STUDENT SOLUTIONS MANUAL
Mark Stevenson

INTERMEDIATE ALGEBRA
FOURTH EDITION

K. Elayn Martin-Gay

PEARSON

Prentice
Hall

Upper Saddle River, NJ 07458

Senior Acquisitions Editor: Paul Murphy
Project Manager: Mary Beckwith
Assistant Editor: Christina Simoneau
Executive Managing Editor: Vince O'Brien
Production Editor: Jeffrey Rydell
Supplement Cover Manager: Paul Gourhan
Supplement Cover Designer: Joanne Alexandris
Manufacturing Buyer: Ilene Kahn

© 2005 Pearson Education, Inc.
Pearson Prentice Hall
Pearson Education, Inc.
Upper Saddle River, NJ 07458

Pearson Prentice Hall® is a trademark of Pearson Education, Inc.

The author and publisher of this book have used their best efforts in preparing this book. These efforts include the development, research, and testing of the theories and programs to determine their effectiveness. The author and publisher make no warranty of any kind, expressed or implied, with regard to these programs or the documentation contained in this book. The author and publisher shall not be liable in any event for incidental or consequential damages in connection with, or arising out of, the furnishing, performance, or use of these programs.

Printed in the United States of America

10 9 8 7 6 5 4

ISBN 0-13-149353-1

Pearson Education Ltd., *London*
Pearson Education Australia Pty. Ltd., *Sydney*
Pearson Education Singapore, Pte. Ltd.
Pearson Education North Asia Ltd., *Hong Kong*
Pearson Education Canada, Inc., *Toronto*
Pearson Educación de Mexico, S.A. de C.V.
Pearson Education—Japan, *Tokyo*
Pearson Education Malaysia, Pte. Ltd.

Table of Contents

Chapter 1

Exercise Set 1.2

1. $5x = 5(7) = 35$

3. $9.8z = 9.8(3.1) = 30.38$

5. $ab = \left(\dfrac{1}{2}\right)\left(\dfrac{3}{4}\right) = \dfrac{3}{8}$

7. $3x + y = 3(6) + (4) = 22$

9. $400t = 400(5) = 2000$ miles

11. $lw = (5.1)(4) = 20.4$
The display needs 20.4 sq. ft. of floor space.

13. $2948t = 2948(3.6) = \$10,612.80$

15. $\{1, 2, 3, 4, 5\}$

17. $\{11, 12, 13, 14, 15, 16\}$

19. $\{0\}$

21. $\{0, 2, 4, 6, 8\}$

23.

25.

27.

29. Answers may vary.

31. $\left\{3, 0, \sqrt{36}\right\}$

33. $\left\{3, \sqrt{36}\right\}$

35. $\left\{\sqrt{7}\right\}$

37. $-11 \in \{x \mid x$ is an integer$\}$

39. $0 \notin \{x \mid x$ is a positive integer$\}$

41. $12 \notin \{1, 3, 5, \dots\}$

43. $0 \notin \{1, 2, 3, \dots\}$

45. True

47. True

49. False

51. False

53. True

55. False; the number $\sqrt{7}$, for example, is a real number, but it is not a rational number.

57. Answers may vary.

59. $-|2| = -2$ (the opposite of $|2|$)

61. $|-4| = 4$ since -4 is located 4 units from 0 on the number line.

63. $|0| = 0$ since 0 is located 0 units from 0 on the number line.

65. $-|-3| = -3$ (the opposite of $|-3|$).

67. Answers may vary.

69. The opposite of -6.2 is $-(-6.2) = 6.2$.

71. The opposite of $\dfrac{4}{7}$ is $-\dfrac{4}{7}$.

73. The opposite of $-\dfrac{2}{3}$ is $\dfrac{2}{3}$.

75. The opposite of 0 is 0.

77. $2x$

79. $2x+5$

81. $x-10$

83. $x+2$

85. $\dfrac{x}{11}$

87. $3x+12$

89. $x-17$

91. $2(x+3)$

93. $\dfrac{5}{4-x}$

95. China = 137
USA = 102
France = 93
Spain = 71
Hong Kong = 59

97. Answers may vary.

Section 1.3

Mental Math

1. B, C

2. A, B

3. B, D

4. A, D

5. B

6. A

Exercise Set 1.3

1. $-3+8=5$

3. $-14+(-10)=-24$

5. $-4.3-6.7=-11$

7. $13-17=-4$

9. $\dfrac{11}{15}-\left(-\dfrac{3}{5}\right)=\dfrac{11}{15}+\dfrac{9}{15}=\dfrac{20}{15}=\dfrac{4}{3}$

11. $19-10-11=9-11=-2$

13. $(-5)(12)=-60$

15. $(-8)(-10)=80$

17. $\dfrac{-12}{-4}=3$

19. $\dfrac{0}{-2}=0$

21. $(-4)(-2)(-1)=8(-1)=-8$

23. $\dfrac{-6}{7}\div 2=\dfrac{-6}{7}\cdot\dfrac{1}{2}=\dfrac{-3}{7}=-\dfrac{3}{7}$

25. $\left(-\dfrac{2}{7}\right)\left(-\dfrac{1}{6}\right)=\dfrac{2}{42}=\dfrac{1}{21}$

27. $-7^2=-(7\cdot 7)=-49$

29. $(-6)^2=(-6)(-6)=36$

31. $(-2)^3=(-2)(-2)(-2)=4(-2)=-8$

33. $\sqrt{49}=7$ since 7 is positive and $7^2=49$.

35. $-\sqrt{\dfrac{1}{9}}=-\dfrac{1}{3}$ since $\dfrac{1}{3}$ is positive and $\left(\dfrac{1}{3}\right)^2=\dfrac{1}{9}$.

37. $\sqrt[3]{64}=4$ since $4^3=64$.

39. $\sqrt[4]{81}=3$ since $3^4=81$.

41. $\sqrt{-100}$ is not a real number.

43. $3(5-7)^4=3(-2)^4=3(16)=48$

45. $-3^2 + 2^3 = -9 + 8 = -1$

47. $\dfrac{3-(-12)}{-5} = \dfrac{3+12}{-5} = \dfrac{15}{-5} = -3$

49. $|3.6 - 7.2| + |3.6 + 7.2| = |-3.6| + |10.8|$
$$= 3.6 + 10.8$$
$$= 14.4$$

51. $\dfrac{(3-\sqrt{9})-(-5-1.3)}{-3} = \dfrac{(3-3)-(-6.3)}{-3}$
$$= \dfrac{0+6.3}{-3}$$
$$= \dfrac{6.3}{-3}$$
$$= -2.1$$

53. $\dfrac{|3-9|-|-5|}{-3} = \dfrac{6-5}{-3} = \dfrac{1}{-3} = -\dfrac{1}{3}$

55. $(-3)^2 + 2^3 = 9 + 8 = 17$

57. $4[8-(2-4)] = 4[8-(-2)] = 4(10) = 40$

59. $2-[(7-6)+(9-19)] = 2-[1+(-10)]$
$$= 2-(-9)$$
$$= 11$$

61. $\dfrac{(-9+6)(-1^2)}{-2-2} = \dfrac{(-3)(-1)}{-4} = \dfrac{3}{-4} = -\dfrac{3}{4}$

63. $(\sqrt[3]{8})(-4)-(\sqrt{9})(-5) = (2)(-4)-(3)(-5)$
$$= -8-(-15)$$
$$= -8+15$$
$$= 7$$

65. $25-[(3-5)+(14-18)]^2$
$$= 25-[(-2)+(-4)]^2$$
$$= 25-(-6)^2$$
$$= 25-36$$
$$= -11$$

67. $\dfrac{\frac{1}{3}\cdot 9 - 7}{3+\frac{1}{2}\cdot 4} = \dfrac{3-7}{3+2} = \dfrac{-4}{5} = -\dfrac{4}{5}$

69. $\dfrac{3(-2+1)}{5} - \dfrac{-7(2-4)}{1-(-2)} = \dfrac{3(-1)}{5} - \dfrac{-7(-2)}{3}$
$$= \dfrac{-3}{5} - \dfrac{14}{3}$$
$$= \dfrac{-9}{15} - \dfrac{70}{15}$$
$$= -\dfrac{79}{15}$$

71. $\dfrac{\frac{-3}{10}}{\frac{42}{50}} = \dfrac{-3}{10}\cdot\dfrac{50}{42} = -\dfrac{5}{14}$

73. $x^2 + z^2 = (-2)^2 + (3)^2 = 4+9 = 13$

75. $-5(-x+3y) = -5(-(-2)+3(-5))$
$$= -5(2-15)$$
$$= -5(-13)$$
$$= 65$$

77. $\dfrac{3z-y}{2x-z} = \dfrac{3(3)-(-5)}{2(-2)-(3)}$
$$= \dfrac{9+5}{-4-3}$$
$$= \dfrac{14}{-7}$$
$$= -2$$

79. $\dfrac{y_2-y_1}{x_2-x_1} = \dfrac{2-(-3)}{4-2} = \dfrac{2+3}{2} = \dfrac{5}{2}$

81. a.

y	5	7	10	100
$8+2y$	18	22	28	208

b. Increase

83. a.

x	10	100	1000
$\dfrac{100x + 5000}{x}$	600	150	105

b. Decrease

85. $1 - \dfrac{1}{5} - \dfrac{3}{7} = \dfrac{35}{35} - \dfrac{7}{35} - \dfrac{15}{35} = \dfrac{13}{35}$

87. $10{,}203 - 5998 = 4205$ meters

89.

Year	Increase
1950	$71.1 - 65.7 = 5.4$
1960	$73.2 - 71.1 = 2.1$
1970	$74.9 - 73.2 = 1.7$
1980	$77.5 - 74.9 = 2.6$
1990	$78.9 - 77.5 = 1.4$
2000	$79.6 - 78.9 = 0.7$

91. $6 - 5 \cdot 2 + 2 = 6 - (5 \cdot 2 + 2)$

93. Answers may vary.

95. $\sqrt{273} \approx 16.5227$

97. $\sqrt{19.6} \approx 4.4272$

99. $\dfrac{(-5.161)(3.222)}{7.955 - 19.676} = \dfrac{-16.628742}{-11.721}$
≈ 1.4187

101. 13.2%

103. $12.9\% - 2.1\% = 10.8\%$

105. b

107. d

109. Yes. Two players have 6 points each (third player has 0 points), or two players have 5 points each (third has 2 points).

Exercise Set 1.4

1. $0 > -2$

3. $7.4 = 7.40$

5. $-7.9 < -7.09$

7. $2x + 5 = -14$

9. $3(x + 1) = 7$

11. $\dfrac{n}{5} = 4n$

13. $z - 2 = 2z$

15. $7x \leq -21$

17. $-2 + x \neq 10$

19. $2(x - 6) > \dfrac{1}{11}$

21. $5y - 7 = 6$

23. $2(x - 6) = -27$

	Number	Opposite	Reciprocal
25.		-5	$\dfrac{1}{5}$
27.	-8		$-\dfrac{1}{8}$
29.		$\dfrac{1}{7}$	-7
31.		0	Undefined
33.	$\dfrac{7}{8}$	$-\dfrac{7}{8}$	

4

35. Zero. For every real number x, $0 \cdot x \neq 1$, so 0 has no reciprocal. It is the only real number that has no reciprocal because if $x \neq 0$, then $x \cdot \dfrac{1}{x} = 1$ by definition.

37. $7x + y = y + 7x$

39. $z \cdot w = w \cdot z$

41. $\dfrac{1}{3} \cdot \dfrac{x}{5} = \dfrac{x}{5} \cdot \dfrac{1}{3}$

43. No, subtraction is not commutative. Answers may vary (for example, $8 - 5 \neq 5 - 8$).

45. $5 \cdot (7x) = (5 \cdot 7)x$

47. $(x + 1.2) + y = x + (1.2 + y)$

49. $(14z) \cdot y = 14(z \cdot y)$

51. $12 - (5 - 3) = 10$; $(12 - 5) - 3 = 4$; Subtraction is not associative.

53. $3(x + 5) = 3x + 15$

55. $-(2a + b) = -2a - b$

57. $2(6x + 5y + 2z) = 12x + 10y + 4z$

59. $-4(x - 2y + 7) = -4x + 8y - 28$

61. $0.5x(6y - 3) = 3xy - 1.5x$

63. $6 + 3x$

65. 0

67. 7

69. $(10 \cdot 2)y$

71. $a(b + c) = ab + ac$

73. In words:

$\boxed{\text{Value of a dime}} \cdot \boxed{\text{Number of dimes}}$

Translate: $0.1 \cdot d$ or $0.1d$

75. If two numbers have a sum of 112 and one number is x, then the other number is the "rest of 112." So, in other words, we have

$\boxed{\text{One hundred twelve}} - \boxed{x}$

Translate: $112 - x$

77. In words:

$\boxed{\text{Ninety}} - \boxed{5x}$

Translate: $90 - 5x$

79. In words:

$\boxed{\text{Cost of a book}} \cdot \boxed{\text{Number of books}}$

Translate: $\$35.61y$

81. The next even integer would be 2 more than the given even integer. In words:

$\boxed{\text{Even integer}} + \boxed{\text{Two}}$

Translate: $2x + 2$

83. $\begin{aligned} 5y - 14 + 7y - 20y &= 5y + 7y - 20y - 14 \\ &= (5 + 7 - 20)y - 14 \\ &= -8y - 14 \end{aligned}$

85. $\begin{aligned} -11c - (4 - 2c) &= -11c - 4 + 2c \\ &= -11c + 2c - 4 \\ &= (-11 + 2)c - 4 \\ &= -9c - 4 \end{aligned}$

87. $\begin{aligned} (8 - 5y) - (4 + 3y) &= 8 - 5y - 4 - 3y \\ &= -5y - 3y + 8 - 4 \\ &= (-5 - 3)y + 4 \\ &= -8y + 4 \\ &\text{or } 4 - 8y \end{aligned}$

89. $-4(yz+3)-7yz+1+y^2$
$= -4yz-12-7yz+1+y^2$
$= y^2-4yz-7yz-11$
$= y^2+(-4-7)yz-11$
$= y^2-11yz-11$

91. $-(8-t)+(2t-6)=-8+t+2t-6$
$\qquad\qquad\qquad\quad = t+2t-8-6$
$\qquad\qquad\qquad\quad = (1+2)t-14$
$\qquad\qquad\qquad\quad = 3t-14$

93. $5(2z^3-6)+10(3-z^3)$
$= 10z^3-30+30-10z^3$
$= 10z^3-10z^3-30+30$
$= (10-10)z^3+0$
$= 0$

95. $7n+3(2n-6)-2=7n+6n-18-2$
$\qquad\qquad\qquad\qquad = (7+6)n-20$
$\qquad\qquad\qquad\qquad = 13n-20$

97. $6.3y-9.7+2.2y-11.1$
$= 6.3y+2.2y-9.7-11.1$
$= (6.3+2.2)y-20.8$
$= 8.5y-20.8$

99. $\dfrac{7}{8}a-\dfrac{11}{12}-\dfrac{1}{2}a+\dfrac{5}{6}=\dfrac{7}{8}a-\dfrac{1}{2}a-\dfrac{11}{12}+\dfrac{5}{6}$
$\qquad\qquad\qquad\qquad\quad = \left(\dfrac{7}{8}-\dfrac{1}{2}\right)a-\dfrac{11}{12}+\dfrac{10}{12}$
$\qquad\qquad\qquad\qquad\quad = \left(\dfrac{7}{8}-\dfrac{4}{8}\right)a-\dfrac{1}{12}$
$\qquad\qquad\qquad\qquad\quad = \dfrac{3}{8}a-\dfrac{1}{12}$

101. $4(5y+12)=20y+48$

103. $\dfrac{1}{2}(10x-2)-\dfrac{1}{6}(60x-5y)$
$= 5x-1-10x+\dfrac{5}{6}y$
$= 5x-10x+\dfrac{5}{6}y-1$
$= -5x+\dfrac{5}{6}y-1$

105. $\dfrac{1}{3}(6x-33y)-\dfrac{1}{8}(24x-40y+1)-\dfrac{1}{3}$
$= 2x-11y-3x+5y-\dfrac{1}{8}-\dfrac{1}{3}$
$= 2x-3x-11y+5y-\dfrac{3}{24}-\dfrac{8}{24}$
$= (2-3)x+(-11+5)y-\dfrac{11}{24}$
$= -x-6y-\dfrac{11}{24}$

107. $5.8(-9.6-31.2y)-18.65$
$= -55.68-180.96y-18.65$
$= -180.96y-55.98-18.65$
$= 180.96y-74.33$

109. $6.5y-4.4(1.8x-3.3)+10.95$
$= 6.5y-7.92x+14.52+10.95$
$= 6.5y-7.92x+25.47$

111. It is not the case that two rectangles with the same perimeter will necessarily have the same area. Take a rectangle that is 5 in. by 5 in. and another that is 8 in. by 2 in. Both have the same perimeter, but the areas are 25 square inches and 16 square inches, respectively.

113. 80 million

115. 35 million

117. $(2.5)(8.1\%)=20.25\%$

Chapter 1 Review Exercises

1. $7x = 7(3) = 21$

2. $st = (1.6)(5) = 8$

3. One hour is $60(60) = 3600$ seconds.
   $90t = 90(3600) = 324,000$
   324,000 wing beats per hour

4. $\{-1, 1, 3\}$

5. $\{-2, 0, 2, 4, 6\}$

6. $\varnothing$

7. $\varnothing$

8. $\{6, 7, 8, \ldots\}$

9. $\{\ldots, -1, 0, 1, 2\}$

10. True

11. False

12. True, since $\sqrt{169} = 13$.

13. True, since zero is not an element of the empty set.

14. False, since π is irrational.

15. True, since π is a real number.

16. False, since $\sqrt{4} = 2$.

17. True, since -9 is a rational number.

18. True, since every element of A is also an element of D.

19. True, since C is not a subset of B.

20. False, since all integers are rational numbers.

21. True, since the empty set is a subset of all sets.

22. True, since every set is a subset of itself.

23. True, since every element of D is also an element of C.

24. True, since every integer is a real number.

25. True, since every irrational number is also a real number.

26. False, since B does not contain the set $\{5\}$.

27. True, since $\{5\}$ is a subset of B.

28. $\left\{5, \dfrac{8}{2}, \sqrt{9}\right\}$

29. $\left\{5, \dfrac{8}{2}, \sqrt{9}\right\}$

30. $\left\{5, -\dfrac{2}{3}, \dfrac{8}{2}, \sqrt{9}, 0.3, 1\dfrac{5}{8}, -1\right\}$

31. $\left\{\sqrt{7}, \pi\right\}$

32. $\left\{5, -\dfrac{2}{3}, \dfrac{8}{2}, \sqrt{9}, 0.3, \sqrt{7}, 1\dfrac{5}{8}, -1, \pi\right\}$

33. $\left\{5, \dfrac{8}{2}, \sqrt{9}, -1\right\}$

34. The opposite of $-\dfrac{3}{4}$ is $-\left(-\dfrac{3}{4}\right) = \dfrac{3}{4}$.

35. The opposite of 0.6 is -0.6.

36. The opposite of 0 is $-0 = 0$.

37. The opposite of 1 is -1.

38. The reciprocal of $-\dfrac{3}{4}$ is $\dfrac{1}{\left(-\dfrac{3}{4}\right)} = -\dfrac{4}{3}$.

39. The reciprocal of 0.6 is $\dfrac{1}{0.6}$.

40. The reciprocal of 0 is $\dfrac{1}{0}$ which is undefined.

41. The reciprocal of 1 is $\dfrac{1}{1} = 1$.

42. $-7 + 3 = -4$

43. $-10 + (-25) = -35$

44. $5(-0.4) = -2$

45. $(-3.1)(-0.1) = 0.31$

46. $-7 - (-15) = -7 + 15 = 8$

47. $9 - (-4.3) = 9 + 4.3 = 13.3$

48. $(-6)(-4)(0)(-3) = 0$

49. $(-12)(0)(-1)(-5) = 0$

50. $(-24) \div 0$ is undefined.

51. $0 \div (-45) = 0$

52. $(-36) \div (-9) = 4$

53. $60 \div (-12) = -5$

54. $\left(-\dfrac{4}{5}\right) - \left(-\dfrac{2}{3}\right) = -\dfrac{4}{5} + \dfrac{2}{3} = -\dfrac{12}{15} + \dfrac{10}{15} = -\dfrac{2}{15}$

55. $\left(\dfrac{5}{4}\right) - \left(-2\dfrac{3}{4}\right) = \dfrac{5}{4} + \dfrac{11}{4} = \dfrac{16}{4} = 4$

56. $1 - \dfrac{1}{4} - \dfrac{1}{3} = \dfrac{12}{12} - \dfrac{3}{12} - \dfrac{4}{12} = \dfrac{5}{12}$

57. $-5 + 7 - 3 - (-10) = 2 - 3 + 10$
$$= -1 + 10$$
$$= 9$$

58. $8 - (-3) + (-4) + 6 = 8 + 3 - 4 + 6$
$$= 11 - 4 + 6$$
$$= 7 + 6$$
$$= 13$$

59. $3(4-5)^4 = 3(-1)^4 = 3(1) = 3$

60. $6(7-10)^2 = 6(-3)^2 = 6(9) = 54$

61. $\left(-\dfrac{8}{15}\right) \cdot \left(-\dfrac{2}{3}\right)^2 = -\dfrac{8}{15} \cdot \dfrac{4}{9} = -\dfrac{32}{135}$

62. $\left(-\dfrac{3}{4}\right)^2 \cdot \left(-\dfrac{10}{21}\right) = \left(\dfrac{9}{16}\right)\left(-\dfrac{10}{21}\right) = -\dfrac{15}{56}$

63. $-\dfrac{6}{15} \div \dfrac{8}{25} = -\dfrac{6}{15} \cdot \dfrac{25}{8} = -\dfrac{150}{120} = -\dfrac{5}{4}$

64. $\dfrac{4}{9} \div \left(-\dfrac{8}{45}\right) = \dfrac{4}{9} \cdot \left(-\dfrac{45}{8}\right) = -\dfrac{180}{72} = -\dfrac{5}{2}$

65. $-\dfrac{3}{8} + 3(2) \div 6 = -\dfrac{3}{8} + 6 \div 6$
$$= -\dfrac{3}{8} + 1$$
$$= -\dfrac{3}{8} + \dfrac{8}{8}$$
$$= \dfrac{5}{8}$$

66. $5(-2) - (-3) - \dfrac{1}{6} + \dfrac{2}{3} = -10 + 3 - \dfrac{1}{6} + \dfrac{2}{3}$
$$= -7 - \dfrac{1}{6} + \dfrac{2}{3}$$
$$= -\dfrac{42}{6} - \dfrac{1}{6} + \dfrac{4}{6}$$
$$= -\dfrac{39}{6}$$
$$= -6\dfrac{1}{2}$$

67. $\left|2^3 - 3^2\right| - \left|5 - 7\right| = \left|8 - 9\right| - \left|-2\right|$
$$= \left|-1\right| - 2$$
$$= 1 - 2$$
$$= -1$$

68. $\left|5^2 - 2^2\right| + \left|9 \div (-3)\right| = \left|25 - 4\right| + \left|-3\right|$
$$= \left|21\right| + 3$$
$$= 21 + 3$$
$$= 24$$

69. $(2^3 - 3^2) - (5 - 7) = (8 - 9) - (-2)$
$$= -1 + 2$$
$$= 1$$

70. $(5^2 - 2^4) + [9 \div (-3)] = (25 - 16) + (-3)$
$$= 9 + (-3)$$
$$= 6$$

71. $\dfrac{(8-10)^3 - (-4)^2}{2 + 8(2) \div 4} = \dfrac{(-2)^3 - 16}{2 + 16 \div 4}$
$$= \dfrac{-8 - 16}{2 + 4}$$
$$= \dfrac{-24}{6}$$
$$= -4$$

72. $\dfrac{(2+4)^2 + (-1)^5}{12 \div 2 \cdot 3 - 3} = \dfrac{(6)^2 + (-1)}{6 \cdot 3 - 3}$
$$= \dfrac{36 - 1}{18 - 3}$$
$$= \dfrac{35}{15}$$
$$= \dfrac{7}{3}$$

73. $\dfrac{(4-9) + 4 - 9}{10 - 12 \div 4 \cdot 8} = \dfrac{(-5) + 4 - 9}{10 - 3 \cdot 8}$
$$= \dfrac{-1 - 9}{10 - 24}$$
$$= \dfrac{-10}{-14}$$
$$= \dfrac{5}{7}$$

74. $\dfrac{3 - 7 - (7 - 3)}{15 + 30 \div 6 \cdot 2} = \dfrac{-4 - (4)}{15 + 5 \cdot 2}$
$$= \dfrac{-8}{15 + 10}$$
$$= \dfrac{-8}{25}$$
$$= -\dfrac{8}{25}$$

75. $\dfrac{\sqrt{25}}{4 + 3 \cdot 7} = \dfrac{5}{4 + 21} = \dfrac{5}{25} = \dfrac{1}{5}$

76. $\dfrac{\sqrt{64}}{24 - 8 \cdot 2} = \dfrac{8}{24 - 16} = \dfrac{8}{8} = 1$

77. $x^2 - y^2 + z^2 = (0)^2 - (3)^2 + (-2)^2$
$$= 0 - 9 + 4$$
$$= -5$$

78. $\dfrac{5x + z}{2y} = \dfrac{5(0) + (-2)}{2(3)} = \dfrac{0 - 2}{6} = \dfrac{-2}{6} = -\dfrac{1}{3}$

79. $\dfrac{-7y - 3z}{-3} = \dfrac{-7(3) - 3(-2)}{-3}$
$$= \dfrac{-21 + 6}{-3}$$
$$= \dfrac{-15}{-3}$$
$$= 5$$

80. $(x - y + z)^2 = (0 - 3 + (-2))^2$
$$= (-3 - 2)^2$$
$$= (-5)^2$$
$$= 25$$

81. a. When $r = 1$,
$\quad 2\pi r = 2\pi(1) = 2(3.14) = 62.8$.
When $r = 10$,
$\quad 2\pi r = 2\pi(10) = 20(3.14) = 62.8$.
When $r = 100$,
$\quad 2\pi r = 2\pi(100) = 200(3.14) = 628$.

r	1	10	100
$2\pi r$	6.28	62.8	628

b. As the radius increases, the circumference increases.

82. $5xy - 7xy + 3 - 2 + xy$
$$= 5xy - 7xy + xy + 3 - 2$$
$$= (5 - 7 + 1)xy + (3 - 2)$$
$$= (-1)xy + 1$$
$$= -xy + 1$$

83. $4x + 10x - 19x + 10 - 19$
$= (4 + 10 - 19)x + (10 - 19)$
$= -5x + (-9)$
$= -5x - 9$

84. $6x^2 + 2 - 4(x^2 + 1) = 6x^2 + 2 - 4x^2 - 4$
$= 6x^2 - 4x^2 + 2 - 4$
$= (6 - 4)x^2 + (2 - 4)$
$= 2x^2 + (-2)$
$= 2x^2 - 2$

85. $-7(2x^2 - 1) - x^2 - 1$
$= -14x^2 + 7 - x^2 - 1$
$= -14x^2 - x^2 + 7 - 1$
$= (-14 - 1)x^2 + (7 - 1)$
$= -15x^2 + 6$

86. $(3.2x - 1.5) - (4.3x - 1.2)$
$= 3.2x - 1.5 - 4.3x + 1.2$
$= 3.2x - 4.3x - 1.5 + 1.2$
$= (3.2 - 4.3)x - 0.3$
$= -1.1x - 0.3$

87. $(7.6x + 4.7) - (1.9x + 3.6)$
$= 7.6x + 4.7 - 1.9x - 3.6$
$= 7.6x - 1.9x + 4.7 - 3.6$
$= (7.6 - 1.9)x + 4.7 - 3.6$
$= 5.7x + 1.1$

88. $12 = -4x$

89. $n + 2n = -15$

90. $4(y + 3) = -1$

91. $6(t - 5) = 4$

92. $z - 7 = 6$

93. $9x - 10 = 5$

94. $x - 5 \geq 12$

95. $-4 < 7y$

96. $\frac{2}{3} \neq 2\left(n + \frac{1}{4}\right)$

97. $t + 6 \leq -12$

98. Associative Property of Addition

99. Distributive Property

100. Additive Inverse Property

101. Commutative Property of Addition

102. Associative and Commutative Properties of Multiplication
To see this: $(XY)Z = X(YZ) = (YZ)X$

103. Multiplicative Inverse Property

104. Multiplication Property of Zero

105. Associative Property of Multiplication

106. Additive Identity Property

107. Multiplicative Identity Property

108. $5x - 15z = 5(x - 3z)$

109. $(7 + y) + (3 + x) = (3 + x) + (7 + y)$

110. $0 = 2 + (-2)$, for example

111. $1 = 2 \cdot \frac{1}{2}$, for example

112. $[(3.4)(0.7)]5 = (3.4)[(0.7)(5)]$

113. $7 = 7 + 0$

114. $-9 > -12$

115. $0 > -6$

116. $-3 < -1$

117. $7 = |7|$

118. $-5 < -(-5)$, since $-(-5) = 5$

119. $-(-2) > -2$, since $-(-2) = 2$

Chapter 1 Test

1. True; -2.3 lies to the right of -2.33 on the number line.

2. False; $-6^2 = -36$, while $(-6)^2 = 36$.

3. False; $-5 - 8 = -13$, while $-(5-8) = -(-3) = 3$.

4. False; $(-2)(-3)(0) = 0$, while $\dfrac{(-4)}{0}$ is undefined.

5. True

6. False; for example, $\dfrac{1}{2}$ is a rational number that is not an integer.

7. $5 - 12 \div 3(2) = 5 - 4(2) = 5 - 8 = -3$

8. $5^2 - 3^4 = 25 - 81 = -56$

9. $(4-9)^3 - |-4-6|^2 = (-5)^3 - |-10|^2$
$$= -125 - 10^2$$
$$= -125 - 100$$
$$= -225$$

10. $12 + \{6 - [5 - 2(-5)]\} = 12 + \{6 - [5 + 10]\}$
$$= 12 + (6 - 15)$$
$$= 12 + (-9)$$
$$= 12 - 9$$
$$= 3$$

11. $\dfrac{6(7-9)^3 + (-2)}{(-2)(-5)(-5)} = \dfrac{6(-2)^3 - 2}{10(-5)}$
$$= \dfrac{6(-8) - 2}{-50}$$
$$= \dfrac{-48 - 2}{-50}$$
$$= \dfrac{-50}{-50}$$
$$= 1$$

12. $\dfrac{(4 - \sqrt{16}) - (-7 - 20)}{-2(1-4)^2} = \dfrac{(4-4) - (-27)}{-2(-3)^2}$
$$= \dfrac{0 + 27}{-2(9)}$$
$$= \dfrac{27}{-18}$$
$$= -\dfrac{3}{2}$$

13. $q^2 - r^2 = (4)^2 - (-2)^2 = 16 - 4 = 12$

14. $\dfrac{5t - 3q}{3r - 1} = \dfrac{5(1) - 3(4)}{3(-2) - 1} = \dfrac{5 - 12}{-6 - 1} = \dfrac{-7}{-7} = 1$

15. **a.** When $x = 1$, $5.75x = 5.75(1) = 5.75$.
    When $x = 3$, $5.75x = 5.75(3) = 17.25$.
    When $x = 10$,
    $5.75x = 5.75(10) = 57.50$.
    When $x = 20$,
    $5.75x = 5.75(20) = 115.00$.

x	1	3	10	20
$5.75x$	5.75	17.25	57.50	115.00

    b. As the number of adults increases the total cost increases.

16. $2(x + 5) = 30$

17. $\dfrac{(6 - y)^2}{7} < -2$

18. $\dfrac{9z}{|-12|} \neq 10$

19. $3\left(\dfrac{n}{5}\right) = -n$

20. $20 = 2x - 6$

21. $-2 = \dfrac{x}{x + 5}$

22. Distributive Property

23. Associative Property of Addition

24. Additive Inverse Property

25. Multiplication Property of Zero

26. $0.05n = 0.1d$

27. $-2(3x + 7) = -6x - 14$

28. $\dfrac{1}{3}a - \dfrac{3}{8} + \dfrac{1}{6}a - \dfrac{3}{4} = \dfrac{1}{3}a + \dfrac{1}{6}a - \dfrac{3}{8} - \dfrac{3}{4}$

$$= \left(\dfrac{1}{3} + \dfrac{1}{6}\right)a - \dfrac{3}{8} - \dfrac{3}{4}$$

$$= \left(\dfrac{2}{6} + \dfrac{1}{6}\right)a - \dfrac{3}{8} - \dfrac{6}{8}$$

$$= \left(\dfrac{3}{6}\right)a - \dfrac{9}{8}$$

$$= \dfrac{1}{2}a - \dfrac{9}{8}$$

29. $4y + 10 - 2(y + 10) = 4y + 10 - 2y - 20$

$$= 4y - 2y + 10 - 20$$

$$= (4 - 2)y - 10$$

$$= 2y - 10$$

30. $(8.3x - 2.9) - (9.6x - 4.8)$

$$= 8.3x - 2.9 - 9.6x + 4.8$$

$$= 8.3x - 9.6x - 2.9 + 4.8$$

$$= (8.3 - 9.6)x + 1.9$$

$$= -1.3x + 1.9$$

Chapter 2

Section 2.1

Mental Math

1. $8x + 21$

2. $11y + 18$

3. $6n - 7$

4. $3m - 4$

5. $-4x - 1$

6. $-6x - 3$

7. Expression

8. Equation

9. Equation

10. Expression

11. $2x + 3 = 2x + 3$
    $3 = 3;$ True
    All real numbers

12. $2x + 1 = 2x + 3$
    $1 = 3;$ False
    No solution

13. $5x - 2 = 5x - 7$
    $-2 = -7;$ False
    No solution

14. $5x - 3 = 5x - 3$
    $-3 = -3;$ True
    All real numbers

Exercise Set 2.1

1. $-3x = 36$
   $$\frac{-3x}{-3} = \frac{36}{-3}$$
   $x = -12$

Check: $\quad -3x = 36$
$$-3(-12) = 36$$
$$36 = 36 \quad \text{True}$$
The solution is -12.

3. $\quad x + 2.8 = 1.9$
   $$x + 2.8 - 2.8 = 1.9 - 2.8$$
   $$x = -0.9$$
   Check: $\quad x + 2.9 = 1.9$
   $$(-0.9) + 2.8 = 1.9$$
   $$1.9 = 1.9 \quad \text{True}$$
   The solution is -0.9.

5. $\quad 5x - 4 = 26$
   $$5x - 4 + 4 = 26 + 4$$
   $$5x = 30$$
   $$\frac{5x}{5} = \frac{30}{5}$$
   $$x = 6$$
   Check: $\quad 5x - 4 = 26$
   $$5(6) - 4 = 26$$
   $$30 - 6 = 4$$
   $$4 = 4 \quad \text{True}$$
   The solution is 6.

7. $\quad -4 = 3x + 11$
   $$-4 - 11 = 3x + 11 - 11$$
   $$-15 = 3x$$
   $$\frac{-15}{3} = \frac{3x}{3}$$
   $$-5 = x$$
   Check: $-4 = 3x + 11$
   $$-4 = 3(-5) + 11$$
   $$-4 = -15 + 11$$
   $$-4 = -4 \quad \text{True}$$
   The solution is -5.

9. $\quad -4.1 - 7z = 3.6$
   $$-4.1 - 7z + 4.1 = 3.6 + 4.1$$
   $$-7z = 7.7$$
   $$\frac{-7z}{-7} = \frac{7.7}{-7}$$
   $$z = -1.1$$

Check: $-4.1 - 7z = 3.6$
$-4.1 - 7(-1.1) = 3.6$
$-4.1 + 7.7 = 3.6$
$3.6 = 3.6$ True

The solution is –1.1.

11. $5y + 12 = 2y - 3$
$5y + 12 - 2y = 2y - 3 - 2y$
$3y + 12 = -3$
$3y + 12 - 12 = -3 - 12$
$3y = -15$
$\dfrac{3y}{3} = \dfrac{-15}{3}$
$y = -5$

Check: $5y + 12 = 2y - 3$
$5(-5) + 12 = 2(-5) - 3$
$-25 + 12 = -10 - 3$
$-13 = -13$ True

The solution is –5.

13. $8x - 5x + 3 = x - 7 + 10$
$3x + 3 = x + 3$
$2x = 0$
$x = 0$

Check: $8x - 5x + 3 = x + 7 - 10$
$8(0) - 5(0) + 3 = (0) + 7 - 10$
$0 + 3 = 3$
$3 = 3$ True

The solution is 0.

15. $5x + 12 = 2(2x + 7)$
$5x + 12 = 4x + 14$
$x + 12 = 14$
$x = 2$

Check: $5x + 12 = 2(2x + 7)$
$5(2) + 12 = 2(2(2) + 7)$
$10 + 12 = 2(4 + 7)$
$22 = 2(11)$
$22 = 22$ True

The solution is 2.

17. $3(x - 6) = 5x$
$3x - 18 = 5x$
$-18 = 2x$
$-9 = x$

Check: $3(x - 6) = 5x$
$3((-9) - 6) = 5(-9)$
$3(-15) = -45$
$-45 = -45$ True

The solution is –9.

19. $-2(5y - 1) - y = -4(y - 3)$
$-10y + 2 = -4y + 12$
$-7y + 2 = 12$
$-7y = 10$
$y = -\dfrac{10}{7}$

Check:

$-2(5y - 1) - y = -4(y - 3)$

$-2\left(5\left(-\dfrac{10}{7}\right) - 1\right) - \left(-\dfrac{10}{7}\right) = -4\left(\left(-\dfrac{10}{7}\right) - 3\right)$

$-2\left(-\dfrac{50}{7} - 1\right) + \dfrac{10}{7} = -4\left(-\dfrac{31}{7}\right)$

$-2\left(-\dfrac{57}{7}\right) + \dfrac{10}{7} = \dfrac{124}{7}$

$\dfrac{114}{7} + \dfrac{10}{7} = \dfrac{124}{7}$

$\dfrac{124}{7} = \dfrac{124}{7}$ True

The solution is $-\dfrac{10}{7}$.

21. a. $4(x + 1) + 1 = 4x + 4 + 1 = 4x + 5$

b. $4(x + 1) + 1 = -7$
$4x + 4 + 1 = -7$
$4x + 5 = -7$
$4x = -12$
$x = -3$

The solution is –3.

c. Answers may vary.

23. $\dfrac{x}{2} + \dfrac{2}{3} = \dfrac{3}{4}$
$12\left(\dfrac{x}{2} + \dfrac{2}{3}\right) = 12\left(\dfrac{3}{4}\right)$
$6x + 8 = 9$
$6x = 1$
$x = \dfrac{1}{6}$

Check: $\dfrac{x}{2} + \dfrac{2}{3} = \dfrac{3}{4}$

$\dfrac{(1/6)}{2} + \dfrac{2}{3} = \dfrac{3}{4}$

$\dfrac{1}{12} + \dfrac{2}{3} = \dfrac{3}{4}$

$\dfrac{1}{12} + \dfrac{8}{12} = \dfrac{3}{4}$

$\dfrac{9}{12} = \dfrac{3}{4}$

$\dfrac{3}{4} = \dfrac{3}{4}$ True

The solution is $\dfrac{1}{6}$.

25. $\dfrac{3t}{4} - \dfrac{t}{2} = 1$

$4\left(\dfrac{3t}{4} - \dfrac{t}{2}\right) = 4(1)$

$3t - 2t = 4$

$t = 4$

Check: $\dfrac{3t}{4} - \dfrac{t}{2} = 1$

$\dfrac{3(4)}{4} - \dfrac{(4)}{2} = 1$

$3 - 2 = 1$

$1 = 1$ True

The solution is 4.

27. $\dfrac{n-3}{4} + \dfrac{n+5}{7} = \dfrac{5}{14}$

$28\left(\dfrac{n-3}{4}\right) + 28\left(\dfrac{n+5}{7}\right) = 28\left(\dfrac{5}{14}\right)$

$7(n-3) + 4(n+5) = 2(5)$

$7n - 21 + 4n + 20 = 10$

$11n - 1 = 10$

$11n = 11$

$n = 1$

Check:

$\dfrac{n-3}{4} + \dfrac{n+5}{7} = \dfrac{5}{14}$

$\dfrac{(1)-3}{4} + \dfrac{(1)+5}{7} = \dfrac{5}{14}$

$\dfrac{-2}{4} + \dfrac{6}{7} = \dfrac{5}{14}$

$\dfrac{-1}{2} + \dfrac{6}{7} = \dfrac{5}{14}$

$-\dfrac{7}{14} + \dfrac{12}{14} = \dfrac{5}{14}$

$\dfrac{5}{14} = \dfrac{5}{14}$ True

The solution is 1.

29. $0.6x - 10 = 1.4x - 14$

$-0.8x - 10 = -14$

$-0.8x = -4$

$x = 5$

Check: $0.6x - 10 = 1.4x - 14$

$0.6(5) - 10 = 1.4(5) - 14$

$3 - 10 = 7 - 14$

$-7 = -7$ True

The solution is 5.

31. $4(n+3) = 2(6+2n)$

$4n + 12 = 12 + 4n$

$4n + 12 = 4n + 12$

This is true for all n.
Therefore, all real numbers are solutions.

33. $3(x-1) + 5 = 3x + 7$

$3x - 3 + 5 = 3x + 7$

$3x + 2 = 3x + 7$

$2 = 7$

This is false for any x.
Therefore, the solution set is $\varnothing$.

35. Answers may vary.

37. $-9x = -72$

$x = 8$

The solution is 8.

39. $x - 1.7 = -7.6$

$x = -5.9$

The solution is -5.9.

41. $6x + 9 = 51$

$6x = 42$

$x = 7$

The solution is 7.

43. $-5x+1.5=-19.5$
$$-5x=-21$$
$$x=4.2$$
The solution is 4.2.

45. $x-10=-6x+4$
$$7x=14$$
$$x=2$$
The solution is 2.

47. $3x-4-5x=x+4+x$
$$-2x-4=2x+4$$
$$-4x=8$$
$$x=-2$$
The solution is -2.

49. $5(y+4)=4(y+5)$
$$5y+20=4y+20$$
$$y=0$$
The solution is 0.

51. $0.6x-10=1.4x-14$
$$-0.8x=-4$$
$$x=5$$
The solution is 5.

53. $6x-2(x-3)=4(x+1)+4$
$$6x-2x+6=4x+4+4$$
$$4x+6=4x+8$$
$$6=8$$
There is no solution.
Therefore, the solution set is $\varnothing$.

55. $$\frac{3}{8}+\frac{b}{3}=\frac{5}{12}$$
$$24\left(\frac{3}{8}\right)+24\left(\frac{b}{3}\right)=24\left(\frac{5}{12}\right)$$
$$9+8b=10$$
$$8b=1$$
$$b=\frac{1}{8}$$
The solution is $\frac{1}{8}$.

57. $z+3(2+4z)=6(z+1)+5z$
$$z+6+12z=6z+6+5z$$
$$13z+6=11z+6$$
$$2z=0$$
$$z=0$$
The solution is 0.

59. $$\frac{3t+1}{8}=\frac{5+2t}{7}+2$$
$$56\left(\frac{3t+1}{8}\right)=56\left(\frac{5+2t}{7}\right)+56(2)$$
$$7(3t+1)=8(5+2t)+112$$
$$21t+7=40+16t+112$$
$$21t+7=16t+152$$
$$5t=145$$
$$t=29$$
The solution is 29.

61. $$\frac{m-4}{3}-\frac{3m-1}{5}=1$$
$$15\left(\frac{m-4}{3}\right)-15\left(\frac{3m-1}{5}\right)=15(1)$$
$$5(m-4)-3(3m-1)=15$$
$$5m-20-9m+3=15$$
$$-4m-17=15$$
$$-4m=32$$
$$m=-8$$
The solution is -8.

63. $5(x-2)+2x=7(x+4)-38$
$$5x-10+2x=7x+28-38$$
$$7x-10=7x-10$$
$$0=0$$
This is true for all x.
Therefore, all real numbers are solutions.

65. $y+0.2=0.6(y+3)$
$$y+0.2=0.6y+1.8$$
$$0.4y=1.6$$
$$y=4$$
The solution is 4.

67.
$$-(3x-5)-(2x-6)+1=-5(x-1)-(3x+2)+3$$
$$-3x+5-2x+6+1=-5x+5-3x-2+3$$
$$-5x+12=-8x+6$$
$$3x=-6$$
$$x=-2$$
The solution is -2.

69. $2(x-8)+x=3(x-6)+2$
$$2x-16+x=3x-18+2$$
$$3x-16=3x-16$$
$$0=0$$
This is true for all x.
Therefore, all real numbers are solutions.

71.
$$\frac{3x-1}{9}+x=\frac{3x+1}{3}+4$$
$$9\left(\frac{3x-1}{9}+x\right)=9\left(\frac{3x+1}{3}+4\right)$$
$$(3x-1)+9x=3(3x+1)+36$$
$$3x-1+9x=9x+3+36$$
$$12x-9=9x+40$$
$$3x=40$$
$$x=\frac{40}{3}$$
The solution is $\frac{40}{3}$.

73.
$$1.5(4-x)=1.3(2-x)$$
$$10[1.5(4-x)]=10[1.3(2-x)]$$
$$15(4-x)=13(2-x)$$
$$60-15x=26-13x$$
$$-2x=-34$$
$$x=17$$
The solution is 17.

75. $-2(b-4)-(3b-1)=5b+3$
$$-2b+8-3b+1=5b+3$$
$$-5b+9=5b+3$$
$$-10b=-6$$
$$b=\frac{6}{10}=\frac{3}{5}$$
The solution is $\frac{3}{5}$.

77.
$$\frac{1}{3}(y+4)+5=\frac{1}{4}(3y-1)-2$$
$$12\left[\frac{1}{3}(y+4)+5\right]=12\left[\frac{1}{4}(3y-1)-2\right]$$
$$4(y+4)+60=3(3y-1)-24$$
$$4y+16+60=9y-3-24$$
$$4y+76=9y-27$$
$$-5y=-103$$
$$y=\frac{103}{5}$$
The solution is $\frac{103}{5}$ or 20.6.

79. $\frac{8}{x}$

81. $8x$

83. $3x+2$

85.
$$3.2x+4=5.4x-7$$
$$3.2x+4-4=5.4x-7-4$$
$$3.2x=5.4x-11$$
From this we see that $K=-11$.

87.
$$\frac{x}{6}+4=\frac{x}{3}$$
$$6\left(\frac{x}{6}+4\right)=6\left(\frac{x}{3}\right)$$
$$x+24=2x$$
From this we see that $K=24$.

89. $x(x-6)+7=x(x+1)$
$$x^2-6x+7=x^2+x$$
$$-6x+7=x$$
$$7=7x$$
$$1=x$$

The solution is 1.

91. $3x(x+5)-12 = 3x^2 +10x+3$

$3x^2 +15x-12 = 3x^2 +10x+3$

$15x-12 = 10x+3$

$5x = 15$

$x = 3$

The solution is 3.

93. $2.569x = -12.48534$

$\dfrac{2.569x}{2.569} = \dfrac{-12.48534}{2.569}$

$x = -4.86$

The solution is -4.86.

95. $2.86z - 8.1258 = -3.75$

$2.86z = 4.3758$

$\dfrac{2.86z}{2.86} = \dfrac{4.3758}{2.86}$

$z = 1.53$

The solution is 1.53.

97. Not a fair game.

Exercise Set 2.2

1. $y+y+y+y = 4y$

3. $z+(z+1)+(z+2) = 3z+3$

5. $5x+10(x+3) = 5x+10x+30$

$\qquad\qquad\qquad = (15x+30)$ cents

7. $4x+3(2x+1) = 4x+6x+3 = 10x+3$

9. Let x = the number.

$4(x-2) = 2+6x$

$4x-8 = 2+6x$

$-2x = 10$

$x = -5$

The number is -5.

11. Let x = 1st number;

then $5x$ = 2nd number.

$x+5x = 270$

$6x = 270$

$x = 45$

$5x = 5(45) = 225$

The numbers are 45 and 225.

13. $30\% \cdot 260 = 0.30 \cdot 260 = 78$

15. $12\% \cdot 16 = 0.12 \cdot 16 = 1.92$

17. $29\% \cdot 2271 = 0.29 \cdot 2271 = 658.59$;

$2271 - 658.59 = 1612.41$.

Approximately 1612.41 million acres are not federally owned.

19. $85\% \cdot 2342 = 0.85 \cdot 2342 = 1990.7$;

Approximately 1991 minor earthquakes occurred.

21. $33\frac{1}{3}\% \cdot 1290 \approx 0.3333 \cdot 1290 = 429.957$

$1290 - 429.957 = 860.043$.

About 860 shoppers would be expected to not spend more than they intended.

23. 17%

25. $6\% \cdot 112,500 = 0.06 \cdot 112,500 = 6750$

You would expect 6750 users to check their e-mail about once a week.

27. $3x+21.1 = 205.9$

$3x = 184.8$

$x = 61.6$

$x+15.3 = 61.6+15.3 = 76.9$

$x+5.8 = 61.6+5.8 = 67.4$

The Los Angeles airport has 61.6 million annual arrivals and departures. The Atlanta airport has 76.9 million annual arrivals and departures. The Chicago airport has 67.4 million annual arrivals and departures.

29. Let x = no. of seats in the 737-200;
then $x + 21$ = no. in the 737-300
and $2x - 36$ = no. in the 757-200.
$$x + (x + 21) + (2x - 36) = 437$$
$$4x - 15 = 437$$
$$4x = 452$$
$$x = 113$$
$$x + 21 = 113 + 21 = 134$$
$$2x - 33 = 2(113) - 36 = 190$$
The 737-200 has 113 seats. The 737-300 has 134 seats. The 757-200 has 190 seats.

31. Let x = price before taxes.
$$x + 0.08x = 464.4$$
$$1.08x = 464.40$$
$$x = 430$$
The price was $430 before taxes.

33. Let x = number seats in Heinz Field;
then $x + 11,675$ = no. seats in Mile High.
$$x + (x + 11,675) = 140,575$$
$$2x + 11,675 = 140,575$$
$$2x = 128900$$
$$x = 64450$$
$$x + 11,675 = 64,450 + 11,675 = 76,125$$
Mile High stadium has 76,125 seats and Heinz Field has 64,450 seats.

35. Let x = number of subscribers to MSN;
Then $x + 700,000$ = no. sub. to Earthlink
and $5x + 3,700,000$ = no. sub. to AOL.
$$x + (x + 700,000) + (5x + 3,700,000)$$
$$= 32,400,000$$
$$7x + 4,400,000 = 32,400,000$$
$$7x = 280,000,000$$
$$x = 4,000,000$$
$$x + 700,000 = (4,000,000) + 700,000$$
$$= 4,700,000$$
$$5x + 3,700,000 = 5(4,000,000)$$
$$+ 3,700,000$$
$$= 23,700,000$$

MSN had 4,000,000 subscribers.
Earthlink had 4,700,000 subscribers.
AOL had 23,700,000 subscribers.

37. Let x = population in 2000.
$$x + 0.091x = 31.2$$
$$1.091x = 31.2$$
$$x \approx 28.6$$
The population of Morocco in 2000 was approximately 28.6 million.

39. a. Let x = no. of operators in 1998
$$x - 0.139x = 185,000$$
$$0.861x = 185,000$$
$$x \approx 214,866$$
There were approximately 214,866 switchboard operators in 1998.

b. Answers may vary.

41. Let x = first integer;
then $x + 1$ = next integer
and $x + 2$ = third integer.
$$x + (x + 1) + (x + 2) = 228$$
$$3x + 3 = 228$$
$$3x = 225$$
$$x = 75$$
$$x + 1 = x + 75 = 76$$
$$x + 2 = 75 + 2 = 77$$
The integers are 75, 76, and 77.

43. Let x = measure of second angle;
then $2x$ = measure of first angle
and $3x - 12$ = measure of third angle.
$$x + 2x + (3x - 12) = 180$$
$$6x - 12 = 180$$
$$6x = 192$$
$$x = 32$$
$$2x = 2(32) = 64$$
$$3x - 12 = 3(32) - 12 = 84$$
The angles measure 64°, 32°, and 84°.

45. Let x = height of sign;
then $2x + 12$ = length of sign.
$$2x + 2(2x + 12) = 312$$
$$2x + 4x + 24 = 312$$
$$6x + 24 = 312$$
$$6x = 288$$
$$x = 48$$
$$2x + 12 = 2(48) + 12 = 108$$
The height is 48 inches and the length is 108 inches.

47. Let x = width of room;
then $2x + 2$ = length of room.
$$2x + 2(2x + 2) = 40$$
$$2x + 4x + 4 = 40$$
$$6x + 4 = 40$$
$$6x = 36$$
$$x = 6$$
$$2x + 2 = 2(6) + 2 = 14$$
The width is 6 centimeters and the length is 14 centimeters.

49.
$$x + (x + 20) = 180$$
$$2x + 20 = 180$$
$$2x = 160$$
$$x = 80$$
$$x + 20 = (80) + 20 = 100$$
The angles measure 80° and 100°.

51.
$$x + 5x = 90$$
$$6x = 90$$
$$x = 15$$
$$5x = 5(15) = 75$$
The angles measure 15° and 75°.

53. Let x = measure of the angle; then
$180 - x$ = measure of its supplement
$$x = 3(180 - x) + 20$$
$$x = 540 - 3x + 20$$
$$4x = 560$$
$$x = 140$$
$$180 - x = 180 - 140 = 40$$
The angles measure 140° and 40°.

55. Let x = width of tank;
then $5(x + 1)$ = height of tank.
$$x + 5(x + 1) = 55.4$$
$$x + 5x + 5 = 55.4$$
$$6x + 5 = 55.4$$
$$6x = 50.4$$
$$x = 8.4$$
$$5(x + 1) = 5[(8.4) + 1] = 5(9.4) = 47$$
The width is 8.4 meters and the height is 47 meters.

57. Let x = hours for halogen;
then $25x$ = hours for fluorescent
and $x - 2500$ = hours for incandescent.
$$x + 25x + (x - 2500) = 105,500$$
$$27x - 2500 = 105,500$$
$$27x = 108,000$$
$$x = 4000$$
$$25x = 100,000;\ x - 2500 = 1500$$
The halogen has 4000 bulb hours. The fluorescent has 100,000 bulb hours. The incandescent has 1500 bulb hours.

59. Let x = number of returns filed electronically in 1999; then
$x + 1.564x$ = number in 2002
$$x + 1.564x = 54.1$$
$$2.564x = 54.1$$
$$x \approx 21.1$$
Approximately 21.1 million returns were filed electronically in 1999.

61. Let x = home runs by Sexson;
then $x + 2$ = home runs by Palmeiro
and $x + 4$ = home runs by Thome.
$$x + (x + 2) + (x + 4) = 141$$
$$3x + 6 = 141$$
$$3x = 135$$
$$x = 45$$
$$x + 2 = 47;\ x + 4 = 49$$
Sexson hit 45 home runs, Palmeiro hit 47 home runs, and Thome hit 49 home runs during the 2001 season.

63. $2a + b - c = 2(5) + (-1) - (3)$
$$= 10 - 1 - 3$$
$$= 6$$

65. $4ab - 3bc = 4(-5)(-8) - 3(-8)(2)$
$$= 160 + 48$$
$$= 208$$

67. $n^2 - m^2 = (-3)^2 - (-8)^2 = 9 - 64 = -55$

69. $P + PRT = 3000 + 3000(0.0325)(2)$
$$= 3195$$

71. Let x = trees worth of newsprint recycled
then $x + 30$ = trees worth of discarded
$$x = 0.27(x + 30)$$
$$x = 0.27x + 8.1$$
$$0.73x = 8.1$$
$$x = 11$$
About 11 million trees' worth of newsprint is recycled.

73. a.
$$y = -64.45x + 2795.5$$
$$0 = -64.45x + 2795.5$$
$$64.45x = 2795.5$$
$$x \approx 43;$$
$$1990 + 43 = 2003$$
The average annual number of cigarettes smoked will be 0 during the year 2033.

b. $y = -64.45x + 2795.5$
$$y = -64.45(15) + 2795.5$$
$$y = 1828.75$$
Americans will smoke about 1828.75 cigarettes annually in 2005.

c. $1828.75 \div 365 \approx 5$
An American adult will smoke an average of 5 cigarettes a day in 2005. No; this is the daily number of cigarettes for all American adults – smokers and non-smokers.

75. Let x = first odd integer;
then $x + 2$ = next odd integer
$$7x = 5(x + 2) + 54$$
$$7x = 5x + 10 + 54$$
$$7x = 5x + 64$$
$$2x = 64$$
$$x = 32, \text{ which is not an odd interger.}$$
Therefore, no such odd integers exist.

77. $R = C$
$$60x = 50x + 5000$$
$$10x = 5000$$
$$x = 500$$
$$R = 60x = 60(500) = 30,000$$
$$C = 50x + 5000 = 50(500) + 5000 = 30,000$$
To break even, 500 boards must be sold. You need $30,000 to produce the 500 boards.

79. The company makes a profit.

Section 2.3

Mental Math

1. $2x + y = 5$
$$y = 5 - 2x$$

2. $7x - y = 3$
$$-y = 3 - 7x$$
$$y = 7x - 3$$

3. $a - 5b = 8$
$$a = 5b + 8$$

4. $7r + s = 10$
$$s = 10 - 7r$$

5. $5j + k - h = 6$
$$k = h - 5j + 6$$

6. $w - 4y + z = 0$
$$z = 4y - w$$

Exercise Set 2.3

1. $D = rt$

$$\frac{D}{r} = \frac{rt}{r}$$

$$\frac{D}{r} = t$$

$$t = \frac{D}{r}$$

3. $I = PRT$

$$\frac{I}{PT} = \frac{PRT}{PR}$$

$$\frac{I}{PT} = R$$

$$R = \frac{I}{PT}$$

5. $9x - 4y = 16$

$$9x - 4y - 9x = 16 - 9x$$

$$-4y = 16 - 9x$$

$$\frac{-4y}{-4} = \frac{16 - 9x}{-4}$$

$$y = \frac{9x - 16}{4}$$

7. $P = 2L + 2W$

$$P - 2L = 2W$$

$$\frac{P - 2L}{2} = \frac{2W}{2}$$

$$\frac{P - 2L}{2} = W$$

$$W = \frac{P - 2L}{2}$$

9. $J = AC - 3$

$$J + 3 = AC$$

$$\frac{J + 3}{C} = \frac{AC}{C}$$

$$\frac{J + 3}{C} = A$$

$$A = \frac{J + 3}{C}$$

11. $W = gh - 3gt^2$

$$W = g(h - 3t^2)$$

$$\frac{W}{h - 3t^2} = \frac{g(h - 3t^2)}{h - 3t^2}$$

$$\frac{W}{h - 3t^2} = g$$

$$g = \frac{W}{h - 3t^2}$$

13. $T = C(2 + AB)$

$$T = 2C + ABC$$

$$T - 2C = 2C + ABC - 2C$$

$$T - 2C = ABC$$

$$\frac{T - 2C}{AC} = \frac{ABC}{AC}$$

$$\frac{T - 2C}{AC} = B$$

$$B = \frac{T - 2C}{AC}$$

15. $C = 2\pi r$

$$\frac{C}{2\pi} = \frac{2\pi r}{2\pi}$$

$$\frac{C}{2\pi} = r$$

$$r = \frac{C}{2\pi}$$

17. $E = I(r + R)$

$$E = Ir + IR$$

$$E - IR = Ir + IR - IR$$

$$E - IR = Ir$$

$$\frac{E - IR}{I} = \frac{Ir}{I}$$

$$\frac{E - IR}{I} = r$$

$$r = \frac{E - IR}{I}$$

19.

$$s = \frac{n}{2}(a+L)$$

$$2s = 2 \cdot \frac{n}{2}(a+L)$$

$$2s = n(a+L)$$

$$2s = na + nL$$

$$2s - na = na + nL - an$$

$$2s - na = nL$$

$$\frac{2s - na}{n} = \frac{nL}{n}$$

$$\frac{2s - na}{n} = L$$

$$L = \frac{2s - na}{n}$$

21.

$$N = 3st^4 - 5sv$$

$$N - 3st^4 = 3st^4 - 5sv - 3st^4$$

$$N - 3st^4 = -5sv$$

$$\frac{N - 3st^4}{-5s} = \frac{-5sv}{-5s}$$

$$\frac{3st^4 - N}{5s} = v$$

$$v = \frac{3st^4 - N}{5s}$$

23.

$$S = 2LW + 2LH + 2WH$$

$$S - 2LW = 2LW + 2LH + 2WH - 2LW$$

$$S - 2LW = 2LH + 2WH$$

$$S - 2LW = H(2L + 2W)$$

$$\frac{S - 2LW}{2L + 2W} = \frac{H(2L + 2W)}{2L + 2W}$$

$$\frac{S - 2LW}{2L + 2W} = H$$

$$H = \frac{S - 2LW}{2L + 2W}$$

25. $A = P\left(1 + \frac{r}{n}\right)^{nt} = 3500\left(1 + \frac{0.03}{n}\right)^{10t}$

n	1	2	4
A	\$4703.71	\$4713.99	\$4719.22

n	12	365
A	\$4722.74	\$4724.45

27. $A = P\left(1 + \frac{r}{n}\right)^{nt} = 6000\left(1 + \frac{0.04}{n}\right)^{5n}$

a. $n = 2$

$$A = 6000\left(1 + \frac{0.04}{2}\right)^{5 \cdot 2} \approx 7313.97$$

\$7313.97

b. $n = 4$

$$A = 6000\left(1 + \frac{0.04}{2}\right)^{5 \cdot 4} \approx 7321.14$$

\$7321.14

c. $n = 12$

$$A = 6000\left(1 + \frac{0.04}{2}\right)^{5 \cdot 12} \approx 7325.98$$

\$7325.98

29. $C = \frac{5}{9}(F - 32)$

$$C = \frac{5}{9}(104 - 32)$$

$$C = \frac{5}{9}(72)$$

$$C = 40°$$

The day's high temperature was 40°C.

31.
$$d = rt$$
$$2(90 \text{ mi}) = 50t$$
$$180 = 50t$$
$$\frac{180}{50} = t$$
$$t = 3.6$$
She takes 3.6 hours or 3 hours, 36 minutes to make the round trip.

33. $A = s^2 = (64)^2 = 4096 \text{ ft}^2$; $\dfrac{4096}{24} \approx 171$

There should be 171 packages of tiles bought.

35. $A = \dfrac{1}{2}bh$
$$18 = \dfrac{1}{2}(4)h$$
$$18 = 2h$$
$$9 = h$$
The height is 9 feet.

37. The area of one pair of walls is $2 \cdot 14 \cdot 8 = 224 \text{ ft}^2$ and the area of the other walls is $2 \cdot 16 \cdot 8 = 256 \text{ ft}^2$ for a total of 480 ft^2. Multiplying by 2, the number of coats, yields 960 ft^2. Dividing this by 500 yields 1.92. Thus, 2 gallons should be purchased.

39. a. $V = \pi r^2 h$
$$V = \pi(4.2)^2(2.12)$$
$$V = 1174.86$$
The volume of the cylinder is 1174.86 cubic meters.

b. $V = \dfrac{4}{3}\pi r^3$
$$V = \dfrac{4}{3}\pi(4.2)^3$$
$$V = 310.34$$
The volume of the sphere is 310.34 cubic meters.

c. $V = 1174.86 + 310.34 = 1485.20$
The volume of the tank is 1485.20 cubic meters.

41. Note that the radius of the circle is equal to $22,248 + 4000 = 26,248$.
$$C = 2\pi r$$
$$C = 2\pi(26,248)$$
$$C = 52,496\pi$$
$$C \approx 164,921.0479$$
The "length" of the Clarke belt is approximately 164,921 miles.

43. $V = \pi r^2 h$
1 mile = 5280 feet
1.3 miles = 6864 feet
$$3800 = \pi(4.2)^2(6864)$$
$$0.42 = r$$
The radius of the hole is 0.42 feet.

45. $C = \pi d$
$$= \pi(41.125)$$
$$= 41.125\pi \text{ ft}$$
$$\approx 129.1325 \text{ ft}$$
The circumference of Eartha is $41.125\pi \approx 129.1325$ feet.

47. $A = P\left(1 + \dfrac{r}{n}\right)^{nt}$
$$= 10,000\left(1 + \dfrac{0.085}{4}\right)^{4 \cdot 2}$$
$$= 10,000(1 + 0.02125)^{4 \cdot 2}$$
$$= \$11,831.96$$
$$\$11,831.96 - \$10,000 = \$1831.96$$

49.
$$C = 4h + 9f + 4p$$
$$C - 4h - 4p = 9f$$
$$\dfrac{C - 4h - 4p}{9} = f$$
$$f = \dfrac{C - 4h - 4p}{9}$$

51. $C = 4h + 9f + 4p$
$C = 4(7) + 9(14) + 4(6)$
$C = 178$
There are 178 calories in this serving.

53. $C = 4h + 9f + 4p$
$130 = 4(31) + 9(0) + 4p$
$130 = 124 + 4p$
$6 = 4p$
$\dfrac{6}{4} = p$
$p = 1.5$
There are 1.5 g of protein provided by this serving of raisins.

55. $\{-3, -2, -1\}$

57. $\{-3, -2, -1, 0, 1\}$

59. Answers may vary.

61.

Planet	AU from Sun
Mercury	0.388
Venus	0.723
Earth	1.00
Mars	1.523
Jupiter	5.202
Saturn	9.538
Uranus	19.193
Neptune	30.065
Pluto	39.505

63. Answers may vary.
$\dfrac{1,700,000,000}{250,000,000} = 6.8$
It cost \$6.80 per person to build the *Endeavour*.

65. Answers may vary.

67. $\dfrac{168 \text{ mi}}{1 \text{ hr}} \cdot \dfrac{5280 \text{ ft}}{1 \text{ mi}} \cdot \dfrac{1 \text{ hr}}{60 \text{ min}} \cdot \dfrac{1 \text{ min}}{60 \text{ sec}}$
$= 246.4 \text{ ft/sec}$
$d = rt$
$60.5 \text{ ft} = 246.6 \text{ ft/sec} \cdot t$
$0.25 \text{ sec} = t$
The ball would reach the plate in approximately 0.25 second.

69. $\dfrac{1}{4}$

71. $\dfrac{3}{8}$

73. $\dfrac{3}{8}$

75. $\dfrac{3}{4}$

77. 1

79. 1

Section 2.4

Mental Math

1. $x - 2 < 4$
$x < 6$
$\{x \mid x < 6\}$

2. $x - 1 > 6$
$x > 7$
$\{x \mid x > 7\}$

3. $x + 5 \geq 15$
$x \geq 10$
$\{x \mid x \geq 10\}$

4. $x + 1 \leq 8$
$x \leq 7$
$\{x \mid x \leq 7\}$

5. $3x > 12$

$\quad x > 4$

$\quad \{x \mid x > 4\}$

6. $5x < 20$

$\quad x < 4$

$\quad \{x \mid x < 4\}$

7. $\frac{x}{2} \le 1$

$\quad x \le 2$

$\quad \{x \mid x \le 2\}$

8. $\frac{x}{4} \ge 2$

$\quad x \ge 8$

$\quad \{x \mid x \ge 8\}$

Exercise Set 2.4

1. $(-\infty, -3)$

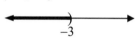

$\qquad\qquad -3$

3. $[0.3, \infty)$

$\qquad\qquad 0.3$

5. $\left(\frac{5}{9}, \infty\right)$

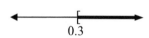

$\qquad\qquad \frac{5}{9}$

7. $(-2, 5)$

$\qquad\quad 2 \qquad 5$

9. $(-1, 5)$

$\qquad\quad -1 \qquad 5$

11. Answers may vary

13. D

15. B

17. $x - 7 \ge -9$

$\quad x \ge -2$

$\quad [-2, \infty)$

$\qquad\qquad -2$

19. $7x < 6x + 1$

$\quad x < 1$

$\quad (-\infty, 1)$

$\qquad\qquad 1$

21. $8x - 7 \le 7x - 5$

$\quad x - 7 \le -5$

$\quad x \le 2$

$\quad (-\infty, 2]$

$\qquad\qquad 2$

23. $2 + 4x > 5x + 6$

$\quad -4 > x$

$\quad x < -4$

$\quad (-\infty, -4)$

$\qquad\qquad -4$

25. $\frac{3}{4}x \ge 2$

$\quad x \ge \frac{8}{3}$

$\quad \left[\frac{8}{3}, \infty\right)$

$\qquad\qquad \frac{8}{3}$

27. $5x < -23.5$

$\quad x < -4.7$

$\quad (-\infty, -4.7)$

$\qquad\qquad -4.7$

29. $-3x \geq 9$

$\quad x \leq -3$

$\quad (-\infty, -3]$

$$\xleftarrow{\qquad\qquad}\underset{-3}{\rule{0pt}{0pt}]}\xrightarrow{\qquad\qquad}$$

31. $-x < -4$

$\quad x > 4$

$\quad (4, \infty)$

$$\xleftarrow{\qquad\qquad}\underset{4}{\rule{0pt}{0pt}(}\xrightarrow{\qquad\qquad}$$

33. $-2x + 7 \geq 9$

$\quad -2x \geq 2$

$\quad\quad x \leq -1$

$\quad (-\infty, -1]$

35. $15 + 2x \geq 4x - 7$

$\quad\quad 15 \geq 2x - 7$

$\quad\quad 22 \geq 2x$

$\quad\quad 11 \geq x \text{ or } x \leq 11$

$\quad (-\infty, 11]$

37. $3(x - 5) < 2(2x - 1)$

$\quad 3x - 15 < 4x - 2$

$\quad\quad -15 < x - 2$

$\quad\quad -13 < x$

$\quad\quad\quad x > -13$

$\quad (-13, \infty)$

39. $\dfrac{1}{2} + \dfrac{2}{3} \geq \dfrac{x}{6}$

$\quad 6\left(\dfrac{1}{2} + \dfrac{2}{3}\right) \geq 6\left(\dfrac{x}{6}\right)$

$\quad\quad 3 + 4 \geq x$

$\quad\quad\quad 7 \geq x$

$\quad\quad\quad x \leq 7$

$\quad (-\infty, 7]$

41. $4(x - 1) \geq 4x - 8$

$\quad 4x - 4 \geq 4x - 8$

$\quad\quad -4 \geq -8 \text{ (True for all } x)$

All real numbers

$(-\infty, \infty)$

43. $7x < 7(x - 2$

$\quad 7x < 7x - 14$

$\quad\quad 0 < -14 \text{ (False)}$

No solution; $\varnothing$

45. $4(2x + 1) > 4$

$\quad 8x + 4 > 4$

$\quad\quad 8x > 0$

$\quad\quad\; x > 0$

$\quad (0, \infty)$

47. $\quad \dfrac{x + 7}{5} > 1$

$\quad 5\left(\dfrac{x + 7}{5}\right) > 5(1)$

$\quad\quad x + 7 > 5$

$\quad\quad\quad x > -2$

$\quad (-2, \infty)$

49. $\quad \dfrac{-5x + 11}{2} \leq 7$

$\quad 2\left(\dfrac{-5x + 11}{2}\right) \leq 2(7)$

$\quad\quad -5x + 11 \leq 14$

$\quad\quad -5x \leq 3$

$\quad\quad\quad x \geq -\dfrac{3}{5}$

$\quad \left[-\dfrac{3}{5}, \infty\right)$

51. $8x - 16.4 \leq 10x + 2.8$

$\quad\quad -16.4 \leq 2x + 2.8$

$\quad\quad -19.2 \leq 2x$

$\quad\quad -9.6 \leq x$

$\quad\quad\quad x \geq -9.6$

$\quad [-9.6, \infty)$

53. $2(x - 3) > 70$

$\quad 2x - 6 > 70$

$\quad\quad 2x > 76$

$\quad\quad\; x > 38$

$\quad (38, \infty)$

55. Answers may vary.

57. $-5x + 4 \le -4(x-1)$
$-5x + 4 \le -4x + 4$
$-x \le 0$
$x \ge 0$
$[0, \infty)$

59. $\frac{1}{4}(x-7) \ge x+2$
$4 \cdot \frac{1}{4}(x-7) \ge 4(x+2)$
$x - 7 \ge 4x + 8$
$-7 \ge 3x + 8$
$-15 \ge 3x$
$-5 \ge x$
$x \le -5$
$(-\infty, -5]$

61. $\frac{2}{3}(x+2) < \frac{1}{5}(2x+7)$
$15\left[\frac{2}{3}(x+2)\right] < 15\left[\frac{1}{5}(2x+7)\right]$
$10(x+2) < 3(2x+7)$
$10x + 20 < 6x + 21$
$4x < 1$
$x < \frac{1}{4}$
$\left(-\infty, \frac{1}{4}\right)$

63. $4(x-6) + 2x - 4 \ge 3(x-7) + 10x$
$4x - 24 + 2x - 4 \ge 3x - 21 + 10x$
$6x - 28 \ge 13x - 21$
$-28 \ge 7x - 21$
$-7 \ge 7x$
$-1 \ge x$
$x \le -1$
$(-\infty, -1]$

65. $\frac{5x+1}{7} - \frac{2x-6}{4} \ge -4$
$28\left(\frac{5x+1}{7} - \frac{2x-6}{4}\right) \ge 28(-4)$
$4(5x+1) - 7(2x-6) \ge -112$
$20x + 4 - 14x + 42 \ge -112$
$6x + 46 \ge -112$
$6x \ge -158$
$x \ge -\frac{79}{3}$
$\left[-\frac{79}{3}, \infty\right)$

67. $\frac{-x+2}{2} - \frac{1-5x}{8} < -1$
$8\left(\frac{-x+2}{2} - \frac{1-5x}{8}\right) < 8(-1)$
$4(-x+2) - (1-5x) < -8$
$-4x + 8 - 1 + 5x < -8$
$x + 7 < -8$
$x < -15$
$(-\infty, -15)$

69. $0.8x + 0.6x \ge 4.2$
$1.4x \ge 4.2$
$x \ge 3$
$[3, \infty)$

71. $\frac{x+5}{5} - \frac{3+x}{8} \ge -\frac{3}{10}$
$40\left(\frac{x+5}{5} - \frac{3+x}{8}\right) \ge 40\left(-\frac{3}{10}\right)$
$8(x+5) - 5(3+x) \ge -12$
$8x + 40 - 15 - 5x \ge -12$
$3x + 25 \ge -12$
$3x \ge -37$
$x \ge -\frac{37}{3}$
$\left[-\frac{37}{3}, \infty\right)$

73.
$$\frac{x+3}{12}+\frac{x-5}{15}<\frac{2}{3}$$
$$60\left(\frac{x+3}{12}+\frac{x-5}{15}\right)<60\left(\frac{2}{3}\right)$$
$$5(x+3)+4(x-5)<20(2)$$
$$5x+15+4x-20<40$$
$$9x-5<40$$
$$9x<45$$
$$x<5$$
$$(-\infty,5)$$

75. Let x = her score on the final.
Then $\dfrac{72+67+82+79+2x}{6}\ge 60$
$$300+2x\ge 360$$
$$2x\ge 360$$
$$x\ge 30$$
She must score at least 30 on the final exam.

77. Let x = weight of the luggage and cargo.
Then $6(160)+x\le 2000$
$$960+x\le 2000$$
$$x\le 1040$$
The plane can carry a maximum of 1040 pound of luggage and cargo.

79.
$$0.37+0.23(x-1)\le 4.00$$
$$100[0.37+0.23(x-1)]\le 100(4.00)$$
$$37+23(x-1)\le 400$$
$$37+23x-23\le 400$$
$$23x+14\le 400$$
$$23x\le 386$$
$$x\le 16.78$$
At most 16 ounces can be mailed for $4.00.

81. Let n = number of calls made in a given month
Then $25<13+0.06n$
$$12<0.06n$$
$$200<n \text{ or } n>200$$
Plan 1 is more economical than Plan 2 when 200 calls are made.

83. $F\ge\dfrac{9}{5}C+32$
$$F\ge\frac{9}{5}(500)+32$$
$$F\ge 932°$$
Glass is a liquid at temperatures of 932°F or higher.

85. $s=2806.6t+32,558$

  a. $60,000<2806.6t+32,558$
  $$27,442<2806.6t$$
  $$9.78<t$$
  Beginning salaries will be greater than $60,000 in $1995+9=2004$.

  b. Answers may vary.

87. The consumption of whole milk is decreasing. The graph of the line is going down over time.

89. $t=2008-1997=11$
$$w=-0.13t+8$$
$$w=-0.13(11)+8$$
$$w=6.57$$
The consumption of whole milk will be about 6.57 gallons per person per year.

91. $-0.13t+8\le 7$
$$-0.13t\le -1$$
$$t\ge 7.69$$
$$1997+7=2004$$
Consumption of whole mike will be less than 7 gallons per person per year beginning in 2004.

93. Answers may vary.

95. $x<5$ and $x>1$
The integers are 2, 3, 4 or {2, 3, 4}.

97. $x\ge -2$ and $x\ge 2$
The integers are {2, 3, 4, ...}.

99. $\{x \mid 0 \le x \le 5\}$
$[0, 5]$

101. $\left\{x \mid -\dfrac{1}{2} < x < \dfrac{3}{2}\right\}$
$\left(-\dfrac{1}{2}, \dfrac{3}{2}\right)$

103. $4(x-1) \ge 4x - 8$
$4x - 4 \ge 4x - 8$
$\qquad -4 \ge -8$ (True for all x)
All real numbers.
$(-\infty, \infty)$

105. $7x < 7(x-2)$
$\quad 7x < 7x - 14$
$\qquad 0 < -14$ (False)
$\varnothing$

Integrated Review

1. $-4x = 20$
$\dfrac{-4x}{-4} = \dfrac{20}{-4}$
$\quad x = -5$
The solution is -5.

2. $-4x < 20$
$\dfrac{-4x}{-4} > \dfrac{20}{-4}$
$\quad x > -5$
$(-5, \infty)$

3. $\dfrac{3x}{4} \ge 2$
$4\left(\dfrac{3x}{4}\right) \ge 4(2)$
$\quad 3x \ge 8$
$\qquad x \ge \dfrac{8}{3}$
$\left[\dfrac{8}{3}, \infty\right)$

4. $5x + 3 \ge 2 + 4x$
$\quad x + 3 \ge 2$
$\qquad x \ge -1$
$[-1, \infty)$

5. $6(y - 4) = 3(y - 8)$
$\quad 6y - 24 = 3y - 24$
$\qquad 3y = 0$
$\qquad y = 0$
The solution is 0.

6. $-4x \le \dfrac{2}{5}$
$\quad -20x \le 2$
$\qquad x \ge -\dfrac{1}{10}$
$\left[-\dfrac{1}{10}, \infty\right)$

7. $-3x \ge \dfrac{1}{2}$
$2(-3x) \ge 2\left(\dfrac{1}{2}\right)$
$\quad -6x \ge 1$
$\qquad x \le -\dfrac{1}{6}$
$\left(-\infty, -\dfrac{1}{6}\right]$

8. $5(y + 4) = 4(y + 5)$
$\quad 5y + 20 = 4y + 20$
$\qquad y = 0$
The solution is 0.

9. $7x < 7(x-2)$
$7x < 7x - 2$
$0 < -2$ (False)
No Solution; $\varnothing$

10. $\dfrac{-5x+11}{2} \le 7$
$2\left(\dfrac{-5x+11}{2}\right) \le 2(7)$
$-5x + 11 \le 14$
$-5x \le 3$
$x \ge -\dfrac{3}{5}$
$\left[-\dfrac{3}{5}, \infty\right)$

11. $-5x + 1.5 = -19.5$
$10(-5x + 1.5) = 10(-19.5)$
$-50x + 15 = -195$
$-50x = -210$
$x = 4.2$

The solution is 4.2.

12. $-5x + 4 = -26$
$-5x = -30$
$x = 6$
The solution is 6.

13. $5 + 2x - x = -x + 3 - 14$
$5 + x = -x - 11$
$5 + 2x = -11$
$2x = -16$
$x = -8$
The solution is -8.

14. $12x + 14 < 11x - 2$
$x + 14 < -2$
$x < -16$
$(-\infty, -16)$

15. $\dfrac{x}{5} - \dfrac{x}{4} = \dfrac{x-2}{2}$
$20\left(\dfrac{x}{5} - \dfrac{x}{4}\right) = 20\left(\dfrac{x-2}{2}\right)$
$4x - 5x = 10(x-2)$
$-x = 10x - 20$
$-11x = -20$
$x = \dfrac{20}{11}$

The solution is $\dfrac{20}{11}$.

16. $12x - 12 = 8(x-1)$
$12x - 12 = 8x - 8$
$4x - 12 = -8$
$4x = 4$
$x = 1$
The solution is 1.

17. $2(x-3) > 70$
$2x - 6 > 70$
$2x > 76$
$x > 38$
$(38, \infty)$

18. $-3x - 4.7 = 11.8$
$10(-3x - 4.7) = 10(11.8)$
$-30x - 47 = 118$
$-30x = 165$
$x = -\dfrac{11}{2} = -5.5$

The solution is -5.5.

19. $-2(b-4) - (3b-1) = 5b + 3$
$-2b + 8 - 3b + 1 = 5b + 3$
$-5b + 9 = 5b + 3$
$-10b = -6$
$x = \dfrac{3}{5}$

The solution is $\dfrac{3}{5}$.

20. $8(x+3) < 7(x+5) + x$
$8x + 24 < 7x + 35 + x$
$8x + 24 < 8x + 35$
$24 < 35$ (True for all x)
All real numbers; $(-\infty, \infty)$

21. $\dfrac{3t+1}{8} = \dfrac{5+2t}{7} + 2$
$56\left(\dfrac{3t+1}{8}\right) = 56\left(\dfrac{5+2t}{7}\right) + 56(2)$
$7(3t+1) = 8(5+2t) + 112$
$21t + 7 = 40 + 16t + 112$
$21t + 7 = 16t + 152$
$5t = 145$
$t = 29$
The solution is 29.

22. $4(x-6) - x = 8(x-3) - 5x$
$4x - 24 - x = 8x - 24 - 5x$
$3x - 24 = 3x - 24$
$-24 = -24$ (True for all x)
The solution is all real numbers.

23. $\dfrac{x+3}{12} + \dfrac{x-5}{15} < \dfrac{2}{3}$
$60\left(\dfrac{x+3}{12} + \dfrac{x-5}{15}\right) < 60\left(\dfrac{2}{3}\right)$
$5(x+3) + 4(x-5) < 20(2)$
$5x + 15 + 4x - 20 < 40$
$9x - 5 < 40$
$9x < 45$
$x < 5$
$(-\infty, 5)$

24. $\dfrac{y}{3} + \dfrac{y}{5} = \dfrac{y+3}{10}$
$30\left(\dfrac{y}{3}\right) + 30\left(\dfrac{y}{5}\right) = 30\left(\dfrac{y+3}{10}\right)$
$10y + 6y = 3(y+3)$
$16y = 3y + 9$
$13y = 9$
$y = \dfrac{9}{13}$
The solution is $\dfrac{9}{13}$.

25. $5(x-6) + 2x > 3(2x-1) - 4$
$5x - 30 + 2x > 6x - 3 - 4$
$7x - 30 > 6x - 7$
$x > 23$
$(23, \infty)$

26. $14(x-1) - 7x \le 2(3x-6) + 4$
$14x - 14 - 7x \le 6x - 12 + 4$
$7x - 14 \le 6x - 8$
$x \le 6$
$(-\infty, 6]$

27. $\dfrac{1}{4}(3x+2) - x \ge \dfrac{3}{8}(x-5) + 2$
$8\left[\dfrac{1}{4}(3x+2) - x\right] \ge 8\left[\dfrac{3}{8}(x-5) + 2\right]$
$2(3x+2) - 8x \ge 3(x-5) + 16$
$6x + 4 - 8x \ge 3x - 15 + 16$
$-2x + 4 \ge 3x + 1$
$3 \ge 5x$
$\dfrac{3}{5} \ge x$ or $x \le \dfrac{3}{5}$
$\left(-\infty, \dfrac{3}{5}\right]$

28. $\dfrac{1}{3}(x-10) - 4x > \dfrac{5}{6}(2x+1) - 1$
$6\left[\dfrac{1}{3}(x-10) - 4x\right] > 6\left[\dfrac{5}{6}(2x+1) - 1\right]$
$2(x-10) - 24x > 5(2x+1) - 6$
$2x - 20 - 24x > 10x + 5 - 6$
$-22x - 20 > 10x - 1$
$-19 > 32x$
$-\dfrac{19}{32} > x$ or $x < -\dfrac{19}{32}$
$\left(-\infty, -\dfrac{19}{32}\right)$

Exercise Set 2.5

1. $C \cup D = \{2, 3, 4, 5, 6, 7\}$

3. $A \cap D = \{4, 6\}$

5. $A \cup B = \{..., -2, -1, 0, 1, ...\}$

7. $B \cap D = \{5, 7\}$

9. $B \cup C = \{x \mid x \text{ is an odd interger}$
$\qquad\qquad \text{or } x = 2 \text{ or } x = 4\}$

11. $A \cap C = \{2, 4\}$

13. $x < 5$ and $x > -2$
$\quad -2 < x < 5$
$\quad (-2, 5)$

15. $x + 1 \geq 7$ and $3x - 1 \geq 5$
$\quad\; x \geq 6$ and $\quad\; 3x \geq 6$
$\qquad\qquad\qquad\qquad x \geq 2$
$\quad x \geq 6$
$\quad [6, \infty)$

17. $4x + 2 \leq -10$ and $2x \leq 0$
$\quad\; 4x \leq -12$ and $\quad x \leq 0$
$\qquad\; x \leq -3$
$\quad x \leq -3$
$\quad (-\infty, -3]$

19. $5 < x - 6 < 11$
$\quad 11 < x < 17$
$\quad (11, 17)$

21. $-2 \leq 3x - 5 \leq 7$
$\quad\; 3 \leq 3x \leq 12$
$\quad\; 1 \leq x \leq 4$
$\quad\; [1, 4]$

23. $1 \leq \dfrac{2}{3}x + 3 \leq 4$
$\quad -2 \leq \dfrac{2}{3}x \leq 1$
$\quad -3 \leq x \leq \dfrac{3}{2}$
$\quad \left[-3, \dfrac{3}{2}\right]$

25. $-5 \leq \dfrac{x+1}{4} \leq -2$
$\quad -20 \leq x + 1 \leq -8$
$\quad -21 \leq x \leq -9$
$\quad [-21, -9]$

27. $x < -1$ or $x > 0$
$\quad (-\infty, -1) \cup (0, \infty)$

29. $-2x \leq -4$ or $5x - 20 \geq 5$
$\quad\; x \geq 2$ or $\qquad 5x \geq 25$
$\qquad\qquad\qquad\qquad x \geq 5$
$\quad x \geq 2$
$\quad [2, \infty)$

31. $3(x - 1) < 12$ or $x + 7 > 10$
$\quad 3x - 3 < 12$ or $\qquad x > 3$
$\quad\; 3x < 15$
$\qquad x < 5$
All real numbers.
$(-\infty, \infty)$

33. Answers may very.

35. $x < 2$ and $x > -1$
$-1 < x < 2$
$(-1, 2)$

37. $x < 2$ or $x > -1$
All real numbers.
$(-\infty, \infty)$

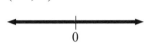

39. $x \geq -5$ and $x \geq -1$
$x \geq -1$
$[-1, \infty)$

41. $x \geq -5$ or $x \geq -1$
$x \geq -5$
$[-5, \infty)$

43. $0 \leq 2x - 3 \leq 9$
$3 \leq 2x \leq 12$
$\dfrac{3}{2} \leq x \leq 6$
$\left[\dfrac{3}{2}, 6\right]$

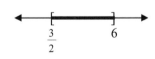

45. $\dfrac{1}{2} < x - \dfrac{3}{4} < 2$
$4\left(\dfrac{1}{2}\right) < 4\left(x - \dfrac{3}{4}\right) < 4(2)$
$2 < 4x - 3 < 8$
$5 < 4x < 11$
$\dfrac{5}{4} < x < \dfrac{11}{4}$
$\left(\dfrac{5}{4}, \dfrac{11}{4}\right)$

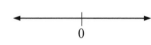

47. $x + 3 \geq 3$ and $x + 3 \leq 2$
     $x \geq 0$ and $x \leq -1$
No solution exist.
$\varnothing$

49. $3x \geq 5$ or $-x - 6 < 1$
$x \geq \dfrac{5}{3}$ or $-x < 7$
                   $x > -7$
$x > -7$
$(-7, \infty)$

51. $0 < \dfrac{5 - 2x}{3} < 5$
$0 < 5 - 2x < 15$
$\dfrac{-5}{-2} > \dfrac{-2x}{-2} > \dfrac{10}{-2}$
$\dfrac{5}{2} > x > -5$
$-5 < x < \dfrac{5}{2}$
$\left(-5, \dfrac{5}{2}\right)$

53. $-6 < 3(x-2) \le 8$
$-6 < 3x - 6 \le 8$
$0 < 3x \le 14$
$0 < x \le \dfrac{14}{3}$
$\left(0, \dfrac{14}{3}\right]$

55. $-x + 5 > 6$ and $1 + 2x \le -5$
$-x > 1$ and $2x \le -6$
$x < -1$ and $x \le -3$
$x \le -3$
$(-\infty, -3]$

57. $3x + 2 \le 5$ or $7x > 29$
$3x \le 3$ or $x > \dfrac{29}{7}$
$x \le 1$ or $x > \dfrac{29}{7}$
$(-\infty, 1] \cup \left(\dfrac{29}{7}, \infty\right)$

59. $5 - x > 7$ and $2x + 3 \ge 13$
$-x > 2$ and $2x \ge 10$
$x < -2$ and $x \ge 5$
No solution exist.
$\varnothing$

61. $-\dfrac{1}{2} \le \dfrac{4x-1}{6} < \dfrac{5}{6}$
$6\left(-\dfrac{1}{2}\right) \le 6\left(\dfrac{4x-1}{6}\right) < 6\left(\dfrac{5}{6}\right)$
$-3 \le 4x - 1 < 5$
$-2 \le 4x < 6$
$-\dfrac{1}{2} \le x < \dfrac{3}{2}$
$\left[-\dfrac{1}{2}, \dfrac{3}{2}\right)$

63. $\dfrac{1}{15} < \dfrac{8-3x}{15} < \dfrac{4}{5}$
$15\left(\dfrac{1}{15}\right) < 15\left(\dfrac{8-3x}{15}\right) < 15\left(\dfrac{4}{5}\right)$
$1 < 8 - 3x < 12$
$-7 < -3x < 4$
$-\dfrac{4}{3} < x < \dfrac{7}{3}$
$\left(-\dfrac{4}{3}, \dfrac{7}{3}\right)$

65. $0.3 < 0.2x - 0.9 < 1.5$
$1.2 < 0.2x < 2.4$
$6 < x < 12$
$(6, 12)$

67. $|-7| - |19| = 7 - 19 = -12$

69. $-(-6) - |-10| = 6 - 10 = -4$

71. $|x| = 7$
$x = -7, 7$

73. $|x| = 0$
$x = 0$

75. $-29 \le C \le 35$
$-29 \le \dfrac{5}{9}(F - 32) \le 35$
$-52.5 \le F - 32 \le 63$
$-20.2 \le F \le 95$
$-20.2° \le F \le 95°$

77. $70 \le \dfrac{68 + 65 + 75 + 78 + 2x}{6} \le 79$
$420 \le 286 + 2x \le 474$
$134 \le 2x \le 188$
$67 \le x \le 94$

If Christian scores between 67 and 94 inclusive on his final exam, he will receive a C in the course.

79. The years that the consumption of pork was greater than 48 pounds per person were 1994, 1995, 1998, and 1999. The years that the consumption of chicken was greater than 48 pounds per person were 1994, 1995, 1996, 1997, 1998, 1999, 2000, and 2001. The years in common are 1994, 1995, 1998, and 1999.

81. $2x - 3 < 3x + 1 < 4x - 5$
$2x - 3 < 3x + 1$ and $3x + 1 < 4x - 5$
$-x < 4$ and $-x < -6$
$x > -4$ and $x > 6$
$x > 6$
$(6, \infty)$

83. $-3(x - 2) \le 3 - 2x \le 10 - 3x$
$-3x + 6 \le 3 - 2x$ and $3 - 2x \le 10 - 3x$
$-x \le -3$ and $x \le 7$
$x \ge 3$
$3 \le x \le 7$
$[3, 7]$

85. $5x - 8 < 2(2 + x) < -2(1 + 2x)$
$5x - 8 < 4 + 2x$ and $4 + 2x < -2 - 4x$
$3x < 12$ and $6x < -6$
$x < 4$ and $x < -1$
$x < -1$
$(-\infty, -1)$

Section 2.6

Mental Math

1. $|-7| = 7$

2. $|-8| = 8$

3. $-|5| = -5$

4. $-|10| = -10$

5. $-|-6| = -6$

6. $-|-3| = -3$

7. $|-3| + |-2| + |-7| = 3 + 2 + 7 = 12$

8. $|-1| + |-6| + |-8| = 1 + 6 + 8 = 15$

Exercise Set 2.6

1. $|x| = 7$
$x = 7$ or $x = -7$
The solution set is $\{-7, 7\}$.

3. $|3x| = 12.6$
$3x = 12.6$ or $3x = -12.6$
$x = 4.2$ or $x = -4.2$
The solution set is $\{-4.2, 4.2\}$.

5. $|2x - 5| = 9$
$2x - 5 = 9$ or $2x - 5 = -9$
$2x = 14$ or $2x = -4$
$x = 7$ or $x = -2$
The solution set is $\{-2, 7\}$.

7. $\left|\dfrac{x}{2} - 3\right| = 1$

$\dfrac{x}{2} - 3 = 1$ or $\dfrac{x}{2} - 3 = -1$

$2\left(\dfrac{x}{2} - 3\right) = 2(1)$ or $2\left(\dfrac{x}{2} - 3\right) = 2(-1)$

$x - 6 = 2$ or $x - 6 = -2$
$x = 8$ or $x = 4$
The solution set is $\{4, 8\}$.

9. $|z| + 4 = 9$
$|z| = 5$
$z = -5$ or $z = -5$
The solution set is $\{-5, 5\}$.

11. $|3x| + 5 = 14$
$|3x| = 9$
$3x = 9$ or $3x = -9$
$x = 3$ or $x = -3$
The solution set is $\{-3, 3\}$.

13. $|2x| = 0$
$2x = 0$
$x = 0$
The solution set is $\{0\}$.

15. $|4n + 1| + 10 = 4$
$|4n + 1| = -6$ which is impossible.
The solution set is $\varnothing$.

17. $|5x - 1| = 0$
$5x - 1 = 0$
$5x = 1$
$x = \dfrac{1}{5}$
The solution set is $\left\{\dfrac{1}{5}\right\}$.

19. $|x| = 5$

21. $|5x - 7| = |3x + 11|$
$5x - 7 = 3x + 11$ or $5x - 7 = -(3x + 11)$
$2x = 18$ or $5x - 7 = -3x - 11$
$x = 9$ or $8x = -4$
$x = -\dfrac{1}{2}$
The solution set is $\left\{-\dfrac{1}{2}, 9\right\}$.

23. $|z + 8| = |z - 3|$
$z + 8 = z - 3$ or $z + 8 = -(z - 3)$
$8 = -3$ or $z + 8 = -z + 3$
$2z = -5$
$z = -\dfrac{5}{2}$
The solution set is $\left\{-\dfrac{5}{2}\right\}$.

25. Answers may vary.

27. $|x| = 4$
$x = 4$ or $x = -4$
The solution set is $\{-4, 4\}$.

29. $|y| = 0;\ y = 0$
The solution set is $\{0\}$.

31. $|z| = -2$ is impossible.
The solution set is $\varnothing$.

33. $|7-3x|=7$

$7-3x=7$ or $7-3x=-7$

$-3x=0$ or $-3x=-14$

$x=0$ or $x=\dfrac{14}{3}$

The solution set is $\left\{0, \dfrac{14}{3}\right\}$.

35. $|6x|-1=11$

$|6x|=12$

$6x=12$ or $6x=-12$

$x=2$ or $x=-2$

The solution set is $\{-2, 2\}$.

37. $|4p|=-8$ is impossible.

The solution set is $\varnothing$.

39. $|x-3|+3=7$

$|x-3|=4$

$x-3=4$ or $x-3=-4$

$x=7$ or $x=-1$

The solution set is $\{-1, 7\}$.

41. $\left|\dfrac{z}{4}+5\right|=-7$ is impossible.

The solution set is $\varnothing$.

43. $|9v-3|=-8$ is impossible.

The solution set is $\varnothing$.

45. $|8n+1|=0$

$8n+1=0$

$8n=-1$ so $n=-\dfrac{1}{8}$

The solution set is $\left\{-\dfrac{1}{8}\right\}$.

47. $|1-6c|-7=-3$

$|1-6c|=4$

$1-6c=4$ or $1-6c=-4$

$6c=3$ or $6c=-5$

$c=\dfrac{1}{2}$ or $c=-\dfrac{5}{6}$

The solution set is $\left\{-\dfrac{5}{6}, \dfrac{1}{2}\right\}$.

49. $|5x+1|=11$

$5x+1=11$ or $5x+1=-11$

$5x=10$ or $5x=-12$

$x=2$ or $x=-\dfrac{12}{5}$

The solution set is $\left\{-\dfrac{12}{5}, 2\right\}$.

51. $|4x-2|=|-10|$

$|4x-2|=10$

$4x-2=10$ or $4x-2=-10$

$4x=12$ or $4x=-8$

$x=3$ or $x=-2$

The solution set is $\{-2, 3\}$.

53. $|5x+1|=|4x-7|$

$5x+1=4x-7$ or $5x+1=-(4x-7)$

$x=-8$ or $5x+1=-4x+7$

$9x=6$

$x=\dfrac{2}{3}$

The solution set is $\left\{-8, \dfrac{2}{3}\right\}$.

55. $|6+2x|=-|-7|$

$|6+2x|=-7$ which is impossible.

The solution set is $\varnothing$.

57. $|2x-6|=|10-2x|$

$2x-6=10-2x$ or $2x-6=-(10-2x)$

$4x=16$ or $2x-6=-10+2x$

$x=4$ or $-6=-10$

$-6=-10$ is impossible.

The solution set is $\{4\}$.

59. $\left|\dfrac{2x-5}{3}\right| = 7$

$\dfrac{2x-5}{3} = 7$ or $\dfrac{2x-5}{3} = -7$

$2x - 5 = 21$ or $2x - 5 = -21$

$2x = 26$ or $2x = -16$

$x = 13$ or $x = -8$

The solution set is $\{-8, 13\}$.

61. $2 + |5n| = 17$

$|5n| = 15$

$5n = 15$ or $5n = -15$

$n = 3$ or $n = -3$

The solution set is $\{-3, 3\}$.

63. $\left|\dfrac{2x-1}{3}\right| = |-5|$

$\left|\dfrac{2x-1}{3}\right| = 5$

$\dfrac{2x-1}{3} = 5$ or $\dfrac{2x-1}{3} = -5$

$2x - 1 = 15$ or $2x - 1 = -15$

$2x = 16$ or $2x = -14$

$x = 8$ or $x = -7$

The solution set is $\{-7, 8\}$.

65. $|2y - 3| = |9 - 4y|$

$2y - 3 = 9 - 4y$ or $2y - 3 = -(9 - 4y)$

$6y = 12$ or $2y - 3 = -9 + 4y$

$y = 2$ or $-2y = -6$

$y = 3$

The solution set is $\{2, 3\}$.

67. $\left|\dfrac{3n+2}{8}\right| = |-1|$

$\left|\dfrac{3n+2}{8}\right| = 1$

$\dfrac{3n+2}{8} = 1$ or $\dfrac{3n+2}{8} = -1$

$3n + 2 = 8$ or $3n + 2 = -8$

$3n = 6$ or $3n = -10$

$n = 2$ or $n = -\dfrac{10}{3}$

The solution set is $\left\{-\dfrac{10}{3}, 2\right\}$.

69. $|x + 4| = |7 - x|$

$x + 4 = 7 - x$ or $x + 4 = -(7 - x)$

$2x = 3$ or $x + 4 = -7 + x$

$x = \dfrac{3}{2}$ or $4 = -7$

$4 = -7$ is impossible.

The solution set is $\left\{\dfrac{3}{2}\right\}$.

71. $\left|\dfrac{8c-7}{3}\right| = -|-5|$

$\left|\dfrac{8c-7}{3}\right| = -5$ which is impossible.

The solution set is $\varnothing$.

73. Answers may vary.

75. 33% of cheese consumption came from cheddar cheese.

77. $32\% \cdot (120 \text{ pounds}) = 0.32(120 \text{ pounds})$
$= 38.4 \text{ pounds}$

We might expect they consumed 38.4 pounds.

79. Answers may vary.

81. $|y| < 0$
No solution.

83. $|x| = 2$

85. $|2x - 1| = 4$

Section 2.7

Mental Math

1. D

2. E

3. C

4. B

5. A

Exercise Set 2.7

1. $|x| \le 4$
$-4 \le x \le 4$
$[-4, 4]$

3. $|x - 3| < 2$
$-2 < x - 3 < 2$
$1 < x < 5$
$(1, 5)$

5. $|x + 3| < 2$
$-2 < x + 3 < 2$
$-5 < x < -1$
$(-5, -1)$

7. $|2x + 7| \le 13$
$-13 \le 2x + 7 \le 13$
$-20 \le 2x \le 6$
$-10 \le x \le 3$
$[-10, 3]$

9. $|x| + 7 \le 12$
$|x| \le 5$
$-5 \le x \le 5$
$[-5, 5]$

11. $|3x - 1| < -5$
No real solutions; $\varnothing$

13. $|x - 6| - 7 \le -1$
$|x - 6| \le 6$
$-6 \le x - 6 \le 6$
$0 \le x \le 12$
$[0, 12]$

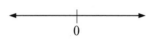

15. $|x| > 3$
$x < -3$ or $x > 3$
$(-\infty, -3) \cup (3, \infty)$

17. $|x + 10| \ge 14$
$x + 10 \le -14$ or $x + 10 \ge 14$
$x \le -24$ or $\quad x \ge 4$
$(-\infty, -24] \cup [4, \infty)$

19. $|x| + 2 > 6$
$|x| > 4$
$x < -4$ or $x > 4$
$(-\infty, -4) \cup (4, \infty)$

21. $|5x| > -4$
All real numbers.
$(-\infty, \infty)$

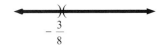

23. $|6x - 8| + 3 > 7$
$|6x - 8| > 4$
$6x - 8 < -4$ or $6x - 8 > 4$
$6x < 4$ or $6x > 12$
$x < \dfrac{2}{3}$ or $x > 2$
$\left(-\infty, \dfrac{2}{3}\right) \cup (2, \infty)$

25. $|x| \le 0$
$|x| = 0$
$x = 0$

27. $|8x + 3| > 0$ only excludes $|8x + 3| = 0$
$8x + 3 = 0$
$8x = -3$
$x = -\dfrac{3}{8}$

All real numbers except $-\dfrac{3}{8}$.
$\left(-\infty, -\dfrac{3}{8}\right) \cup \left(-\dfrac{3}{8}, \infty\right)$

29. $|x| \le 2$
$-2 \le x \le 2$
$[-2, 2]$

31. $|y| > 1$
$y < -1$ or $y > 1$
$(-\infty, -1) \cup (1, \infty)$

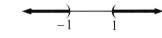

33. $|x - 3| < 8$
$-8 < x - 3 < 8$
$-5 < x < 11$
$(-5, 11)$

35. $|0.6x - 3| > 0.6$
$0.6x - 3 < -0.6$ or $0.6x - 3 > 0.6$
$0.6x < -2.4$ or $0.6x > 3.6$
$x < 4$ or $x > 6$
$(-\infty, 4) \cup (6, \infty)$

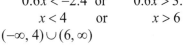

37. $5 + |x| \le 2$
$|x| \le -3$
No real solution.
$\varnothing$

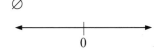

39. $|x| > -4$
All real numbers.
$(-\infty, \infty)$

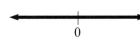

41. $|2x-7| \le 11$
$-11 \le 2x-7 \le 11$
$-4 \le 2x \le 18$
$-2 \le x \le 9$
$[-2, 9]$

43. $|x+5|+2 \ge 8$
$|x+5| \ge 6$
$x+5 \le -6$ or $x+5 \ge 6$
$x \le -11$ or $x \ge 1$
$(-\infty, -11] \cup [1, \infty)$

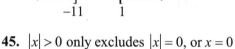

45. $|x| > 0$ only excludes $|x| = 0$, or $x = 0$
All real numbers except $x = 0$.
$(-\infty, 0) \cup (0, \infty)$

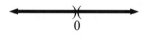

47. $9+|x| > 7$
$|x| > -2$
All real numbers.
$(-\infty, \infty)$

49. $6+|4x-1| \le 9$
$|4x-1| \le 3$
$-3 \le 4x-1 \le 3$
$-2 \le 4x \le 4$
$-\dfrac{1}{2} \le x \le 1$
$\left[-\dfrac{1}{2}, 1\right]$

51. $\left|\dfrac{2}{3}x+1\right| > 1$
$\dfrac{2}{3}x+1 < -1$ or $\dfrac{2}{3}x+1 > 1$
$\dfrac{2}{3}x < -2$ or $\dfrac{2}{3}x > 0$
$x < -3$ or $x > 0$
$(-\infty, -3) \cup (0, \infty)$

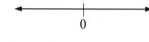

53. $|5x+3| < -6$
No real solution.
$\varnothing$

55. $|8x+3| > 0$
All real numbers.
$(-\infty, \infty)$

57. $|1+3x|+4 < 5$
$|1+3x| < 1$
$-1 < 1+3x < 1$
$-2 < 3x < 0$
$-\dfrac{2}{3} < x < 0$
$\left(-\dfrac{2}{3}, 0\right)$

59. $\left|\dfrac{x+6}{3}\right| > 2$

$\dfrac{x+6}{3} < -2$ or $\dfrac{x+6}{3} > 2$

$x + 6 < -6$ or $x + 6 > 6$

$\quad x < -12$ or $\qquad x > 0$

$(-\infty, -12) \cup (0, \infty)$

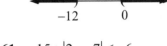

61. $-15 + |2x - 7| \le -6$

$\qquad |2x - 7| \le 9$

$-9 \le 2x - 7 \le 9$

$-2 \le 2x \le 16$

$-1 \le x \le 8$

$[-1, 8]$

63. $\left|2x + \dfrac{3}{4}\right| - 7 \le -2$

$\left|2x + \dfrac{3}{4}\right| \le 5$

$-5 \le 2x + \dfrac{3}{4} \le 5$

$-20 \le 8x + 3 \le 20$

$-23 \le 8x \le 17$

$-\dfrac{23}{8} \le x \le \dfrac{17}{8}$

$\left[-\dfrac{23}{8}, \dfrac{17}{8}\right]$

65. $|2x - 3| < 7$

$-7 < 2x - 3 < 7$

$-4 < 2x < 10$

$-2 < x < 5$

The solution set is $(-2, 5)$.

67. $|2x - 3| = 7$

$2x - 3 = 7$ or $2x - 3 = -7$

$2x = 10$ or $\quad 2x = -4$

$x = 5$ or $\qquad x = -2$

The solution set is $\{-2, 5\}$.

69. $|x - 5| \ge 12$

$x - 5 \le -12$ or $x - 5 \ge 12$

$x \le -7$ or $\qquad x \ge 17$

The solution set is $(-\infty, -7] \cup [17, \infty)$.

71. $|9 + 4x| = 0$

$9 + 4x = 0$

$4x = -9$

$x = -\dfrac{9}{4}$

The solution set is $\left\{-\dfrac{9}{4}\right\}$.

73. $|2x + 1| + 4 < 7$

$\qquad |2x + 1| < 3$

$-3 < 2x + 1 < 3$

$-4 < 2x < 2$

$-2 < x < 1$

The solution set is $(-2, 1)$.

75. $|3x - 5| + 4 = 5$

$\qquad |3x - 5| = 1$

$3x - 5 = 1$ or $3x - 5 = -1$

$3x = 6$ or $\quad 3x = 4$

$x = 2$ or $\qquad x = \dfrac{4}{3}$

The solution set is $\left\{\dfrac{4}{3}, 2\right\}$.

77. $|x + 11| = -1$ is impossible.

The solution set is $\varnothing$.

20.
$$\frac{2t-1}{3} = \frac{3t+2}{15}$$
$$15\left(\frac{2t-1}{3}\right) = 15\left(\frac{3t+2}{15}\right)$$
$$5(2t-1) = 3t+2$$
$$10t-5 = 3t+2$$
$$7t = 7$$
$$t = 1$$

21.
$$\frac{2(t+1)}{3} = \frac{2(t-1)}{3}$$
$$3\left[\frac{2(t+1)}{3}\right] = 3\left[\frac{2(t-1)}{3}\right]$$
$$2(t+1) = 2(t-1)$$
$$2t+2 = 2t-2$$
$$2 = -2$$
No solution

22.
$$\frac{3a-3}{6} = \frac{4a+1}{15} + 2$$
$$30\left(\frac{3a-3}{6}\right) = 30\left(\frac{4a+1}{15} + 2\right)$$
$$5(3a-3) = 2(4a+1) + 30(2)$$
$$15a-15 = 8a+2+60$$
$$15a-15 = 8a+62$$
$$7a = 77$$
$$a = 11$$

23.
$$\frac{x-2}{5} + \frac{x+2}{2} = \frac{x+4}{3}$$
$$30\left(\frac{x-2}{5} + \frac{x+2}{2}\right) = 30\left(\frac{x+4}{3}\right)$$
$$6(x-2) + 15(x+2) = 10(x+4)$$
$$6x-12+15x+30 = 10x+40$$
$$21x+18 = 10x+40$$
$$11x = 22$$
$$x = 2$$

24.
$$\frac{2z-3}{4} - \frac{4-z}{2} = \frac{z+1}{3}$$
$$12\left(\frac{2z-3}{4} - \frac{4-z}{2}\right) = 12\left(\frac{z+1}{3}\right)$$
$$3(2z-3) - 6(4-z) = 4(z+1)$$
$$6z-9-24+6z = 4z+4$$
$$12z-33 = 4z+4$$
$$8z = 37$$
$$z = \frac{37}{8}$$

25. Let $x =$ the number.
$$2(x-3) = 3x+1$$
$$2x-6 = 3x+1$$
$$-7 = x$$
The number is -7.

26. Let $x =$ smaller number,
then $x+5 =$ larger number.
$$x+x+5 = 285$$
$$2x = 280$$
$$x = 140$$
$$x+5 = 145$$
The numbers are 140 and 145.

27. $40\% \cdot 130 = 0.40 \cdot 130 = 52$

28. $1.5\% \cdot 8 = 0.015 \cdot 8 = 0.12$

29. Let $x =$ viewers in 1995.
$$x - 40\% \text{ of } x = 33 \text{ million}$$
$$x - 0.40x = 33$$
$$0.60x = 33$$
$$x = 55$$
There were 55 million Oscar viewers in 1995.

30. Let $n =$ the first integer, then
$n+1 =$ the second integer,
$n+2 =$ the third integer, and
$n+3 =$ the fourth integer.

59. $\left|\dfrac{x+6}{3}\right| > 2$

$\dfrac{x+6}{3} < -2$ or $\dfrac{x+6}{3} > 2$

$x+6 < -6$ or $x+6 > 6$

    $x < -12$ or $x > 0$

$(-\infty, -12) \cup (0, \infty)$

61. $-15 + |2x - 7| \le -6$

         $|2x - 7| \le 9$

$-9 \le 2x - 7 \le 9$

$-2 \le 2x \le 16$

$-1 \le x \le 8$

$[-1, 8]$

63. $\left|2x + \dfrac{3}{4}\right| - 7 \le -2$

$\left|2x + \dfrac{3}{4}\right| \le 5$

$-5 \le 2x + \dfrac{3}{4} \le 5$

$-20 \le 8x + 3 \le 20$

$-23 \le 8x \le 17$

$-\dfrac{23}{8} \le x \le \dfrac{17}{8}$

$\left[-\dfrac{23}{8}, \dfrac{17}{8}\right]$

65. $|2x - 3| < 7$

$-7 < 2x - 3 < 7$

$-4 < 2x < 10$

$-2 < x < 5$

The solution set is $(-2, 5)$.

67. $|2x - 3| = 7$

$2x - 3 = 7$ or $2x - 3 = -7$

    $2x = 10$ or $2x = -4$

     $x = 5$ or $x = -2$

The solution set is $\{-2, 5\}$.

69. $|x - 5| \ge 12$

$x - 5 \le -12$ or $x - 5 \ge 12$

   $x \le -7$ or $x \ge 17$

The solution set is $(-\infty, -7] \cup [17, \infty)$.

71. $|9 + 4x| = 0$

$9 + 4x = 0$

    $4x = -9$

     $x = -\dfrac{9}{4}$

The solution set is $\left\{-\dfrac{9}{4}\right\}$.

73. $|2x + 1| + 4 < 7$

     $|2x + 1| < 3$

$-3 < 2x + 1 < 3$

$-4 < 2x < 2$

$-2 < x < 1$

The solution set is $(-2, 1)$.

75. $|3x - 5| + 4 = 5$

     $|3x - 5| = 1$

$3x - 5 = 1$ or $3x - 5 = -1$

    $3x = 6$ or $3x = 4$

     $x = 2$ or $x = \dfrac{4}{3}$

The solution set is $\left\{\dfrac{4}{3}, 2\right\}$.

77. $|x + 11| = -1$ is impossible.

The solution set is $\varnothing$.

79. $\left|\dfrac{2x-1}{3}\right| = 6$

$\dfrac{2x-1}{3} = 6$ or $\dfrac{2x-1}{3} = -6$

$2x-1 = 18$ or $2x-1 = -18$

$2x = 19$ or $2x = -17$

$x = \dfrac{19}{2}$ or $x = -\dfrac{17}{2}$

The solution set is $\left\{-\dfrac{17}{2}, \dfrac{19}{2}\right\}$.

81. $\left|\dfrac{3x-5}{6}\right| > 5$

$\dfrac{3x-5}{6} < -5$ or $\dfrac{3x-5}{6} > 5$

$3x-5 < -30$ or $3x-5 > 30$

$3x < -25$ or $3x > 35$

$x < -\dfrac{25}{3}$ or $x > \dfrac{35}{3}$

$\left(-\infty, -\dfrac{25}{3}\right) \cup \left(\dfrac{35}{3}, \infty\right)$

83. $P(\text{rolling a } 2) = \dfrac{1}{6}$

85. $P(\text{rolling a } 7) = 0$

87. $P(\text{rolling a 1 or } 3) = \dfrac{1}{3}$

89. $3x - 4y = 12$

$3(2) - 4y = 12$

$6 - 4y = 12$

$-4y = 6$

$y = -\dfrac{3}{2} = -1.5$

91. $3x - 4y = 12$

$3x - 4(-3) = 12$

$3x + 12 = 12$

$3x = 0$

$x = 0$

93. $|x| < 7$

95. $|x| \le 5$

97. Answers may vary.

99. $|3.5 - x| < 0.05$

$-0.05 < 3.5 - x < 0.05$

$-3.55 < -x < -3.45$

$3.55 > x > 3.45$

$3.45 < x < 3.55$

Chapter 2 Review Exercises

1. $4(x-5) = 2x - 14$

$4x - 20 = 2x - 14$

$2x = 6$

$x = 3$

2. $x + 7 = -2(x+8)$

$x + 7 = -2x - 16$

$3x = -23$

$x = -\dfrac{23}{3}$

3. $3(2y-1) = -8(6+y)$

$6y - 3 = -48 - 8y$

$14y = -45$

$y = -\dfrac{45}{14}$

4. $-(z+12) = 5(2z-1)$

$-z - 12 = 10z - 5$

$-11z = 7$

$z = -\dfrac{7}{11}$

5. $n - (8+4n) = 2(3n-4)$

$n - 8 - 4n = 6n - 8$

$-3n = 6n$

$-9n = 0$

$n = 0$

6. $4(9v+2) = 6(1+6v) - 10$

$36v + 8 = 6 + 36v - 10$

$36v + 8 = 36v - 4$

$8 = -4$

No solution, or $\varnothing$

7. $0.3(x-2) = 1.2$
$10[0.3(x-2) = 10(1.2)$
$3(x-2) = 12$
$3x - 6 = 12$
$3x = 18$
$x = 6$

8. $1.5 = 0.2(c-0.3)$
$1.5 = 0.2c - 0.06$
$100(1.5) = 100(0.2c - 0.06)$
$150 = 20c - 6$
$156 = 20c$
$7.8 = c$

9. $-4(2-3h) = 2(3h-4) + 6h$
$-8 + 12h = 6h - 8 + 6h$
$-8 + 12h = 12h - 8$
$-8 = -8$
All real numbers

10. $6(m-1) + 3(2-m) = 0$
$6m - 6 + 6 - 3m = 0$
$3m = 0$
$m = 0$

11. $6 - 3(2g+4) - 4g = 5(1-2g)$
$6 - 6g - 12 - 4g = 5 - 10g$
$-6 - 10g = 5 - 10g$
$-6 = 5$

No solution

12. $20 - 5(p+1) + 3p = -(2p-15)$
$20 - 5p - 5 + 3p = -2p + 15$
$-15 - 2p = -2p + 15$
$15 = 15$

All real numbers

13. $\frac{x}{3} - 4 = x - 2$
$3\left(\frac{x}{3} - 4\right) = 3(x-2)$
$x - 12 = 3x - 6$
$-2x = 6$
$x = -3$

14. $\frac{9}{4}y = \frac{2}{3}y$
$12\left(\frac{9}{4}y\right) = 12\left(\frac{2}{3}y\right)$
$27y = 8y$
$19y = 0$
$y = 0$

15. $\frac{3n}{8} - 1 = 3 + \frac{n}{6}$
$24\left(\frac{3n}{8} - 1\right) = 24\left(3 + \frac{n}{6}\right)$
$9n - 24 = 72 + 4n$
$5n = 96$
$n = \frac{96}{5}$

16. $\frac{z}{6} + 1 = \frac{z}{2} + 2$
$6\left(\frac{z}{6} + 1\right) = 6\left(\frac{z}{2} + 2\right)$
$z + 6 = 3z + 12$
$-2z = 6$
$z = -3$

17. $\frac{y}{4} - \frac{y}{2} = -8$
$4\left(\frac{y}{4} - \frac{y}{2}\right) = 4(-8)$
$y - 2y = -32$
$-y = -32$
$y = 32$

18. $\frac{2x}{3} - \frac{8}{3} = x$
$3\left(\frac{2x}{3} - \frac{8}{3}\right) = 3 \cdot x$
$2x - 8 = 3x$
$-8 = x$

19. $\frac{b-2}{3} = \frac{b+2}{5}$
$5(b-2) = 3(b+2)$
$5b - 10 = 3b + 6$
$2b = 16$
$b = 8$

20.
$$\frac{2t-1}{3}=\frac{3t+2}{15}$$
$$15\left(\frac{2t-1}{3}\right)=15\left(\frac{3t+2}{15}\right)$$
$$5(2t-1)=3t+2$$
$$10t-5=3t+2$$
$$7t=7$$
$$t=1$$

21.
$$\frac{2(t+1)}{3}=\frac{2(t-1)}{3}$$
$$3\left[\frac{2(t+1)}{3}\right]=3\left[\frac{2(t-1)}{3}\right]$$
$$2(t+1)=2(t-1)$$
$$2t+2=2t-2$$
$$2=-2$$
No solution

22.
$$\frac{3a-3}{6}=\frac{4a+1}{15}+2$$
$$30\left(\frac{3a-3}{6}\right)=30\left(\frac{4a+1}{15}+2\right)$$
$$5(3a-3)=2(4a+1)+30(2)$$
$$15a-15=8a+2+60$$
$$15a-15=8a+62$$
$$7a=77$$
$$a=11$$

23.
$$\frac{x-2}{5}+\frac{x+2}{2}=\frac{x+4}{3}$$
$$30\left(\frac{x-2}{5}+\frac{x+2}{2}\right)=30\left(\frac{x+4}{3}\right)$$
$$6(x-2)+15(x+2)=10(x+4)$$
$$6x-12+15x+30=10x+40$$
$$21x+18=10x+40$$
$$11x=22$$
$$x=2$$

24.
$$\frac{2z-3}{4}-\frac{4-z}{2}=\frac{z+1}{3}$$
$$12\left(\frac{2z-3}{4}-\frac{4-z}{2}\right)=12\left(\frac{z+1}{3}\right)$$
$$3(2z-3)-6(4-z)=4(z+1)$$
$$6z-9-24+6z=4z+4$$
$$12z-33=4z+4$$
$$8z=37$$
$$z=\frac{37}{8}$$

25. Let $x=$ the number.
$$2(x-3)=3x+1$$
$$2x-6=3x+1$$
$$-7=x$$
The number is -7.

26. Let $x=$ smaller number,
then $x+5=$ larger number.
$$x+x+5=285$$
$$2x=280$$
$$x=140$$
$$x+5=145$$
The numbers are 140 and 145.

27. $40\%\cdot130=0.40\cdot130=52$

28. $1.5\%\cdot8=0.015\cdot8=0.12$

29. Let $x=$ viewers in 1995.
$$x-40\%\text{ of }x=33\text{ million}$$
$$x-0.40x=33$$
$$0.60x=33$$
$$x=55$$
There were 55 million Oscar viewers in 1995.

30. Let $n=$ the first integer, then
$n+1=$ the second integer,
$n+2=$ the third integer, and
$n+3=$ the fourth integer.

$$(n+1)+(n+2)+(n+3)-2n=16$$
$$n+6=16$$
$$n=10$$

Therefore, the integers are 10, 11, 12, and 13.

31. Let x = smaller odd integer, then $x+2$ = larger odd integer.

$$5x=3(x+2)+54$$
$$5x=3x+6+54$$
$$2x=60$$
$$x=30$$

Since this is not odd, no such consecutive odd integers exist.

32. Let w = width of the playing field, then $2w-5$ = length of the playing field.

$$2w+2(2w+5)=230$$
$$2w+4w-10=230$$
$$6w=240$$
$$w=40$$

Then $2w-5=2(40)-5=75$

The field is 75 meters long and 40 meters wide.

33. Let m = number of miles driven.

$$2(39.95)+0.15(m-200)=103.6$$
$$79.9+0.15m-30=103.6$$
$$49.9+0.15m=103.6$$
$$0.15m=53.7$$
$$m=358$$

The customer drove 358 miles.

34. Solve $R = C$.

$$16.50x=4.50x+3000$$
$$12x=3000$$
$$x=250$$

Thus, 250 calculators must be produced and sold in order to break even.

35. Solve $R = C$.

$$40x=20x+100$$
$$20x=100$$
$$x=5$$
$$R=40x=40\cdot5=200$$

She will break even if she sells 5 plants. The revenue will be \$200.

36. $V=lwh$

$$w=\frac{V}{lh}$$

37. $C=2\pi r$

$$\frac{C}{2\pi}=r$$

38. $5x-4y=-12$

$$5x+12=4y$$
$$y=\frac{5x+12}{4}$$

39. $5x-4y=-12$

$$5x=4y-12$$
$$x=\frac{4y-12}{5}$$

40. $y-y_1=m(x-x_1)$

$$m=\frac{y-y_1}{x-x_1}$$

41.
$$y-y_1=m(x-x_1)$$
$$y-y_1=mx-mx_1$$
$$y-y_1+mx_1=mx$$
$$\frac{y-y_1+mx_1}{m}=x$$

42.
$$E=I(R+r)$$
$$E=IR+Ir$$
$$E-IR=Ir$$
$$\frac{E-IR}{I}=r$$

43.
$$S = vt + gt^2$$
$$S - vt = gt^2$$
$$\frac{S - vt}{t^2} = g$$

44.
$$T = gr + gvt$$
$$T = g(r + vt)$$
$$g = \frac{T}{r + vt}$$

45.
$$I = Prt + P$$
$$I = P(rt + 1)$$
$$\frac{I}{rt + 1} = P$$

46.
$$A = \frac{h}{2}(B + b)$$
$$2A = hB + hb$$
$$2A - hb = hB$$
$$\frac{2A - hb}{h} = B$$

47.
$$V = \frac{1}{3}\pi r^2 h$$
$$3V = \pi r^2 h$$
$$\frac{3V}{\pi r^2} = h$$

48.
$$R = \frac{r_1 + r_2}{2}$$
$$2R = r_1 + r_2$$
$$2R - r_2 = r_1$$

49.
$$\frac{V_1}{T_1} = \frac{V_2}{T_2}$$
$$T_2 V_1 = T_1 V_2$$
$$T_2 = \frac{T_1 V_2}{V_1}$$

50. $A = P\left(1 + \frac{r}{n}\right)^{nt} = 3000\left(1 + \frac{0.03}{n}\right)^{7n}$

　　a. $A = 3000\left(1 + \frac{0.03}{2}\right)^{14} \approx \$3695.27$

b. $A = 3000\left(1 + \frac{0.03}{52}\right)^{364} \approx \$3700.81$

51. $C = \frac{5}{9}(F - 32)$
$$C = \frac{5}{9}(90 - 32)$$
$$C = \frac{5}{9}(58)$$
$$C = \frac{290}{9} \approx 32.2$$

90°F is $\left(\frac{290}{9}\right)^\circ$ C $\approx 32.2°$C.

52. Let $x =$ original width,
then $x + 2 =$ original length.
$$(x + 4)(x + 2 + 4) = x(x + 2) + 88$$
$$(x + 4)(x + 6) = x^2 + 2x + 88$$
$$x^2 + 10x + 24 = x^2 + 2x + 88$$
$$8x = 64$$
$$x = 8$$
$$x + 2 = 10$$
The original width is 8 in. and the
original length is 10 in.

53. Area $= 18 \times 21 = 378 \text{ ft}^2$
Packages $= \frac{378}{24} = 15.75$
There are 16 packages needed.

54. $V_{box} = lwh = 8 \cdot 5 \cdot 3 = 120 \text{ in}^3$, while
$V_{cyl} = \pi r^2 h = \pi \cdot 3^2 \cdot 6 = 54\pi \approx 170 \text{ in}^3$
Therefore, the cylinder holds more ice
cream.

55. $d = rt$ or $r = \frac{d}{t}$
11:00 A.M. to 1:15 P.M. is 2.25 hours.
$r = \frac{130}{2.25} \approx 58$
His average speed was 58 mph.

56. $3(x-5) > -(x+3)$
$$3x - 15 > -x - 3$$
$$4x > 12$$
$$x > 3$$
$$(3, \infty)$$

57. $-2(x+7) \geq 3(x+2)$
$$-2x - 14 \geq 3x + 6$$
$$-5x \geq 20$$
$$x \leq -4$$
$$(-\infty, -4]$$

58. $4x - (5 + 2x) < 3x - 1$
$$4x - 5 - 2x < 3x - 1$$
$$2x - 5 < 3x - 1$$
$$-x < 4$$
$$x > -4$$
$$(-4, \infty)$$

59. $3(x-8) < 7x + 2(5-x)$
$$3x - 24 < 7x + 10 - 2x$$
$$3x - 24 < 5x + 10$$
$$-2x < 34$$
$$x > -17$$
$$(-17, \infty)$$

60. $24 \geq 6x - 2(3x - 5) + 2x$
$$24 \geq 6x - 6x + 10 + 2x$$
$$24 \geq 10 + 2x$$
$$14 \geq 2x$$
$$7 \geq x$$
$$(-\infty, 7]$$

61. $48 + x \geq 5(2x + 4) - 2x$
$$48 + x \geq 10x + 20 - 2x$$
$$48 + x \geq 8x + 20$$
$$28 \geq 7x$$
$$4 \geq x$$
$$(-\infty, 4]$$

62. $\dfrac{x}{3} + \dfrac{1}{2} > \dfrac{2}{3}$
$$6\left(\dfrac{x}{3} + \dfrac{1}{2}\right) > 6\left(\dfrac{2}{3}\right)$$
$$2x + 3 > 4$$
$$2x > 1$$
$$x > \dfrac{1}{2}$$
$$\left(\dfrac{1}{2}, \infty\right)$$

63. $x + \dfrac{3}{4} < -\dfrac{x}{2} + \dfrac{9}{4}$
$$4\left(x + \dfrac{3}{4}\right) < 4\left(-\dfrac{x}{2} + \dfrac{9}{4}\right)$$
$$4x + 3 < 2x + 9$$
$$6x < 6$$
$$x < 1$$
$$(-\infty, 1)$$

64. $\dfrac{x-5}{2} \leq \dfrac{3}{8}(2x + 6)$
$$8\left(\dfrac{x-5}{2}\right) \leq 8\left[\dfrac{3}{8}(2x + 6)\right]$$
$$4(x - 5) \leq 3(2x + 6)$$
$$4x - 20 \leq 6x + 18$$
$$-2x \leq 38$$
$$x \geq -19$$
$$[-19, \infty)$$

65. $\dfrac{3(x-2)}{5} > \dfrac{-5(x-2)}{3}$
$$15\left[\dfrac{3(x-2)}{5}\right] > 15\left[\dfrac{-5(x-2)}{3}\right]$$
$$9(x - 2) > -25(x - 2)$$
$$9x - 18 > -25x + 50$$
$$34x > 68$$
$$x > 2$$
$$(2, \infty)$$

66. Let n = number of pounds of laundry

$$25 < 0.9(10) + 0.8(n - 10)$$
$$25 < 9 + 0.8n - 8$$
$$25 < 1 + 0.8n$$
$$24 < 0.8n$$
$$30 < n$$

It is more economical to use the housekeeper for more than 30 pounds of laundry per week.

67. $500 \leq F \leq 1000$

$$500 \leq \frac{9}{5}C + 32 \leq 1000$$
$$468 \leq \frac{9}{5}C \leq 968$$
$$260 \leq C \leq 538$$

Rounded to the nearest degree, firing temperatures range from 260°C to 538°C.

68. Let x = minimum score the last judge can give

$$\frac{9.5 + 9.7 + 9.9 + 9.7 + 9.7 + 9.6 + 9.5 + x}{8} \geq 9.65$$
$$67.6 + x \geq 77.2$$
$$x \geq 9.6$$

The last judge must give Nana at least a 9.6 for her to win a silver medal.

69. Let x = the amount saved each summer

$$4000 \leq 2x + 500 \leq 8000$$
$$3500 \leq 2x \leq 7500$$
$$1750 \leq x \leq 3750$$

She must save between \$1750 and \$3750 each summer.

70. $1 \leq 4x - 7 \leq 3$

$$8 \leq 4x \leq 10$$
$$2 \leq x \leq \frac{5}{2}$$
$$\left[2, \frac{5}{2}\right]$$

71. $-2 \leq 8 + 5x < -1$

$$-10 \leq 5x < -9$$
$$-2 \leq x < -\frac{9}{5}$$
$$\left[-2, \frac{9}{5}\right)$$

72. $-3 < 4(2x - 1) < 12$

$$-3 < 8x - 4 < 12$$
$$1 < 8x < 16$$
$$\frac{1}{8} < x < 2$$
$$\left(\frac{1}{8}, 2\right)$$

73. $-6 < x - (3 - 4x) < -3$

$$-6 < x - 3 + 4x < -3$$
$$-6 < 5x - 3 < -3$$
$$-3 < 5x < 0$$
$$-\frac{3}{5} < x < 0$$
$$\left(-\frac{3}{5}, 0\right)$$

74. $\dfrac{1}{6} < \dfrac{4x - 3}{3} \leq \dfrac{4}{5}$

$$30\left(\frac{1}{6}\right) < 30\left(\frac{4x - 3}{3}\right) \leq 30\left(\frac{4}{5}\right)$$
$$5 < 10(4x - 3) \leq 24$$
$$5 < 40x - 30 \leq 24$$
$$35 < 40x \leq 54$$
$$\frac{7}{8} < x \leq \frac{27}{20}$$
$$\left(\frac{7}{8}, \frac{27}{20}\right]$$

75. $0 \le \dfrac{2(3x+4)}{5} \le 3$

$5(0) \le 5\left[\dfrac{2(3x+4)}{5}\right] \le 5(3)$

$0 \le 2(3x+4) \le 15$

$0 \le 6x+8 \le 15$

$-8 \le 6x \le 7$

$-\dfrac{4}{3} \le x \le \dfrac{7}{6}$

$\left[-\dfrac{4}{3}, \dfrac{7}{6}\right]$

76. $x \le 2$ and $x > -5$

$-5 < x \le 2$

$(-5, 2]$

77. $x \le 2$ or $x > -5$

$(-\infty, \infty)$

78. $3x-5 > 6$ or $-x < -5$

$3x > 11$ or $x < 5$

$x > \dfrac{11}{3}$ or $x < 5$

$x > \dfrac{11}{3}$

$\left(\dfrac{11}{3}, \infty\right)$

79. $-2x \le 6$ and $-2x+3 < -7$

$x \ge -3$ and $-2x < -10$

$x \ge -3$ and $x > 5$

$x > 5$

$(5, \infty)$

80. $|x-7| = 9$

$x-7 = 9$ or $x-7 = -9$

$x = 16$ or $x = -2$

The solution set is $\{-2, 16\}$.

81. $|8-x| = 3$

$8-x = 3$ or $8-x = -3$

$-x = -5$ or $-x = -11$

$x = 5$ or $x = 11$

The solution set is $\{5, 11\}$.

82. $|2x+9| = 9$

$2x+9 = 9$ or $2x+9 = -9$

$2x = 0$ or $2x = -18$

$x = 0$ or $x = -9$

The solution set is $\{-9, 0\}$.

83. $|-3x+4| = 7$

$-3x+4 = 7$ or $-3x+4 = -7$

$-3x = 3$ or $-3x = -11$

$x = -1$ or $x = \dfrac{11}{3}$

The solution set is $\left\{-1, \dfrac{11}{3}\right\}$.

84. $|3x-2|+6 = 10$

$|3x-2| = 4$

$3x-2 = 4$ or $3x-2 = -4$

$3x = 6$ or $3x = -2$

$x = 2$ or $x = -\dfrac{2}{3}$

The solution set is $\left\{-\dfrac{2}{3}, 2\right\}$.

85. $5+|6x+1| = 5$

$|6x+1| = 0$

$6x+1 = 0$

$6x = -1$

$x = -\dfrac{1}{6}$

The solution set is $\left\{-\dfrac{1}{6}\right\}$.

86. $-5 = |4x-3|$

The solution set is $\varnothing$.

87. $|5-6x|+8 = 3$

$|5-6x| = -5$

The solution set is $\varnothing$.

88. $|7x| - 26 = -5$

$|7x| = 21$

$7x = 21$ or $7x = -21$

$x = 3$ or $x = -3$

The solution set is $\{-3, 3\}$.

89. $-8 = |x - 3| - 10$

$2 = |x - 3|$

$x - 3 = 2$ or $x - 3 = -2$

$x = 5$ or $x = 1$

The solution set is $\{1, 5\}$.

90. $\left|\dfrac{3x - 7}{4}\right| = 2$

$\dfrac{3x - 7}{4} = 2$ or $\dfrac{3x - 7}{4} = -2$

$3x - 7 = 8$ or $3x - 7 = -8$

$3x = 15$ or $3x = -1$

$x = 5$ or $x = -\dfrac{1}{3}$

The solution set is $\left\{-\dfrac{1}{3}, 5\right\}$.

91. $\left|\dfrac{9 - 2x}{5}\right| = -3$

The solution set is $\varnothing$.

92. $|6x + 1| = |15 + 4x|$

$6x + 1 = 15 + 4x$ or $6x + 1 = -(15 + 4x)$

$2x = 14$ or $6x + 1 = -15 - 4x$

$x = 7$ or $10x = -16$

$x = -\dfrac{8}{5}$

The solution set is $\left\{-\dfrac{8}{5}, 7\right\}$.

93. $|x - 3| = |7 + 2x|$

$x - 3 = 7 + 2x$ or $x - 3 = -(7 + 2x)$

$-10 = x$ or $x - 3 = -7 - 2x$

$3x = -4$

$x = -\dfrac{4}{3}$

The solution set is $\left\{-10, -\dfrac{4}{3}\right\}$.

94. $|5x - 1| < 9$

$-9 < 5x - 1 < 9$

$-8 < 5x < 10$

$-\dfrac{8}{5} < x < 2$

$\left(-\dfrac{8}{5}, 2\right)$

95. $|6 + 4x| \geq 10$

$6 + 4x \leq -10$ or $6 + 4x \geq 10$

$4x \leq -16$ or $4x \geq 4$

$x \leq -4$ or $x \geq 1$

$(-\infty, -4] \cup [1, \infty)$

96. $|3x| - 8 > 1$

$|3x| > 9$

$3x < -9$ or $3x > 9$

$x < -3$ or $x > 3$

$(-\infty, -3) \cup (3, \infty)$

97. $9 + |5x| < 24$

$|5x| < 15$

$-15 < 5x < 15$

$-3 < x < 3$

$(-3, 3)$

98. $|6x - 5| \le -1$

The solution set is $\varnothing$.

99. $|6x - 5| \ge -1$

Since $|6x - 5|$ is nonnegative for all numbers x, the solution set is $(-\infty, \infty)$.

100. $\left|3x + \dfrac{2}{5}\right| \ge 4$

$$3x + \frac{2}{5} \le -4 \quad \text{or} \quad 3x + \frac{2}{5} \ge 4$$

$$5\left(3x + \frac{2}{5}\right) \le 5(-4) \quad \text{or} \quad 5\left(3x + \frac{2}{5}\right) \ge 5(4)$$

$$15x + 2 \le -20 \quad \text{or} \quad 15x + 2 \ge 20$$

$$15x \le -22 \quad \text{or} \quad 15x \ge 18$$

$$x \le -\frac{22}{15} \quad \text{or} \quad x \ge \frac{6}{5}$$

$$\left(-\infty, -\frac{22}{15}\right] \cup \left[\frac{6}{5}, \infty\right)$$

101. $\left|\dfrac{4x - 3}{5}\right| < 1$

$$-1 < \frac{4x - 3}{5} < 1$$

$$-5 < 4x - 3 < 5$$

$$-2 < 4x < 8$$

$$-\frac{1}{2} < x < 2$$

$$\left(-\frac{1}{2}, 2\right)$$

102. $\left|\dfrac{x}{3} + 6\right| - 8 > -5$

$$\left|\frac{x}{3} + 6\right| > 3$$

$$\frac{x}{3} + 6 < -3 \quad \text{or} \quad \frac{x}{3} + 6 > 3$$

$$\frac{x}{3} < -9 \quad \text{or} \quad \frac{x}{3} > -3$$

$$x < -27 \quad \text{or} \quad x > -9$$

$$(-\infty, -27) \cup (-9, \infty)$$

103. $\left|\dfrac{4(x - 1)}{7}\right| + 10 < 2$

$$\left|\frac{4(x - 1)}{7}\right| < -8$$

The solution set is $\varnothing$.

Chapter 2 Test

1. $8x + 14 = 5x + 44$

$$3x = 30$$

$$x = 10$$

2. $3(x + 2) = 11 - 2(2 - x)$

$$3x + 6 = 11 - 4 + 2x$$

$$3x + 6 = 2x + 7$$

$$x = 1$$

3. $3(y - 4) + y = 2(6 + 2y)$

$$3y - 12 + y = 12 + 4y$$

$$4y - 12 = 12 + 4y$$

$$-12 = 12$$

No solution

4. $7n - 6 + n = 2(4n - 3)$

$$8n - 6 = 8n - 6$$

$$-6 = -6$$

All real numbers

5. $\dfrac{7w}{4} + 5 = \dfrac{3w}{10} + 1$

$20\left(\dfrac{7w}{4} + 5\right) = 20\left(\dfrac{3w}{10} + 1\right)$

$35w + 100 = 6w + 20$

$29w = -80$

$w = -\dfrac{80}{29}$

6. $|6x - 5| - 3 = -2$

$|6x - 5| = 1$

$6x - 5 = 1 \ \text{ or } \ 6x - 5 = -1$

$6x = 6 \ \text{ or } \ \ \ \ 6x = 4$

$x = 1 \ \text{ or } \ \ \ \ \ x = \dfrac{2}{3}$

The solution set is $\left\{\dfrac{2}{3}, 1\right\}$.

7. $|8 - 2t| = -6$

No solution.

8. $|x - 5| = |x + 2|$

$x - 5 = x + 2 \ \text{ or } \ x - 5 = -(x + 2)$

$-5 = 2 \ \ \ \ \text{ or } \ x - 5 = -x - 2$

$2x = 3$

$x = \dfrac{3}{2}$

Since $-5 = 2$ is not possible, the solution

set is $\left\{\dfrac{3}{2}\right\}$.

9. $3x - 4y = 8$

$3x - 8 = 4y$

$y = \dfrac{3x - 8}{4}$

10. $S = gt^2 + gvt$

$S = g(t^2 + vt)$

$g = \dfrac{S}{t^2 + vt}$

11. $F = \dfrac{9}{5}C + 32$

$F - 32 = \dfrac{9}{5}C$

$C = \dfrac{5}{9}(F - 32)$

12. $3(2x - 7) - 4x > -(x + 6)$

$6x - 21 - 4x > -x - 6$

$2x - 21 > -x - 6$

$3x > 15$

$x > 5$

$(5, \infty)$

13. $8 - \dfrac{x}{2} \geq 7$

$2\left(8 - \dfrac{x}{2}\right) \geq 2(7)$

$16 - x \geq 14$

$-x \geq -2$

$x \leq 2$

$(-\infty, 2]$

14. $-3 < 2(x - 3) \leq 4$

$-3 < 2x - 6 \leq 4$

$3 < 2x \leq 10$

$\dfrac{3}{2} < x \leq 5$

$\left(\dfrac{3}{2}, 5\right]$

15. $|3x + 1| > 5$

$3x + 1 < -5 \ \text{ or } \ 3x + 1 > 5$

$3x < -6 \ \text{ or } \ \ \ \ \ 3x > 4$

$x < -2 \ \text{ or } \ \ \ \ \ \ x > \dfrac{4}{3}$

$(-\infty, -2) \cup \left(\dfrac{4}{3}, \infty\right)$

16. $|x - 5| - 4 < -2$

$|x - 5| < 2$

$-2 < x - 5 < 2$

$3 < x < 7$

$(-3, 7)$

17. $-x > 1$ and $3x + 3 \geq x - 3$
   $x < -1$ and $\quad 2x \geq -6$
   $\qquad\qquad\qquad x \geq -3$

   $-3 \leq x < -1$
   $[-3, -1)$

18. $6x + 1 > 5x + 4 \quad$ or $\quad 1 - x > -4$
   $\quad x > 3 \qquad\quad$ or $\qquad 5 > x$

   $(-\infty, \infty)$

19. $12\% \cdot 80 = 0.12 \cdot 80 = 9.6$

20. Let x = employees in 1996
   $x + 1.8x = 461,000$
   $2.18x = 461,000$
   $\qquad x = 211,468$
   The number of people employed in these occupations in 1996 was 211,468.

21. Recall that $C = 2\pi r$. Here $C = 78.5$.
   $78.5 = 2\pi r$
   $r = \dfrac{78.5}{2\pi} = \dfrac{39.25}{\pi}$
   Also, recall that $A = \pi r^2$.
   $A = \pi \left(\dfrac{39.25}{\pi}\right)^2 = \dfrac{39.25^2}{3.14} \approx 490.63$
   Dividing this by 60 yields approximately 8.18. Therefore, about 8 hunting dogs could safely be kept in the pen.

22. Solve $R > C$
   $7.4x > 3910 + 2.8x$
   $4.6x > 3910$
   $\quad x > 850$
   Therefore, more than 850 sunglasses must be produced and sold in order for them to yield a profit.

23. $A = P\left(1 + \dfrac{r}{n}\right)^{nt}$
   $= 2500\left(1 + \dfrac{0.035}{4}\right)^{4 \cdot 10}$
   $\approx \$3542.27$

24. Let x = population of NY, then
   $x + 13.2$ = population of Mexico City
   and $2x + 0.7$ = population of Tokyo
   $x + (x + 13.2) + (2x + 0.7) = 72.3$
   $\qquad\qquad 4x + 13.9 = 72.3$
   $\qquad\qquad\qquad 4x = 58.4$
   $\qquad\qquad\qquad\quad x = 14.6$ million
   $x + 13.2 = 14.6 + 13.2 = 27.8$ million
   $2x + 0.7 = 2(14.6) + 0.7 = 29.9$ million
   The population of New York is 14.6 million. The population of Mexico City is 27.8 million. The population of Tokyo is 29.9 million.

Chapter 2 Cumulative Review

1. a. $\{2, 3, 4, 5\}$

   b. $\{101, 102, 103, \ldots\}$

2. a. $\{-2, -1, 0, 1, 2, 3, 4\}$

   b. $\{4\}$

3. a. $|3| = 3$

   b. $|-5| = 5$

   c. $-|2| = -2$

   d. $-|-8| = -8$

   e. $|0| = 0$

4. a. The opposite of $\dfrac{2}{3}$ is $-\dfrac{2}{3}$.

   b. The opposite of -9 is 9.

   c. The opposite of 1.5 is -1.5.

5. a. $-3 + (-11) = -14$

   b. $3 + (-7) = -4$

c. $-10+15=5$

d. $-8.3+(-1.9)=-10.2$

e. $-\dfrac{2}{3}+\dfrac{3}{7}=-\dfrac{14}{21}+\dfrac{9}{21}=-\dfrac{5}{21}$

6. a. $-2-(-10)=8$

b. $1.7-8.9=-7.2$

c. $-\dfrac{1}{2}-\dfrac{1}{4}=-\dfrac{2}{4}-\dfrac{1}{4}=-\dfrac{3}{4}$

7. a. $\sqrt{9}=3$

b. $\sqrt{25}=5$

c. $\sqrt{\dfrac{1}{4}}=\dfrac{1}{2}$

d. $-\sqrt{36}=-6$

e. $\sqrt{-36}$ is not a real number

8. a. $-3(-2)=6$

b. $-\dfrac{3}{4}\left(-\dfrac{4}{7}\right)=\dfrac{3}{7}$

c. $\dfrac{0}{-2}=0$

d. $\dfrac{-20}{-2}=10$

9. a. $z-y=(-3)-(-1)=-2$

b. $z^2=(-3)^2=9$

c. $\dfrac{2x+y}{z}=\dfrac{2(2)+(-1)}{-3}=\dfrac{3}{-3}=-1$

10. a. $\sqrt[4]{1}=1$

b. $\sqrt[3]{8}=2$

c. $\sqrt[4]{81}=3$

11. a. $x+5=20$

b. $2(3+y)=4$

c. $x-8=2x$

d. $\dfrac{z}{9}=3(z-5)$

12. a. $-3>-5$

b. $\dfrac{-12}{-4}=3$

c. $0>-2$

13. $7x+5=5+7x$

14. $5\cdot(7x)=(5\cdot7)x=35x$

15. $2x+5=9$
$2x=4$
$x=2$

16. $11.2=1.2-5x$
$10=-5x$
$-2=x$

17. $6x-4=2+6(x-1)$
$6x-4=2+6x-6$
$6x-4=6x-4$
$-4=-4,$ which is always true.
All real numbers

18. $2x+1.5=-0.2+1.6x$
$0.4x=-1.7$
$x=-4.25$

19. a. Let $x=$ the first integer. Then
$x+1=$ the second integer and
$x+2=$ the third integer.
$x+(x+1)+(x+2)=3x+3$

b. $x+(5x)+(6x-3)=12x-3$

20. a. See Exercise #19, part a.

b. $4(3x+1)=12x+4$

21. Let x = first number,
then $2x + 3$ = second number
$$x + (2x + 3) = 72$$
$$3x + 3 = 72$$
$$3x = 69$$
$$x = 23$$
$$2x + 3 = 2(23) + 3 = 49$$
The two numbers are 23 and 49.

22. Let x = first number,
then $3x + 2$ = second number
$$(3x + 2) - x = 24$$
$$2x + 2 = 24$$
$$2x = 22$$
$$x = 11$$
$$3x + 2 = 3(11) + 2 = 35$$
The two numbers are 11 and 35.

23. $3y - 2x = 7$
$$3y = 2x + 7$$
$$y = \frac{2x + 7}{3}, \text{ or } y = \frac{2x}{3} + \frac{7}{3}$$

24. $7x - 4y = 10$
$$7x = 4y + 10$$
$$x = \frac{4y + 10}{7}, \text{ or } x = \frac{4y}{7} + \frac{10}{7}$$

25.
$$A = \frac{1}{2}(B + b)h$$
$$2A = (B + b)h$$
$$2A = Bh + bh$$
$$2A - Bh = bh$$
$$\frac{2A - Bh}{h} = b$$

26.
$$P = 2\ell + 2w$$
$$P - 2w = 2\ell$$
$$\frac{P - 2w}{2} = \ell$$

27. a. $\{x \mid x \geq 2\}$

$[2, \infty)$

b. $\{x \mid x < -1\}$

$(-\infty, -1)$

c. $\{x \mid 0.5 < x \leq 3\}$

$(0.5, 3]$

28. a. $\{x \mid x \leq -3\}$

$(-\infty, -3]$

b. $\{x \mid -2 \leq x < 0.1\}$

$[-2, 0.1)$

29. $-(x - 3) + 2 \leq 3(2x - 5) + x$
$$-x + 3 + 2 \leq 6x - 15 + x$$
$$-x + 5 \leq 7x - 15$$
$$20 \leq 8x$$
$$\frac{5}{2} \leq x$$
$$\left[\frac{5}{2}, \infty \right)$$

30. $2(7x-1)-5x > -(-7x)+4$
$14x-2-5x > 7x+4$
$9x-2 > 7x+4$
$2x > 6$
$x > 3$
$(3, \infty)$

31. $2(x+3) > 2x+1$
$2x+6 > 2x+1$
$6 > 1;$ True for all real numbers x.
$(-\infty, \infty)$

32. $4(x+1)-3 < 4x+1$
$4x+4-3 < 4x+1$
$4x+1 < 4x+1$
$1 < 1$ Never true
$\varnothing$

33. $\{4, 6\}$

34. $\{-2, -1, 0, 1, 2, 3, 4, 5\}$

35. $x-7 < 2$ and $2x+1 < 9$
$x < 9$ and $\quad 2x < 8$
$x < 4$
$x < 4$
$(-\infty, 4)$

36. $x+3 \le 1$ or $3x-1 < 8$
$x \le -2$ or $\quad 3x < 9$
$x < 3$
$x < 3$
$(-\infty, 3)$

37. $\{2, 3, 4, 5, 6, 8\}$

38. $\varnothing$

39. $-2x-5 < -3$ or $6x < 0$
$-2x < 2$ or $\quad x < 0$
$x > -1$
All real numbers
$(-\infty, \infty)$

40. $-2x-5 < -3$ and $6x < 0$
$-2x < 2$ and $\quad x < 0$
$x > -1$
$-1 < x < 0$
$(-1, 0)$

41. $|p| = 2$
$p = 2$ or $p = -2$
The solution set is $\{-2, 2\}$.

42. $|x| = 5$
$x = 5$ or $x = -5$
The solution set is $\{-5, 5\}$.

43. $\left|\dfrac{x}{2}-1\right| = 11$
$\dfrac{x}{2}-1 = 11$ or $\dfrac{x}{2}-1 = -11$
$\dfrac{x}{2} = 12$ or $\quad \dfrac{x}{2} = -10$
$x = 24$ or $\quad x = -20$
The solution set is $\{-20, 24\}$.

44. $\left|\dfrac{y}{3}+2\right| = 10$
$\dfrac{y}{3}+2 = 10$ or $\dfrac{y}{3}+2 = -10$
$\dfrac{y}{3} = 8$ or $\quad \dfrac{y}{3} = -12$
$y = 24$ or $\quad y = -36$
The solution set is $\{-36, 24\}$.

45. $|x-3| = |5-x|$
$x-3 = 5-x$ or $x-3 = -(5-x)$
$2x = 8$ or $x-3 = -5+x$
$x = 4$ or $-3 = -5$
Since $-3 = -5$ is not possible, the solution set is $\{4\}$.

46. $|x+3| = |7-x|$

$\quad x+3 = 7-x \quad \text{or} \quad x+3 = -(7-x)$

$\quad\quad 2x = 4 \quad\quad \text{or} \quad x-3 = -7+x$

$\quad\quad x = 2 \quad\quad \text{or} \quad -3 = -7$

Since $-3 = -7$ is not possible, the solution set is $\{2\}$.

47. $|x| \le 3$

$-3 \le x \le 3$

$[-3, 3]$

48. $|x| > 1$

$x < -1 \quad \text{or} \quad x > 1$

$(-\infty, -1) \cup (1, \infty)$

49. $|2x+9| + 5 > 3$

$\quad\quad |2x+9| > -2$

Since $|2x+9|$ is nonnegative for all numbers x, the solution set is $(-\infty, \infty)$.

50. $|3x+1| + 9 < 1$

$\quad\quad |3x+1| < -8$

The solution set is $\varnothing$.

Chapter 3

Section 3.1

Graphing Calculator Explorations

1.

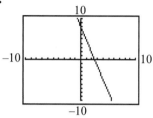

3.

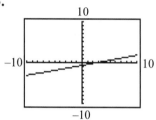

5.

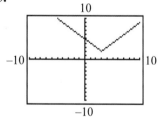

7.

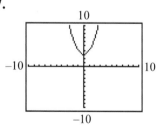

Mental Math

1. Point A is (5, 2).

2. Point B is (2, 5).

3. Point C is (3, 0).

4. Point D is (–1, 3).

5. Point E is (–5, –2).

6. Point F is (–3, 5).

7. Point G is (–1, 0).

8. Point H is (0, –3).

9. (2, 3); QI

10. (0, 5); y-axis

11. (–2, 7); QII

12. (–3, 0); x-axis

13. (–1, –4); QIII

14. (4, –2); QIV

15. (0, –100); y-axis

16. (10, 30); QI

17. (–10, –30); QIII

18. (0, 0); x- and y-axis

19. (–87, 0); x-axis

20. (–42, 17); QII

Exercise Set 3.1

1. (3, 2) is in quadrant I

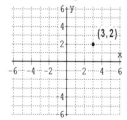

60

3. $(-5, 3)$ is in quadrant II.

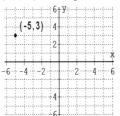

5. $\left(5\frac{1}{2}, -4\right)$ is in quadrant IV.

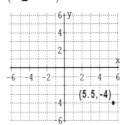

7. $(0, 3.5)$ is on the y-axis.

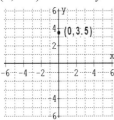

9. $(-2, -4)$ is in quadrant III.

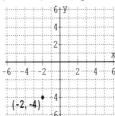

11. quadrant IV

13. x-axis

15. quadrant III

17. Let $x = 0$, $y = 5$
$$y = 3x - 5$$
$$5 = 3 \cdot 0 - 5$$
$$5 = -5$$
False; No

Let $x = -1$, $y = -8$
$$y = 3x - 5$$
$$-8 = 3 \cdot (-1) - 5$$
$$-8 = -8$$
True; Yes

19. Let $x = 1$, $y = 0$
$$-6x + 5y = -6$$
$$-6(1) + 5(0) = -6$$
$$-6 = -6$$
True; Yes

Let $x = 2$, $y = \frac{6}{5}$
$$-6x + 5y = -6$$
$$-6(2) + 5\left(\frac{6}{5}\right) = -6$$
$$-6 = -6$$
True; Yes

21. Let $x = 1$, $y = 2$
$$y = 2x^2$$
$$2 = 2(1)^2$$
$$2 = 2$$
True; Yes

Let $x = 3$, $y = 18$
$$y = 2x^2$$
$$18 = 2(3)^2$$
$$18 = 18$$
True; Yes

23. Let $x = 2$, $y = 8$
$$y = x^3$$
$$8 = (2)^3$$
$$8 = 8$$
True; Yes

Let $x = 3$, $y = 9$

$y = x^3$

$9 = (3)^3$

$9 = 27$

False; No

25. Let $x = 1$, $y = 3$

$y = \sqrt{x} + 2$

$3 = \sqrt{1} + 2$

$3 = 3$

True; Yes

Let $x = 4$, $y = 4$

$y = \sqrt{x} + 2$

$4 = \sqrt{4} + 2$

$4 = 4$

True; Yes

27. Linear

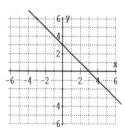

29. Linear

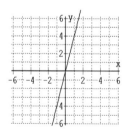

31. Linear

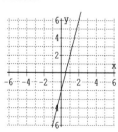

33. Not linear

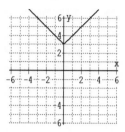

35. Linear

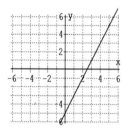

37. Not linear

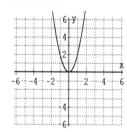

39. Not linear

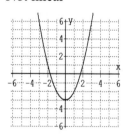

41. Linear

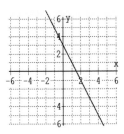

43. Linear

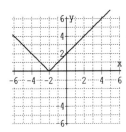

45. Not linear

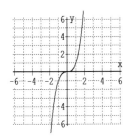

47. Not linear

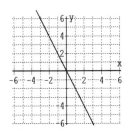

49. Not linear

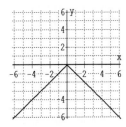

51. Linear

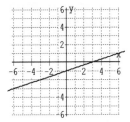

53. Linear

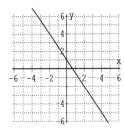

55.
$$3(x-2)+5x = 6x-16$$
$$3x-6+5x = 6x-16$$
$$8x-6 = 6x-16$$
$$2x = -10$$
$$x = -5$$
The solution is -5.

57.
$$3x+\frac{2}{5} = \frac{1}{10}$$
$$30x+4 = 1$$
$$30x = -3$$
$$x = -\frac{1}{10}$$
The solution is $-\frac{1}{10}$.

59. $3x \le -15$
$x \le -5$
$(-\infty, -5]$

61. $2x - 5 > 4x + 3$
$-2x > 8$
$x < -4$
$(-\infty, -4)$

63. B

65. C

67. 1991: In April it rose to $4.25.

69. Answers may vary.

71. $y = x^2 - 4x + 7$

x	y
0	7
1	4
2	3
3	4
4	7

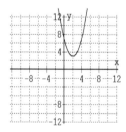

73. a.

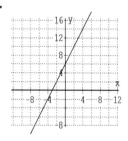

    b. 14 inches

75. $7000

77. $7000 - 6500 = \$500$

79. Depreciation is the same from year to year.

81. They are parallel.

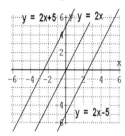

83. Answers may vary.

85. $y = -3 - 2x$

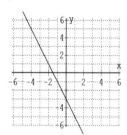

87. $y = 5 - x^2$

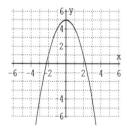

89.

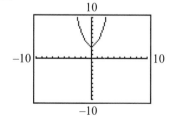

91.

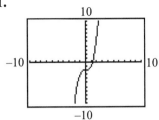

Section 3.2

Graphing Calculator Explorations

1.

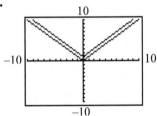

3.

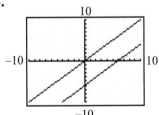

5.

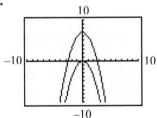

Exercise Set 3.2

1. Domain = {−1, 0, −2, 5}
   Range = {7, 6, 2}
   Function

3. Domain = {−2, 6, −7}
   Range = {4, −3, −8}
   The relation is not a function since −2 is paired with both 4 and −3.

5. Domain = {1}
   Range = {1, 2, 3, 4}
   The relation is not a function since 1 is paired with both 1 and 2 for example.

7. Domain = $\left\{\dfrac{3}{2}, 0\right\}$

   Range = $\left\{\dfrac{1}{2}, -7, \dfrac{4}{5}\right\}$

   The relation is not a function since $\dfrac{3}{2}$ is paired with both $\dfrac{1}{2}$ and −7.

9. Domain = {−3, 0, 3}
   Range = {−3, 0, 3}
   Function

11. Domain = {−1, 1, 2, 3}
    Range = {2, 1}
    Function

13. Domain = {Colorado, Alaska, Delaware, Illinois, Connecticut, Texas}
    Range = {6, 1, 20, 30}
    Function

15. Domain = {32°, 104°, 212°, 50°}
    Range = {0°, 40°, 10°, 100°}
    Function

17. Domain = {0}
    Range = {2, −1, 5, 100}
    Not a function

19. Function

21. Not a function

23. Function

25. Not a function

27. Function

29. Domain = $[0, \infty)$
Range = $(-\infty, \infty)$
The relation is not a function since it fails the vertical line test (try $x = 1$).

31. Domain = $[-1, 1]$
Range = $(-\infty, \infty)$
The relation is not a function since it fails the vertical line test (try $x = 0$).

33. Domain = $(-\infty, \infty)$
Range = $(-\infty, -3] \cup [3, \infty)$
The relation is not a function since it fails the vertical line test (try $x = 2$).

35. Domain = $[2, 7]$
Range = $[1, 6]$
The relation is not a function since it fails the vertical line test (try $x = 4$)

37. Domain = $\{-2\}$
Range = $(-\infty, \infty)$
The relation is not a function since it fails the vertical line test.

39. Domain = $(-\infty, \infty)$
Range = $(-\infty, 3]$
Function

41. Answers may vary.

43. Yes

45. No

47. Yes

49. Yes

51. Yes

53. No

55. $f(x) = 3x + 3$
$f(4) = 3(4) + 3 = 12 + 3 = 15$

57. $h(x) = 5x^2 - 7$
$$\begin{aligned} h(-3) &= 5(-3)^2 - 7 \\ &= 5(9) - 7 \\ &= 45 - 7 \\ &= 38 \end{aligned}$$

59. $g(x) = 4x^2 - 6x + 3$
$$\begin{aligned} g(2) &= 4(2)^2 - 6(2) + 3 \\ &= 4(4) - 12 + 3 \\ &= 16 - 12 + 3 \\ &= 7 \end{aligned}$$

61. $g(x) = 4x^2 - 6x + 3$
$$\begin{aligned} g(0) &= 4(0)^2 - 6(0) + 3 \\ &= 4(0) - 0 + 3 \\ &= 0 - 0 + 3 \\ &= 3 \end{aligned}$$

63. $f(x) = \dfrac{1}{2}x$

   a. $f(0) = \dfrac{1}{2}(0) = 0$

   b. $f(2) = \dfrac{1}{2}(2) = 1$

   c. $f(-2) = \dfrac{1}{2}(-2) = -1$

65. $g(x) = 2x^2 + 4$

   a. $\begin{aligned} g(-11) &= 2(-11)^2 + 4 \\ &= 2(121) + 4 \\ &= 242 + 4 \\ &= 246 \end{aligned}$

   b. $\begin{aligned} g(-1) &= 2(-1)^2 + 4 \\ &= 2(1) + 4 \\ &= 2 + 4 \\ &= 6 \end{aligned}$

c. $g\left(\dfrac{1}{2}\right) = 2\left(\dfrac{1}{2}\right)^2 + 4$

$= 2\left(\dfrac{1}{4}\right) + 4$

$= \dfrac{1}{2} + \dfrac{8}{2}$

$= \dfrac{9}{2}$

67. $f(x) = -5$

    a. $f(2) = -5$

    b. $f(0) = -5$

    c. $f(606) = -5$

69. $f(x) = 1.3x^2 - 2.6x + 5.1$

    a. $f(2) = 1.3(2)^2 - 2.6(2) + 5.1$

        $= 1.3(4) - 5.2 + 5.1$

        $= 5.2 - 5.2 + 5.1$

        $= 5.1$

    b. $f(-2) = 1.3(-2)^2 - 2.6(-2) + 5.1$

        $= 1.3(4) + 5.2 + 5.1$

        $= 5.2 + 5.2 + 5.1$

        $= 15.5$

    c. $f(3.1) = 1.3(3.1)^2 - 2.6(3.1) + 5.1$

        $= 1.3(9.61) - 8.06 + 5.1$

        $= 12.493 - 8.06 + 5.1$

        $= 9.533$

71. $(1, -10)$

73. $(4, 56)$

75. $f(-1) = -2$

77. $g(2) = 0$

79. $-4, 0$

81. 3

83. Infinite number

The reason is that a graph is a function as long as it passes the vertical line test. So it does not matter if the equation of the graph takes on the value 0 many times.

85. a. \$17.1 billion

    b. \$16.21 billion

87. \$36.464 billion

89. $f(x) = x + 7$

91. $A(r) = \pi r^2$

$A(5) = \pi(5)^2 = 25\pi$ square centimeters

93. $V(x) = x^3$

$V(14) = (14)^3 = 2744$ cubic inches

95. $H(f) = 2.59f + 47.24$

$H(46) = 2.59(46) + 47.24$

$\qquad = 166.38$ centimeters

97. $D(x) = \dfrac{136}{25}x$

$D(30) = \dfrac{136}{25}(30) = 163.2$ milligrams

99. $C(x) = 1.69x + 87.54$

    a. $C(5) = 1.69(5) + 87.54 = 95.99$

The per capita consumption of poultry was about 95.99 lb in 2000.

    b. 2002 gives $x = 7$

$C(7) = 1.69(7) + 87.54$

$\qquad = 99.37$ pounds

101. $x - y = 5$

x	0	−5	1
y	5	0	6

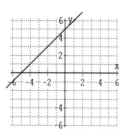

103. $7x + 4y = 8$

x	0	$\frac{8}{7}$	$\frac{12}{7}$
y	2	0	−1

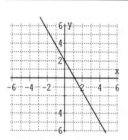

105. $y = 6x$

x	0	0	−1
y	0	0	−6

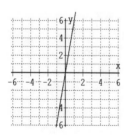

107. Yes; 170 meters

109. $g(x) = -3x + 12$

    a. $g(s) = -3s + 12$

    b. $g(r) = -3r + 12$

111. $f(x) = x^2 - 12$

    a. $f(12) = (12)^2 - 12 = 144 - 12 = 132$

    b. $f(a) = a^2 - 12$

113. answers may vary

Section 3.3

Graphing Calculator Explorations

1. $y = \dfrac{x}{3.5}$

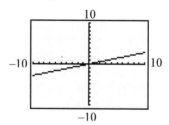

3. $y = -\dfrac{5.78}{2.31}x + \dfrac{10.98}{2.31}$

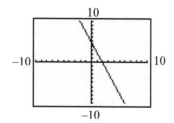

5. $y = |x| + 3.78$

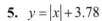
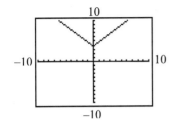

7. $y = 5.6x^2 + 7.7x + 1.5$

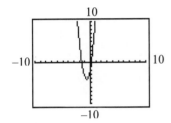

Exercise Set 3.3

1. $f(x) = -2x$

x	0	−1	1
y	0	2	−2

Plot the points to obtain the graph.

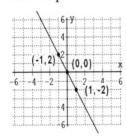

3. $f(x) = -2x + 3$

x	0	1	−1
y	3	1	5

Plot the points to obtain the graph.

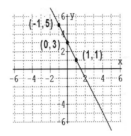

5. $f(x) = \dfrac{1}{2}x$

x	0	2	−2
y	0	1	−1

Plot the points to obtain the graph.

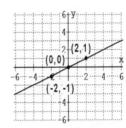

7. $f(x) = \dfrac{1}{2}x - 4$

x	0	2	4
y	−4	−3	−2

Plot the points to obtain the graph.

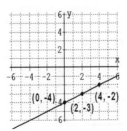

9. C

11. D

13. $x - y = 3$

Let $x = 0$	Let $y = 0$	Let $x = 2$
$0 - y = 3$	$x - 0 = 3$	$2 - y = 3$
$y = -3$	$x = 3$	$y = -1$

x	0	3	2
y	-3	0	-1

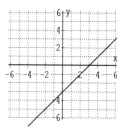

$f(x) = x - 3$

15. $x = 5y$

Let $x = 0$	Let $x = 5$	Let $x = -5$
$0 = 5y$	$5 = 5y$	$-5 = 5y$
$y = 0$	$y = 1$	$y = -1$

x	0	5	-5
y	0	1	-1

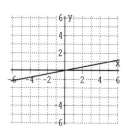

$f(x) = \dfrac{x}{5}$

17. $-x + 2y = 6$

Let $x = 0$	Let $y = 0$	Let $x = 2$
$-0 + 2y = 6$	$-x + 2(0) = 6$	$-2 + 2y = 6$
$y = 3$	$x = -6$	$y = 2$

x	0	-6	2
y	3	0	2

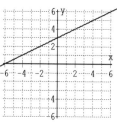

$f(x) = \dfrac{1}{2}x + 3$

19. $2x - 4y = 8$

Let $x = 0$	Let $y = 0$
$2(0) - 4y = 8$	$2x - 4(0) = 8$
$y = -2$	$x = 4$

Let $x = 2$
$2(2) - 4y = 8$
$y = -1$

x	0	4	2
y	-2	0	-1

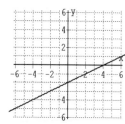

$f(x) = \dfrac{1}{2}x - 2$

21. Answers may vary.

23. $x = -1$

Vertical line with x-intercept at -1

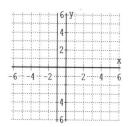

70

25. $y = 0$

Horizontal line with y-intercept at 0

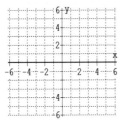

27. $y + 7 = 0$

$y = -7$

Horizontal line with y-intercept at -7

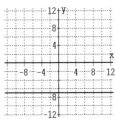

29. C

31. A

33. The vertical line $x = 0$ has a y-intercept.

35. $x + 2y = 8$

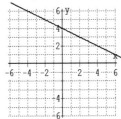

37. $3x + 5y = 7$

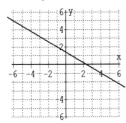

39. $x + 8y = 8$

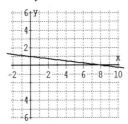

41. $5 = 6x - y$

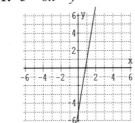

43. $-x + 10y = 11$

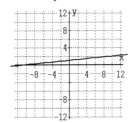

45. $y = \dfrac{3}{2}$

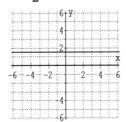

47. $2x + 3y = 6$

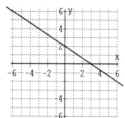

49. $x + 3 = 0$, or $x = -3$

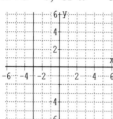

51. $f(x) = \dfrac{3}{4}x + 2$

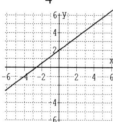

53. $f(x) = x$

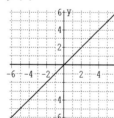

55. $f(x) = \dfrac{1}{2}x$

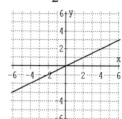

57. $f(x) = 4x - \dfrac{1}{3}$

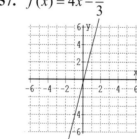

59. $x = -3$

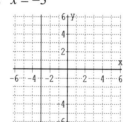

61. $|x - 3| = 6$
$x - 3 = 6$ or $x - 3 = -6$
$x = 9$ or $\quad x = -3$
The solution set is $\{-3, 9\}$.

63. $|2x + 5| > 3$
$2x + 5 < -3$ or $2x + 5 > 3$
$2x < -8$ or $\quad 2x > -2$
$x < -4$ or $\quad x > -1$
$(-\infty, -4) \cup (-1, \infty)$

65. $|3x-4| \le 2$
$-2 \le 3x-4 \le 2$
$2 \le 3x \le 6$
$\dfrac{2}{3} \le x \le 2$
$\left[\dfrac{2}{3}, 2\right]$

67. $\dfrac{-6-3}{2-8} = \dfrac{-9}{-6} = \dfrac{3}{2}$

69. $\dfrac{-8-(-2)}{-3-(-2)} = \dfrac{-8+2}{-3+2} = \dfrac{-6}{-1} = 6$

71. $\dfrac{0-6}{5-0} = \dfrac{-6}{5} = -\dfrac{6}{5}$

73. $2x+3y = 1500$

 a. $2(0)+3y = 1500$
 $3y = 1500$
 $y = 500$
 (0, 500); If no tables are produced, 500 chairs can be produced.

 b. $2x+3(0) = 1500$
 $2x = 1500$
 $x = 750$
 (750, 0); If no chairs are produced, 750 tables can be produced.

 c. $2(50)+3y = 1500$
 $100+3y = 1500$
 $3y = 1400$
 $y = 466.7$
 466 chairs

75. $C(x) = 0.2x + 24$

 a. $C(200) = 0.2(200)+24$
 $= 40+24$
 $= 64$
 $64

 b.

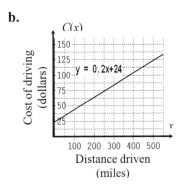

 c. The line moves upward from left to right.

77. $f(x) = 53.6x + 849.88$

 a. $f(20) = 53.6(20)+849.88$
 $= 1072+849.88$
 $= 1921.88$
 $1921.88

 b. $2000 = 53.6x + 849.88$
 $1150.12 = 53.6x$
 $21.46 = x$
 $1990 + 22 = 2012$

 c. Answers may vary.

79.

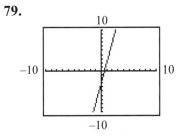

81.

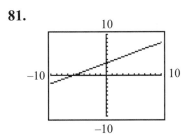

83. a. $y = -4x + 2$ is a line parallel to
$y = -4x$ but with *y*-intercept $(0, 2)$.

   b. $y = -4x - 5$ is a line parallel to
$y = -4x$ but with *y*-intercept $(0, -5)$.

85. B

87. A

Section 3.4

Graphing Calculator Explorations

1.

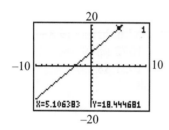

$y = 18.4$

3.

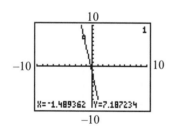

$x = -1.5$

5.

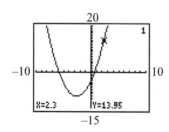

$y = 14.0$

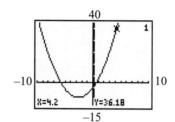

$x = 4.2$

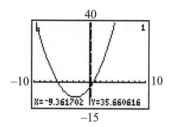

$x = -9.4$

Mental Math

1. $m = \dfrac{7}{6}$ slants upward.

2. $m = -3$ slants downward.

3. $m = 0$ slants horizontally.

4. m is undefined; slants vertically.

Exercise Set 3.4

1. $m = \dfrac{11-2}{8-3} = \dfrac{9}{5}$

3. $m = \dfrac{8-1}{1-3} = \dfrac{7}{-2} = -\dfrac{7}{2}$

5. $m = \dfrac{3-8}{4-(-2)} = \dfrac{-5}{6} = -\dfrac{5}{6}$

7. $m = \dfrac{-4-(-6)}{4-(-2)} = \dfrac{-4+6}{4+2} = \dfrac{2}{6} = \dfrac{1}{3}$

9. $m = \dfrac{11-(-1)}{-12-(-3)} = \dfrac{12}{-9} = -\dfrac{4}{3}$

11. $m = \dfrac{5-5}{3-(-2)} = \dfrac{0}{5} = 0$

13. $m = \dfrac{-5-1}{-1-(-1)} = \dfrac{-6}{0}$

undefined slope

15. $m = \dfrac{0-6}{-3-0} = \dfrac{-6}{-3} = 2$

17. $m = \dfrac{4-2}{-3-(-1)} = \dfrac{2}{-2} = -1$

19. ℓ_2

21. ℓ_2

23. ℓ_2

25. $f(x) = 5x - 2$
$m = 5, b = -2$

27. $2x + y = 7$
$\qquad y = -2x + 7$
$\quad m = -2, b = 7$

29. $2x - 3y = 10$
$\qquad -3y = -2x + 10$
$\qquad\quad y = \dfrac{2}{3}x - \dfrac{10}{3}$
$\quad m = \dfrac{2}{3}, b = -\dfrac{10}{3}$

31. $f(x) = \dfrac{1}{2}x$
$\quad m = \dfrac{1}{2}, b = 0$

33. A

35. B

37. $x = 1$
m is undefined

39. $y = -3$
$m = 0$

41. $x + 2 = 0$
$\quad x = -2$
m is undefined

43. Answers may vary.

45. $f(x) = -x + 5$
$m = -1, b = 5$

47. $-6x + 5y = 30$
$\qquad\quad 5y = 6x + 30$
$\qquad\quad\ y = \dfrac{6}{5}x + 6$
$\quad m = \dfrac{6}{5}, b = 6$

49. $3x + 9 = y$
$\qquad y = 3x + 9$
$\quad m = 3, b = 9$

51. $y = 4$
$\quad m = 0, b = 4$

53. $f(x) = 7x$
$\quad m = 7, b = 0$

55. $6 + y = 0$
$\qquad y = -6$
$\quad m = 0, b = -6$

57. $2 - x = 3$
$\qquad x = -1$
m is undefined. There is no y-intercept.

59. $f(x) = -3x + 6 \qquad g(x) = 3x + 5$
$\quad m = -3 \qquad\qquad m = 3$
Neither, since their slopes are not equal nor does their product equal -1.

61. $-4x + 2y = 5$ $2x - y = 7$

$$y = 2x + \frac{5}{2} \qquad y = 2x + 7$$
$$m = 2$$
$$m = 2$$

Parallel, since they have the same slope.

63. $-2x + 3y = 1$ $3x + 2y = 12$

$$y = \frac{2}{3}x + \frac{1}{3} \qquad y = -\frac{3}{2}x + 6$$
$$m = \frac{2}{3} \qquad\qquad m = -\frac{3}{2}$$

Perpendicular, since the product of their slopes is -1.

65. Answers may vary.

67. Two points on the line: $(0, 0)$, $(2, 3)$

$$m = \frac{3 - 0}{2 - 0} = \frac{3}{2}$$

69. Two points on the line: $(4, 0)$, $(0, 2)$

$$m = \frac{2 - 0}{0 - 4} = \frac{2}{-4} = -\frac{1}{2}$$

71. $m = \dfrac{8}{12} = \dfrac{2}{3}$

73. $m = \dfrac{-1600 \text{ ft.}}{2.5 \text{ mi}}$

$$= \frac{-1600 \text{ ft.}}{2.5(5280 \text{ ft.})}$$
$$= \frac{-1600}{13,200} \approx -0.12$$

75. $y = 1545.4x + 33,858.4$

a. $x = 2005 - 1997 = 8$

$y = 1545.4(8) + 33,858.4 = 46,221.6$

His average income will be about $46,221.60.

b. $m = 1545.4$; The annual income increases $1545.40 every year.

c. $b = 33,858.4$; at year $x = 0$, or 1997, the annual average income of an American man with an associate degree was about $33,858.40.

77. $-245x + 10y = 59$

a. $y = 24.5x + 5.9$

$m = 24.5$; $b = 5.9$

b. The number of public wireless internet access points is projected to increase 24.5 thousand every year.

c. There were about 5.9 thousand wireless internet access points in 2002.

79. $f(x) = 174.4x + 2074.38$

a. $m = 174.4$; The total cost of tuition and fees increase $174.40 every year.

b. $b = 2074.38$; The total cost of tuition and fees was $2074.38 in 1990.

81. $f(x) = -\dfrac{7}{2}x - 6$

$$m = -\frac{7}{2}$$

The slope of a parallel line is $-\dfrac{7}{2}$.

83. $f(x) = -\dfrac{7}{2}x - 6$

$$m = -\frac{7}{2}$$

The slope of a perpendicular line is $\dfrac{2}{7}$.

85. $5x - 2y = 6 \Rightarrow y = \dfrac{5}{2}x - 3$

$$m = \frac{5}{2}$$

The slope of a parallel line is $\dfrac{5}{2}$.

87. $5x - 2y = 6 \Rightarrow y = \dfrac{5}{2}x - 3$

$$m = \frac{5}{2}$$

The slope of a perpendicular line is $-\dfrac{2}{5}$.

89. $P(B) = \dfrac{2}{11}$

91. $P(I \text{ or } T) = \dfrac{3}{11}$

93. $P(\text{vowel}) = \dfrac{4}{11}$

95. $y - 0 = -3[x - (-10)]$
$y = -3(x + 10)$
$y = -3x - 30$

97. $y - 9 = -8[x - (-4)]$
$y - 9 = -8(x + 4)$
$y - 9 = -8x - 32$
$y = -8x - 23$

99. a. $(6, 20)$

b. $(10, 13)$

c. $m = \dfrac{13 - 20}{10 - 6} = \dfrac{-7}{4} = -\dfrac{7}{4} = -1.75$
The rate of change is -1.75 yd per sec.

d. $F(22, 2)$, $G(26, 8)$
$m = \dfrac{8 - 2}{26 - 22} = \dfrac{6}{4} = \dfrac{3}{2} = 1.5$
The rate of change is 1.5 yd per sec.

101. $-4x + 2y = 5$
$2x - y = 7$

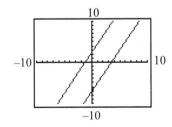

103. a. $y = \dfrac{1}{2}x + 1$
$y = x + 1$
$y = 2x + 1$

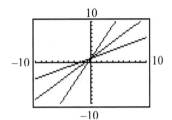

b. $y = -\dfrac{1}{2}x + 1$
$y = -x + 1$
$y = -2x + 1$

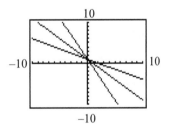

c. True

Section 3.5

Mental Math

1. $m = -4$, $b = 12$

2. $m = \dfrac{2}{3}$, $b = -\dfrac{7}{2}$

3. $m = 5$, $b = 0$

4. $m = -1$, $b = 0$

5. $m = \dfrac{1}{2}$, $b = 6$

6. $m = -\dfrac{2}{3}$, $b = 5$

7. Parallel

8. Parallel

9. Neither

10. Neither

Exercise Set 3.5

1. $y = -x + 1$

3. $y = 2x + \dfrac{3}{4}$

5. $y = \dfrac{2}{7}x$

7. $y = 5x$

possible points: $(0, 0)$, $(1, 5)$

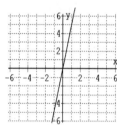

9. $x + y = 7$
$y = -x + 7$
possible points: $(0, 7)$, $(1, 6)$

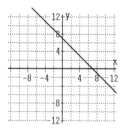

11. $-3x + 2y = 3$
$2y = 3x + 3$
$y = \dfrac{3}{2}x + \dfrac{3}{2}$

possible points: $\left(0, \dfrac{3}{2}\right), \left(2, \dfrac{9}{2}\right)$

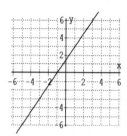

13. $y - y_1 = m(x - x_1)$
$y - 2 = 3(x - 1)$
$y - 2 = 3x - 3$
$y = 3x - 1$

15. $y - y_1 = m(x - x_1)$
$y - (-3) = -2(x - 1)$
$y + 3 = -2x + 2$
$y = -2x - 1$

17. $y - y_1 = m(x - x_1)$
$y - 2 = \dfrac{1}{2}[x - (-6)]$
$y - 2 = \dfrac{1}{2}(x + 6)$
$y - 2 = \dfrac{1}{2}x + 3$
$y = \dfrac{1}{2}x + 5$

19. $y - y_1 = m(x - x_1)$
$y - 0 = -\dfrac{9}{10}[x - (-3)]$
$y = -\dfrac{9}{10}(x + 3)$
$y = -\dfrac{9}{10}x - \dfrac{27}{10}$

21. $(0, 3)\ (1, 1)$
$m = \dfrac{1 - 3}{1 - 0} = \dfrac{-2}{1} = -2$
$b = 3$
$y = -2x + 3$
$2x + y = 3$

23. $(-2, 1) (4, 5)$

$$m = \frac{5-1}{4-(-2)} = \frac{4}{6} = \frac{2}{3}$$
$$y - 1 = \frac{2}{3}(x+2)$$
$$3y - 3 = 2(x+2)$$
$$3y - 3 = 2x + 4$$
$$2x - 3y = -7$$

25. $m = \frac{6-0}{4-2} = \frac{6}{2} = 3$

$$y - 0 = 3(x-2)$$
$$y = 3x - 6$$
$$f(x) = 3x - 6$$

27. $m = \frac{13-5}{-6-(-2)} = \frac{8}{-4} = -2$

$$y - 5 = -2[x-(-2)]$$
$$y - 5 = -2(x+2)$$
$$y - 5 = -2x - 4$$
$$y = -2x + 1$$
$$f(x) = -2x + 1$$

29. $m = \frac{-3-(-4)}{-4-(-2)} = \frac{1}{-2} = -\frac{1}{2}$

$$y - (-4) = -\frac{1}{2}[x-(-2)]$$
$$y + 4 = -\frac{1}{2}(x+2)$$
$$2y + 8 = -(x+2)$$
$$2y + 8 = -x - 2$$
$$2y = -x - 10$$
$$y = -\frac{1}{2}x - 5$$
$$f(x) = -\frac{1}{2}x - 5$$

31. $m = \frac{-9-(-8)}{-6-(-3)} = \frac{-1}{-3} = \frac{1}{3}$

$$y - (-8) = \frac{1}{3}[x-(-3)]$$
$$y + 8 = \frac{1}{3}(x+3)$$
$$3y + 24 = x + 3$$
$$3y = x - 21$$
$$y = \frac{1}{3}x - 7$$
$$f(x) = \frac{1}{3}x - 7$$

33. Answers may vary.

35. $f(0) = -2$

37. $f(2) = 2$

39. $f(x) = -6$
$$f(-2) = -6$$
$$x = -2$$

41. $y = mx + b$
$$-4 = 0(-2) + b$$
$$-4 = b$$
$$y = -4$$

43. Every vertical line is in the form $x = c$. Since the line passes through the point $(4, 7)$, its equation is $x = 4$.

45. Every horizontal line is in the form $y = c$. Since the line passes through the point $(0, 5)$, its equation is $y = 5$.

47. $y = 4x - 2$ so $m = 4$
$$y - 8 = 4(x-3)$$
$$y - 8 = 4x - 12$$
$$y = 4x - 4$$
$$f(x) = 4x - 4$$

49. $3y = x - 6$ or $y = \frac{1}{3}x - 2$

so $m = \frac{1}{3}$ and $m_{\perp} = -3$

$$y-(-5)=-3(x-2)$$
$$y+5=-3x+6$$
$$y=-3x+1$$
$$f(x)=-3x+1$$

51. $3x+2y=5$
$$2y=-3x+5$$
$$y=-\frac{3}{2}x+\frac{5}{2} \text{ so } m=-\frac{3}{2}$$

$$y-(-3)=-\frac{3}{2}[x-(-2)]$$
$$2(y+3)=-3(x+2)$$
$$2y+6=-3(x+2)$$
$$2y+6=-3x-6$$
$$y=-\frac{3}{2}x-6$$
$$f(x)=-\frac{3}{2}x-6$$

53. $\quad y-3=2[x-(-2)]$
$$y-3=2(x+2)$$
$$y-3=2x+4$$
$$2x-y=-7$$

55. $m=\dfrac{2-6}{5-1}=\dfrac{-4}{4}=-1$
$$y-6=-1(x-1)$$
$$y-6=-x+1$$
$$y=-x+7$$
$$f(x)=-x+7$$

57. $\quad y=-\dfrac{1}{2}x+11$
$$2y=-x+22$$
$$x+2y=22$$

59. $m=\dfrac{-6-(-4)}{0-(-7)}=\dfrac{-2}{7}=-\dfrac{2}{7}$
$$y=-\frac{2}{7}x-6$$
$$7y=-2x-42$$
$$2x+7y=-42$$

61. $\quad y-0=-\dfrac{4}{3}[x-(-5)]$
$$3y=-4(x+5)$$
$$3y=-4x-20$$
$$4x+3y=-20$$

63. $x=-2$

65. $2x+4y=8$
$$4y=-2x+8$$
$$y=-\frac{1}{2}x+2 \text{ so } m=-\frac{1}{2}$$
$$y-(-2)=-\frac{1}{2}(x-6)$$
$$2(y+2)=-(x-6)$$
$$2y+4=-x+6$$
$$x+2y=2$$

67. $y=12$

69. $8x-y=9$
$$y=8x-9 \text{ so } m=8$$
$$y-1=8(x-6)$$
$$y-1=8x-48$$
$$8x-y=47$$

71. $x=5$

73. $m=\dfrac{-5-(-8)}{-6-2}=\dfrac{3}{-8}=-\dfrac{3}{8}$
$$y-(-8)=-\frac{3}{8}(x-2)$$
$$8(y+8)=-3(x-2)$$
$$8y+64=-3x+6$$
$$8y=-3x-58$$
$$y=-\frac{3}{8}x-\frac{29}{4}$$
$$f(x)=-\frac{3}{8}x-\frac{29}{4}$$

75. a. $(1, 30,000), (4, 66,000)$
$$m=\frac{66,000-30,000}{4-1}=12,000$$
$$y-30,000=12,000(x-1)$$
$$y=12,000x+18,000$$
$$P(x)=12,000x+18,000$$

b. $P(7) = 12,000(7) + 18,000$
$= \$102,000$

c. $126,000 = 12,000x + 18,000$
$x = \dfrac{126,000 - 18,000}{12,000}$
$x = 9$ years

77. a. $(3, 10,000), (5, 8000)$
$m = \dfrac{8000 - 10,000}{5 - 3} = -1000$
$y - 10,000 = -1000(x - 3)$
$y - 10,000 = -1000x + 3,000$
$y = -1000x + 13,000$

b. $y = -1000(3.5) + 13,000$
$y = 9500$
9500 Fun Noodles

79. a. $(0, 133,300), (3, 147,802)$
$m = \dfrac{147,802 - 133,300}{3 - 0}$
$= \dfrac{14502}{3}$
$= 4834$
$y - 133,300 = 4834(x - 0)$
$y = 4834x + 133,300$

b. $x = 2008 - 1999 = 9$
$y = 4834(9) + 133,300$
$= \$176,806$

c. The median price of a home is rising $4834 every year.

81. a. $(0, 757), (10, 1052)$
$m = \dfrac{1052 - 757}{10 - 0} = \dfrac{295}{10} = 29.5$
$y - 757 = 29.5(x - 0)$
$y = 29.5x + 757$

b. $x = 2004 - 2000 = 4$
$y = 29.5(4) + 757$
$= 875$ thousand people

83. $2x - 7 \le 21$
$2x \le 28$
$x \le 14$
$(-\infty, 14]$

14

85. $5(x - 2) \ge 3(x - 1)$
$5x - 10 \ge 3x - 3$
$2x \ge 7$
$x \ge \dfrac{7}{2}$
$\left[\dfrac{7}{2}, \infty\right)$

$\dfrac{7}{2}$

87. $\dfrac{x}{2} + \dfrac{1}{4} < \dfrac{1}{8}$
$8\left(\dfrac{x}{2} + \dfrac{1}{4}\right) < 8\left(\dfrac{1}{8}\right)$
$4x + 2 < 1$
$4x < -1$
$x < -\dfrac{1}{4}$
$\left(-\infty, -\dfrac{1}{4}\right)$

$-\dfrac{1}{4}$

89. $m = \dfrac{1 - (-1)}{-5 - 3} = \dfrac{2}{-8} = -\dfrac{1}{4}$ so $m_{\perp} = 4$
$M((3, -1), (5, 1)) = \left(\dfrac{3 - 5}{2}, \dfrac{-1 + 1}{2}\right) = (1, 0)$
$y - 0 = 4[x - (-1)]$
$y = 4(x + 1)$
$y = 4x + 4$
$-4x + y = 4$

91. $m = \dfrac{-4 - 6}{-22 - (-2)} = \dfrac{-10}{-20} = \dfrac{1}{2}$ so $m_{\perp} = -2$

$$M((-2, 6), (-22, -4)) = \left(\frac{-2-22}{2}, \frac{6-4}{2}\right)$$
$$= (-12, 1)$$
$$y - 1 = -2[x - (-12)]$$
$$y - 1 = -2(x + 12)$$
$$y - 1 = -2x - 24$$
$$2x + y = -23$$

93. $m = \dfrac{7-3}{-4-2} = \dfrac{4}{-6} = -\dfrac{2}{3}$ so $m_\perp = \dfrac{3}{2}$

$$M((2, 3), (-4, 7)) = \left(\frac{2-4}{2}, \frac{3+7}{2}\right)$$
$$= (-1, 5)$$
$$y - 5 = \frac{3}{2}[x - (-1)]$$
$$2(y - 5) = 3(x + 1)$$
$$2y - 10 = 3x + 3$$
$$3x - 2y = -13$$

95. $f(x) = -x + 7$

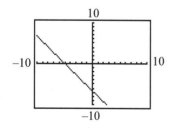

97. $4x + 3y = -20$

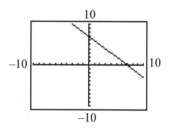

99. True

101. $y = 4x - 2$
    $y = 4x - 4$

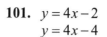
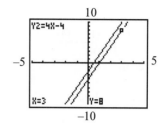

Integrated Review

1. $y = -2x$

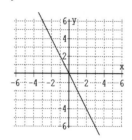

2. $3x - 2y = 6$

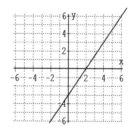

3. $x = -3$

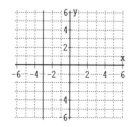

4. $y = 1.5$

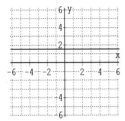

5. $m = \dfrac{-5-(-5)}{3-(-2)} = \dfrac{0}{5} = 0$

6. $m = \dfrac{5-2}{0-5} = \dfrac{3}{-5} = -\dfrac{3}{5}$

7. $y = 3x - 5$
$m = 3; \; (0, -5)$

8. $5x - 2y = 7$
$\quad -2y = -5x + 7$
$\quad\quad y = \dfrac{5}{2}x - \dfrac{7}{2}$
$\quad m = \dfrac{5}{2}; \; \left(0, -\dfrac{7}{2}\right)$

9. $y = 8x - 6 \quad\quad y = 8x + 6$
$m = 8 \quad\quad\quad m = 8$
Parallel, since their slopes are equal.

10. $y = \dfrac{2}{3}x + 1 \quad\quad 2y + 3x = 1$
$m = \dfrac{2}{3} \quad\quad\quad\quad 2y = -3x + 1$
$\quad\quad\quad\quad\quad\quad\quad y = -\dfrac{3}{2}x + \dfrac{1}{2}$
$\quad\quad\quad\quad\quad\quad\quad m = -\dfrac{3}{2}$

Perpendicular, since the product of their slopes is –1.

11. $m = \dfrac{2-6}{5-1} = \dfrac{-4}{4} = -1$
$y - 1 = -1(x - 6)$
$y - 1 = -x + 6$
$\quad y = -x + 7$

12. $x = -2$

13. $y = 0$

14. $m = \dfrac{-5-(-8)}{-6-2} = \dfrac{3}{-8} = -\dfrac{3}{8}$
$y - (-8) = -\dfrac{3}{8}(x - 2)$
$8(y + 8) = -3(x - 2)$
$8y + 64 = -3x + 6$
$8y = -3x - 58$
$\quad y = -\dfrac{3}{8}x - \dfrac{29}{4}$

15. $y - 4 = -5[x - (-2)]$
$y - 4 = -5(x + 2)$
$y - 4 = -5x - 10$
$\quad y = -5x - 6$

16. $y = -4x + \dfrac{1}{3}$

17. $y = \dfrac{1}{2}x - 1$

18. $y - 0 = 3\left(x - \dfrac{1}{2}\right)$
$\quad\quad y = 3x - \dfrac{3}{2}$

19. $3x - y = 5$
$\quad\quad y = 3x - 5$
$m = 3$
$y - (-5) = 3[x - (-1)]$
$\quad y + 5 = 3(x + 1)$
$\quad y + 5 = 3x + 3$
$\quad\quad y = 3x - 2$

20. $4x - 5y = 10$
$\quad -5y = -4x + 10$
$\quad\quad y = \dfrac{4}{5}x - 2; \; m = \dfrac{4}{5} \; \text{ so } \; m_\perp = -\dfrac{5}{4}$

Therefore, $y = -\dfrac{5}{4}x + 4$.

21. $4x + y = \dfrac{2}{3}$

$y = -4x + \dfrac{2}{3};\ \ m = -4\ \ \text{so}\ \ m_{\perp} = \dfrac{1}{4}$

$y - (-3) = \dfrac{1}{4}(x - 2)$

$4(y + 3) = x - 2$

$4y + 12 = x - 2$

$4y = x - 14$

$y = \dfrac{1}{4}x - \dfrac{7}{2}$

22. $5x + 2y = 2$

$2y = -5x + 2$

$y = -\dfrac{5}{2}x + 1$

$m = -\dfrac{5}{2}$

$y - 0 = -\dfrac{5}{2}[x - (-1)]$

$y = -\dfrac{5}{2}(x + 1)$

$2y = -5(x + 1)$

$2y = -5x - 5$

$y = -\dfrac{5}{2}x - \dfrac{5}{2}$

23. A line having undefined slope is vertical. Therefore, the equation is $x = -1$.

24. $y - 3 = 0[x - (-1)]$

$y - 3 = 0$

$y = 3$

Exercise Set 3.6

1. $x < 2$

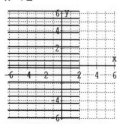

3. $x - y \geq 7$

Test: $(0, 0)$

$0 - 0 \geq 7$

$0 \geq 7$ False

Shade the half-plane that does not contain $(0, 0)$.

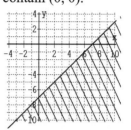

5. $3x + y > 6$

Test: $(0, 0)$

$3(0) + 0 > 6$

$0 > 6$ False

Shade the half-plane that does not contain $(0, 0)$.

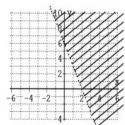

7. $y \leq -2x$

Test: $(1, 1)$

$1 \leq -2(1)$

$1 \leq -2$ False

Shade the half-plane that does not contain $(1, 1)$.

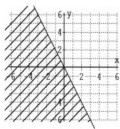

9. $2x + 4y \geq 8$

Test: (0, 0)

$2(0) + 4(0) \geq 8$

$\qquad 0 \geq 8$ False

Shade the half-plane that does not contain (0, 0).

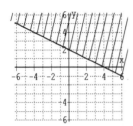

11. $5x + 3y > -15$

Test: (0, 0)

$5(0) + 3(0) > -15$

$\qquad 0 > -15$ True

Shade the half-plane that contains (0, 0).

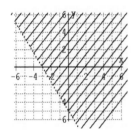

13. Answers may vary. A dashed boundary line should be used when the inequality contains a < or >.

15. $x \geq 3$ and $y \leq -2$

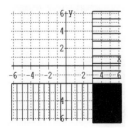

17. $x \leq -2$ or $y \geq 4$

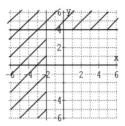

19. $x - y < 3$ and $x > 4$

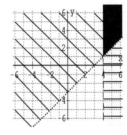

21. $x + y \leq 3$ or $x - y \geq 5$

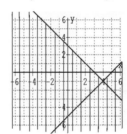

23. $y \geq 2$

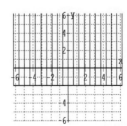

25. $x - 6y < 12$

$\quad -6y < 12 - x$

$\quad\quad y > \dfrac{1}{6}x - 2$

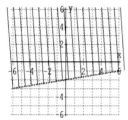

27. $x > 5$

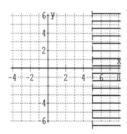

29. $-2x + y \le 4$

$\quad\quad y \le 2x + 4$

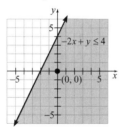

31. $x - 3y < 0$

$\quad -3y < -x$

$\quad\quad y > \dfrac{x}{3}$

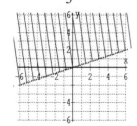

33. $3x - 2y \le 12$

$\quad -2y \le -3x + 12$

$\quad\quad y \ge \dfrac{3}{2}x - 6$

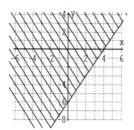

35. $x - y > 2 \quad$ or $y < 5$

$\quad\ y < x - 2 \ $ or $y < 5$

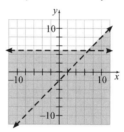

37. $x + y \le 1 \quad$ and $y \le -1$

$\quad\ y \le -x + 1 \ $ and $y \le -1$

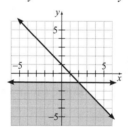

39. $2x + y > 4 \quad$ or $x \ge 1$

$\quad\ y > -2x + 4 \ $ or $x \ge 1$

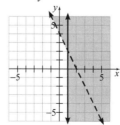

41. $-2 \le x \le 1$

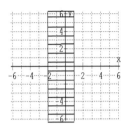

43. $x + y \le 0$ or $3x - 6y \ge 12$

$\quad\quad y \le -x$ or $-6y \ge -3x + 12$

$\quad\quad y \le -x$ or $y \le \dfrac{1}{2}x - 2$

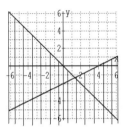

45. $2x - y > 3$ and $x \ge 0$

$\quad\quad y < 2x - 3$ and $x \ge 0$

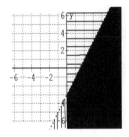

47. D

49. A

51. $x \ge 2$

53. $y \le -3$

55. $y > 4$

57. $x < 1$

59. $2^3 = 8$

61. $-5^2 = -(5 \cdot 5) = -25$

63. $(-2)^4 = 16$

65. $\left(\dfrac{3}{5}\right)^3 = \dfrac{27}{125}$

67. Domain: $[1, 5]$
Range: $[1, 3]$
Not a function

69. $x \le 20$ and $y \ge 10$

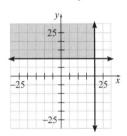

71. $\begin{cases} x \ge 0 \\ y \ge 0 \\ 2x + 4y \le 40 \end{cases}$

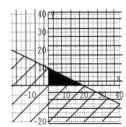

Chapter 3 Review Exercises

1. $A(2, -1)$, quadrant IV
   $B(-2, 1)$, quadrant II
   $C(0, 3)$, y-axis
   $D(-3, -5)$, quadrant III

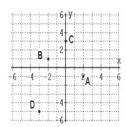

2. $A(-3, 4)$, quadrant II
   $B(4, -3)$, quadrant IV
   $C(-2, 0)$, x-axis
   $D(-4, 1)$, quadrant II

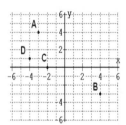

3. $7x - 8y = 56$

   $(0, 56)$; No
   $7(0) - 8(56) = 56$
   $\qquad -448 = 56$, False

   $(8, 0)$; Yes
   $7(8) - 8(0) = 56$
   $\qquad 56 = 56$, True

4. $-2x + 5y = 10$

   $(-5, 0)$; Yes
   $-2(-5) + 5(0) = 10$
   $\qquad\qquad 10 = 10$, True

   $(1, 1)$, No
   $-2(1) + 5(1) = 10$
   $\qquad\qquad 3 = 10$, False

5. $x = 13$
   $(13, 5)$; Yes
   $13 = 13$, True
   $(13, 13)$; Yes
   $13 = 13$, True

6. $y = 2$
   $(7, 2)$; Yes
   $2 = 2$, True
   $(2, 7)$; No
   $7 = 2$, False

7. $y = 3x$; Linear

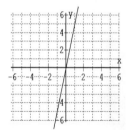

8. $y = 5x$; Linear

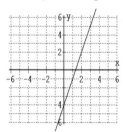

9. $3x - y = 4$; Linear
   Find three ordered pair solutions, or find
   x- and y-intercepts, or find m and b.

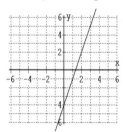

10. $x - 3y = 2$; Linear

Find three ordered pair solutions, or find x- and y-intercepts, or find m and b.

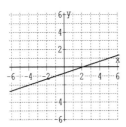

11. $y = |x| + 4$; Nonlinear

x	−3	−2	−1	0	1	2	3
y	7	6	5	4	5	6	7

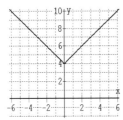

12. $y = x^2 + 4$; Nonlinear

x	−3	−2	−1	0	1	2	3
y	13	8	5	4	5	8	13

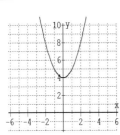

13. $y = -\dfrac{1}{2}x + 2$; Linear

Find three ordered pair solutions, or find x- and y-intercepts, or find m and b.

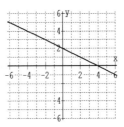

14. $y = -x + 5$; Linear

Find three ordered pair solutions, or find x- and y-intercepts, or find m and b.

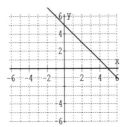

15. $y = 2x - 1$; Linear

Find three ordered pair solutions, or find x- and y-intercepts, or find m and b.

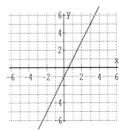

16. $y = \dfrac{1}{3}x + 1$; Linear

Find three ordered pair solutions, or find x- and y-intercepts, or find m and b.

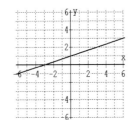

17. $y = -1.36x$; Linear

Find three ordered pair solutions, or find x- and y-intercepts, or find m and b.

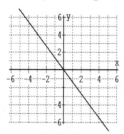

18. $y = 2.1x + 5.9$

Find three ordered pair solutions, or find x- and y-intercepts, or find m and b.

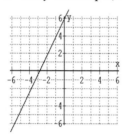

19. Domain: $\left\{-\dfrac{1}{2}, 6, 0, 25\right\}$

Range: $\left\{\dfrac{3}{4} \text{ or } 0.75, -12, 25\right\}$

Function

20. Domain: $\left\{\dfrac{3}{4} \text{ or } 0.75, -12, 25\right\}$

Range: $\left\{-\dfrac{1}{2}, 6, 0, 25\right\}$

Not a function

21. Domain: $\{2, 4, 6, 8\}$
Range: $\{2, 4, 5, 6\}$
Not a function

22. Domain: {Triangle, Square, Rectangle, Parallelogram}
Range: $\{3, 4\}$
Function

23. Domain: $(-\infty, \infty)$
Range: $(-\infty, -1] \cup [1, \infty)$
Not a function

24. Domain: $\{-3\}$
Range: $(-\infty, \infty)$
Not a function

25. Domain: $(-\infty, \infty)$
Range: $\{4\}$
Function

26. Domain: $[-1, 1]$
Range: $[-1, 1]$
Not a function

27. $f(x) = x - 5$
$f(2) = (2) - 5 = -3$

28. $g(x) = -3x$
$g(0) = -3(0) = 0$

29. $g(x) = -3x$
$g(-6) = -3(-6) = 18$

30. $h(x) = 2x^2 - 6x + 1$
$h(-1) = 2(-1)^2 - 6(-1) + 1$
$\qquad = 2(1) + 6 + 1$
$\qquad = 9$

31. $h(x) = 2x^2 - 6x + 1$
$h(1) = 2(1)^2 - 6(1) + 1 = 2 - 6 + 1 = -3$

32. $f(x) = x - 5$
$f(5) = (5) - 5 = 0$

33. $J(x) = 2.54x$
$J(150) = 2.54(150) = 381$ pounds

34. $J(x) = 2.54x$
$J(2000) = 2.54(2000) = 5080$ pounds

35. $f(-1) = 0$

36. $f(1) = -2$

37. $f(x) = 1$
$f(-2) = f(4) = 1$
$x = -2, 4$

38. $f(x) = -1$
$f(0) = f(2) = -1$
$x = 0, 2$

39. $f(x) = x$ or $y = x$
$m = 1, \; b = 0$

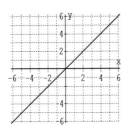

40. $f(x) = -\dfrac{1}{3}x$ or $y = -\dfrac{1}{3}x$
$m = -\dfrac{1}{3}, \; b = 0$

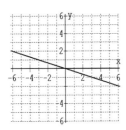

41. $g(x) = 4x - 1$ or $y = 4x - 1$
$m = 4, \; b = -1$

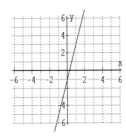

42. C

43. A

44. B

45. D

46. $4x + 5y = 20$

Let $x = 0$ Let $y = 0$
$4(0) + 5y = 20$ $4x + 5(0) = 20$
        $y = 4$ $x = 5$
    $(0, 4)$ $(5, 0)$

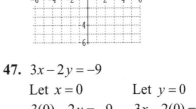

47. $3x - 2y = -9$

Let $x = 0$ Let $y = 0$
$3(0) - 2y = -9$ $3x - 2(0) = -9$
        $y = \dfrac{9}{2}$ $x = -3$
                  $(-3, 0)$
$\left(0, \dfrac{9}{2}\right)$

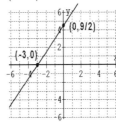

48. $4x - y = 3$

Let $x = 0$ Let $y = 0$
$4(0) - y = 3$ $4x - (0) = 3$
          $y = -3$ $x = \dfrac{3}{4}$
   $(0, -3)$
                  $\left(\dfrac{3}{4}, 0\right)$

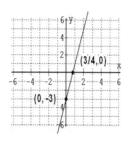

49. $2x + 6y = 9$

Let $x = 0$ Let $y = 0$

$2(0) + 6y = 9$ $2x + 6(0) = 9$

$\quad\quad y = \dfrac{3}{2}$ $x = \dfrac{9}{2}$

$\left(0, \dfrac{3}{2}\right)$ $\left(\dfrac{9}{2}, 0\right)$

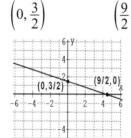

50. $y = 5$

Horizontal line with *y*-intercept 5.

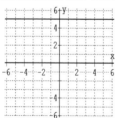

51. $x = -2$

Vertical line with *x*-intercept –2.

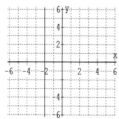

52. $x - 2 = 0$

$\quad\quad x = 2$

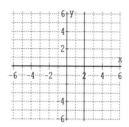

53. $y + 3 = 0$

$\quad\quad y = -3$

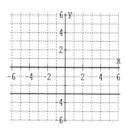

54. $C(x) = 0.3x + 42$

a. $C(150) = 0.3(150) + 42$

$\quad\quad\quad\quad\quad = 45 + 42$

$\quad\quad\quad\quad\quad = 87$

$\quad\quad\quad$ $\$87$

b. $m = 0.3, \; b = 42$

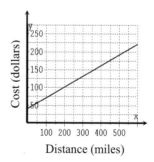

55. $m = \dfrac{-4 - 8}{6 - 2} = \dfrac{-12}{4} = -3$

56. $m = \dfrac{13 - 9}{5 - (-3)} = \dfrac{4}{8} = \dfrac{1}{2}$

57. $m = \dfrac{6-(-4)}{-3-(-7)} = \dfrac{10}{4} = \dfrac{5}{2}$

58. $m = \dfrac{7-(-2)}{-5-7} = \dfrac{9}{-12} = -\dfrac{3}{4}$

59. $6x - 15y = 20$

$\qquad -15y = -6x + 20$

$\qquad y = \dfrac{2}{5}x - \dfrac{4}{3}$

$\qquad m = \dfrac{2}{5}, \; b = -\dfrac{4}{3}$

60. $4x + 14y = 21$

$\qquad 14y = -4x + 21$

$\qquad y = -\dfrac{2}{7}x + \dfrac{3}{2}$

$\qquad m = -\dfrac{2}{7}, \; b = \dfrac{3}{2}$

61. $y - 3 = 0$

$\qquad y = 3; \quad \text{Slope} = 0$

62. $x = -5$; Vertical line
Slope is undefined.

63. l_2

64. l_2

65. l_2

66. l_1

67. $y = 0.3x + 42$

   a. $m = 0.3$; the cost increases by $0.30
for each additional mile driven.

   b. $b = 42$; the cost for 0 miles driven is
$42.

68. $f(x) = -2x + 6 \qquad g(x) = 2x - 1$
$m = -2 \qquad\qquad m = 2$
Neither; The slopes are not the same and
their product is not -1.

69. $-x + 3y = 2 \qquad\qquad 6x - 18y = 3$

$\qquad y = \dfrac{1}{3}x + \dfrac{2}{3} \qquad\qquad y = \dfrac{1}{3}x - \dfrac{1}{6}$

$m = \dfrac{1}{3} \qquad\qquad\qquad m = \dfrac{1}{3}$

Parallel, since their slopes are equal.

70. $y = -x + 1$
$m = -1, \; b = 1$

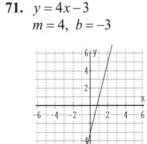

71. $y = 4x - 3$
$m = 4, \; b = -3$

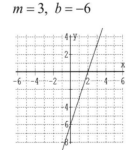

72. $3x - y = 6$

$\qquad y = 3x - 6$

$\qquad m = 3, \; b = -6$

73. $y = -5x$
$m = -5, \ b = 0$

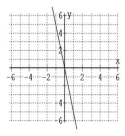

74. $y = -1$

75. $x = -2$

76. $x = -4$

77. $y = 5$

78. $y - 5 = 3[x - (-3)]$
$y - 5 = 3(x + 3)$
$y - 5 = 3x + 9$
$3x - y = -14$

79. $y - (-2) = 2(x - 5)$
$y + 2 = 2x - 10$
$2x - y = 12$

80. $m = \dfrac{-2 - (-1)}{-4 - (-6)} = \dfrac{-1}{2} = -\dfrac{1}{2}$

$y - (-1) = -\dfrac{1}{2}[x - (-6)]$
$2(y + 1) = -(x + 6)$
$2y + 2 = -x - 6$
$x + 2y = -8$

81. $m = \dfrac{-8 - 3}{-4 - (-5)} = \dfrac{-11}{1} = -11$

$y - 3 = -11[x - (-5)]$
$y - 3 = -11(x + 5)$
$y - 3 = -11x - 55$
$11x + y = -52$

82. $x = 4$ has undefined slope
A line perpendicular to $x = 4$ has
slope = 0 and is therefore horizontal.
$y = 3$

83. $y = 8$ has slope = 0
A line parallel to $y = 8$ has slope = 0.
$y = -5$

84. $y = -\dfrac{2}{3}x + 4$
$f(x) = -\dfrac{2}{3}x + 4$

85. $y = -x - 2$
$f(x) = -x - 2$

86. $6x + 3y = 5$
$3y = -6x + 5$
$y = -2x + \dfrac{5}{3}$ so $m = -2$
$y - (-6) = -2(x - 2)$
$y + 6 = -2x + 4$
$y = -2x - 2$
$f(x) = -2x - 2$

87. $3x + 2y = 8$
$2y = -3x + 8$
$y = -\dfrac{3}{2}x + 4$ so $m = -\dfrac{3}{2}$
$y - (-2) = -\dfrac{3}{2}[x - (-4)]$
$2(y + 2) = -3(x + 4)$
$2y + 4 = -3x - 12$
$2y = -3x - 16$
$y = -\dfrac{3}{2}x - 8$
$f(x) = -\dfrac{3}{2}x - 8$

88. $4x + 3y = 5$

$$3y = -4x + 5$$

$$y = -\frac{4}{3}x + \frac{5}{3}$$

so $m = -\frac{4}{3}$ and $m_\perp = \frac{3}{4}$

$$y - (-1) = \frac{3}{4}[x - (-6)]$$

$$4(y + 1) = 3(x + 6)$$

$$4y + 4 = 3x + 18$$

$$4y = 3x + 14$$

$$y = \frac{3}{4}x + \frac{7}{2}$$

$$f(x) = \frac{3}{4}x + \frac{7}{2}$$

89. $2x - 3y = 6$

$$-3y = -2x + 6$$

$$y = \frac{2}{3}x - 2$$

so $m = \frac{2}{3}$ and $m_\perp = -\frac{3}{2}$

$$y - 5 = -\frac{3}{2}[x - (-4)]$$

$$2(y - 5) = -3(x + 4)$$

$$2y - 10 = -3x - 12$$

$$2y = -3x - 2$$

$$y = -\frac{3}{2}x - 1$$

$$f(x) = -\frac{3}{2}x - 1$$

90. a. Use ordered pairs (0, 65) and (12, 81)

$$m = \frac{81 - 65}{12 - 0} = \frac{16}{12} = \frac{4}{3} \text{ and } b = 65$$

$$y = \frac{4}{3}x + 65$$

b. $x = 2009 - 1990 = 19$

$$y = \frac{4}{3}(19) + 65 \approx 90.3$$

About 90% of US drivers will be wearing seat belts.

91. a. Use ordered pairs (0, 43) and (22, 60)

$$m = \frac{60 - 43}{22 - 0} = \frac{17}{22} \text{ and } b = 43$$

$$y = \frac{17}{22}x + 43$$

b. $x = 2010 - 1998 = 12$

$$y = \frac{17}{22}(12) + 43 \approx 52.3$$

There will be about 52 million people reporting arthritis.

92. $3x + y > 4$

$$y > -3x + 4$$

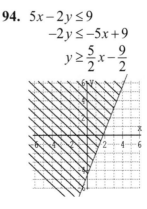

93. $\frac{1}{2}x - y < 2$

$$y > \frac{1}{2}x - 2$$

94. $5x - 2y \le 9$

$$-2y \le -5x + 9$$

$$y \ge \frac{5}{2}x - \frac{9}{2}$$

95. $3y \geq x$

$$y \geq \frac{x}{3}$$

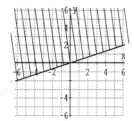

96. $y < 1$

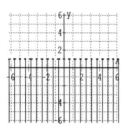

97. $x > -2$

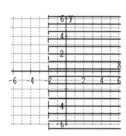

98. $y > 2x + 3$ or $x \leq -3$

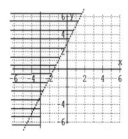

99. $2x < 3y + 8$ and $y \geq -2$

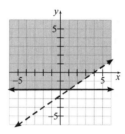

Chapter 3 Test

1.

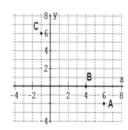

A is in quadrant IV.
B is on the x-axis, no quadrant.
C is in quadrant II.

2. $2x - 3y = -6$

$$-3y = -2x - 6$$

$$y = \frac{2}{3}x + 2$$

$$m = \frac{2}{3}, \ b = 2$$

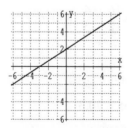

3. $4x + 6y = 7$

$6y = -4x + 7$

$y = -\dfrac{2}{3}x + \dfrac{7}{6}$

$m = -\dfrac{2}{3},\ b = \dfrac{7}{6}$

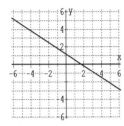

4. $f(x) = \dfrac{2}{3}x$ or $y = \dfrac{2}{3}x$

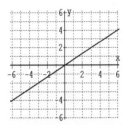

5. $y = -3$

Horizontal line with y-intercept at -3.

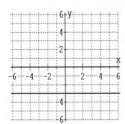

6. $m = \dfrac{10 - (-8)}{-7 - 5} = \dfrac{18}{-12} = -\dfrac{3}{2}$

7. $3x + 12y = 8$

$12y = -3x + 8$

$y = -\dfrac{1}{4}x + \dfrac{2}{3}$

$m = -\dfrac{1}{4},\ b = \dfrac{2}{3}$

8. $f(x) = (x-1)^2$

x	−2	−1	0	1	2	3	4
y	9	4	1	0	1	4	9

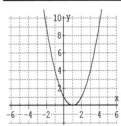

9. $f(x) = |x| + 2$

x	−3	−2	−1	0	1	2	3
y	5	4	3	2	3	4	5

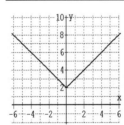

10. $y = -8$

11. $x = -4$

12. $y = -2$

13. $y - (-1) = -3(x - 4)$

$y + 1 = -3x + 12$

$3x + y = 11$

14. $y - (-2) = 5(x - 0)$

$y + 2 = 5x$

$5x - y = 2$

15. $m = \dfrac{-3-(-2)}{6-4} = \dfrac{-1}{2} = -\dfrac{1}{2}$

$y-(-2) = -\dfrac{1}{2}(x-4)$

$2(y+2) = -(x-4)$

$2y+4 = -x+4$

$2y = -x$

$y = -\dfrac{1}{2}x$

$f(x) = -\dfrac{1}{2}x$

16. $3x - y = 4$

$y = 3x - 4$

$m = 3$ so $m_\perp = -\dfrac{1}{3}$

$y-2 = -\dfrac{1}{3}[x-(-1)]$

$3(y-2) = -(x+1)$

$3y-6 = -x-1$

$3y = -x+5$

$y = -\dfrac{1}{3}x + \dfrac{5}{3}$

$f(x) = -\dfrac{1}{3}x + \dfrac{5}{3}$

17. $2y + x = 3$

$2y = -x + 3$

$y = -\dfrac{1}{2}x + 3$ so $m = -\dfrac{1}{2}$

$y-(-2) = -\dfrac{1}{2}(x-3)$

$2(y+2) = -(x-3)$

$2y+4 = -x+3$

$2y = -x-1$

$y = -\dfrac{1}{2}x - \dfrac{1}{2}$

$f(x) = -\dfrac{1}{2}x - \dfrac{1}{2}$

18. $2x - 5y = 8$

$-5y = -2x + 8$

$y = \dfrac{2}{5}x - \dfrac{8}{5}$ so $m_1 = \dfrac{2}{5}$

$m_2 = \dfrac{-1-4}{-1-1} = \dfrac{-5}{-2} = \dfrac{5}{2}$

Therefore, lines L_1 and L_2 are neither parallel nor perpendicular since there slopes are not equal and the product of their slopes is not -1.

19. $x \le -4$

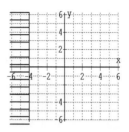

20. $2x - y > 5$

$y < 2x - 5$

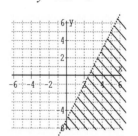

21. $2x + 4y < 6$ and $y \le -4$

$4y < -2x + 6$ and $y \le -4$

$y < -\dfrac{1}{2}x + \dfrac{3}{2}$ and $y \le -4$

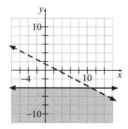

22. Domain: $(-\infty, \infty)$
Range: $\{5\}$
Function

23. Domain: $\{-2\}$
Range: $(-\infty, \infty)$
Not a function

24. Domain: $(-\infty, \infty)$
Range: $[0, \infty)$
Function

25. Domain: $(-\infty, \infty)$
Range: $(-\infty, \infty)$
Function

26. $f(x) = 732x + 21,428$

   a. $x = 1998 - 1996 = 2$
      $f(2) = 732(2) + 21,428 = 22,892$
      The average earnings were $22,892.

   b. $x = 2005 - 1996 = 9$
      $f(9) = 732(9) + 21,428 = 28,016$
      The average earnings will be $28,016.

   c. $\qquad 30,000 \geq 732x + 21,428$
      $\dfrac{30,000 - 21,428}{732} \geq x$
      $\qquad 11.71 \geq x$
      It will be more than 11 years or 2008.

   d. $m = 732$; the yearly earnings for high
      graduates increases by $732 per year.

   e. $b = 21,428$; the yearly earnings for a
      high school graduate in 1996 was
      $21,428.

Chapter 3 Cumulative Review

1. $3x - y = 3(15) - (4) = 45 - 4 = 41$

2. a. $-4 + (-3) = -7$

   b. $\dfrac{1}{2} - \left(-\dfrac{1}{3}\right) = \dfrac{3}{6} + \dfrac{2}{6} = \dfrac{5}{6}$

   c. $7 - 20 = -13$

3. a. True, 3 is a real number

   b. False, $\dfrac{1}{5}$ is not an irrational number.

   c. False, every rational number is not an
      integer, for example, $\dfrac{2}{3}$.

   d. False, since 1 is not in the second set.

4. a. The opposite of -7 is 7.

   b. The opposite of 0 is 0.

   c. The opposite of $\dfrac{1}{4}$ is $-\dfrac{1}{4}$.

5. a. $2 - 8 = -6$

   b. $-8 - (-1) = -8 + 1 = -7$

   c. $-11 - 5 = -16$

   d. $10.7 - (-9.8) = 10.7 + 9.8 = 20.5$

   e. $\dfrac{2}{3} - \dfrac{1}{2} = \dfrac{4}{6} - \dfrac{3}{6} = \dfrac{1}{6}$

   f. $1 - 0.06 = 0.94$

   g. $4 - 7 = -3$

6. a. $\dfrac{-42}{-6} = 7$

   b. $\dfrac{0}{14} = 0$

   c. $-1(-5)(-2) = 5(-2) = -10$

7. a. $3^2 = 9$

  b. $\left(\dfrac{1}{2}\right)^4 = \dfrac{1}{16}$

  c. $-5^2 = -25$

  d. $(-5)^2 = 25$

  e. $-5^3 = -125$

  f. $(-5)^3 = -125$

8. a. Distributive Property

  b. Commutative Property of Addition

9. a. $-1 > -2$

  b. $\dfrac{12}{4} = 3$

  c. $-5 < 0$

  d. $-3.5 < -3.05$

10. $2x^2$

  a. $2(7)^2 = 2(49) = 98$

  b. $2(-7)^2 = 2(49) = 98$

11. a. The reciprocal of 11 is $\dfrac{1}{11}$.

  b. The reciprocal of –9 is $-\dfrac{1}{9}$.

  c. The reciprocal of $\dfrac{7}{4}$ is $\dfrac{4}{7}$.

12. $-2 + 3[5 - (7 - 10)] = -2 + 3[5 - (-3)]$
$$= -2 + 3(8)$$
$$= -2 + 24$$
$$= 22$$

13. $0.6 = 2 - 3.5c$
$$-1.4 = -3.5c$$
$$\dfrac{-1.4}{-3.5} = \dfrac{-3.5c}{-3.5}$$
$$0.4 = c$$
The solution is 0.4.

14. $2(x - 3) = -40$
$$2x - 6 = -40$$
$$2x = -34$$
$$x = -17$$

15. $3x + 5 = 3(x + 2)$
$$3x + 5 = 3x + 6$$
$$5 = 6 \text{ False}$$
The solution is $\varnothing$.

16. $5(x - 7) = 4x - 35 + x$
$$5x - 35 = 5x - 35$$
$$-35 = -35 \text{ True for any number}$$
The solution is all real numbers.

17. 16% of $25 = 0.16(25) = 4$

18. 25% of $16 = 0.25(16) = 4$

19. Let $x =$ first quiz score,
then $x + 2 =$ second quiz score.
And $x + 4 =$ third quiz score
$$x + (x + 2) + (x + 4) = 264$$
$$3x + 6 = 264$$
$$3x = 258$$
$$x = 86$$
$x + 2 = 86 + 2 = 88$
$x + 4 = 86 + 4 = 90$
Benji's scores were 86 ,88, and 90.

20. Let $x =$ first odd integer,
then $x + 2 =$ next odd integer
and $x + 4 =$ third odd integer.
$$x + (x + 2) + (x + 4) = 213$$
$$3x + 6 = 213$$
$$3x = 207$$
$$x = 69$$
$x + 2 = 69 + 2 = 71$
$x + 4 = 69 + 4 = 73$
The integers are 69, 71, and 73.

21. $V = lwh$

$$\frac{V}{lw} = \frac{lwh}{lw}$$

$$\frac{V}{lw} = h$$

22. $7x + 3y = 21$

$$3y = -7x + 21$$

$$y = -\frac{7}{3}x + 7$$

23. $x - 2 < 5$

$$x < 7$$

$$(-\infty, 7)$$

24. $-x - 17 \geq 9$

$$-x \geq 26$$

$$x \leq -26$$

$$(-\infty, -26]$$

25. $\frac{2}{5}(x - 6) \geq x - 1$

$$5\left[\frac{2}{5}(x - 6)\right] \geq 5[x - 1]$$

$$2(x - 6) \geq 5x - 5$$

$$2x - 12 \geq 5x - 5$$

$$-3x \geq 7$$

$$x \leq -\frac{7}{3}$$

$$\left(-\infty, -\frac{7}{3}\right]$$

26. $3x + 10 > \frac{5}{2}(x - 1)$

$$2(3x + 10) > 2\left[\frac{5}{2}(x - 1)\right]$$

$$6x + 20 > 5(x - 1)$$

$$6x + 20 > 5x - 5$$

$$x > -25$$

$$(-25, \infty)$$

27. $2x \geq 0$ and $4x - 1 \leq -9$

$$x \geq 0 \text{ and } \quad 4x \leq -8$$

$$x \geq 0 \text{ and } \quad \quad x \leq -2$$

The solution set is $\varnothing$.

28. $x - 2 < 6$ and $3x + 1 > 1$

$$x < 8 \text{ and } \quad 3x > 0$$

$$x < 8 \text{ and } \quad \quad x > 0$$

$$0 < x < 8$$

$$(0, 8)$$

29. $5x - 3 \leq 10$ or $x + 1 \geq 5$

$$5x \leq 13 \text{ or } \quad x \geq 4$$

$$x \leq \frac{13}{5} \text{ or } \quad x \geq 4$$

$$\left(-\infty, \frac{13}{5}\right] \cup [4, \infty)$$

30. $x - 2 < 6$ or $3x + 1 > 1$

$$x < 8 \text{ or } \quad 3x > 0$$

$$x < 8 \text{ or } \quad \quad x > 0$$

$$(-\infty, \infty)$$

31. $|5w + 3| = 7$

$$5w + 3 = 7 \text{ or } 5w + 3 = -7$$

$$5w = 4 \text{ or } \quad 5w = -10$$

$$w = \frac{4}{5} \text{ or } \quad w = -2$$

The solution set is $\left\{-2, \frac{4}{5}\right\}$.

32. $|5x - 2| = 3$

$$5x - 2 = 3 \text{ or } 5x - 2 = -3$$

$$5x = 5 \text{ or } \quad 5x = -1$$

$$x = 1 \text{ or } \quad \quad x = -\frac{1}{5}$$

The solution set is $\left\{-\frac{1}{5}, 1\right\}$.

33. $|3x + 2| = |5x - 8|$

$$3x + 2 = 5x - 8 \text{ or } 3x + 2 = -(5x - 8)$$

$$-2x = -10 \quad \text{ or } 3x + 2 = -5x + 8$$

$$x = 5 \quad \quad \text{ or } \quad \quad 8x = 6$$

$$x = \frac{3}{4}$$

The solution set is $\left\{\frac{3}{4}, 5\right\}$.

34. $|7x-2|=|7x+4|$

$7x-2=7x+4$ or $7x-2=-(7x+4)$
$-2=4$ or $7x-2=-7x-4$
   False   or $14x=-2$
$$x=-\frac{1}{7}$$

The solution set is $\left\{-\frac{1}{7}\right\}$.

35. $|5x+1|+1\le 10$

$|5x+1|\le 9$
$-9\le 5x+1\le 9$
$-10\le 5x\le 8$
$-2\le x\le \frac{8}{5}$

$\left[-2,\frac{8}{5}\right]$

36. $|-x+8|-2\le 8$

$|-x+8|\le 10$
$-10\le -x+8\le 10$
$-18\le -x\le 2$
$18\ge x\ge -2$
$-2\le x\le 18$
$[-2, 18]$

37. $|y-3|>7$

$y-3<-7$ or $y-3>7$
$y<-4$ or $y>10$
$(-\infty, -4)\cup(10,\infty)$

38. $|x+3|>1$

$x+3<-1$ or $x+3>1$
$x<-4$ or $x>-2$
$(-\infty, -4)\cup(-2,\infty)$

39. $3x-y=12$

$3(0)-(-12)=12$
$12=12$ True
$(0, -12)$ is a solution.

$3(1)-9=12$
$-6=12$ False
$(1, 9)$ is not a solution.

$3(2)-(-6)=12$
$6+6=12$
$12=12$ True
$(2, -6)$ is a solution.

40. $7x+2y=10$

$2y=-7x+10$
$y=-\frac{7}{2}x+5$

$m=-\frac{7}{2},$ y-intercept $=(0, 5)$

41. Yes, $y=2x+1$ is a function (graph the function and use the vertical line test).

42. No, it is not a function (by the vertical line test).

43. **a.** y-intercept $=\left(0,\frac{3}{7}\right)$

b. y-intercept $=(0, -3.2)$

44. $m=\dfrac{9-6}{0-(-1)}=\dfrac{3}{1}=3$

45. $m=\dfrac{2}{3}$

46. $x=-2$

47. y-intercept $=(0, -3)$ means that $b=-3$. Using the equation $y=mx+b$, we have $y=\dfrac{1}{4}x-3$.

48. $y=-\dfrac{3}{4}$

49. $2x - y < 6$

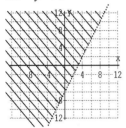

50. $m = \dfrac{7-5}{-4-(-2)} = \dfrac{2}{-4+2} = \dfrac{2}{-2} = -1$

$y - 5 = -1[x - (-2)]$
$y - 5 = -(x + 2)$
$y - 5 = -x - 2$
$x + y = 3$

Chapter 4

Section 4.1

Graphing Calculator Explorations

1.

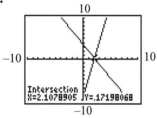

$(2.11, 0.17)$

3.

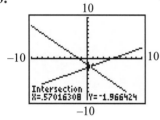

$(0.57, -1.97)$

Mental Math

1. B

2. C

3. A

4. D

Exercise Set 3.1

1. $\begin{cases} x - y = 3 \\ 2x - 4y = 8 \end{cases}$

$x - y = 3$
$2 - (-1) = 3$
$ 3 = 3 \;\; \text{True}$

$2x - 4y = 8$
$2(2) - 4(-1) = 8$
$4 + 4 = 8$
$ 8 = 8 \;\; \text{True}$

Yes, $(2, -1)$ is a solution.

3. $\begin{cases} 2x - 3y = -9 \\ 4x + 2y = -2 \end{cases}$

$2x - 3y = -9$
$2(3) - 3(5) = -9$
$6 - 15 = -9$
$ -9 = -9 \;\; \text{True}$

$4x + 2y = -2$
$4(3) + 2(5) = -2$
$12 + 10 = -2$
$ 22 = -2 \;\; \text{False}$

No, $(3, 5)$ is not a solution.

5. $\begin{cases} y = -5x \\ x = -2 \end{cases}$

$y = -5x \qquad\qquad x = -2$
$10 = -5(-2) \qquad -2 = -2 \;\; \text{True}$
$10 = 10 \;\; \text{True}$

Yes, $(-2, 10)$ is a solution.

7. $\begin{cases} x + y = 1 \\ x - 2y = 4 \end{cases}$

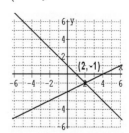

The solution is $(2, -1)$.

9. $\begin{cases} 2y - 4 = 0 \\ x + 2y = 5 \end{cases}$

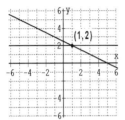

The solution is $(1, 2)$.

11. $\begin{cases} 3x - y = 4 \\ 6x - 2y = 4 \end{cases}$

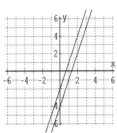

The solution is $\varnothing$.

13. No. Answers may vary.

15. $\begin{cases} x + y = 10 \\ \quad y = 4x \end{cases}$

Replace y with $4x$ in E1.
$x + (4x) = 10$
$\qquad 5x = 10$
$\qquad x = 2$
Replace x with 2 in E2.
$y = 4(2)$
$y = 8$
The solution is $(2, 8)$.

17. $\begin{cases} 4x - y = 9 \\ 2x + 3y = -27 \end{cases}$

Solve E1 for y.
$4x - y = 9$
$\qquad y = 4x - 9$

Replace y with $4x - 9$ in E2.
$2x + 3(4x - 9) = -27$
$2x + 12x - 27 = -27$
$\qquad\quad 14x = 0$
$\qquad\quad\quad x = 0$
Replace x with 0 in E1.
$4(0) - y = 9$
$\qquad\quad y = -9$
The solution is $(0, -9)$.

19. $\begin{cases} \dfrac{1}{2}x + \dfrac{3}{4}y = -\dfrac{1}{4} \\ \dfrac{3}{4}x - \dfrac{1}{4}y = 1 \end{cases}$

Clear fractions by multiplying each equation by 4.
$\begin{cases} 2x + 3y = -1 \\ 3x - y = 4 \end{cases}$
Now solve E2 for y.
$3x - y = 4$
$\qquad y = 3x - 4$
Replace y with $3x - 4$ in E1.
$2x + 3(3x - 4) = -1$
$2x + 9x - 12 = -1$
$\qquad\quad 11x = 11$
$\qquad\quad\quad x = 1$
Replace x with 1 in equation $y = 3x - 4$.
$y = 3(1) - 4$
$y = -1$
The solution is $(1, -1)$.

21. $\begin{cases} \dfrac{x}{3} + y = \dfrac{4}{3} \\ -x + 2y = 11 \end{cases}$

Clear fractions by multiplying the first equation by 3.
$\begin{cases} x + 3y = 4 \\ -x + 2y = 11 \end{cases}$
Solve E2 for x.
$2y - 11 = x$
$\qquad x = 2y - 11$

Replace x with $2y-11$ in E1.

$(2y-11)+3y=4$

$\qquad\qquad 5y=15$

$\qquad\qquad\; y=3$

Replace y with 3 in equation $x=2y-11$.

$x=2(3)-11$

$x=-5$

The solution is $(-5, 3)$.

23. $\begin{cases} 2x-4y=0 \\ \;\; x+2y=5 \end{cases}$

Multiply E2 by 2.

$\begin{cases} 2x-4y=0 \\ 2x+4y=10 \end{cases}$

E1 + E2:
$\begin{array}{r} 2x-4y=0 \\ 2x+4y=10 \\ \hline 4x\qquad=10 \end{array}$

$\qquad x=\dfrac{5}{2}$

Replace x with $\dfrac{5}{2}$ in E1.

$\left(\dfrac{5}{2}\right)-2y=0$

$\qquad \dfrac{5}{2}=2y$

$\qquad\; y=\dfrac{5}{4}$

The solution is $\left(\dfrac{5}{2},\dfrac{5}{4}\right)$.

25. $\begin{cases} 5x+2y=1 \\ \;\; x-3y=7 \end{cases}$

Multiply E2 by -5.

$\begin{cases} \;\; 5x+2y=1 \\ -5x+15y=-35 \end{cases}$

E1 + E2:
$\begin{array}{r} 5x+2y=1 \\ -5x+15y=-35 \\ \hline 17y=-34 \\ y=-2 \end{array}$

Replace y with -2 in E2.

$x-3(-2)=7$

$x+6=7$

$\qquad x=1$

The solution is $(1, -2)$.

27. $\begin{cases} 5x-2y=27 \\ -3x+5y=18 \end{cases}$

Multiply E1 by 3 and E2 by 5.

$\begin{cases} \;\; 15x-6y=81 \\ -15x+25y=90 \end{cases}$

E1 + E2:
$\begin{array}{r} 15x-6y=81 \\ -15x+25y=90 \\ \hline 19y=171 \\ y=9 \end{array}$

Replace y with 9 in E1.

$5x-2(9)=27$

$5x-18=27$

$\qquad 5x=45$

$\qquad\; x=9$

The solution is $(9, 9)$.

29. $\begin{cases} 3x-5y=11 \\ 2x-6y=2 \end{cases}$

Multiply E1 by 2 and E2 by -3.

$\begin{cases} \;\; 6x-10y=22 \\ -6x+18y=-6 \end{cases}$

E1 + E2:
$\begin{array}{r} 6x-10y=22 \\ -6x+18y=-6 \\ \hline 8y=16 \\ y=2 \end{array}$

Replace y with 2 in E2.

$2x-6(2)=2$

$2x-12=2$

$\qquad 2x=14$

$\qquad\; x=7$

The solution is $(7, 2)$.

31. $\begin{cases} \;\; x-2y=4 \\ 2x-4y=4 \end{cases}$

Multiply E1 by -2.

$$\begin{cases} -2x + 4y = -8 \\ 2x - 4y = 4 \end{cases}$$

E1 + E2:
$$\begin{array}{r} -2x + 4y = -8 \\ 2x - 4y = 4 \\ \hline 0 = -4 \quad \text{False} \end{array}$$

Inconsistent system
The solution is $\varnothing$.

33. $\begin{cases} 3x + y = 1 \\ \quad 2y = 2 - 6x \end{cases}$

$\begin{cases} 3x + y = 1 \\ 6x + 2y = 2 \end{cases}$

Multiply E1 by -2.
$$\begin{cases} -6x - 2y = -2 \\ 6x + 2y = 2 \end{cases}$$

E1 + E2:
$$\begin{array}{r} -6x - 2y = -2 \\ 6x + 2y = 2 \\ \hline 0 = 0 \quad \text{True} \end{array}$$

Dependent system
The solution is $\{(x, y) \mid 3x + y = 1\}$.

35. $\begin{cases} 2x + 5y = 8 \\ 6x + y = 10 \end{cases}$

Multiply E1 by -3.
$$\begin{cases} -6x - 15y = -24 \\ 6x + y = 10 \end{cases}$$

E1 + E2:
$$\begin{array}{r} -6x - 15y = -24 \\ 6x + y = 10 \\ \hline -14y = -14 \\ y = 1 \end{array}$$

Replace y with 1 in E2.
$6x + 1 = 10$
$6x = 9$
$x = \dfrac{9}{6} = \dfrac{3}{2}$

The solution is $\left(\dfrac{3}{2}, 1\right)$.

37. $\begin{cases} x + y = 1 \\ x - 2y = 4 \end{cases}$

Multiply E1 by -1.
$$\begin{cases} -x - y = -1 \\ x - 2y = 4 \end{cases}$$

E1 + E2:
$$\begin{array}{r} -x - y = -1 \\ x - 2y = 4 \\ \hline -3y = 3 \\ y = -1 \end{array}$$

Replace y with -1 in E1.
$x + (-1) = 1$
$x - 1 = 1$
$x = 2$
The solution is $(2, -1)$.

39. $\begin{cases} \dfrac{1}{3}x + y = \dfrac{4}{3} \\ -\dfrac{1}{4}x - \dfrac{1}{2}y = -\dfrac{1}{4} \end{cases}$

Clear fractions by multiplying E1 by 3 and E2 by 4.
$$\begin{cases} x + 3y = 4 \\ -x - 2y = -1 \end{cases}$$

E1 + E2:
$$\begin{array}{r} x + 3y = 4 \\ -x - 2y = -1 \\ \hline y = 3 \end{array}$$

Replace y with 3 in the equation
$x + 3y = 4$.
$x + 3(3) = 4$
$x + 9 = 4$
$x = -5$
The solution is $(-5, 3)$.

41. $\begin{cases} 2x + 6y = 8 \\ 3x + 9y = 12 \end{cases}$

Multiply E1 by -3 and E2 by 2.
$$\begin{cases} -6x - 18y = -24 \\ 6x + 18y = 24 \end{cases}$$

E1 + E2:
$$\begin{array}{r} -6x - 18y = -24 \\ 6x + 18y = 24 \\ \hline 0 = 0 \quad \text{True} \end{array}$$

Dependent system
The solution is $\{(x, y) \mid 3x + 9y = 12\}$.

43. $\begin{cases} 4x+2y=5 \\ 2x+y=-1 \end{cases}$

Multiply E2 by –2.

$\begin{cases} 4x+2y=5 \\ -4x-2y=2 \end{cases}$

E1 + E2: $\begin{array}{r} 4x+2y=5 \\ -4x-2y=2 \\ \hline 0=7 \end{array}$ False

Inconsistent system
The solution is $\varnothing$.

45. $\begin{cases} 10y-2x=1 \\ 5y=4-6x \end{cases}$

$\begin{cases} 10y-2x=1 \\ 5y+6x=4 \end{cases}$

Multiply E2 by –2.

$\begin{cases} 10y-2x=1 \\ -10y-12x=-8 \end{cases}$

E1 + E2: $\begin{array}{r} 10y-2x=1 \\ -10y-12x=-8 \\ \hline -14x=-7 \\ x=\dfrac{1}{2} \end{array}$

Replace x with $\dfrac{1}{2}$ in the equation
$5y=4-6x$.

$5y=4-6\left(\dfrac{1}{2}\right)$

$5y=4-3$

$5y=1$

$y=\dfrac{1}{5}$

The solution is $\left(\dfrac{1}{2},\dfrac{1}{5}\right)$.

47. $\begin{cases} \dfrac{3}{4}x+\dfrac{5}{2}y=11 \\ \dfrac{1}{16}x-\dfrac{3}{4}y=-1 \end{cases}$

Clear fractions by multiplying E1 by 4
and E2 by 16.

Page 108

$\begin{cases} 3x+10y=44 \\ x-12y=-16 \end{cases}$

Multiply E2 by –3.

$\begin{cases} 3x+10y=44 \\ -3x+36y=48 \end{cases}$

E1 + E2: $\begin{array}{r} 3x+10y=44 \\ -3x+36y=48 \\ \hline 46y=92 \\ y=2 \end{array}$

Replace y with 3 in the equation
$x-12y=-16$.

$x-12(2)=-16$

$x-24=-16$

$x=8$

The solution is (8, 2).

49. $\begin{cases} x=3y+2 \\ 5x-15y=10 \end{cases}$

Replace x with $3y+2$ in E2.

$5(3y+2)-15y=10$

$15y+10-15y=10$

$10=10$ True

The system is dependent.
The solution is $\{(x,\,y)\mid x=3y+2\}$.

51. $\begin{cases} 2x-y=-1 \\ y=-2x \end{cases}$

Replace y with $-2x$ in E1.

$2x-(-2x)=-1$

$4x=-1$

$x=-\dfrac{1}{4}$

Replace x with $-\dfrac{1}{4}$ in E2.

$y=-2\left(-\dfrac{1}{4}\right)$

$y=\dfrac{1}{2}$

The solution is $\left(-\dfrac{1}{4},\dfrac{1}{2}\right)$.

53. $\begin{cases} 2x = 6 \\ y = 5 - x \end{cases}$

E1 yields $x = 3$.
Replace x with 3 in E2.
$y = 5 - 3$
$y = 2$
The solution is (3, 2).

55. $\begin{cases} \dfrac{x+5}{2} = \dfrac{6-4y}{3} \\ \dfrac{3x}{5} = \dfrac{21-7y}{10} \end{cases}$

Multiply E1 by 6 and E2 by 10.
$\begin{cases} 3x + 15 = 12 - 8y \\ 6x = 21 - 7y \end{cases}$
$\begin{cases} 3x + 8y = -3 \\ 6x + 7y = 21 \end{cases}$
Multiply E1 by –2.
$\begin{cases} -6x - 16y = 6 \\ 6x + 7y = 21 \end{cases}$

E1 + E2: $\begin{array}{r} -6x - 16y = 6 \\ 6x + 7y = 21 \\ \hline -9y = 27 \\ y = -3 \end{array}$

Replace y with –3 in the equation
$3x + 8y = -3$.
$3x + 8(-3) = -3$
$3x - 24 = -3$
$3x = 21$
$x = 7$
The solution is (7, –3).

57. $\begin{cases} 4x - 7y = 7 \\ 12x - 21y = 24 \end{cases}$

Multiply E1 by –3.
$\begin{array}{r} -12x + 21y = -21 \\ 12x - 21y = 24 \\ \hline 0 = 3 \quad \text{False} \end{array}$

Inconsistent system
The solution is $\varnothing$.

59. $\begin{cases} \dfrac{2}{3}x - \dfrac{3}{4}y = -1 \\ -\dfrac{1}{6}x + \dfrac{3}{8}y = 1 \end{cases}$

Multiply E1 by 12 and E2 by 24.
$\begin{cases} 8x - 9y = -12 \\ -4x + 9y = 24 \end{cases}$

E1 + E2: $\begin{array}{r} 8x - 9y = -12 \\ -4x + 9y = 24 \\ \hline 4x = 12 \\ x = 3 \end{array}$

Replace x with 3 in the equation
$-4x + 9y = 24$.
$-4(3) + 9y = 24$
$-12 + 9y = 24$
$9y = 36$
$y = 4$
The solution is (3, 4).

61. $\begin{cases} 0.7x - 0.2y = -1.6 \\ 0.2x - y = -1.4 \end{cases}$

Multiply both equations by 10.
$\begin{cases} 7x - 2y = -16 \\ 2x - 10y = -14 \end{cases}$

Multiply E1 by –5.
$\begin{cases} -35x + 10y = 80 \\ 2x - 10y = -14 \end{cases}$

E1 + E2: $\begin{array}{r} -35x + 10y = 80 \\ 2x - 10y = -14 \\ \hline -33x = 66 \\ x = -2 \end{array}$

Replace x with –2 in the equation
$7x - 2y = -16$.
$7(-2) - 2y = -16$
$-14 - 2y = -16$
$-2y = -2$
$y = 1$
The solution is (–2, 1).

63. $\begin{cases} 4x - 1.5y = 10.2 \\ 2x + 7.8y = -25.68 \end{cases}$

Multiply E2 by –2.

$\begin{cases} 4x - 1.5y = 10.2 \\ -4x - 15.6y = 51.36 \end{cases}$

E1 + E2:
$$\begin{array}{r} 4x - 1.5y = 10.2 \\ -4x - 15.6y = 51.36 \\ \hline -17.1y = 61.56 \\ y = -3.6 \end{array}$$

Replace y with –3.6 in E1.

$4x - 1.5(-3.6) = 10.2$

$4x + 5.4 = 10.2$

$4x = 4.8$

$x = 1.2$

The solution is (1.2, –3.6).

65. $3x - 4y + 2z = 5$

$3(1) - 4(2) + 2(5) = 5$

$3 - 8 + 10 = 5$

$5 = 5$

True

67. $-x - 5y + 3z = 15$

$-(0) - 5(-1) + 3(5) = 15$

$0 + 5 + 15 = 15$

$20 = 15$

False

69. $\begin{array}{r} 3x + 2y - 5z = 10 \\ -3x + 4y + z = 15 \\ \hline 6y - 4z = 25 \end{array}$

71. $\begin{array}{r} 10x + 5y + 6z = 14 \\ -9x + 5y - 6z = -12 \\ \hline x + 10y \quad\;\; = 2 \end{array}$

73. The equilibrium point occurs when $x = 5$ and $y = 21$. Since the x-axis represents the number of DVDs in thousands and the y-axis represents the price of the DVD in dollars, supply equals demand for 5000 DVDs at $21 per DVD.

75. When $x = 7$, the graph of the supply equation is above the graph of the demand equation. Therefore, when $x = 7$, supply is greater than demand.

77. $\begin{cases} y = 2.5x \\ y = 0.9x + 3000 \end{cases}$

Substitute.

$2.5x = 0.9x + 3000$

$1.6x = 3000$

$x = 1875$

$y = 2.5(1875) = 4687.5$

The point of intersection is (1875, 4687.56).

79. Makes money because revenue is greater than cost at $x = 2000$.

81. For values of $x > 1875$ because the x-value at the intersection is 1875.

83. Answers may vary.

85. Red meat: $y = -x + 124.6$

Poultry: $y = 0.9x + 93$

a. $m_{\text{red meat}} = -1$ and $m_{\text{poultry}} = 0.9$.
Consumption of red meat is decreasing while consumption of poultry is increasing.

b. $\begin{cases} y = -x + 124.6 \\ y = 0.9x + 93 \end{cases}$

Substitute.

$-x + 124.6 = 0.9x + 93$

$31.6 = 1.9x$

$16.63 = x$

$y = -(16.63) + 124.6 = 107.97$

The solution is (17, 108).

c. $x = 17 \Rightarrow 1998 + 17 = 2015$

In the year 2015, red meat and poultry consumption will each be about 108 pounds per person.

87. $\begin{cases} \dfrac{1}{x} + y = 12 \\ \dfrac{3}{x} - y = 4 \end{cases}$

Replacing $\dfrac{1}{x}$ with a, we have

$\begin{cases} a + y = 12 \\ 3a - y = 4 \end{cases}$

Adding the two new equations we get
$4a = 16$
$a = 4$

Replace a with 4 in the equation
$a + y = 12$.
$4 + y = 12$
$y = 8$

Since $a = 4$, $x = \dfrac{1}{4}$.

The solution is $\left(\dfrac{1}{4}, 8 \right)$.

89. $\begin{cases} \dfrac{1}{x} + \dfrac{1}{y} = 5 \\ \dfrac{1}{x} - \dfrac{1}{y} = 1 \end{cases}$

Replace $\dfrac{1}{x}$ with a and $\dfrac{1}{y}$ with b.

$\begin{cases} a + b = 5 \\ a - b = 1 \end{cases}$

Adding the two new equations we get
$2a = 6$
$a = 3$

Replace a with 3 in the equation
$a + b = 5$.
$3 + b = 5$
$b = 2$

Since $a = 3$, $x = \dfrac{1}{3}$. Similarly, $y = \dfrac{1}{2}$.

The solution is $\left(\dfrac{1}{3}, \dfrac{1}{2} \right)$.

91. $\begin{cases} \dfrac{2}{x} + \dfrac{3}{y} = -1 \\ \dfrac{3}{x} - \dfrac{2}{y} = 18 \end{cases}$

Replace $\dfrac{1}{x}$ with a and $\dfrac{1}{y}$ with b.

$\begin{cases} 2a + 3b = -1 \\ 3a - 2b = 18 \end{cases}$

Multiply E1 by 2 and E2 by 3 to obtain

$\begin{cases} 4a + 6b = -2 \\ 9a - 6b = 54 \end{cases}$

Adding these last two equations we have
$13a = 52$
$a = 4$

Replace a with 4 in the equation
$2a + 3b = -1$.
$2(4) + 3b = -1$
$3b = -9$
$b = -3$

Since $a = 4$, $x = \dfrac{1}{4}$. Similarly, $y = -\dfrac{1}{3}$.

The solution is $\left(\dfrac{1}{4}, -\dfrac{1}{3} \right)$.

93. $\begin{cases} \dfrac{2}{x} - \dfrac{4}{y} = 5 \\ \dfrac{1}{x} - \dfrac{2}{y} = \dfrac{3}{2} \end{cases}$

Replace $\dfrac{1}{x}$ with a and $\dfrac{1}{y}$ with b.

$\begin{cases} 2a - 4b = 5 \\ a - 2b = \dfrac{3}{2} \end{cases}$

Multiply E2 by 2.
$\begin{cases} 2a - 4b = 5 \\ 2a - 4b = 3 \end{cases}$

This system in inconsistent.
The solution is $\varnothing$.

Exercise Set 4.2

1. $x+y+z=3$ $-x+y+z=5$
    $(-1)+3+1=3$ $-(-1)+3+1=5$
    $\qquad 3=3$ $\qquad 5=5$
    A is true.   B is true.
    $-x+y+2z=0$ $x+2y-3z=2$
    $-(-1)+3+2(1)=0$ $(-1)+2(3)-3(1)=2$
    $\qquad 6=0$ $\qquad 2=2$
    C is false.   D is true.
    Therefore, equations A, B, and D.

3. Yes. Answers may vary.

5. $\begin{cases} x-y+z=-4 & (1) \\ 3x+2y-z=5 & (2) \\ -2x+3y-z=15 & (3) \end{cases}$

    Add E1 and E2.
    $4x+y=1$
    Add E1 and E3
    $-x+2y=11$
    Solve the new system:
    $\begin{cases} 4x+y=1 \\ -x+2y=11 \end{cases}$
    Multiply the second equation by 4.
    $\begin{cases} 4x+y=1 \\ -4x+8y=44 \end{cases}$
    Add the equations.
    $\quad 4x+y=1$
    $\underline{-4x+8y=44}$
    $\qquad 9y=45$
    $\qquad\ y=5$
    Replace y with 5 in the equation
    $4x+y=1$.
    $4x+5=1$
    $\quad 4x=-4$
    $\qquad x=-1$
    Replace x with -1 and y with 5 in E1.
    $(-1)-(5)+z=-4$
    $\qquad -6+z=-4$
    $\qquad\qquad z=2$
    The solution is $(-1, 5, 2)$.

7. $\begin{cases} x+y=3 & (1) \\ 2y=10 & (2) \\ 3x+2y-3z=1 & (3) \end{cases}$

    Solve E2 for y: $y=5$
    Replace y with 5 in E1.
    $x+5=3$
    $\quad x=-2$
    Replace x with -1 and y with 5 in E3.
    $3(-2)+2(5)-3z=1$
    $\qquad -6+10-3z=1$
    $\qquad\qquad 4-3z=1$
    $\qquad\qquad\ -3z=-3$
    $\qquad\qquad\quad z=1$
    The solution is $(-2, 5, 1)$.

9. $\begin{cases} 2x+2y+z=1 & (1) \\ -x+y+2z=3 & (2) \\ x+2y+4z=0 & (3) \end{cases}$

    Add E2 and E3.
    $3y+6z=3$ or $y+2z=1$
    Multiply E2 by 2 and add to E1.
    $-2x+2y+4z=6$
    $\underline{2x+2y+z=1}$
    $\qquad 4y+5z=7$
    Solve the new system:
    $\begin{cases} y+2z=1 \\ 4y+5z=7 \end{cases}$
    Multiply the first equation by -4.
    $\begin{cases} -4y-8z=-4 \\ 4y+5z=7 \end{cases}$
    Add the equations.
    $-4y-8z=-4$
    $\underline{\ 4y+5z=7}$
    $\qquad -3z=3$
    $\qquad\quad z=-1$
    Replace z with -1 in the equation
    $y+2z=1$.
    $y+2(-1)=1$
    $\quad y-2=1$
    $\qquad y=3$

Replace y with 3 and z with -1 in E3.
$$x + 2(3) + 4(-1) = 0$$
$$x + 6 - 4 = 0$$
$$x + 2 = 0$$
$$x = -2$$
The solution is $(-2, 3, -1)$.

11. $\begin{cases} x - 2y + z = -5 & (1) \\ -3x + 6y - 3z = 15 & (2) \\ 2x - 4y + 2z = -10 & (3) \end{cases}$

Multiply E2 by $-\dfrac{1}{3}$ and E3 by $\dfrac{1}{2}$.

$\begin{cases} x - 2y + z = -5 \\ x - 2y + z = -5 \\ x - 2y + z = -5 \end{cases}$

All three equations are identical. There are infinitely many solutions.
The solution is $\{(x, y, z) \mid x - 2y + z = -5\}$.

13. $\begin{cases} 4x - y + 2z = 5 & (1) \\ 2y + z = 4 & (2) \\ 4x + y + 3z = 10 & (3) \end{cases}$

Multiply E1 by -1 and add to E3.
$$-4x + y - 2z = -5$$
$$\underline{4x + y + 3z = 10}$$
$$2y + z = 5 \quad (4)$$
Multiply E4 by -1 and add to E2.
$$-2y - z = -5$$
$$\underline{2y + z = 4}$$
$$0 = -1 \quad \text{False}$$
Inconsistent system.
The solution is $\varnothing$.

15. $\begin{cases} x + 5z = 0 & (1) \\ 5x + y = 0 & (2) \\ y - 3z = 0 & (3) \end{cases}$

Multiply E3 by -1 and add to E2.
$$-y + 3z = 0$$
$$\underline{5x + y = 0}$$
$$5x + 3z = 0 \quad (4)$$
Multiply E1 by -5 and add to E4.

$$-5x - 25z = 0$$
$$\underline{5x + 3z = 0}$$
$$-22z = 0$$
$$z = 0$$
Replace z with 0 in E4.
$$5x + 3(0) = 0$$
$$5x = 0$$
$$x = 0$$
Replace x with 0 in E2.
$$5(0) + y = 0$$
$$y = 0$$
The solution is $(0, 0, 0)$.

17. $\begin{cases} 6x - 5z = 17 & (1) \\ 5x - y + 3z = -1 & (2) \\ 2x + y = -41 & (3) \end{cases}$

Add E2 and E3.
$$7x + 3z = -42 \quad (4)$$
Multiply E4 by 5, multiply E1 by 3, and add.
$$35x + 15z = -210$$
$$\underline{18x - 15z = 51}$$
$$53x = -159$$
$$x = -3$$
Replace x with -3 in E1.
$$6(-3) - 5z = 17$$
$$-18 - 5z = 17$$
$$-5z = 35$$
$$z = -7$$
Replace x with -3 in E3.
$$2(-3) + y = -41$$
$$-6 + y = -41$$
$$y = -35$$
The solution is $(-3, -35, -7)$.

19. $\begin{cases} x + y + z = 8 & (1) \\ 2x - y - z = -1 & (2) \\ x - 2y - 3z = 22 & (3) \end{cases}$

Add E1 and E2.
$$3x = 18 \quad \text{or} \quad x = 6$$
Add twice E1 to E3.

$$2x+2y+2z=16$$
$$\underline{x-2y-3z=22}$$
$$3x-\quad\ z=38$$

Replace x with 6 in this equation.
$$3(6)-z=38$$
$$18-z=38$$
$$-z=20$$
$$z=-20$$

Replace x with 6 and z with -20 in E1.
$$6+y+(-20)=8$$
$$y-14=8$$
$$y=22$$

The solution is $(6, 22, -20)$.

21. $\begin{cases} x+2y-z=5 & (1) \\ 6x+y+z=7 & (2) \\ 2x+4y-2z=5 & (3) \end{cases}$

Add E1 and E2.
$$7x+3y=12 \quad (4)$$
Add twice E2 to E3.
$$12x+2y+2z=14$$
$$\underline{2x+4y-2z=5}$$
$$14x+6y\quad\ =19 \quad (5)$$
Multiply E4 by -2 and add to E5.
$$-14x-6y=-24$$
$$\underline{14x+6y=19}$$
$$0=-5 \quad \text{False}$$

Inconsistent system.
The solution is $\varnothing$.

23. $\begin{cases} 2x-3y+z=2 & (1) \\ x-5y+5z=3 & (2) \\ 3x+y-3z=5 & (3) \end{cases}$

Add -2 times E2 to E1.
$$2x-\ 3y+\ z=2$$
$$\underline{-2x+10y-10z=-6}$$
$$7y-\ 9z=-4 \quad (4)$$
Add -3 times E2 to E3.

$$-3x+15y-15z=-9$$
$$\underline{3x+\ y-\ 3z=5}$$
$$16y-18z=-4$$

Solve the new system:
$\begin{cases} 7y-9z=-4 & (4) \\ 16y-18z=-4 & (5) \end{cases}$

Multiply E4 by -2 and add to E5.
$$-14y+18z=8$$
$$\underline{16y-18z=-4}$$
$$2y\quad\ =4$$
$$y=2$$

Replace y with 2 in E4.
$$7(2)-9z=-4$$
$$-9z=-18$$
$$z=2$$

Replace y with 2 and z with 2 in E1.
$$2x-3(2)+2=2$$
$$x=3$$

The solution is $(3, 2, 2)$.

25. $\begin{cases} -2x-4y+6z=-8 & (1) \\ x+2y-3z=4 & (2) \\ 4x+8y-12z=16 & (3) \end{cases}$

Add 2 times E2 to E1.
$$2x+4y-6z=8$$
$$\underline{-2x-4y+6z=-8}$$
$$0=0$$

Add -4 times E2 to E3.
$$-4x-8y+12z=-16$$
$$\underline{4x+8y-12z=16}$$
$$0=0$$

The system is dependent.
The solution is $\{(x, y, z)\,|\,x+2y-3z=4\}$.

27. $\begin{cases} 2x+2y-3z=1 & (1) \\ y+2z=-14 & (2) \\ 3x-2y=5 & (3) \end{cases}$

Add E1 to E3.
$$5x-3z=0 \quad (4)$$
Add twice E2 to E3.

$2y + 4z = -28$

$\underline{3x - 2y \qquad = -1}$

$3x + \qquad 4z = -29$ (5)

Multiply E4 by 4, multiply E5 by 2, and add.

$20x - 12z = 0$

$\underline{9x - 12z = -87}$

$29x \qquad = -87$

$x = -3$

Replace x with -3 in E4.

$5(-3) - 3z = 0$

$3z = -15$

$z = -5$

Replace z with -5 in E2.

$y + 2(-5) = -14$

$y - 10 = -14$

$y = -4$

The solution is $(-3, -4, -5)$.

29. $\begin{cases} x + 2y - z = 5 & (1) \\ -3x - 2y - 3z = 11 & (2) \\ 4x + 4y + 5z = -18 & (3) \end{cases}$

Add E1 and E2.

$-2x - 4z = 16$ or $x + 2z = -8$ (4)

Add twice E2 to E3.

$-6x - 4y - 6z = 22$

$\underline{4x + 4y + 5z = -18}$

$-2x - \qquad z = 4$ (5)

Solve the new system:

$\begin{cases} x + 2z = -8 & (4) \\ -2x - z = 4 & (5) \end{cases}$

Add twice E4 to E5.

$2x + 4z = -16$

$\underline{-2x - z = 4}$

$3z = -12$

$z = -4$

Replace z with -4 in E4.

$x + 2(-4) = -8$

$x - 8 = -8$

$x = 0$

Replace x with 0 and z with -4 in E1.

$0 + 2y - (-4) = 5$

$2y = 1$

$y = \dfrac{1}{2}$

The solution is $\left(0, \dfrac{1}{2}, -4\right)$.

31. $\begin{cases} \dfrac{3}{4}x - \dfrac{1}{3}y + \dfrac{1}{2}z = 9 & (1) \\[2mm] \dfrac{1}{6}x + \dfrac{1}{3}y - \dfrac{1}{2}z = 2 & (2) \\[2mm] \dfrac{1}{2}x - \quad y + \dfrac{1}{2}z = 2 & (3) \end{cases}$

Multiply E1 by 12, multiply E2 by 6, and multiply E3 by 2.

$\begin{cases} 9x - 4y + 6z = 108 & (4) \\ x + 2y - 3z = 12 & (5) \\ x - 2y + z = 4 & (6) \end{cases}$

Add twice E5 to E4.

$2x + 4y - 6z = 24$

$\underline{9x - 4y + 6z = 108}$

$11x \qquad = 132$

$x = 12$

Add E5 and E6.

$2x - 2z = 16$ or $x - z = 8$

Replace x with 12 in this equation.

$12 - z = 8$

$z = 4$

Replace x with 12 and z with 4 in E6.

$12 - 2y + 4 = 4$

$12 - 2y = 0$

$-2y = -12$

$y = 6$

The solution is $(12, 6, 4)$.

33. Let $x =$ the first number,

then $2x =$ the second number.

$x + 2x = 45$

$3x = 45$

$x = 15$

$2x = 2(15) = 30$

The numbers are 15 and 30.

35. $2(x-1)-3x = x-12$
$2x-2-3x = x-12$
$-x-2 = x-12$
$-2x = -10$
$x = 5$
The solution set is $\{5\}$.

37. $-y-5(y+5) = 3y-10$
$-y-5y-25 = 3y-10$
$-6y-25 = 3y-10$
$-9y = 15$
$y = -\dfrac{15}{9} = -\dfrac{5}{3}$
The solution set is $\left\{-\dfrac{5}{3}\right\}$.

39. Answers may vary.

41. Answers may vary.

43. $\begin{cases} x+\ y+\ z = 1 & (1) \\ 2x-\ y+\ z = 0 & (2) \\ -x+2y+2z = -1 & (3) \end{cases}$

Add E1 and E3.
$3y+3z = 0$ or $y+z = 0$ (4)
Add -2 times E1 to E2.
$-2x-2y-2z = -2$
$\underline{2x-\ y+\ z = 0}$
$-3y-\ z = -2$ (5)
Add E4 and E5.
$-2y = -2$
$y = 1$
Replace y with 1 in E4.
$1+z = 0$
$z = -1$
Replace y with 1 and z with -1 in E1.
$x+1+(-1) = 1$
$x = 1$
The solution is $(1, 1, -1)$, and

$\dfrac{x}{8}+\dfrac{y}{4}+\dfrac{z}{3} = \dfrac{1}{8}+\dfrac{1}{4}-\dfrac{1}{3}$
$\qquad = \dfrac{3}{24}+\dfrac{6}{24}-\dfrac{8}{24}$
$\qquad = \dfrac{1}{24}.$

45. $\begin{cases} x+y\ \ \ \ -w = 0 & (1) \\ y+2z+w = 3 & (2) \\ x\ \ \ -z\ \ \ \ = 1 & (3) \\ 2x-y\ \ \ \ -w = -1 & (4) \end{cases}$

Add E1 and E2.
$x+2y+2z = 3$ (4)
Add E2 and E4.
$2x+z = 2$ (5)
Add E3 and E5.
$x-z = 1$
$\underline{2x+z = 2}$
$3x\ \ \ \ = 3$
$x = 1$
Replace x with 1 in E3.
$1-z = 1$
$z = 0$
Replace x with 1 and z with 0 in E4.
$1+2y+2(0) = 3$
$1+2y = 3$
$2y = 2$
$y = 1$
Replace y with 1, and z with 0 in E2.
$1+2(0)+w = 3$
$1+w = 3$
$w = 2$
The solution is $(1, 1, 0, 2)$.

47. $\begin{cases} x+y+z+w = 5 & (1) \\ 2x+y+z+w = 6 & (2) \\ x+y+z\ \ \ \ = 2 & (3) \\ x+y\ \ \ \ \ \ \ \ = 0 & (4) \end{cases}$

Add -1 times E4 to E3.
$-x-y\ \ \ \ = 0$
$\underline{x+y+z = 2}$
$z = 2$

Replace z with 2 in E1 and E2.

$\begin{cases} x+y+w=3 & (5) \\ 2x+y+w=4 & (6) \end{cases}$

Add -1 times E5 to E6.

$-x-y-w=-3$

$\underline{2x+y+w=4}$

$x \qquad\;\; =1$

Replace x with 1 in E4.

$1+y=0$

$\quad y=-1$

Replace x with 1, y with -1, and z with 2 in E1.

$1+(-1)+2+w=5$

$\qquad\quad 2+w=5$

$\qquad\qquad\;\; w=3$

The solution is $(1, -1, 2, 3)$.

49. Answers may vary.

Exercise Set 4.3

1. Let x = the first number, and y = the second number.

$\begin{cases} x=y+2 & (1) \\ 2x=3y-4 & (2) \end{cases}$

Substitute $x=y+2$ in E2.

$2(y+2)=3y-4$

$2y+4=3y-4$

$\qquad y=8$

Replace y with 8 in E1.

$x=8+2=10$

The numbers are 10 and 8.

3. Let x = length of Enterprise, and y = length of Nimitz.

$\begin{cases} x+y=2193 & (1) \\ x-y=9 & (2) \end{cases}$

Add E1 and E2.

$2x=2202$

$\;\; x=1101$

Replace x with 1102 in E1.

$1101+y=2193$

$\qquad\quad y=1092$

The Enterprise is 1101 feet long and the Nimitz is 1092 feet long.

b. $\dfrac{1 \text{ field}}{100 \text{ yds}} \cdot \dfrac{1 \text{ yd}}{3 \text{ ft}} \cdot 1101 \text{ ft} = \dfrac{1101}{300}$ fields

$\qquad\qquad\qquad\qquad\qquad = 3.67$ fields

The Enterprise is 3.67 football fields in length.

5. Let p = the speed of the plane in still air, and w = the speed of the wind.

$\begin{cases} p+w=560 & (1) \\ p-w=480 & (2) \end{cases}$

Add E1 and E2.

$2p=1040$

$\;\; p=520$

Replace p with 520 in E1.

$520+w=560$

$\qquad\;\; w=40$

The speed of the plane in still air is 520 mph and the speed of the wind is 40 mph.

7. Let x = number of quarts of 4% butterfat milk, and
y = number of quarts of 1% butterfat milk.

$\begin{cases} x+ \qquad y=60 & (1) \\ 0.04x+0.01y=0.02(60) & (2) \end{cases}$

Multiply E2 by -100 and add to E1.

$\quad x+y=60$

$\underline{-4x-y=-120}$

$-3x \qquad =-60$

$\qquad\; x=20$

Replace x with 20 in E1.

$20+y=60$

$\qquad y=40$

There should be 20 quarts of 4% butterfat used and 40 quarts of 1% butterfat used.

9. Let k = number of students studied abroad in the United Kingdom and s = number of students studied abroad in Spain.

$$\begin{cases} k + \quad s = 40{,}012 \quad (1) \\ \quad k = 15{,}428 + s \quad (2) \end{cases}$$

Substitute $k = 15{,}428 + s$ in E1.

$$(15{,}428 + s) + s = 40{,}012$$
$$15{,}428 + 2s = 40{,}012$$
$$2s = 24{,}584$$
$$s = 12{,}292$$

Replace s with 12,292 in E21.

$$k = 15{,}428 + 12{,}292 = 27{,}720$$

The United Kingdom had 27,720 students and Spain had 12,292 students.

11. Let l = the number of large frames, and s = the number of small frames.

$$\begin{cases} l + s = 22 \quad (1) \\ 15l + 8s = 239 \quad (2) \end{cases}$$

Multiply E1 by -8 and add to E2.

$$\begin{array}{r} -8l - 8s = -176 \\ 15l + 8s = 239 \\ \hline 7l \quad\quad = 63 \\ l = 9 \end{array}$$

Replace l with 9 in E1.

$$9 + s = 22$$
$$s = 13$$

She bought 9 large frames and 13 small frames.

13. Let x = the first number, and y = the second number.

$$\begin{cases} x = y - 2 \quad (1) \\ 2x = 3y + 4 \quad (2) \end{cases}$$

Substitute $x = y - 2$ in E2.

$$2(y - 2) = 3y + 4$$
$$2y - 4 = 3y + 4$$
$$y = -8$$

Replace y with -8 in E1.

$$x = -8 + 2 = -10$$

The numbers are -10 and -8.

15. $\begin{cases} y = 7x + 18.7 \quad (1) \\ y = 6x + 27.7 \quad (2) \end{cases}$

Substitute $7x + 18.7$ for y in E2.

$$7x + 18.7 = 6x + 27.7$$
$$x = 9$$
$$1996 + 9 = 2005$$

The year would be 2005.

17. Let x = price of each tablet, and y = the price of each pen.

$$\begin{cases} 7x + 4y = 6.40 \quad (1) \\ 2x + 19y = 5.40 \quad (2) \end{cases}$$

Multiply E1 by 2 and E2 by -7 and add.

$$\begin{array}{r} 14x + 8y = 12.8 \\ -14x - 133y = -37.80 \\ \hline -125y = -25 \\ y = 0.20 \end{array}$$

Replace y with 0.20 in E1.

$$7x + 4(0.20) = 6.40$$
$$7x + 0.80 = 6.40$$
$$7x = 5.60$$
$$x = 0.80$$

Tablets cost $0.80 each and pens cost $0.20 each.

19. Let p = the speed of the plane in still air, and w = the speed of the wind.

First note:

$$\frac{2160 \text{ mi}}{3 \text{ hr}} = 720 \text{ mph} \text{ and}$$
$$\frac{2160 \text{ mi}}{4 \text{ hr}} = 540 \text{ mph}$$

Now,

$$\begin{cases} p + w = 720 \\ p - w = 540 \end{cases}$$

Add E1 and E2.

$$2p = 1260$$
$$p = 630$$

Replace p with 630 in E1.

$$630 + w = 720$$
$$w = 90$$

The speed of the plane in still air is 630 mph and the speed of the wind is 90 mph.

21. a. Answers may vary, but notice the slope of each function.

b. $\begin{cases} y = -0.52x + 64.5 \ (1) \\ y = 0.02x + 7.2 \quad (2) \end{cases}$

Substitute $0.02x + 7.2$ for y in E1.
$$0.02x + 7.2 = -0.52x + 64.5$$
$$0.54x = 57.3$$
$$x = \frac{57.3}{0.54} \approx 106.11$$
$$1990 + 106 = 2006$$

They would be the same in 2006.

23. let x = length of shortest two sides, and y = length of the longest side.

$\begin{cases} 2x + y = 93 \ (1) \\ y = x + 9 \quad (2) \end{cases}$

Replace y with $x + 9$ in E1.
$$2x + (x + 9) = 93$$
$$3x + 9 = 93$$
$$3x = 84$$
$$x = 28$$

Replace x with 28 in E2.
$$y = 28 + 9 = 37$$

The three sides are 28 cm, 28 cm, and 37 cm.

25. Cost for Hertz: $H(x) = 25 + 0.10x$

Cost for Budget: $B(x) = 20 + 0.25x$

We want to find when $B(x) = 2 \cdot H(x)$
$$20 + 0.25x = 2(25 + 0.10x)$$
$$20 + 0.25x = 50 + 0.20x$$
$$0.05x = 30$$
$$x = 600$$

The Budget charge will be twice that of the Hertz charge at 600 miles.

27. $\begin{cases} x + y = 180 \\ x = y - 30 \end{cases}$

Replace x with $y - 30$ in E1.

Replace y with $x + 9$ in E1.
$$(y - 30) + y = 180$$
$$2y - 30 = 180$$
$$2y = 210$$
$$y = 105$$

Replace y with 105 in E2.
$$x = 105 - 30 = 75$$

The value of x is $75°$ and the value of y is $105°$.

29. $C(x) = 30x + 10,000$
$R(x) = 46x$
$$46x = 30x + 10,000$$
$$16x = 10,000$$
$$x = 625$$
625 units

31. $C(x) = 1.2x + 1500$
$R(x) = 1.7x$
$$1.7x = 1.2x + 1500$$
$$0.5x = 1500$$
$$x = 3000 \text{ units}$$
3000 units

33. $C(x) = 75x + 160,000$
$R(x) = 200x$
$$200x = 75x + 160,000$$
$$125x = 160,000$$
$$x = 1280$$
1280 units

35. a. $R(x) = 450x$

b. $C(x) = 200x + 6000$

c. $R(x) = C(x)$
$$450x = 200x + 6000$$
$$250x = 6000$$
$$x = 24 \text{ desks}$$

37. Let x = units of Mix A
y = units of Mix B, and
z = units of Mix C.

$$\begin{cases} 4x + 6y + 4z = 30 \ (1) \\ 6x + y + z = 16 \ (2) \\ 3x + 2y + 12z = 24 \ (3) \end{cases}$$

Multiply E2 by –6 and add to E1.
$$-36x - 6y - 6z = -96$$
$$\underline{4x + 6y + 4z = 30}$$
$$-32x - 2z = -66 \ \text{or} \ 16x + z = 33 \ (4)$$

Multiply E2 by –2 and add to E3.
$$-12x - 2y - 2z = -32$$
$$\underline{3x + 2y + 12z = 24}$$
$$-9x + 10z = -8 \ (5)$$

Multiply E4 by –10 and add to E5.
$$-160x - 10z = -330$$
$$\underline{-9x + 10z = -8}$$
$$-169x = -338$$
$$x = 2$$

Replace x with 2 in E4.
$$16(2) + z = 33$$
$$32 + z = 33$$
$$z = 1$$

Replace x with 2 and z with 1 in E2.
$$6(2) + y + 1 = 16$$
$$y = 3$$

You need 2 units of Mix A, 3 units of Mix B, and 1 unit of Mix C.

39. Let x = length of shortest side,
y = length of longest side, and
z = length of the other two sides

$$\begin{cases} x + y + 2z = 29 \ (1) \\ y = 2x \ (2) \\ z = x + 2 \ (3) \end{cases}$$

Substitute $y = 2x$ and $z = x + 2$ in E1.
$$x + (2x) + 2(x + 2) = 29$$
$$x + 2x + 2x + 4 = 29$$
$$5x = 25$$
$$x = 5$$

Replace x with 5 in E2 and E3.

$$-2(5) + y = 0 \qquad -(5) + z = 2$$
$$y = 10 \qquad\qquad z = 7$$

The sides are 5 in., 7 in., 7 in., and 10 in.

41. Let x = the first number
y = the second number, and
z = the third number.

$$\begin{cases} x + y + z = 40 \\ x = y + 5 \\ x = 2z \end{cases}$$

$$\begin{cases} x + y + z = 40 \ (1) \\ x - y = 5 \ (2) \\ x - 2z = 0 \ (3) \end{cases}$$

Add E1 and E2.
$$2x + z = 45 \ (4)$$

Multiply E3 by –2 and add to E4.
$$-2x + 4z = 0$$
$$\underline{2x + z = 45}$$
$$5z = 45$$
$$z = 9$$

Replace z with 9 in E3.
$$x - 2(9) = 0$$
$$x = 18$$

Replace x with 15 in E2.
$$18 - y = 5$$
$$y = 13$$

The numbers are 18, 13, and 9.

43. Let x = number of free throws,
y = number of two-point field goals, and
z = number of three-point field goals

$$\begin{cases} x + 2y + 3z = 698 \\ y = 6z - 19 \\ x = y - 64 \end{cases}$$

$$\begin{cases} x + 2y + 3z = 698 \ (1) \\ y - 6z = -19 \ (2) \\ x - y = -64 \ (3) \end{cases}$$

Multiply E3 by –1 and add to E1.
$$-x + y = 64$$
$$\underline{x + 2y + 3z = 698}$$
$$3y + 3z = 762 \ \text{or} \ y + z = 254 \ (4)$$

Multiply E4 by –1 and add to E2.

$-y - z = -254$

$\underline{y - 6z = -19}$

$\overline{-7z = -273}$

$z = 39$

Replace z with 39 in E2.

$y - 6(39) = -19$

$y - 234 = -19$

$y = 215$

Replace y with 215 in E3.

$x - 215 = -64$

$x = 151$

She made 151 free throws, 215 two-point field goals, and 39 three-point field goals.

45. $\begin{cases} x + y + z = 180 \\ y + 2x + 5 = 180 \\ z + 2x - 5 = 180 \end{cases}$

$\begin{cases} x + y + z = 180 \quad (1) \\ 2x + y \quad\;\;\; = 175 \quad (2) \\ 2x + \quad\;\; z = 185 \quad (3) \end{cases}$

Multiply E1 by –1 and add to E2.

$-x - y - z = -180$

$\underline{2x + y \quad\quad = 175}$

$\overline{x \quad\quad - z = -5} \quad (4)$

Add E3 and E4.

$3x = 180$

$x = 60$

Replace x with 60 in E4.

$60 - z = -5$

$z = 65$

Replace x with 60 in E2.

$2(60) + y = 175$

$120 + y = 175$

$y = 55$

$x = 60$, $y = 55$, and $z = 65$

47. $\begin{cases} 3x - y + z = 2 \\ -x + 2y + 3z = 6 \end{cases}$

$6x - 2y + 2z = 4$

$\underline{-x + 2y + 3z = 6}$

$\overline{5x \quad\quad\;\; + 5z = 10}$

$5x + 5z = 10$

49. $\begin{cases} x + 2y - z = 0 \\ 3x + y - z = 6 \end{cases}$

$-3x - 6y + 3z = 0$

$\underline{3x + y - z = 2}$

$\overline{-5y + 2z = 2}$

$-5y + 2z = 2$

51. Let x = number filed in 1980 and y = number filed in 2001

$\begin{cases} y = 4x + 200{,}000 \quad (1) \\ y - x = 1{,}100{,}000 \quad (2) \end{cases}$

Substitute $y = 4x + 200{,}000$ in E2.

$(4x + 200{,}000) - x = 1{,}100{,}000$

$3x + 200{,}000 = 1{,}100{,}000$

$3x = 900{,}000$

$x = 300{,}000$

Replace x with 300,000 in E1.

$y = 4(300{,}000) + 200{,}000 = 1{,}400{,}000$

There were 300,000 filed in 1980 and 1,400,000 filed in 2001.

53. $y = ax^2 + bx + c$

$(1, 6): \quad 6 = a + b + c \quad (1)$

$(-1, -2): -2 = a - b + c \quad (2)$

$(0, -1): \quad -1 = c \quad\quad\quad (3)$

Substitute $c = -1$ in E1 and E2 to obtain

$\begin{cases} a + b = 7 \quad \text{(from E1)} \\ a - b = -1 \;\text{(from E2)} \end{cases}$

Add these equations.

$2a = 6$

$a = 3$

Replace a with 3 in $a + b = 7$.

$3 + b = 7$ so $b = 4$

Therefore, $a = 3$, $b = 4$, and $c = -1$.

55. $y = ax^2 + bx + c$

(0, 1065): $1065 = c$ (1)
(1, 1070): $1070 = a + b + c$ (2)
(3, 1175): $1175 = 9a + 3b + c$ (3)

Substitute $c = 1065$ in E2 and E3 to

obtain $\begin{cases} a + b = 5 & \text{(from E2)} \\ 9a + b = 105 & \text{(from E3)} \end{cases}$

Multiply the first equation by -1 and add to the second equation.

$$-3a - 3b = -15$$
$$\underline{9a + 3b = 110}$$
$$6a \quad\quad = 95$$

$$a = \frac{95}{6} = 15\frac{5}{6}$$

Replace a with $15\frac{5}{6}$ in $a + b = 5$.

$$\frac{95}{6} + b = 5$$

$$b = -\frac{65}{6} = -10\frac{5}{6}$$

So, $a = 15\frac{5}{6}$, $b = -10\frac{5}{6}$, and $c = 1065$.

For 2009, we have $x = 9$. Using our

model we get:

$$y = \frac{95}{6}(9)^2 - \frac{65}{6}(9) + 1065 = 2250$$

Our model predicts that 2,250,000

students will take the ACT in 2009

(**note:** remember that y is in thousands).

Integrated Review

1. C

2. D

3. A

4. B

5. $\begin{cases} x + y = 4 & \text{(1)} \\ \quad y = 3x & \text{(2)} \end{cases}$

Substitute $y = 3x$ in E1.

$$x + (3x) = 4$$
$$4x = 4$$
$$x = 1$$
$$y = 3x = 3(1) = 3$$

The solution is (1, 3).

6. $\begin{cases} x - y = -4 & \text{(1)} \\ \quad y = 4x & \text{(2)} \end{cases}$

Substitute $y = 4x$ in E1.

$$x - (4x) = -4$$
$$-3x = -4$$
$$x = \frac{4}{3}$$

$$y = 4x = 4\left(\frac{4}{3}\right) = \frac{16}{3}$$

The solution is $\left(\frac{4}{3}, \frac{16}{3}\right)$.

7. $\begin{cases} x + y = 1 & \text{(1)} \\ x - 2y = 4 & \text{(2)} \end{cases}$

Multiply E1 by -1 and add to E2.

$$-x - y = -1$$
$$\underline{x - 2y = 4}$$
$$-3y = 3$$
$$y = -1$$

Replace y with -1 in E1.

$$x + (-1) = 1$$
$$x - 1 = 1$$
$$x = 2$$

The solution is $(2, -1)$.

8. $\begin{cases} 2x - y = 8 & \text{(1)} \\ x + 3y = 11 & \text{(2)} \end{cases}$

Multiply E1 by 3 and add to E2.

$$6x - 3y = 24$$
$$\underline{x + 3y = 11}$$
$$7x \quad\quad = 35$$
$$x = 5$$

Replace x with 5 in E1.

$2(5) - y = 8$

$10 - y = 8$

$y = 2$

The solution is (5, 2).

9. $\begin{cases} 2x + 5y = 8 & (1) \\ 6x + y = 10 & (2) \end{cases}$

Multiply E2 by –5 and add to E1.

$2x + 5y = 8$

$-30x - 5y = -50$

$\overline{-28x \quad\quad = -42}$

$x = \dfrac{3}{2}$

Replace x with $\dfrac{3}{2}$ in E2.

$6\left(\dfrac{3}{2}\right) + y = 10$

$9 + y = 8$

$y = 1$

The solution is $\left(\dfrac{3}{2}, 1\right)$.

10. $\begin{cases} x - 4y = -5 & (1) \\ -3x - 8y = 0 & (2) \end{cases}$

Multiply E1 by –2 and add to E2.

$-2x + 8y = 10$

$-3x - 8y = 0$

$\overline{-5x \quad\quad = 10}$

$x = -2$

Replace x with –2 in E1.

$-2 - 4y = -5$

$-4y = -3$

$y = \dfrac{3}{4}$

The solution is $\left(-2, \dfrac{3}{4}\right)$.

11. $\begin{cases} 4x - 7y = 7 & (1) \\ 12x - 21y = 24 & (2) \end{cases}$

Multiply E1 by –3 and add to E2.

$-12x + 21y = -21$

$12x - 21y = 24$

$\overline{\quad\quad 0 = 3 \quad \text{False}}$

The system is inconsistent.

The solution set is $\varnothing$.

12. $\begin{cases} 2x - 5y = 3 & (1) \\ -4x + 10y = -6 & (2) \end{cases}$

Multiply E1 by 2 and add to E2.

$4x - 10y = 6$

$-4x + 10y = -6$

$\overline{\quad\quad 0 = 0 \quad \text{True}}$

The system is dependent.

The solution set is $\{(x, y) \mid 2x - 5y = 3\}$.

13. $\begin{cases} x + y = 2 & (1) \\ -3y + z = -7 & (2) \\ 2x + y - z = -1 & (3) \end{cases}$

Add E2 and E3.

$2x - 2y = -8$ or $x - y = -4$ (4)

Add E1 and E4.

$2x = -2$

$x = -1$

Replace x with –1 in E1.

$-1 + y = 2$

$y = 3$

Replace y with 3 in E2.

$-3(3) + z = -7$

$-9 + z = -7$

$z = 2$

The solution is (–1, 3, 2).

14. $\begin{cases} y + 2z = -3 & (1) \\ x - 2y = 7 & (2) \\ 2x - y + z = 5 & (3) \end{cases}$

Multiply E2 by –2 and add to E3.

$-2x + 4y = -14$

$2x - y + z = 5$

$\overline{3y + z = -9 \quad (4)}$

Multiply E4 by –2 and add to E1.

$$-6y - 2z = 18$$
$$\underline{\quad y + 2z = -3\quad}$$
$$-5y = 15$$
$$y = -3$$

Replace y with -3 in E4.
$$3(-3) + z = -9$$
$$z = 0$$

Replace y with -3 in E2.
$$x - 2(-3) = 7$$
$$x + 6 = 7$$
$$x = 1$$

The solution is $(1, -3, 0)$.

15. $\begin{cases} 2x + 4y - 6z = 3 & (1) \\ -x + y - z = 6 & (2) \\ x + 2y - 3z = 1 & (3) \end{cases}$

Multiply E3 by -2 and add to E1.
$$-2x - 4y + 6z = -2$$
$$\underline{\quad 2x + 4y - 6z = 3\quad}$$
$$0 = 1 \text{ False}$$

The system is inconsistent.
The solution set is $\varnothing$.

16. $\begin{cases} x - y + 3z = 2 & (1) \\ -2x + 2y - 6z = -4 & (2) \\ 3x - 3y + 9z = 6 & (3) \end{cases}$

Multiply E1 by 2 and add to E2.
$$2x - 2y + 6z = 4$$
$$\underline{-2x + 2y - 6z = -4}$$
$$0 = 0 \text{ True}$$

The system is dependent.
The solution set is
$\{(x, y) \mid x - y + 3z = 2\}$.

17. $\begin{cases} x + y - 4z = 5 & (1) \\ x - y + 2z = -2 & (2) \\ 3x + 2y + 4z = 18 & (3) \end{cases}$

Add E1 and E2.
$$2x - 2z = 3 \quad (4)$$

Multiply E2 by 2 and add to E3.

$$2x - 2y + 4z = -4$$
$$\underline{3x + 2y + 4z = 18}$$
$$5x \qquad + 8z = 14 \quad (5)$$

Multiply E4 by 4 and add to E5.
$$8x - 8z = 12$$
$$\underline{5x + 8z = 14}$$
$$13x \qquad = 26$$
$$x = 2$$

Replace x with 2 in E4.
$$2(2) - 2z = 3$$
$$-2z = -1$$
$$z = \frac{1}{2}$$

Replace x with 2 and z with $\frac{1}{2}$ in E2.

$$2 + y - 4\left(\frac{1}{2}\right) = 5$$
$$2 + y - 2 = 5$$
$$y = 5$$

The solution is $\left(2, 5, \frac{1}{2}\right)$.

18. $\begin{cases} 2x - y + 3z = 2 & (1) \\ x + y - 6z = 0 & (2) \\ 3x + 4y - 3z = 6 & (3) \end{cases}$

Add E1 and E3.
$$5x + 3y = 8 \quad (4)$$

Multiply E1 by 2 and add to E2.
$$4x - 2y + 6z = 4$$
$$\underline{\quad x + y - 6z = 0\quad}$$
$$5x - y \qquad = 4 \quad (5)$$

Multiply E5 by 3 and add to E4.
$$15x - 3y = 12$$
$$\underline{\quad 5x + 3y = 8\quad}$$
$$20x \qquad = 20$$
$$x = 1$$

Replace x with 1 in E5.
$$5(1) - y = 4$$
$$-y = -1$$
$$y = 1$$

Replace both x and y with 1 in E1.

$$2(1)+(-1)+3z = 2$$
$$1+3z = 2$$
$$3z = 1$$
$$z = \frac{1}{3}$$

The solution is $\left(1, 1, \frac{1}{3}\right)$.

19. Let x = the first number
and y = the second number.
$$\begin{cases} x = y-8 & (1) \\ 2x = y+11 & (2) \end{cases}$$
Substitute $x = y-8$ in E2.
$$2(y-8) = y+11$$
$$2y-16 = y+11$$
$$y = 27$$
Replace y with 27 in E1.
$$x = 27-8 = 19$$
The numbers are 19 and 27.

20. Let x = measure of the two smallest
angles,
y = measure of the third angle, and
z = measure of the fourth angle.
$$\begin{cases} 2x+y+z = 360 \\ y = x+30 \\ z = x+50 \end{cases}$$
Substitute $y = x+30$ and $z = x+50$ in
the first equation.
$$2x+(x+30)+(x+50) = 360$$
$$4x+80 = 360$$
$$4x = 280$$
$$x = 70$$
so $y = 70+30 = 100$
and $z = 70+50 = 120$
The two smallest angles are 70°, the
third angles is 100°, and the fourth
angle is 120°.

Exercise Set 4.4

1. $\begin{cases} x+\ \ y = 1 \\ x-2y = 4 \end{cases}$

$$\left[\begin{array}{cc|c} 1 & 1 & 1 \\ 1 & -2 & 4 \end{array}\right]$$

Multiply R1 by -1 and add to R2.

$$\left[\begin{array}{cc|c} 1 & 1 & 1 \\ 0 & -3 & 3 \end{array}\right]$$

Divide R2 by -3.

$$\left[\begin{array}{cc|c} 1 & 1 & 1 \\ 0 & 1 & -1 \end{array}\right]$$

This corresponds to $\begin{cases} x+y = 1 \\ \quad\ y = -1 \end{cases}$.

$$x+(-1) = 1$$
$$x-1 = 1$$
$$x = 2$$
The solution is $(2, -1)$.

3. $\begin{cases} x+3y = 2 \\ x+2y = 0 \end{cases}$

$$\left[\begin{array}{cc|c} 1 & 3 & 2 \\ 1 & 2 & 0 \end{array}\right]$$

Multiply R1 by -1 and add to R2.

$$\left[\begin{array}{cc|c} 1 & 3 & 2 \\ 0 & -1 & -2 \end{array}\right]$$

Multiply R2 by -1.

$$\left[\begin{array}{cc|c} 1 & 3 & 2 \\ 0 & 1 & 2 \end{array}\right]$$

This corresponds to $\begin{cases} x+3y = 2 \\ \quad\ y = 2 \end{cases}$.

$$x+3(2) = 6$$
$$x+6 = 2$$
$$x = -4$$
The solution is $(-4, 2)$.

5. $\begin{cases} x-2y = 4 \\ 2x-4y = 4 \end{cases}$

$$\left[\begin{array}{cc|c} 1 & -2 & 4 \\ 2 & -4 & 4 \end{array}\right]$$

Multiply R1 by -2 and add to R2.

$$\begin{bmatrix} 1 & -2 & | & 4 \\ 0 & 0 & | & -4 \end{bmatrix}$$

This corresponds to $\begin{cases} x - 2y = 4 \\ \quad\quad 0 = -4 \end{cases}$.

This is an inconsistent system.
The solution is $\varnothing$.

7. $\begin{cases} 3x - 3y = 9 \\ 2x - 2y = 6 \end{cases}$

$$\begin{bmatrix} 3 & -3 & | & 9 \\ 2 & -2 & | & 6 \end{bmatrix}$$

Divide R1 by 3.

$$\begin{bmatrix} 1 & -1 & | & 3 \\ 2 & -2 & | & 6 \end{bmatrix}$$

Multiply R1 by –2 and add to R2.

$$\begin{bmatrix} 1 & -1 & | & 3 \\ 0 & 0 & | & 0 \end{bmatrix}$$

This corresponds to $\begin{cases} x - y = 3 \\ \quad\quad 0 = 0 \end{cases}$.

This is a dependent system.
The solution is $\{(x, y) \mid x - y = 3\}$.

9. $\begin{cases} x + y \quad\quad = 3 \\ \quad\quad 2y \quad\quad = 10 \\ 3x + 2y - 4z = 12 \end{cases}$

$$\begin{bmatrix} 1 & 1 & 0 & | & 3 \\ 0 & 2 & 0 & | & 10 \\ 3 & 2 & -4 & | & 12 \end{bmatrix}$$

Multiply R1 by –3 and add to R3.

$$\begin{bmatrix} 1 & 1 & 0 & | & 3 \\ 0 & 2 & 0 & | & 10 \\ 0 & -1 & -4 & | & 3 \end{bmatrix}$$

Divide R2 by 2.

$$\begin{bmatrix} 1 & 1 & 0 & | & 3 \\ 0 & 1 & 0 & | & 5 \\ 0 & -1 & -4 & | & 3 \end{bmatrix}$$

Add R2 to R3.

$$\begin{bmatrix} 1 & 1 & 0 & | & 3 \\ 0 & 1 & 0 & | & 5 \\ 0 & 0 & -4 & | & 8 \end{bmatrix}$$

Divide R3 by –4.

$$\begin{bmatrix} 1 & 1 & 0 & | & 3 \\ 0 & 1 & 0 & | & 5 \\ 0 & 0 & 1 & | & -2 \end{bmatrix}$$

This corresponds to $\begin{cases} x + y = 3 \\ \quad\quad y = 5 \\ \quad\quad z = -2 \end{cases}$.

$x + 5 = 3$
$\quad x = -2$
The solution is $(-2, 5, -2)$.

11. $\begin{cases} \quad\quad 2y - z = -7 \\ x + 4y + z = -4 \\ 5x - y + 2z = 13 \end{cases}$

$$\begin{bmatrix} 0 & 2 & -1 & | & -7 \\ 1 & 4 & 1 & | & -4 \\ 5 & -1 & 2 & | & 13 \end{bmatrix}$$

Interchange R1 and R2.

$$\begin{bmatrix} 1 & 4 & 1 & | & -4 \\ 0 & 2 & -1 & | & -7 \\ 5 & -1 & 2 & | & 13 \end{bmatrix}$$

Multiply R1 by –5 and add to R3.

$$\begin{bmatrix} 1 & 4 & 1 & | & -4 \\ 0 & 2 & -1 & | & -7 \\ 0 & -21 & -3 & | & 33 \end{bmatrix}$$

Divide R2 by 2.

$$\begin{bmatrix} 1 & 4 & 1 & | & -4 \\ 0 & 1 & -\frac{1}{2} & | & -\frac{7}{2} \\ 0 & -21 & -3 & | & 33 \end{bmatrix}$$

Multiply R2 by 21 and add to R3.

$$\begin{bmatrix} 1 & 4 & 1 & | & -4 \\ 0 & 1 & -\frac{1}{2} & | & -\frac{7}{2} \\ 0 & 0 & -\frac{27}{2} & | & -\frac{81}{2} \end{bmatrix}$$

Multiply R2 by $-\dfrac{2}{27}$.

$$\begin{bmatrix} 1 & 4 & 1 & | & -4 \\ 0 & 1 & -\frac{1}{2} & | & -\frac{7}{2} \\ 0 & 0 & 1 & | & 3 \end{bmatrix}$$

This corresponds to $\begin{cases} x + 4y + z = 4 \\ \quad y - \frac{1}{2}z = -\frac{7}{2} \\ \qquad\quad z = 3 \end{cases}$.

$$y - \frac{1}{2}(3) = -\frac{7}{2}$$
$$y - \frac{3}{2} = -\frac{7}{2}$$
$$y = -2$$
$$x + 4(-2) + 3 = -4$$
$$x - 8 + 3 = -4$$
$$x = 1$$

The solution is $(1, -2, 3)$.

13. $\begin{cases} x - 4 = 0 \\ x + y = 1 \end{cases}$ or $\begin{cases} x \quad\;\; = 4 \\ x + y = 1 \end{cases}$

$$\begin{bmatrix} 1 & 0 & | & 4 \\ 1 & 1 & | & 1 \end{bmatrix}$$

Multiply R1 by -1 and add to R2.

$$\begin{bmatrix} 1 & 0 & | & 4 \\ 0 & 1 & | & -3 \end{bmatrix}$$

This corresponds to $\begin{cases} x = 4 \\ y = -3 \end{cases}$

The solution is $(4, -3)$.

15. $\begin{cases} x + y + z = 2 \\ 2x \quad\;\; - z = 5 \\ \quad\;\; 3y + z = 2 \end{cases}$

$$\begin{bmatrix} 1 & 1 & 1 & | & 2 \\ 2 & 0 & -1 & | & 5 \\ 0 & 3 & 1 & | & 2 \end{bmatrix}$$

Multiply R1 by -2 and add to R2.

$$\begin{bmatrix} 1 & 1 & 1 & | & 2 \\ 0 & -2 & -3 & | & 1 \\ 0 & 3 & 1 & | & 2 \end{bmatrix}$$

Divide R2 by -2.

$$\begin{bmatrix} 1 & 1 & 1 & | & 2 \\ 0 & 1 & \frac{3}{2} & | & -\frac{1}{2} \\ 0 & 3 & 1 & | & 2 \end{bmatrix}$$

Multiply R2 by -3 and add to R3.

$$\begin{bmatrix} 1 & 1 & 1 & | & 2 \\ 0 & 1 & \frac{3}{2} & | & -\frac{1}{2} \\ 0 & 0 & -\frac{7}{2} & | & \frac{7}{2} \end{bmatrix}$$

Multiply R3 by $-\dfrac{2}{7}$.

$$\begin{bmatrix} 1 & 1 & 1 & | & 2 \\ 0 & 1 & \frac{3}{2} & | & -\frac{1}{2} \\ 0 & 0 & 1 & | & -1 \end{bmatrix}$$

This corresponds to $\begin{cases} x + y + z = 2 \\ \quad y + \frac{3}{2}z = -\frac{1}{2} \\ \qquad\quad z = -1 \end{cases}$.

$$y + \frac{3}{2}(-1) = -\frac{1}{2}$$
$$y - \frac{3}{2} = -\frac{1}{2}$$
$$y = 1$$
$$x + 1 + (-1) = 2$$
$$x = 2$$

The solution is $(2, 1, -1)$.

17. $\begin{cases} 5x - 2y = 27 \\ -3x + 5y = 18 \end{cases}$

$$\begin{bmatrix} 5 & -2 & | & 27 \\ -3 & 5 & | & 18 \end{bmatrix}$$

Divide R1 by 5.

$$\begin{bmatrix} 1 & -\frac{2}{5} & | & \frac{27}{5} \\ -3 & 5 & | & 18 \end{bmatrix}$$

Multiply R1 by 3 and add to R2.

$$\begin{bmatrix} 1 & -\dfrac{2}{5} & \bigm| & \dfrac{27}{5} \\ 0 & \dfrac{19}{5} & \bigm| & \dfrac{171}{5} \end{bmatrix}$$

Multiply R2 by $\dfrac{5}{19}$.

$$\begin{bmatrix} 1 & -\dfrac{2}{5} & \bigm| & \dfrac{27}{5} \\ 0 & 1 & \bigm| & 9 \end{bmatrix}$$

This corresponds to $\begin{cases} x - \dfrac{2}{5}y = \dfrac{27}{5} \\ \qquad y = 9 \end{cases}$.

$x - \dfrac{2}{5}(9) = \dfrac{27}{5}$

$x - \dfrac{18}{5} = \dfrac{27}{5}$

$x = 9$

The solution is (9, 9).

19. $\begin{cases} 4x - 7y = 7 \\ 12x - 21y = 24 \end{cases}$

$$\begin{bmatrix} 4 & -7 & | & 7 \\ 12 & -21 & | & 24 \end{bmatrix}$$

Divide R1 by 4.

$$\begin{bmatrix} 1 & -\dfrac{7}{4} & \bigm| & \dfrac{7}{4} \\ 12 & -21 & \bigm| & 24 \end{bmatrix}$$

Multiply R1 by –12 and add to R2.

$$\begin{bmatrix} 1 & -\dfrac{7}{4} & \bigm| & \dfrac{7}{4} \\ 0 & 0 & \bigm| & 3 \end{bmatrix}$$

This corresponds to $\begin{cases} x - \dfrac{7}{4}y = \dfrac{7}{4} \\ \qquad 0 = 3 \end{cases}$.

This is an inconsistent system.
The solution set is $\varnothing$.

21. $\begin{cases} 4x - y + 2z = 5 \\ 2y + z = 4 \\ 4x + y + 3z = 10 \end{cases}$

$$\begin{bmatrix} 4 & -1 & 2 & | & 5 \\ 0 & 2 & 1 & | & 4 \\ 4 & 1 & 3 & | & 10 \end{bmatrix}$$

Divide R1 by 4.

$$\begin{bmatrix} 1 & -\dfrac{1}{4} & \dfrac{1}{2} & \bigm| & \dfrac{5}{4} \\ 0 & 2 & 1 & \bigm| & 4 \\ 0 & 2 & 1 & \bigm| & 5 \end{bmatrix}$$

Divide R2 by 2.

$$\begin{bmatrix} 1 & -\dfrac{1}{4} & \dfrac{1}{2} & \bigm| & \dfrac{5}{4} \\ 0 & 1 & \dfrac{1}{2} & \bigm| & 2 \\ 0 & 2 & 1 & \bigm| & 5 \end{bmatrix}$$

Multiply R2 by –2 and add to R3.

$$\begin{bmatrix} 1 & -\dfrac{1}{4} & \dfrac{1}{2} & \bigm| & \dfrac{5}{4} \\ 0 & 1 & \dfrac{1}{2} & \bigm| & 2 \\ 0 & 0 & 0 & \bigm| & 1 \end{bmatrix}$$

This corresponds to $\begin{cases} x - \dfrac{1}{4}y + \dfrac{1}{2}z = \dfrac{5}{4} \\ \qquad y + \dfrac{1}{2}z = 2 \\ \qquad\qquad 0 = 1 \end{cases}$.

This is an inconsistent system.
The solution set is $\varnothing$.

23. $\begin{cases} 4x + y + z = 3 \\ -x + y - 2z = -11 \\ x + 2y + 2z = -1 \end{cases}$

$\begin{bmatrix} 4 & 1 & 1 & | & 3 \\ -1 & 1 & -2 & | & -11 \\ 1 & 2 & 2 & | & -1 \end{bmatrix}$

Interchange R1 and R3.

$\begin{bmatrix} 1 & 2 & 2 & | & -1 \\ -1 & 1 & -2 & | & -11 \\ 4 & 1 & 1 & | & 3 \end{bmatrix}$

Add R1 to R2.
Multiply R1 by –4 and add to R3.

$\begin{bmatrix} 1 & 2 & 2 & | & -1 \\ 0 & 3 & 0 & | & -12 \\ 0 & -7 & -7 & | & 7 \end{bmatrix}$

Divide R2 by 3.

$\begin{bmatrix} 1 & 2 & 2 & | & -1 \\ 0 & 1 & 0 & | & -4 \\ 0 & -7 & -7 & | & 7 \end{bmatrix}$

Multiply R2 by 7 and add to R3.

$\begin{bmatrix} 1 & 2 & 2 & | & -1 \\ 0 & 1 & 0 & | & -4 \\ 0 & 0 & -7 & | & -21 \end{bmatrix}$

Divide R3 by –7.

$\begin{bmatrix} 1 & 2 & 2 & | & -1 \\ 0 & 1 & 0 & | & -4 \\ 0 & 0 & 1 & | & 3 \end{bmatrix}$

This corresponds to $\begin{cases} x + 2y + 2z = -1 \\ y = -4 \\ z = 3 \end{cases}$

$x + 2(-4) + 2(3) = -1$
$x - 8 + 6 = -1$
$x = 1$

The solution is (1, –4, 3).

25. Function

27. Not a function

29. $(-1)(-5) - (6)(3) = 5 - 18 = -13$

31. $(4)(-10) - (2)(-2) = -40 + 4 = -36$

33. $(-3)(-3) - (-1)(-9) = 9 - 9 = 0$

35. a. Solve the system $\begin{cases} 2.3x + y = 52 \\ -5.4x + y = 14 \end{cases}$.

$\begin{bmatrix} 2.3 & 1 & | & 52 \\ -5.4 & 1 & | & 14 \end{bmatrix}$

Since getting 1 in the first column would lead to repeating decimals, we multiply R1 by –1 and add to R2.

$\begin{bmatrix} 2.3 & 1 & | & 52 \\ -7.7 & 0 & | & -38 \end{bmatrix}$

This corresponds to $\begin{cases} 2.3x + y = 52 \\ -7.7x = -38 \end{cases}$.

From the second equation,
$x = \dfrac{-3.8}{-7.7} \approx 4.935$.

Thus, the percent of U.S. households owning black-and-white television sets was the same as the percent of U.S. households owning a microwave oven in the end of 1984 (about 4.9 years after 1980).

b. Solve the television equation for y:
$y = -2.3x + 52$. Thus, for 1980,
$y = -2.3(0) + 52 = 52$, and for 1993,
$y = -2.3(13) + 52 = 22.1$.
Solve the microwave equation for y:
$y = 5.4x + 14$. Thus, for 1980
$y = 5.4(0) + 14 = 14$, and for 1993,
$y = 5.4(13) + 14 = 84.2$.
In 1980, a greater percent of (and hence more) U.S. households owned black-and-white television sets. In 1993, more households owned a microwave oven. The percent owning black-and-white television sets is

decreasing and the percent owning a microwave oven is increasing. Answers may vary.

c. Let $y = 0$ in the television equation.
$$2.3x + y = 52$$
$$2.3x + 0 = 52$$
$$x = \frac{52}{2.3} \approx 22.6$$

According to this model, the percent of U.S. households owning a black-and-white television set will be 0% about 22.6 years after 1980, or sometime in 2002.

d. Answers may vary. The answer to part **c** is not accurate since there were still many black and white television sets in 2002.

37. Answers may vary. The matrix does not take into account the negative y coefficient in the first equation, nor the implied x coefficient of 1 in the second equation.

Exercise Set 4.5

1. $\begin{vmatrix} 3 & 5 \\ -1 & 7 \end{vmatrix} = 3(7) - 5(-1) = 21 + 5 = 26$

3. $\begin{vmatrix} 9 & -2 \\ 4 & -3 \end{vmatrix} = 39(-3) - 4(-2) = -27 + 8 = -19$

5. $\begin{vmatrix} -2 & 9 \\ 4 & -18 \end{vmatrix} = -2(-18) - 9(4) = 36 - 36 = 0$

7. $\begin{cases} 2y - 4 = 0 \\ x + 2y \quad = 5 \end{cases}$ or $\begin{cases} 2y = 4 \\ x + 2y = 5 \end{cases}$

$D = \begin{vmatrix} 0 & 2 \\ 1 & 2 \end{vmatrix} = 0(2) - 2(1) = 0 - 2 = -2$

$D_x = \begin{vmatrix} 4 & 2 \\ 5 & 2 \end{vmatrix} = 4(2) - 2(5) = 8 - 10 = -2$

$D_y = \begin{vmatrix} 0 & 4 \\ 1 & 5 \end{vmatrix} = 0(5) - 4(1) = 0 - 4 = -4$

$x = \frac{-2}{-2} = 1$ and $y = \frac{-4}{-2} = 2$

The solution is $(1, 2)$.

9. $\begin{cases} 3x + y = 1 \\ \quad 2y = 2 - 6x \end{cases}$ or $\begin{cases} 3x + y = 1 \\ 6x + 2y = 2 \end{cases}$

$D = \begin{vmatrix} 3 & 1 \\ 6 & 2 \end{vmatrix} = 3(2) - 1(6) = 6 - 6 = 0$

Thus, the system cannot be solved by Cramer's rule. Since E2 is 2 times E1, the system is dependent.
The solution is $\{(x, y) \mid 3x + y = 1\}$.

11. $\begin{cases} 5x - 2y = 27 \\ -3x + 5y = 5 \end{cases}$

$D = \begin{vmatrix} 5 & -2 \\ -3 & 5 \end{vmatrix}$
$= 5(5) - (-2)(-3)$
$= 25 - 6$
$= 19$

$D_x = \begin{vmatrix} 27 & -2 \\ 18 & 5 \end{vmatrix}$
$= 27(5) - (-2)(18)$
$= 135 + 36$
$= 171$

$D_y = \begin{vmatrix} 5 & 27 \\ -3 & 18 \end{vmatrix}$
$= 5(18) - 27(-3)$
$= 90 + 81$
$= 171$

$x = \frac{D_x}{D} = \frac{171}{19} = 9$ and $y = \frac{D_y}{D} = \frac{171}{19} = 9$

The solution is $(9, 9)$.

13.
$$\begin{vmatrix} 2 & 1 & 0 \\ 0 & 5 & -3 \\ 4 & 0 & 2 \end{vmatrix}$$

$$= 2\begin{vmatrix} 5 & -3 \\ 0 & 2 \end{vmatrix} - 1\begin{vmatrix} 0 & -3 \\ 4 & 2 \end{vmatrix} + 0\begin{vmatrix} 0 & 5 \\ 4 & 0 \end{vmatrix}$$

$$= 2[5(2)-(-3)(0)]-[0(2)-4(-3)]-0$$

$$= 2(10)-12$$

$$= 8$$

15.
$$\begin{vmatrix} 4 & -6 & 0 \\ -2 & 3 & 0 \\ 4 & -6 & 1 \end{vmatrix}$$

$$= 0\begin{vmatrix} -2 & 3 \\ 4 & -6 \end{vmatrix} - 0\begin{vmatrix} 4 & -6 \\ 4 & -6 \end{vmatrix} + 1 - 2\begin{vmatrix} 4 & -6 \\ 4 & 3 \end{vmatrix}$$

$$= 0-0+[4(3)-(-6)(-2)]$$

$$= 0$$

17.
$$\begin{vmatrix} 3 & 6 & -3 \\ -1 & -2 & 3 \\ 4 & -1 & 6 \end{vmatrix}$$

$$= 3\begin{vmatrix} -2 & 3 \\ -1 & 6 \end{vmatrix} - 6\begin{vmatrix} -1 & 3 \\ 4 & 6 \end{vmatrix} + (-3)\begin{vmatrix} -1 & -2 \\ 4 & -1 \end{vmatrix}$$

$$= 3[-2(-6)-3(-1)]-6[-1(6)-3(4)]$$

$$\quad -3[(-1)(-1)-(-2)(4)]$$

$$= 3(-9)-6(-18)-3(9)$$

$$= -27+108-27$$

$$= 54$$

19. $\begin{cases} 3x \quad\;\; +z = -1 \\ -x-3y+z = 7 \\ \quad\;\; 3y+z = 5 \end{cases}$

$$D = \begin{vmatrix} 3 & 0 & 1 \\ -1 & -3 & 1 \\ 0 & 3 & 1 \end{vmatrix}$$

$$= 3\begin{vmatrix} -3 & 1 \\ 3 & 1 \end{vmatrix} - 0\begin{vmatrix} -1 & 1 \\ 0 & 1 \end{vmatrix} + 1\begin{vmatrix} -1 & -3 \\ 0 & 3 \end{vmatrix}$$

$$= 3[(-3)(1)-1(3)]-0$$

$$\quad +[(-1)(3)-(-3)(0)]$$

$$= 3(-6)-3$$

$$= -21$$

$$D_x = \begin{vmatrix} -1 & 0 & 1 \\ 7 & -3 & 1 \\ 5 & 3 & 1 \end{vmatrix}$$

$$= -1\begin{vmatrix} -3 & 1 \\ 3 & 1 \end{vmatrix} - 0\begin{vmatrix} 7 & 1 \\ 5 & 1 \end{vmatrix} + 1\begin{vmatrix} 7 & -3 \\ 5 & 3 \end{vmatrix}$$

$$= -[(-3)(1)-1(3)]-0$$

$$\quad +[(7)(3)-(-3)(5)]$$

$$= 6+36$$

$$= 42$$

$$D_y = \begin{vmatrix} 3 & -1 & 1 \\ -1 & 7 & 1 \\ 0 & 5 & 1 \end{vmatrix}$$

$$= 3\begin{vmatrix} 7 & 1 \\ 5 & 1 \end{vmatrix} - (-1)\begin{vmatrix} -1 & 1 \\ 0 & 1 \end{vmatrix} + 1\begin{vmatrix} -1 & 7 \\ 0 & 5 \end{vmatrix}$$

$$= 3[7(1)-1(5)]+1[(-1)(1)-1(0)]$$

$$\quad +[(-1)(5)-7(0)]$$

$$= 3(2)+(-1)+(-5)$$

$$= 0$$

$$D_z = \begin{vmatrix} 3 & 0 & -1 \\ -1 & -3 & 7 \\ 0 & 3 & 5 \end{vmatrix}$$

$$= 3\begin{vmatrix} -3 & 7 \\ 3 & 5 \end{vmatrix} - 0\begin{vmatrix} -1 & 7 \\ 0 & 5 \end{vmatrix} + 1\begin{vmatrix} -1 & -3 \\ 0 & 3 \end{vmatrix}$$

$$= 3[(-3)(5)-7(3)]-0$$

$$\quad -[(-1)(3)-(-3)(0)]$$

$$= 3(-36)-(-3)$$

$$= -105$$

$$x = \frac{D_x}{D} = \frac{42}{-21} = -2, \quad y = \frac{D_y}{D} = \frac{0}{-21} = 0,$$

$$z = \frac{D_z}{D} = \frac{-105}{-21} = 5$$

The solution is $(-2, 0, 5)$.

21. $\begin{cases} x+\;\; y+\;\; z = 8 \\ 2x-\;\; y-\;\; z = 10 \\ x-2y+3z = 22 \end{cases}$

$$D = \begin{vmatrix} 1 & 1 & 1 \\ 2 & -1 & -1 \\ 1 & -2 & 3 \end{vmatrix}$$

$$= 1\begin{vmatrix} -1 & -1 \\ -2 & 3 \end{vmatrix} - 1\begin{vmatrix} 2 & -1 \\ 1 & 3 \end{vmatrix} + 1\begin{vmatrix} 2 & -1 \\ 1 & -2 \end{vmatrix}$$

$$= (-3-2) - [6-(-1)] + [-4-(-1)]$$

$$= -5 - 7 - 3$$

$$= -15$$

$$D_x = \begin{vmatrix} 8 & 1 & 1 \\ 10 & -1 & -1 \\ 22 & -2 & 3 \end{vmatrix}$$

$$= 8\begin{vmatrix} -1 & -1 \\ -2 & 3 \end{vmatrix} - 1\begin{vmatrix} 10 & -1 \\ 22 & 3 \end{vmatrix} + 1\begin{vmatrix} 10 & -1 \\ 22 & -2 \end{vmatrix}$$

$$= 8(-3-2) - [30-(-22)]$$
$$\quad + [-20-(-22)]$$

$$= 8(-5) - 52 + 2$$

$$-40 - 52 + 2$$

$$= -90$$

$$D_y = \begin{vmatrix} 1 & 8 & 1 \\ 2 & 10 & -1 \\ 1 & 22 & 3 \end{vmatrix}$$

$$= 1\begin{vmatrix} 10 & -1 \\ 22 & 3 \end{vmatrix} - 8\begin{vmatrix} 2 & -1 \\ 1 & 3 \end{vmatrix} + 1\begin{vmatrix} 2 & 10 \\ 1 & 22 \end{vmatrix}$$

$$= [30-(-22)] - 8[6-(-1)] + (44-10)$$

$$= 52 - 8(7) + 34$$

$$= 52 - 56 + 34$$

$$= 30$$

$$D_z = \begin{vmatrix} 1 & 1 & 8 \\ 2 & -1 & 10 \\ 1 & -2 & 22 \end{vmatrix}$$

$$= 1\begin{vmatrix} -1 & 10 \\ -2 & 22 \end{vmatrix} - 1\begin{vmatrix} 2 & 10 \\ 1 & 22 \end{vmatrix} + 8\begin{vmatrix} 2 & -1 \\ 1 & -2 \end{vmatrix}$$

$$= [-22-(20)] - (44-10)$$
$$\quad + 8[-4-(-1)]$$

$$= -2 - 34 + 8(-3)$$

$$= -36 - 24$$

$$= -60$$

$$x = \frac{D_x}{D} = \frac{-90}{-15} = 6, \quad y = \frac{D_y}{D} = \frac{30}{-15} = -2,$$

$$z = \frac{D_z}{D} = \frac{-60}{-15} = 4$$

The solution is $(6, -2, 4)$.

23. $\begin{vmatrix} 10 & -1 \\ -4 & 2 \end{vmatrix} = 10(2) - (-1)(-4) = 20 - 4 = 16$

25. $\begin{vmatrix} 1 & 0 & 4 \\ 1 & -1 & 2 \\ 3 & 2 & 1 \end{vmatrix}$

$$= 1\begin{vmatrix} -1 & 2 \\ 2 & 1 \end{vmatrix} - 0\begin{vmatrix} 1 & 2 \\ 3 & 1 \end{vmatrix} + 4\begin{vmatrix} 1 & -1 \\ 3 & 2 \end{vmatrix}$$

$$= 1(-1-4) - 1 + 4[2-(-3)]$$

$$= -5 + 4(5)$$

$$= -5 + 20$$

$$= 15$$

27. $\begin{vmatrix} \dfrac{3}{4} & \dfrac{5}{2} \\ -\dfrac{1}{6} & \dfrac{7}{3} \end{vmatrix} = \dfrac{3}{4}\left(\dfrac{7}{3}\right) - \dfrac{5}{2}\left(-\dfrac{1}{6}\right)$

$$= \frac{21}{12} + \frac{5}{12}$$

$$= \frac{26}{12}$$

$$= \frac{13}{6}$$

29. $\begin{vmatrix} 4 & -2 & 2 \\ 6 & -1 & 3 \\ 2 & 1 & 1 \end{vmatrix}$

$$= 4\begin{vmatrix} -1 & 3 \\ 1 & 1 \end{vmatrix} - (-2)\begin{vmatrix} 6 & 3 \\ 2 & 1 \end{vmatrix} + 2\begin{vmatrix} 6 & -1 \\ 2 & 1 \end{vmatrix}$$

$$= 4(-1-3) + 2(6-6) + 2[6-(-2)]$$

$$= 4(-4) + 2(0) + 2(8)$$

$$= -16 + 0 + 16$$

$$= 0$$

31. $\begin{vmatrix} -2 & 5 & 4 \\ 5 & -1 & 3 \\ 4 & 1 & 2 \end{vmatrix}$

$$= -2\begin{vmatrix} -1 & 3 \\ 1 & 2 \end{vmatrix} - 5\begin{vmatrix} 5 & 3 \\ 4 & 2 \end{vmatrix} + 4\begin{vmatrix} 5 & -1 \\ 4 & 1 \end{vmatrix}$$

$$= -2(-2-3) - 5(10-12) + 4[5-(-4)]$$

$$= -2(-5) - 5(-2) + 4(9)$$

$$= 10 + 10 + 36$$

$$= 56$$

33. $\begin{cases} 2x - 5y = 4 \\ x + 2y = -7 \end{cases}$

$D = \begin{vmatrix} 2 & -5 \\ 1 & 2 \end{vmatrix}$

$= 2(2) - (-5)(1) = 4 + 5 = 9$

$D_x = \begin{vmatrix} 4 & -5 \\ -7 & 2 \end{vmatrix}$

$= 4(2) - (-5)(-7)$

$= 8 - 35$

$= -27$

$D_y = \begin{vmatrix} 2 & 4 \\ 1 & -7 \end{vmatrix}$

$= 2(-7) - 4(1)$

$= -14 - 4$

$= -18$

$x = \dfrac{D_x}{D} = \dfrac{-27}{9} = -3$

$y = \dfrac{D_y}{D} = \dfrac{-18}{9} = -2$

The solution is $(-3, -2)$.

35. $\begin{cases} 4x + 2y = 5 \\ 2x + y = -1 \end{cases}$

$D = \begin{vmatrix} 4 & 2 \\ 2 & 1 \end{vmatrix}$

$= 4(1) - (2)(2) = 4 + 4 = 0$

Thus, the system cannot be solved by Cramer's rule. Multiply E2 by 2 yielding the new system:

$\begin{cases} 4x + 2y = 5 \\ 4x + 2y = -2 \end{cases}$

Therefore, the system is inconsistent. The solution is $\varnothing$.

37. $\begin{cases} 2x + 2y + z = 1 \\ -x + y + 2z = 3 \\ x + 2y + 4z = 0 \end{cases}$

$D = \begin{vmatrix} 2 & 2 & 1 \\ -1 & 1 & 2 \\ 1 & 2 & 4 \end{vmatrix}$

$= 2\begin{vmatrix} 1 & 2 \\ 2 & 4 \end{vmatrix} - 2\begin{vmatrix} -1 & 2 \\ 1 & 4 \end{vmatrix} + 1\begin{vmatrix} -1 & 1 \\ 1 & 2 \end{vmatrix}$

$= 2(4 - 4) - 2(-4 - 2) + (-2 - 1)$

$= 2(0) - 2(-6) + (-3)$

$= 0 + 12 - 3$

$= 9$

$D_x = \begin{vmatrix} 1 & 2 & 1 \\ 3 & 1 & 2 \\ 0 & 2 & 4 \end{vmatrix}$

$= 1\begin{vmatrix} 1 & 2 \\ 2 & 4 \end{vmatrix} - 3\begin{vmatrix} 2 & 1 \\ 2 & 4 \end{vmatrix} + 0\begin{vmatrix} 2 & 1 \\ 1 & 2 \end{vmatrix}$

$= (4 - 4) - 3(8 - 2) + 0$

$= 0 - 3(6)$

$= -18$

$D_y = \begin{vmatrix} 2 & 1 & 1 \\ -1 & 3 & 2 \\ 1 & 0 & 4 \end{vmatrix}$

$= 1\begin{vmatrix} 1 & 1 \\ 3 & 2 \end{vmatrix} - 0\begin{vmatrix} 2 & 1 \\ -1 & 2 \end{vmatrix} + 4\begin{vmatrix} 2 & 1 \\ -1 & 3 \end{vmatrix}$

$= (2 - 3) - 0 + 4[6 - (-1)]$

$= -1 + 4(7)$

$= -1 + 28$

$= -27$

$D_z = \begin{vmatrix} 2 & 2 & 1 \\ -1 & 1 & 3 \\ 1 & 2 & 0 \end{vmatrix}$

$= 1\begin{vmatrix} 2 & 1 \\ 1 & 3 \end{vmatrix} - 2\begin{vmatrix} 2 & 1 \\ -1 & 4 \end{vmatrix} 3 + 0\begin{vmatrix} 2 & 2 \\ -1 & 1 \end{vmatrix}$

$= (6 - 1) - 2[6 - (-1)] + 0$

$= 5 - 2(7)$

$= 5 - 14$

$= -9$

$x = \dfrac{D_x}{D} = \dfrac{-18}{9} = -2, \quad y = \dfrac{D_y}{D} = \dfrac{27}{9} = 3,$

$z = \dfrac{D_z}{D} = \dfrac{-9}{9} = -1$

The solution is $(-2, 3, -1)$.

39. $\begin{cases} \dfrac{2}{3}x - \dfrac{3}{4}y = -1 \\ -\dfrac{1}{6}x + \dfrac{3}{4}y = \dfrac{5}{2} \end{cases}$

$D = \begin{vmatrix} \dfrac{2}{3} & -\dfrac{3}{4} \\ -\dfrac{1}{6} & \dfrac{3}{4} \end{vmatrix}$

$\quad = \dfrac{2}{3}\left(\dfrac{3}{4}\right) - \left(-\dfrac{3}{4}\right)\left(-\dfrac{1}{6}\right)$

$\quad = \dfrac{1}{2} - \dfrac{1}{8}$

$\quad = \dfrac{3}{8}$

$D_x = \begin{vmatrix} -1 & -\dfrac{3}{4} \\ \dfrac{5}{2} & \dfrac{3}{4} \end{vmatrix}$

$\quad = (-1)\left(\dfrac{3}{4}\right) - \left(-\dfrac{3}{4}\right)\left(\dfrac{5}{2}\right)$

$\quad = -\dfrac{3}{4} + \dfrac{15}{8}$

$\quad = \dfrac{9}{8}$

$D_y = \begin{vmatrix} \dfrac{2}{3} & -1 \\ -\dfrac{1}{6} & \dfrac{5}{2} \end{vmatrix}$

$\quad = \dfrac{2}{3}\left(\dfrac{5}{2}\right) - (-1)\left(-\dfrac{1}{6}\right)$

$\quad = \dfrac{5}{3} - \dfrac{1}{6}$

$\quad = \dfrac{3}{2}$

$x = \dfrac{D_x}{D} = \dfrac{\frac{9}{8}}{\frac{3}{8}} = 3$ and $y = \dfrac{D_y}{D} = \dfrac{\frac{3}{2}}{\frac{3}{8}} = 4$

The solution is (3, 4).

41. $\begin{cases} 0.7x - 0.2y = -1.6 \\ 0.2x - y = -1.4 \end{cases}$

$D = \begin{vmatrix} 0.7 & -0.2 \\ 0.2 & -1 \end{vmatrix}$

$\quad = 0.7(-1) - (-0.2)(0.2)$

$\quad = -0.7 + 0.04$

$\quad = -0.66$

$D_x = \begin{vmatrix} -1.6 & -0.2 \\ -1.4 & -1 \end{vmatrix}$

$\quad = -1.6(-1) - (-0.2)(-1.4)$

$\quad = 1.6 - 2.8$

$\quad = 1.32$

$D_y = \begin{vmatrix} 0.7 & -1.6 \\ 0.2 & -1.4 \end{vmatrix}$

$\quad = 0.7(-1.4) - (-1.6)(0.2)$

$\quad = -0.98 + 0.32$

$\quad = -0.66$

$x = \dfrac{D_x}{D} = \dfrac{1.32}{-0.66} = -2$ and

$y = \dfrac{D_y}{D} = \dfrac{-0.66}{-0.66} = 1$

The solution is (–2, 1).

43. $\begin{cases} -2x + 4y - 2z = 6 \\ x - 2y + z = -3 \\ 3x - 6y + 3z = -9 \end{cases}$

$D = \begin{vmatrix} -2 & 4 & -2 \\ 1 & -2 & 1 \\ 3 & -6 & 3 \end{vmatrix}$

$\quad = -2\begin{vmatrix} -2 & 1 \\ -6 & 3 \end{vmatrix} - 4\begin{vmatrix} 1 & 1 \\ 3 & 3 \end{vmatrix} + (-2)\begin{vmatrix} 1 & -2 \\ 3 & -6 \end{vmatrix}$

$\quad = -2[-6 - (-6)] - 4(3 - 3) - 2[-6 - (-6)]$

$\quad = 2(0) - 4(0) - 2(0)$

$\quad = 0$

Therefore, Cramer's rule will not provide the solution. Note that E1 is –2 times E2 and that E3 is 3 times E2. Thus, the system is dependent. The solution is $\{(x, y, z) \mid x - 2y + z = -3\}$.

45. $\begin{cases} x - 2y + z = -5 \\ -3y + 2z = 4 \\ 3x - y = -2 \end{cases}$

$D = \begin{vmatrix} 1 & -2 & 1 \\ 0 & 3 & 2 \\ 3 & -1 & 0 \end{vmatrix}$

$= 1\begin{vmatrix} 3 & 2 \\ -1 & 0 \end{vmatrix} - 0\begin{vmatrix} -2 & 1 \\ -1 & 0 \end{vmatrix} + 3\begin{vmatrix} -2 & 1 \\ 3 & 2 \end{vmatrix}$

$= [0 - (-2)] - 0 + 3(-4 - 3)$

$= 2 + 3(-7)$

$= -19$

$D_x = \begin{vmatrix} -5 & -2 & 1 \\ 4 & 3 & 2 \\ -2 & -1 & 0 \end{vmatrix}$

$= 1\begin{vmatrix} 4 & 3 \\ -2 & -1 \end{vmatrix} - 2\begin{vmatrix} -5 & -2 \\ -2 & -1 \end{vmatrix} + 0\begin{vmatrix} 5 & -2 \\ 4 & 3 \end{vmatrix}$

$= [4 - (-6)] - 2(5 - 4) + 0$

$= 2 - 2(1)$

$= 0$

$D_y = \begin{vmatrix} 1 & -5 & 1 \\ 0 & 4 & 2 \\ 3 & -2 & 0 \end{vmatrix}$

$= 1\begin{vmatrix} 4 & 2 \\ -2 & 0 \end{vmatrix} - 0\begin{vmatrix} -5 & 1 \\ -2 & 0 \end{vmatrix} + 3\begin{vmatrix} -5 & 1 \\ 4 & 2 \end{vmatrix}$

$= [0 - (-4)] - 0 + 3(-10 - 4)$

$= 4 + 3(-14)$

$= 4 - 42$

$= -38$

$D_z = \begin{vmatrix} 1 & -2 & -5 \\ 0 & 3 & 4 \\ 3 & -1 & -2 \end{vmatrix}$

$= 1\begin{vmatrix} 3 & 4 \\ -1 & -2 \end{vmatrix} - 0\begin{vmatrix} -2 & -5 \\ -1 & -2 \end{vmatrix} + 3\begin{vmatrix} -2 & -5 \\ 3 & 4 \end{vmatrix}$

$= [-6 - (-4)] - 0 + 3[-8 - (-15)]$

$= -2 + 3(7)$

$= 19$

$x = \dfrac{D_x}{D} = \dfrac{0}{-19} = 0, \quad y = \dfrac{D_y}{D} = \dfrac{-38}{-19} = 2,$

$z = \dfrac{D_z}{D} = \dfrac{19}{-19} = -1$

The solution is $(0, 2, -1)$.

47. $5x - 6 + x - 12 = 6x - 18$

49. $2(3x - 6) + 3(x - 1) = 6x - 12 + 3x - 3$
$\qquad\qquad\qquad\qquad\quad = 9x - 15$

51. $f(x) = 5x - 6$ or $y = 5x - 6$

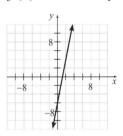

53. $h(x) = 3$ or $y = 3$

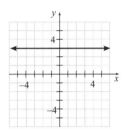

55. $\begin{vmatrix} 1 & x \\ 2 & 7 \end{vmatrix} = -3$

$1(7) - 2x = -3$

$7 - 2x = -3$

$-2x = -10$

$x = 5$

57. If the elements of a single row (or column) of a determinant are all zero, the value of the determinant will be zero. To see this, consider expanding on that row (or column) containing all zeros.

59. The array of signs for use with a 4×4 matrix is

$\begin{array}{cccc} + & - & + & - \\ - & + & - & + \\ + & - & + & - \\ - & + & - & + \end{array}$

61.
$$\begin{vmatrix} 5 & 0 & 0 & 0 \\ 0 & 4 & 2 & -1 \\ 1 & 3 & -2 & 0 \\ 0 & -3 & 1 & 2 \end{vmatrix}$$

$$= 5\begin{vmatrix} 4 & 2 & -1 \\ 3 & -2 & 0 \\ -3 & 1 & 2 \end{vmatrix} - 0\begin{vmatrix} 0 & 2 & -1 \\ 1 & -2 & 0 \\ 0 & 1 & 2 \end{vmatrix}$$

$$+ 0\begin{vmatrix} 0 & 4 & -1 \\ 1 & 3 & 0 \\ 0 & -3 & 2 \end{vmatrix} - 0\begin{vmatrix} 0 & 4 & 2 \\ 1 & 3 & -2 \\ 0 & -3 & 1 \end{vmatrix}$$

$$= 5\left[(-1)\begin{vmatrix} 3 & -2 \\ -3 & 1 \end{vmatrix} - 0\begin{vmatrix} 4 & 2 \\ -3 & 1 \end{vmatrix} + 2\begin{vmatrix} 4 & 2 \\ 3 & -2 \end{vmatrix} \right]$$

$$= 5[-(3-6) - 0 + 2(-8-6)]$$
$$= 5[3 + 2(-14)]$$
$$= 5(3 - 28)$$
$$= 5(-25)$$
$$= -125$$

63.
$$\begin{vmatrix} 4 & 0 & 2 & 5 \\ 0 & 3 & -1 & 1 \\ 0 & 0 & 2 & 0 \\ 0 & 0 & 0 & 1 \end{vmatrix} = 4\begin{vmatrix} 3 & -1 & 1 \\ 0 & 2 & 0 \\ 0 & 0 & 1 \end{vmatrix} - 0\begin{vmatrix} 0 & 2 & 5 \\ 0 & 2 & 0 \\ 0 & 0 & 1 \end{vmatrix}$$

$$+ 0\begin{vmatrix} 0 & 2 & 5 \\ 3 & -1 & 1 \\ 0 & 0 & 1 \end{vmatrix} - 0\begin{vmatrix} 0 & 2 & 5 \\ 3 & -1 & 1 \\ 0 & 2 & 0 \end{vmatrix}$$

$$= 4\left[3\begin{vmatrix} 2 & 0 \\ 0 & 1 \end{vmatrix} - 0\begin{vmatrix} -1 & 1 \\ 0 & 1 \end{vmatrix} + 0\begin{vmatrix} -1 & 1 \\ 2 & 0 \end{vmatrix} \right]$$
$$= 4[3(2-0) - 0 + 0]$$
$$= 4(6)$$
$$= 24$$

Chapter 4 Review

1. $\begin{cases} 3x + 10y = 1 & (1) \\ x + 2y = -1 & (2) \end{cases}$

(1)

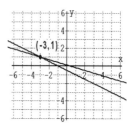

(2) From E2: $x = -2y - 1$
Replace x with $-2y - 1$ in E1.
$$3(-2y-1) + 10y = 1$$
$$-6y - 3 + 10y = 1$$
$$4y = 4$$
$$y = 1$$
Replace y with 1 in the equation
$x = -2y - 1$.
$$x = -2(1) - 1 = -3$$
The solution is $(-3, 1)$.

(3) Multiply E2 by -3 and add to E1.
$$\begin{array}{r} 3x + 10y = 1 \\ -3x - 6y = 3 \\ \hline 4y = 4 \\ y = 1 \end{array}$$
Replace y with 1 in E2.
$$x + 2(1) = -1$$
$$x + 2 = -1$$
$$x = -3$$
The solution is $(-3, 1)$.

2. $\begin{cases} y = \frac{1}{2}x + \frac{2}{3} & (1) \\ 4x + 6y = 4 & (2) \end{cases}$

(1)

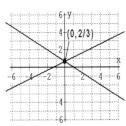

(2) Replace y with $\frac{1}{2}x + \frac{2}{3}$ in E2.

$$4x + 6\left(\frac{1}{2}x + \frac{2}{3}\right) = 4$$
$$4x + 3x + 4 = 4$$
$$x = 0$$

Replace x with 0 in E1.

$$y = \frac{1}{2}(0) + \frac{2}{3} = \frac{2}{3}$$

The solution is $\left(0, \frac{2}{3}\right)$.

(3) Rewrite the system: $\begin{cases} -\frac{1}{2}x + y = \frac{2}{3} \\ 4x + 6y = 4 \end{cases}$.

Multiply the first equation by –6.

$$\begin{cases} 3x - 6y = -4 \\ 4x + 6y = 4 \end{cases}$$

Add these equations.

$$7x = 0$$
$$x = 0$$

Replace x with 0 in second equation.

$$4(0) + 6y = 4$$
$$6y = 4$$
$$y = \frac{4}{6} = \frac{2}{3}$$

The solution is $\left(0, \frac{2}{3}\right)$.

3. $\begin{cases} 2x - 4y = 22 & (1) \\ 5x - 10y = 16 & (2) \end{cases}$

(1)

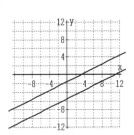

(2) Solve E1 for x.

$$2x - 4y = 22$$
$$2x = 4y + 22$$
$$x = 2y + 11$$

Replace x with $2y + 11$ in E2.

$$5(2y + 11) - 10y = 15$$
$$10y + 55 - 10y = 15$$
$$55 = 15 \quad \text{False}$$

This is an inconsistent system.
The solution is $\varnothing$.

(3) Multiply E1 by 5 and E2 by –2.

$$\begin{cases} 10x - 20y = 110 \\ -10x + 20y = -30 \end{cases}$$

Add these equations.

$$\begin{array}{r} 10x - 20y = 110 \\ -10x + 20y = -30 \\ \hline 0 = 80 \quad \text{False} \end{array}$$

This is an inconsistent system.
The solution is $\varnothing$.

4. $\begin{cases} 3x - 6y = 12 & (1) \\ 2y = x - 4 & (2) \end{cases}$

(1)

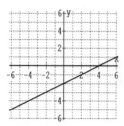

(2) Solve E2 for x.

$x = 2y + 4$

Replace x with $2y + 4$ in E1.

$3(2y + 4) - 6y = 12$
$6y + 12 - 6y = 12$
$\qquad\qquad 12 = 12$ True

This is a dependent system.
The solution is $\{(x, y) \mid 3x - 6y = 12\}$.

(3) $\begin{cases} 3x - 6y = 12 & (1) \\ -x + 2y = -4 & (2) \end{cases}$

Multiply E2 by 3.

$\begin{cases} 3x - 6y = 12 \\ -3x + 6y = -12 \end{cases}$

Add these equations.

$\quad 3x - 6y = 12$
$-3x + 6y = -12$
$\overline{\qquad\qquad\quad 0 = 0}$ True

This is a dependent system.
The solution is $\{(x, y) \mid 3x - 6y = 12\}$.

5. $\begin{cases} \dfrac{1}{2}x - \dfrac{3}{4}y = -\dfrac{1}{2} & (1) \\ \dfrac{1}{8}x + \dfrac{3}{4}y = \dfrac{19}{8} & (2) \end{cases}$

(1)

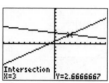

(2) Clear fractions by multiplying E1 by 4 and E2 by 8.

$\begin{cases} 2x - 3y = -2 & (1) \\ x + 6y = 19 & (2) \end{cases}$

Solve the new E2 for x.

$x = -6y + 19$

Replace x with $-6y + 19$ in new E1.

$2(-6y + 19) - 3y = -2$
$\quad -12y + 38 - 3y = -2$
$\qquad\qquad\qquad -15y = -40$
$\qquad\qquad\qquad\quad y = \dfrac{-40}{-15} = \dfrac{8}{3}$

Replace y with $\dfrac{8}{3}$ in the equation

$x = 6y + 19$.

$x = -6\left(\dfrac{8}{3}\right) + 19$
$x = -16 + 19$
$x = 3$

The solution is $\left(3, \dfrac{8}{3}\right)$.

(3) Add the equations.

$$\frac{1}{2}x - \frac{3}{4}y = -\frac{1}{2}$$
$$\frac{1}{8}x + \frac{3}{4}y = \frac{19}{8}$$
$$\overline{\quad\frac{5}{8}x \qquad = \frac{15}{8}\quad}$$
$$5x = 15$$
$$x = 3$$

Replace x with 3 in E1.

$$\frac{1}{2}(3) - \frac{3}{4}y = -\frac{1}{2}$$
$$-\frac{3}{4}y = -2$$
$$-3y = -8$$
$$y = \frac{8}{3}$$

The solution is $\left(3, \frac{8}{3}\right)$.

6. $\begin{cases} y = 32x & (1) \\ y = 15x + 25{,}500 & (2) \end{cases}$

Multiply E1 by -1 and add to E2.

$$-y = -32$$
$$\underline{y = 15x + 25{,}500}$$
$$0 = -17x + 25{,}500$$
$$17x = 25{,}500$$
$$x = 1500$$

Replace x with 1500 in E1.

$$y = 32(1500) = 48{,}000$$

The number of backpacks that the company must sell is 1500.

7. $\begin{cases} x + \quad z = 4 & (1) \\ 2x - y \quad = 4 & (2) \\ x + y - z = 0 & (3) \end{cases}$

Adding E2 and E3 gives $3x - z = 4$ (4)
Adding E1 and E4 gives $4x = 8$ or $x = 2$
Replace x with 2 in E1.

$$2 + z = 4$$
$$z = 2$$

Replace x with 2 and z with 2 in E3.

$$2 + y - 2 = 0$$
$$y = 0$$

The solution is (2, 0, 2).

8. $\begin{cases} 2x + 5y \quad = 4 & (1) \\ x - 5y + z = -1 & (2) \\ 4x \quad - z = 11 & (3) \end{cases}$

Add E2 and E3.

$$5x - 5y = 10 \ (4)$$

Add E1 and E4.

$$7x = 14$$
$$x = 2$$

Replace x with 2 in E1.

$$2(2) + 5y = 4$$
$$4 + 5y = 4$$
$$5y = 0$$
$$y = 0$$

Replace x with 2 in E3.

$$4(2) - z = 11$$
$$8 - z = 11$$
$$z = -3$$

The solution is (2, 0, -3).

9. $\begin{cases} \quad 4y + 2z = 5 & (1) \\ 2x + 8y \quad = 5 & (2) \\ 6x \quad + 4z = 1 & (3) \end{cases}$

Multiply E1 by -2 and add to E2.

$$-8y - 4z = -10$$
$$\underline{2x + 8y \qquad = 5}$$
$$2x \qquad - 4z = -5 \ (4)$$

Add E3 and E4.

$$8x = -4$$
$$x = -\frac{1}{2}$$

Replace x with $-\frac{1}{2}$ in E2.

$$2\left(-\frac{1}{2}\right) + 8y = 5$$
$$-1 + 8y = 5$$
$$8y = 6$$
$$y = \frac{3}{4}$$

Replace x with $-\frac{1}{2}$ in E3.

$$6\left(-\frac{1}{2}\right)+4z=1$$
$$-3+4z=1$$
$$4z=4$$
$$z=1$$

The solution is $\left(-\frac{1}{2},\frac{3}{4},1\right)$.

10. $\begin{cases} 5x+\ 7y\ \ \ \ \ \ =9\ \ (1) \\ \ \ \ \ \ \ 14y-\ z=28\ (2) \\ 4x\ \ \ \ \ \ +2z=-4\ (3) \end{cases}$

Dividing E3 by 2 gives $2x+z=-2$.
Add this equation to E2.

$$\begin{array}{r} 2x\ \ \ \ \ +z=-2 \\ 14y-z=28 \\ \hline 2x+14y\ \ \ =26\ \text{ or }\ x+7y=13\ (4) \end{array}$$

Multiply E4 by -1 and add to E1.

$$\begin{array}{r} -x-7y=-13 \\ 5x+7y=9 \\ \hline 4x\ \ \ \ \ \ =-4 \\ x=-1 \end{array}$$

Replace x with -1 in E4.

$$1+7y=13$$
$$7y=14$$
$$y=2$$

Replace x with -1 in E3.

$$4(-1)+2z=-4$$
$$-4+2z=-4$$
$$2z=0$$
$$z=0$$

The solution is $(-1, 2, 0)$.

11. $\begin{cases} 3x-\ 2y+2z=5\ \ \ (1) \\ -x+\ 6y+\ z=4\ \ \ (2) \\ 3x+14y+7z=20\ (3) \end{cases}$

Multiply E2 by 3 and add to E1.

$$\begin{array}{r} 3x-\ 2y+2z=5 \\ -3x+18y+3z=12 \\ \hline 16y+5z=17\ (4) \end{array}$$

Multiply E3 by -1 and add to E1.

$$\begin{array}{r} 3x-\ 2y+2z=5 \\ -3x-14y-7z=-20 \\ \hline -16y-5z=-15\ (5) \end{array}$$

Add E4 and E5.

$$\begin{array}{r} 16x+5z=17 \\ -16x-5z=-15 \\ \hline 0=2\ \ \text{False} \end{array}$$

The system is inconsistent.
The solution is $\varnothing$.

12. $\begin{cases} x+\ 2y+3z=11\ \ (1) \\ \ \ \ \ \ \ \ \ y+2z=3\ \ \ \ (2) \\ 2x\ \ \ \ \ +2z=10\ \ (3) \end{cases}$

Multiply E2 by -2 and add to E1.

$$\begin{array}{r} x+2y+3z=11 \\ -2y-4z=-6 \\ \hline x\ \ \ \ \ \ -z=5\ \ (4) \end{array}$$

Multiply E4 by 2 and add to E3.

$$\begin{array}{r} 2x+2z=10 \\ 2x-2z=10 \\ \hline 4x\ \ \ \ \ =20 \\ x=5 \end{array}$$

Replace x with 5 in E3.

$$2(5)+2z=10$$
$$10+2z=10$$
$$2z=0$$
$$z=0$$

Replace z with 0 in E2.

$$y+2(0)=3$$
$$y+0=3$$
$$y=3$$

The solution is $(5, 3, 0)$.

13. $\begin{cases} 7x-3y+2z=0\ (1) \\ 4x-4y-\ z=2\ (2) \\ 5x+2y+3z=1\ (3) \end{cases}$

Multiply E2 by 2 and add to E1.

$$\begin{array}{r} 7x-3y+2z=0 \\ 8x-8y-2z=4 \\ \hline 15x-11y\ \ \ \ =4\ (4) \end{array}$$

Multiply E2 by 3 and add to E3.

$12x - 12y - 3z = 6$
$\underline{5x + 2y + 3z = 1}$
$17x - 10y \quad = 7 \ (5)$

Solve the new system.

$\begin{cases} 15x - 11y = 4 \ (4) \\ 17x - 10y = 7 \ (5) \end{cases}$

Multiply E4 by –10, multiply E5 by 11, and add.

$-150x + 110y = -40$
$\underline{187x - 110y = 77}$
$37x \qquad = 37$
$\qquad x = 1$

Replace x with 1 in E4.

$15(1) - 11y = 4$
$15 - 11y = 4$
$-11y = -11$
$y = 1$

Replace x with 1 and y with 1 in E1.

$7(1) - 3(1) + 2z = 0$
$4 + 2z = 0$
$2z = -4$
$z = -2$

The solution is $(1, 1, -2)$.

14. $\begin{cases} x - 3y - 5z = -5 \ (1) \\ 4x - 2y + 3z = 13 \ (2) \\ 5x + 3y + 4z = 22 \ (3) \end{cases}$

Multiply E1 by –4 and add to E2.

$-4x + 12y + 20z = 20$
$\underline{4x - 2y + 3z = 13}$
$10y + 23z = 33 \ (4)$

Multiply E1 by –5 and add to E3.

$-5x + 15y + 20z = 25$
$\underline{5x + 3y + 4z = 22}$
$18y + 29z = 47 \ (5)$

Solve the new system.

$\begin{cases} 10y + 23z = 33 \ (4) \\ 18y + 29z = 47 \ (5) \end{cases}$

Multiply E4 by 9, multiply E5 by –5

and add.

$90y + 207z = 297$
$\underline{-90y - 145z = -235}$
$62z = 62$
$z = 1$

Replace z with 1 in E4.

$10y + 23(1) = 33$
$10y = 10$
$y = 1$

Replace y with 1 and z with 1 in E1.

$x - 3(1) - 5(1) = -5$
$x - 8 = -5$
$x = 3$

The solution is $(3, 1, 1)$.

15. Let x = the first number,
    y = the second number, and
    z = the third number.

$\begin{cases} x + y + z = 98 & (1) \\ x + y = z + 2 & (2) \\ y = 4x & (3) \end{cases}$

Replace y with $4x$ in E1 and E2.

$x + 4x + z = 98$
$5x + z = 98 \ (4)$
$x + 4x \quad = z + 2$
$5x - z = 2 \ (5)$

Add E4 and E5.

$5x + z = 98$
$\underline{5x - z = 2}$
$10x \quad = 100$
$x = 10$

Replace x with 10 in E3.

$y = 4(10) = 40$

Replace x with 10 and y with 40 in E2.

$10 + 40 = z + 2$
$50 = z + 2$
$48 = z$

The numbers are 10, 40, and 48.

16. Let x = the first number and
    y = the second number.

$$\begin{cases} x = 3y & (1) \\ 2(x+y) = 168 & (2) \end{cases}$$

Replace x with $3y$ in E2.

$$2(3y+y) = 168$$
$$8y = 168$$
$$y = 21$$

Replace y with 21 in E1.

$$x = 3(21) = 63$$

The numbers are 63 and 21.

17. Let x = speed of first car and
y = speed of the second car.

$$\begin{cases} 4x + 4y = 492 & (1) \\ y = x + 7 & (2) \end{cases}$$

Replace y with $x+7$ in E1.

$$4x + 4(x+7) = 492$$
$$8x + 28 = 492$$
$$8x = 464$$
$$x = 58$$

Replace x with 58 in E2.

$$y = 58 + 7 = 65$$

The cars are going 58 and 65 miles per hour.

18. Let w = the width of the foundation and
l = the length of the foundation.

$$\begin{cases} l = 3w & (1) \\ 2w + 2l = 296 & (2) \end{cases}$$

Replace l with $3w$ in E2.

$$2w + 2(3w) = 296$$
$$2w + 6w = 296$$
$$8w = 296$$
$$w = 37$$

Replace w with 37 in E1.

$$l = 3(37) = 111$$

The foundation is 37 feet wide and 111 feet long.

19. Let x = liters of 10% solution and
y = liters of 60% solution.

$$\begin{cases} x + y = 50 & (1) \\ 0.10x + 0.60y = 0.40(50) & (2) \end{cases}$$

Solve E1 for y.

$$y = 50 - x$$

Replace y with $50 - x$ in E2.

$$0.10x + 0.60(50 - x) = 0.40(50)$$
$$10[0.10x + 0.60(50 - x)] = 10[0.40(50)]$$
$$x + 6(50 - x) = 4(50)$$
$$x + 300 - 6x = 200$$
$$-5x = -100$$
$$x = 20$$

Replace x with 20 in the equation
$y = 50 - x$.

$$y = 50 - 20 = 30$$

He should use 20 liters of 10% solution and 30 liters of 60% solution.

20. Let c = pounds of chocolate used,
n = pounds of nuts used, and
r = pounds of raisins used.

$$\begin{cases} r = 2n & (1) \\ c + n + r = 45 & (2) \\ 3.00c + 2.70n + 2.25r = 2.80(45) & (3) \end{cases}$$

Replace r with $2n$ in E2.

$$c + n + 2n = 45$$
$$c + 3n = 45$$
$$c = -3n + 45$$

Replace r with $2n$ and c with $-3n + 45$ in E3.

$$3.00(-3n + 45) + 2.70n + 2.25(2n) = 126$$
$$-9n + 135 + 2.7n + 4.5n = 126$$
$$-1.8n + 135 = 126$$
$$-1.8n = -9$$
$$n = 5$$

Replace n with 5 in E1.

$$r = 2(5) = 10$$

Replace n with 5 and r with 10 in E2.

$$c + 5 + 10 = 45$$
$$c + 15 = 45$$
$$c = 30$$

She should use 30 pounds of creme-filled chocolates, 5 pounds of chocolate-covered nuts, and 10 pounds of chocolate-covered raisins.

21. Let x = the number of pennies,
$\quad\quad$ y = the number of nickels, and
$\quad\quad$ z = the number dimes.

$$\begin{cases} x + y + z = 53 & (1) \\ 0.01x + 0.05y + 0.10z = 2.77 & (2) \\ y = z + 4 & (3) \end{cases}$$

Clear the decimals from E2 by multiplying by 100.

$x + 5y + 10z = 277$ (4)

Replace y with $z + 4$ in E1.

$x + z + 4 + z = 53$
$\quad\quad x + 2z = 49$ (5)

Replace y with $z + 4$ in E4.

$x + 5(z + 4) + 10z = 277$
$\quad\quad\quad x + 15z = 257$ (6)

Solve the new system.

$$\begin{cases} x + 2z = 49 & (5) \\ x + 15z = 257 & (6) \end{cases}$$

Multiply E5 by -1 and add to E6.

$$\begin{array}{r} -x - 2z = -49 \\ x + 15z = 257 \\ \hline 13z = 208 \\ z = 16 \end{array}$$

Replace z with 16 in E3.

$x + 2(16) = 49$
$\quad x + 32 = 49$
$\quad\quad\quad x = 17$

Replace z with 16 in E3.

$y = 16 + 4 = 20$

He has 17 pennies, 20 nickels, and 16 dimes in his jar.

22. Let l = rate of interest on the larger investment and s = the rate of interest on the smaller investment, both expressed as decimals.

$$\begin{cases} 10,000l + 4000s = 1250 & (1) \\ l = s + 0.02 & (2) \end{cases}$$

Replace l with $s + 0.02$ in E1.

$10,000(s + 0.02) + 4000s = 1250$
$10,000s + 200 + 4000s = 1250$
$\quad\quad\quad\quad 14,000s = 1050$
$$s = \frac{1050}{14,000} = 0.075$$

and $l = 0.075 + 0.02 = 0.095$.

The interest rate on the larger investment is 9.5% and the rate on the smaller investment is 7.5%.

23. Let x = length of the equal side and y = length of the third side.

$$\begin{cases} 2x + y = 73 & (1) \\ y = x + 7 & (2) \end{cases}$$

Replace y with $x + 7$ in E1.

$2x + x + 7 = 73$
$\quad\quad\quad 3x = 66$
$\quad\quad\quad x = 22$

Replace x with 22 in E2.

$y = 22 + 7 = 29$

Two sides of the triangle have length 22 cm and the third side has length 29 cm.

24. Let f = the first number,
$\quad\quad$ s = the second number, and
$\quad\quad$ t = the third number.

$$\begin{cases} f + s + t = 295 & (1) \\ f = s + 5 & (2) \\ f = 2t & (3) \end{cases}$$

Solve E2 for s and E3 for t.

$s = f - 5$
$t = \dfrac{f}{2}$

Replace s with $f - 5$ and t with $\dfrac{f}{2}$ in E1.

$f + f - 5 + \dfrac{f}{2} = 295$
$\quad\quad\quad \dfrac{5}{2}f = 300$
$\quad\quad\quad\quad f = 120$

Replace f with 300 in the equation $s = f - 5$.

$s = 120 - 5 = 115$

Replace f with 120 the equation $\dfrac{f}{2}$.

$t = \dfrac{120}{2} = 60$

The first number is 120, the second number is 115, and the third number is 60.

25. $\begin{cases} 3x + 10y = 1 \\ x + 2y = -1 \end{cases}$

$\begin{bmatrix} 3 & 10 & | & 1 \\ 1 & 2 & | & -1 \end{bmatrix}$

Interchange R1 and R2.

$\begin{bmatrix} 1 & 2 & | & -1 \\ 3 & 10 & | & 1 \end{bmatrix}$

Multiply R1 by -3 and add to R2.

$\begin{bmatrix} 1 & 2 & | & -1 \\ 0 & 4 & | & 4 \end{bmatrix}$

Divide R2 by 4.

$\begin{bmatrix} 1 & 2 & | & -1 \\ 0 & 1 & | & 1 \end{bmatrix}$

This corresponds to $\begin{cases} x + 2y = -1 \\ y = 1. \end{cases}$

$x + 2(1) = -1$

$\qquad x = -3$

The solution is $(-3, 1)$.

26. $\begin{cases} 3x - 6y = 12 \\ 2y = x - 4 \end{cases}$, or $\begin{cases} 3x - 6y = 12 \\ -x + 2y = -4 \end{cases}$

$\begin{bmatrix} 3 & -6 & | & 12 \\ -1 & 2 & | & -4 \end{bmatrix}$

Divide R1 by 3.

$\begin{bmatrix} 1 & -2 & | & 4 \\ -1 & 2 & | & -4 \end{bmatrix}$

Add R1 to R2.

$\begin{bmatrix} 1 & -2 & | & 4 \\ 0 & 0 & | & 0 \end{bmatrix}$

This corresponds to $\begin{cases} x - 2y = 4 \\ 0 = 0 \end{cases}$

This is a dependent system.
The solution is $\{x, y) \mid x - 2y = 4\}$.

27. $\begin{cases} 3x - 2y = -8 \\ 6x + 5y = 11 \end{cases}$

$\begin{bmatrix} 3 & -2 & | & -8 \\ 6 & 5 & | & 11 \end{bmatrix}$

Divide R1 by 3.

$\begin{bmatrix} 1 & -\dfrac{2}{3} & | & -\dfrac{8}{3} \\ 6 & 5 & | & 11 \end{bmatrix}$

Multiply R1 by -6 and add to R2.

$\begin{bmatrix} 1 & -\dfrac{2}{3} & | & -\dfrac{8}{3} \\ 0 & 9 & | & 27 \end{bmatrix}$

Divide R2 by 9.

$\begin{bmatrix} 1 & -\dfrac{2}{3} & | & -\dfrac{8}{3} \\ 0 & 1 & | & 3 \end{bmatrix}$

This corresponds to $\begin{cases} x - \dfrac{2}{3}y = -\dfrac{8}{3} \\ y = 3. \end{cases}$

$x - \dfrac{2}{3}(3) = -\dfrac{8}{3}$

$\qquad x - 2 = -\dfrac{8}{3}$

$\qquad\qquad x = -\dfrac{2}{3}$

The solution is $\left(-\dfrac{2}{3}, 3\right)$.

28. $\begin{cases} 6x - 6y = -5 \\ 10x - 2y = 1 \end{cases}$

$\begin{bmatrix} 6 & -6 & | & -5 \\ 10 & -2 & | & 1 \end{bmatrix}$

Divide R1 by 6.

$$\begin{bmatrix} 1 & -1 & \Big| & -\frac{5}{6} \\ 10 & -2 & \Big| & 1 \end{bmatrix}$$

Multiply R1 by -10 and add to R2.

$$\begin{bmatrix} 1 & -1 & \Big| & -\frac{5}{6} \\ 0 & 8 & \Big| & \frac{28}{3} \end{bmatrix}$$

Divide R2 by 8.

$$\begin{bmatrix} 1 & -1 & \Big| & -\frac{5}{6} \\ 0 & 1 & \Big| & \frac{7}{6} \end{bmatrix}$$

Add R2 to R1.

$$\begin{bmatrix} 1 & 0 & \Big| & \frac{1}{3} \\ 0 & 1 & \Big| & \frac{7}{6} \end{bmatrix}$$

This corresponds to $\begin{cases} x = \dfrac{1}{3} \\ y = \dfrac{7}{6} \end{cases}$.

The solution is $\left(\dfrac{1}{3}, \dfrac{7}{6} \right)$

29. $\begin{cases} 3x - 6y = 0 \\ 2x + 4y = 5 \end{cases}$

$$\begin{bmatrix} 3 & -6 & \Big| & 0 \\ 2 & 4 & \Big| & 5 \end{bmatrix}$$

Divide R1 by 3.

$$\begin{bmatrix} 1 & -2 & \Big| & 0 \\ 2 & 4 & \Big| & 5 \end{bmatrix}$$

Multiply R1 by -2 and add to R2.

$$\begin{bmatrix} 1 & -2 & \Big| & 0 \\ 0 & 8 & \Big| & 5 \end{bmatrix}$$

Divide R2 by 8.

$$\begin{bmatrix} 1 & -2 & \Big| & 0 \\ 0 & 1 & \Big| & \frac{5}{8} \end{bmatrix}$$

This corresponds to $\begin{cases} x - 2y = 0 \\ y = \dfrac{5}{8} \end{cases}$.

$$x - 2\left(\frac{5}{8} \right) = 0$$

$$x - \frac{5}{4} = 0$$

$$x = \frac{5}{4}$$

The solution is $\left(\dfrac{5}{4}, \dfrac{5}{8} \right)$.

30. $\begin{cases} 5x - 3y = 10 \\ -2x + y = -1 \end{cases}$

$$\begin{bmatrix} 5 & -3 & \Big| & 10 \\ -2 & 1 & \Big| & -1 \end{bmatrix}$$

Divide R1 by 5.

$$\begin{bmatrix} 1 & -\frac{3}{5} & \Big| & 2 \\ -2 & 1 & \Big| & -1 \end{bmatrix}$$

Multiply R1 by 2 and add to R2.

$$\begin{bmatrix} 1 & -\frac{3}{5} & \Big| & 2 \\ 0 & -\frac{1}{5} & \Big| & 3 \end{bmatrix}$$

Multiply R2 by -5.

$$\begin{bmatrix} 1 & -\frac{3}{5} & \Big| & 2 \\ 0 & 1 & \Big| & -15 \end{bmatrix}$$

This corresponds to $\begin{cases} x - \dfrac{3}{5}y = 2 \\ y = -15 \end{cases}$.

$$x - \frac{3}{5}(-15) = 2$$

$$x + 9 = 2$$

$$x = -7$$

The solution is $(-7, -15)$.

31. $\begin{cases} 0.2x - 0.3y = -0.7 \\ 0.5x + 0.3y = 1.4 \end{cases}$

$$\left[\begin{array}{cc|c} 0.2 & -0.3 & -0.7 \\ 0.5 & 0.3 & 1.4 \end{array}\right]$$

Multiply both rows by 10 to clear decimals.

$$\left[\begin{array}{cc|c} 2 & -3 & -7 \\ 5 & 3 & 14 \end{array}\right]$$

Divide R1 by 2.

$$\left[\begin{array}{cc|c} 1 & -\dfrac{3}{2} & -\dfrac{7}{2} \\ 5 & 3 & 14 \end{array}\right]$$

Multiply R1 by –5 and add to R2.

$$\left[\begin{array}{cc|c} 1 & -\dfrac{3}{2} & -\dfrac{7}{2} \\ 0 & \dfrac{21}{2} & \dfrac{63}{2} \end{array}\right]$$

Multiply R2 by $\dfrac{2}{21}$.

$$\left[\begin{array}{cc|c} 1 & -\dfrac{3}{2} & -\dfrac{7}{2} \\ 0 & 1 & 3 \end{array}\right]$$

This corresponds to $\begin{cases} x - \dfrac{3}{2}y = -\dfrac{7}{2} \\ y = 3. \end{cases}$

$$x - \frac{3}{2}(3) = -\frac{7}{2}$$
$$x - \frac{9}{2} = -\frac{7}{2}$$
$$x = 1$$

The solution is (1, 3).

32. $\begin{cases} 3x + 2y = 8 \\ 3x - y = 5 \end{cases}$

$$\left[\begin{array}{cc|c} 3 & 2 & -8 \\ 3 & -1 & 5 \end{array}\right]$$

Divide R1 by 3.

$$\left[\begin{array}{cc|c} 1 & \dfrac{2}{3} & \dfrac{8}{3} \\ 3 & -1 & 5 \end{array}\right]$$

Multiply R1 by –3 and add to R2.

$$\left[\begin{array}{cc|c} 1 & \dfrac{2}{3} & \dfrac{8}{3} \\ 0 & -3 & -3 \end{array}\right]$$

Divide R2 by –3.

$$\left[\begin{array}{cc|c} 1 & -\dfrac{2}{3} & \dfrac{8}{3} \\ 0 & 1 & 1 \end{array}\right]$$

This corresponds to $\begin{cases} x + \dfrac{2}{3}y = \dfrac{8}{3} \\ y = 1. \end{cases}$

$$x - \frac{2}{3}(1) = \frac{8}{3}$$
$$x = 2$$

The solution is (2, 1).

33. $\begin{cases} x \phantom{{}+y} + z = 4 \\ 2x - y \phantom{{}+z} = 0 \\ x + y - z = 0 \end{cases}$

$$\left[\begin{array}{ccc|c} 1 & 0 & 1 & 4 \\ 2 & -1 & 0 & 0 \\ 1 & 1 & -1 & 0 \end{array}\right]$$

Multiply R1 by –2 and add to R2.
Multiply R1 by –1 and add to R3.

$$\left[\begin{array}{ccc|c} 1 & 0 & 1 & 4 \\ 0 & -1 & -2 & -8 \\ 0 & 1 & -2 & -4 \end{array}\right]$$

Multiply R2 by –1.

$$\left[\begin{array}{ccc|c} 1 & 0 & 1 & 4 \\ 0 & 1 & 2 & 8 \\ 0 & 1 & -2 & -4 \end{array}\right]$$

Multiply R2 by –1 and add to R3.

$$\left[\begin{array}{ccc|c} 1 & 0 & 1 & 4 \\ 0 & 1 & 2 & 8 \\ 0 & 0 & -4 & -12 \end{array}\right]$$

Divide R3 by –4.

$$\left[\begin{array}{ccc|c} 1 & 0 & 1 & 4 \\ 0 & 1 & 2 & 8 \\ 0 & 0 & 1 & 3 \end{array}\right]$$

This corresponds to $\begin{cases} x + z = 4 \\ y + 2z = 8 \\ \quad\quad z = 3. \end{cases}$

$y + 2(3) = 8$
$\quad y + 6 = 8$
$\quad\quad y = 2$
$\quad x + 3 = 4$
$\quad\quad x = 1$

The solution is $(1, 2, 3)$.

34. $\begin{cases} 2x + 5y \quad\;\; = 4 \\ x - 5y + z = -1 \\ 4x \quad\quad - z = 11 \end{cases}$

$\begin{bmatrix} 2 & 5 & 0 & | & 4 \\ 1 & -5 & 1 & | & -1 \\ 4 & 0 & -1 & | & 11 \end{bmatrix}$

Interchange R1 and R2.

$\begin{bmatrix} 1 & -5 & 1 & | & -1 \\ 2 & 5 & 0 & | & 4 \\ 4 & 0 & -1 & | & 11 \end{bmatrix}$

Multiply R1 by -2 and add to R2.
Multiply R1 by -4 and add to R3.

$\begin{bmatrix} 1 & -5 & 1 & | & -1 \\ 0 & 15 & -2 & | & 6 \\ 0 & 20 & -5 & | & 15 \end{bmatrix}$

Divide R2 by 15.

$\begin{bmatrix} 1 & -5 & 1 & | & -1 \\ 0 & 1 & -\frac{2}{15} & | & \frac{2}{5} \\ 0 & 20 & -5 & | & 15 \end{bmatrix}$

Multiply R2 by -20 and add to R3.

$\begin{bmatrix} 1 & -5 & 1 & | & -1 \\ 0 & 1 & -\frac{2}{15} & | & \frac{2}{5} \\ 0 & 0 & -\frac{7}{3} & | & 7 \end{bmatrix}$

Multiply R3 by $-\frac{3}{7}$.

$\begin{bmatrix} 1 & -5 & 1 & | & -1 \\ 0 & 1 & -\frac{2}{15} & | & \frac{2}{5} \\ 0 & 0 & 1 & | & -3 \end{bmatrix}$

This corresponds to $\begin{cases} x - 5y + z = -1 \\ y - \frac{2}{15}z = \frac{2}{5} \\ \quad\quad z = -3. \end{cases}$

$y - \frac{2}{15}(-3) = \frac{2}{15}$
$\quad\quad y + \frac{2}{5} = \frac{2}{5}$
$\quad\quad\quad\quad y = 0$
$x - 5(0) + (-3) = -1$
$\quad\quad\quad x - 3 = -1$
$\quad\quad\quad\quad\quad x = 2$

The solution is $(2, 0, -3)$.

35. $\begin{cases} 3x - y \quad\quad = 11 \\ x \quad\; + 2z = 13 \\ \quad y - z = -7 \end{cases}$

$\begin{bmatrix} 3 & -1 & 0 & | & 11 \\ 1 & 0 & 2 & | & 13 \\ 0 & 1 & -1 & | & -7 \end{bmatrix}$

Interchange R1 and R2.

$\begin{bmatrix} 1 & 0 & 2 & | & 13 \\ 3 & -1 & 0 & | & 11 \\ 0 & 1 & -1 & | & -7 \end{bmatrix}$

Interchange R2 and R3.

$\begin{bmatrix} 1 & 0 & 2 & | & 13 \\ 0 & 1 & -1 & | & -7 \\ 3 & -1 & 0 & | & 11 \end{bmatrix}$

Multiply R1 by -3 and add to R3.

$\begin{bmatrix} 1 & 0 & 2 & | & 13 \\ 0 & 1 & -1 & | & -7 \\ 0 & -1 & 6 & | & -28 \end{bmatrix}$

Add R2 to R3.

$$\begin{bmatrix} 1 & 0 & 2 & | & 13 \\ 0 & 1 & -1 & | & -7 \\ 0 & 0 & -7 & | & -35 \end{bmatrix}$$

Divide R3 by –7.

$$\begin{bmatrix} 1 & 0 & 2 & | & 13 \\ 0 & 1 & -1 & | & -7 \\ 0 & 0 & 1 & | & 5 \end{bmatrix}$$

This corresponds to $\begin{cases} x + 2z = 13 \\ y - z = -7 \\ z = 5. \end{cases}$

$$y - 5 = -7$$
$$y = -2$$
$$x + 2(5) = 13$$
$$x = 3$$

The solution is $(3, -2, 5)$.

36. $\begin{cases} 5x + 7y + 3z = 9 \\ 14y - z = 28 \\ 4x + 2z = -4 \end{cases}$

$$\begin{bmatrix} 5 & 7 & 3 & | & 9 \\ 0 & 14 & -1 & | & 28 \\ 4 & 0 & 2 & | & -4 \end{bmatrix}$$

Divide R1 by 5.

$$\begin{bmatrix} 1 & \frac{7}{5} & \frac{3}{5} & | & \frac{9}{5} \\ 0 & 14 & -1 & | & 28 \\ 4 & 0 & 2 & | & -4 \end{bmatrix}$$

Multiply R1 by –4 and add to R3.

$$\begin{bmatrix} 1 & \frac{7}{5} & \frac{3}{5} & | & \frac{9}{5} \\ 0 & 14 & -1 & | & 28 \\ 0 & -\frac{28}{5} & -\frac{2}{5} & | & -\frac{56}{5} \end{bmatrix}$$

Divide R2 by 14.

$$\begin{bmatrix} 1 & \frac{7}{5} & \frac{3}{5} & | & \frac{9}{5} \\ 0 & 1 & -\frac{1}{14} & | & 2 \\ 0 & -\frac{28}{5} & -\frac{2}{5} & | & -\frac{56}{5} \end{bmatrix}$$

Multiply R2 by $\frac{28}{5}$ and add to R3.

$$\begin{bmatrix} 1 & \frac{7}{5} & \frac{3}{5} & | & \frac{9}{5} \\ 0 & 1 & -\frac{1}{14} & | & 2 \\ 0 & 0 & -\frac{4}{5} & | & 0 \end{bmatrix}$$

Multiply R3 by $-\frac{5}{4}$.

$$\begin{bmatrix} 1 & \frac{7}{5} & \frac{3}{5} & | & \frac{9}{5} \\ 0 & 1 & -\frac{1}{14} & | & 2 \\ 0 & 0 & 1 & | & 0 \end{bmatrix}$$

This corresponds to $\begin{cases} x + \frac{7}{5}y + \frac{3}{5}z = \frac{9}{5} \\ y - \frac{1}{14}z = 2 \\ z = 0. \end{cases}$

$$y - \frac{1}{14}(0) = 2$$
$$y = 2$$
$$x + \frac{7}{5}(2) + \frac{3}{5}(0) = \frac{9}{5}$$
$$x + \frac{14}{5} = \frac{9}{5}$$
$$x = -1$$

The solution is $(-1, 2, 0)$.

37. $\begin{cases} 7x - 3y + 2z = 0 \\ 4x - 4y - z = 2 \\ 5x + 2y + 3z = 1 \end{cases}$

$$\begin{bmatrix} 7 & -3 & 2 & | & 0 \\ 4 & -4 & -1 & | & 2 \\ 5 & 2 & 3 & | & 1 \end{bmatrix}$$

Interchange R1 and R2.

$$\begin{bmatrix} 4 & -4 & -1 & | & 2 \\ 7 & -3 & 2 & | & 0 \\ 5 & 2 & 3 & | & 1 \end{bmatrix}$$

Divide R1 by 4.

$$\begin{bmatrix} 1 & -1 & -\frac{1}{4} & \bigg| & \frac{1}{2} \\ 7 & -3 & 2 & \bigg| & 0 \\ 5 & 2 & 3 & \bigg| & 1 \end{bmatrix}$$

Multiply R1 by −7 and add to R2.
Multiply R1 by −5 and add to R3.

$$\begin{bmatrix} 1 & -1 & -\frac{1}{4} & \bigg| & \frac{1}{2} \\ 0 & 4 & \frac{15}{4} & \bigg| & -\frac{7}{2} \\ 0 & 7 & \frac{17}{4} & \bigg| & -\frac{3}{2} \end{bmatrix}$$

Divide R2 by 4.

$$\begin{bmatrix} 1 & -1 & -\frac{1}{4} & \bigg| & \frac{1}{2} \\ 0 & 1 & \frac{15}{16} & \bigg| & -\frac{7}{8} \\ 0 & 7 & \frac{17}{4} & \bigg| & -\frac{3}{2} \end{bmatrix}$$

Multiply R2 by −7 and add to R3.

$$\begin{bmatrix} 1 & -1 & -\frac{1}{4} & \bigg| & \frac{1}{2} \\ 0 & 1 & \frac{15}{16} & \bigg| & -\frac{7}{8} \\ 0 & 0 & -\frac{37}{16} & \bigg| & -\frac{37}{8} \end{bmatrix}$$

Multiply R3 by $-\frac{16}{37}$.

$$\begin{bmatrix} 1 & -1 & -\frac{1}{4} & \bigg| & \frac{1}{2} \\ 0 & 1 & \frac{15}{16} & \bigg| & -\frac{7}{8} \\ 0 & 0 & 1 & \bigg| & -2 \end{bmatrix}$$

This corresponds to $\begin{cases} x - y - \frac{1}{4}z = \frac{1}{2} \\ y + \frac{15}{16}z = -\frac{7}{8} \\ z = -2. \end{cases}$

$$y + \frac{15}{16}(-2) = -\frac{7}{8}$$
$$y - \frac{15}{8} = -\frac{7}{8}$$
$$y = 1$$

$$x - 1 - \frac{1}{4}(-2) = \frac{1}{2}$$
$$x - 1 + \frac{1}{2} = \frac{1}{2}$$
$$x = 1$$

The solution is $(1, 1, -2)$.

38. $\begin{cases} x + 2y + 3z = 14 \\ y + 2z = 3 \\ 2x \quad\quad - 2z = 10 \end{cases}$

$$\begin{bmatrix} 1 & 2 & 3 & \big| & 14 \\ 0 & 1 & 2 & \big| & 3 \\ 2 & 0 & -2 & \big| & 10 \end{bmatrix}$$

Multiply R1 by −2 and add to R3.

$$\begin{bmatrix} 1 & 2 & 3 & \big| & 14 \\ 0 & 1 & 2 & \big| & 3 \\ 0 & -4 & -8 & \big| & -18 \end{bmatrix}$$

Multiply R2 by 4 and add to R3.

$$\begin{bmatrix} 1 & 2 & 3 & \big| & 14 \\ 0 & 1 & 2 & \big| & 3 \\ 0 & 0 & 0 & \big| & 6 \end{bmatrix}$$

This corresponds to $\begin{cases} x + 2y + 3z = 14 \\ y + 2z = 3 \\ 0 = 6. \end{cases}$

This system is inconsistent.
The solution is $\varnothing$.

39. $\begin{vmatrix} -1 & 3 \\ 5 & 2 \end{vmatrix} = -1(2) - 3(5) = -2 - 15 = -17$

40. $\begin{vmatrix} 3 & -1 \\ 2 & 5 \end{vmatrix} = 3(5) - (-1)(2) = 15 + 2 = 17$

41. $\begin{vmatrix} 2 & -1 & -3 \\ 1 & 2 & 0 \\ 3 & -2 & 2 \end{vmatrix}$

$= 2\begin{vmatrix} 2 & 0 \\ -2 & 2 \end{vmatrix} - (-1)\begin{vmatrix} 1 & 0 \\ 3 & 2 \end{vmatrix} + (-3)\begin{vmatrix} 1 & 2 \\ 3 & -2 \end{vmatrix}$

$= 2(4-0) + (2-0) - 3(-2-6)$

$= 2(4) + (2) - 3(-8)$

$= 34$

42. $\begin{vmatrix} -2 & 3 & 1 \\ 4 & 4 & 0 \\ 1 & -2 & 3 \end{vmatrix}$

$= 1\begin{vmatrix} 4 & 4 \\ 1 & -2 \end{vmatrix} - 0\begin{vmatrix} -2 & 3 \\ 1 & -2 \end{vmatrix} + 3\begin{vmatrix} -2 & 3 \\ 4 & 4 \end{vmatrix}$

$= (-8-4) - 0 + 3(-8-12)$

$= -12 + 3(-20)$

$= -12 - 60$

$= -72$

43. $\begin{cases} 3x - 2y = -8 \\ 6x + 5y = 11 \end{cases}$

$D = \begin{vmatrix} 3 & -2 \\ 6 & 5 \end{vmatrix} = 15 + 12 = 27$

$D_x = \begin{vmatrix} -8 & -2 \\ 11 & 5 \end{vmatrix} = -40 + 22 = -18$

$D_y = \begin{vmatrix} 3 & -8 \\ 6 & 11 \end{vmatrix} = 33 + 48 = 81$

$x = \dfrac{D_x}{D} = \dfrac{-18}{27} = -\dfrac{2}{3}$

$y = \dfrac{D_y}{D} = \dfrac{81}{27} = 3$

The solution is $\left(-\dfrac{2}{3}, 3\right)$.

44. $\begin{cases} 6x - 6y = -5 \\ 10x - 2y = 1 \end{cases}$

$D = \begin{vmatrix} 6 & -6 \\ 10 & -2 \end{vmatrix} = -12 - (-60) = 48$

$D_x = \begin{vmatrix} -5 & -6 \\ 1 & -2 \end{vmatrix} = 10 - (-6) = 16$

$D_y = \begin{vmatrix} 6 & -5 \\ 10 & 1 \end{vmatrix} = 6 - (-50) = 56$

$x = \dfrac{D_x}{D} = \dfrac{16}{48} = \dfrac{1}{3}$

$y = \dfrac{D_y}{D} = \dfrac{56}{48} = \dfrac{7}{6}$

The solution is $\left(\dfrac{1}{3}, \dfrac{7}{6}\right)$.

45. $\begin{cases} 3x + 10y = 1 \\ x + 2y = -1 \end{cases}$

$D = \begin{vmatrix} 3 & 10 \\ 1 & 2 \end{vmatrix} = 6 - 10 = -4$

$D_x = \begin{vmatrix} 1 & 10 \\ -1 & 2 \end{vmatrix} = 2 - (-10) = 12$

$D_y = \begin{vmatrix} 3 & 1 \\ 1 & -1 \end{vmatrix} = -3 - 1 = -4$

$x = \dfrac{D_x}{D} = \dfrac{12}{-4} = -3$

$y = \dfrac{D_y}{D} = \dfrac{-4}{-4} = 1$

The solution is $(-3, 1)$.

46. $\begin{cases} y = \dfrac{1}{2}x + \dfrac{2}{3} \\ 4x + 6y = 4 \end{cases}$ or $\begin{cases} -\dfrac{1}{2}x + y = \dfrac{2}{3} \\ 4x + 6y = 4 \end{cases}$

$D = \begin{vmatrix} -\dfrac{1}{2} & 1 \\ 4 & 6 \end{vmatrix} = -3 - 4 = -7$

$D_x = \begin{vmatrix} \dfrac{2}{3} & 1 \\ 4 & 6 \end{vmatrix} = 4 - 4 = 0$

$D_y = \begin{vmatrix} -\dfrac{1}{2} & \dfrac{2}{3} \\ 4 & 4 \end{vmatrix} = -2 - \dfrac{8}{3} = -\dfrac{14}{3}$

$x = \dfrac{D_x}{D} = \dfrac{0}{-7} = 0$

$$y = \frac{D_y}{D} = \frac{-\frac{14}{3}}{-7} = \frac{2}{3}$$

The solution is $\left(0, \frac{2}{3}\right)$.

47. $\begin{cases} 2x - 4y = 22 \\ 5x - 10y = 16 \end{cases}$

$$D = \begin{vmatrix} 2 & -4 \\ 5 & -10 \end{vmatrix} = -20 - (-20) = 0$$

This cannot be solved by Cramer's rule.
Multiply E1 by –5, multiply E2 by 2,
and add.

$$-10x + 20y = -110$$
$$\underline{10x - 20y = 32}$$
$$0 = -78 \quad \text{False}$$

This system is inconsistent.
The solution is $\varnothing$.

48. $\begin{cases} 3x - 6y = 12 \\ 2y = x - 4 \end{cases}$ or $\begin{cases} 3x - 6y = 12 \\ -x + 2y = -4 \end{cases}$

$$D = \begin{vmatrix} 3 & -6 \\ -1 & 2 \end{vmatrix} = 6 - 6 = 0$$

This cannot be solved by Cramer's rule.
Since E1 is –3 times E2, the system is
dependent.
The solution is $\{(x, y) \mid -x + 2y = -4\}$.

49. $\begin{cases} x \qquad + z = 4 \\ 2x - y \qquad = 0 \\ x + y - z = 0 \end{cases}$

$$D = \begin{vmatrix} 1 & 0 & 1 \\ 2 & -1 & 0 \\ 1 & 1 & -1 \end{vmatrix}$$

$$= 1 \begin{vmatrix} -1 & 0 \\ 1 & -1 \end{vmatrix} - 0 \begin{vmatrix} 2 & 0 \\ 1 & -1 \end{vmatrix} + 1 \begin{vmatrix} 2 & -1 \\ 1 & 1 \end{vmatrix}$$

$$= (1 - 0) - 0 + [2 - (-1)]$$
$$= 1 + 3$$
$$= 4$$

$$D_x = \begin{vmatrix} 4 & 0 & 1 \\ 0 & -1 & 0 \\ 0 & 1 & -1 \end{vmatrix}$$

$$= 4 \begin{vmatrix} -1 & 0 \\ 1 & -1 \end{vmatrix} - 0 \begin{vmatrix} 0 & 1 \\ 1 & -1 \end{vmatrix} + 0 \begin{vmatrix} 0 & 1 \\ 1 & 0 \end{vmatrix}$$

$$= 4(1 - 0) - 0 + 0$$
$$= 4$$

$$D_y = \begin{vmatrix} 1 & 4 & 1 \\ 2 & 0 & 0 \\ 1 & 0 & -1 \end{vmatrix}$$

$$= -4 \begin{vmatrix} 2 & 0 \\ 1 & -1 \end{vmatrix} + 0 \begin{vmatrix} 1 & 1 \\ 1 & -1 \end{vmatrix} - 0 \begin{vmatrix} 1 & 1 \\ 2 & 0 \end{vmatrix}$$

$$= -4(-2 - 0) + 0 - 0$$
$$= 8$$

$$D_z = \begin{vmatrix} 1 & 0 & 4 \\ 2 & -1 & 0 \\ 1 & 1 & 0 \end{vmatrix}$$

$$= 4 \begin{vmatrix} 2 & -1 \\ 1 & 1 \end{vmatrix} - 0 \begin{vmatrix} 1 & 0 \\ 1 & 1 \end{vmatrix} + 0 \begin{vmatrix} 1 & 0 \\ 2 & -1 \end{vmatrix}$$

$$= 4[2 - (-1)] - 0 + 0$$
$$= 4(3)$$
$$= 12$$

$$x = \frac{D_x}{D} = \frac{4}{4} = 1, \quad y = \frac{D_y}{D} = \frac{8}{4} = 2,$$

$$z = \frac{D_z}{D} = \frac{12}{4} = 3$$

The solution is (1, 2, 3).

50. $\begin{cases} 2x + 5y \quad = 4 \\ x - 5y + z = -1 \\ 4x \qquad - z = 11 \end{cases}$

$$D = \begin{vmatrix} 2 & 5 & 0 \\ 1 & -5 & 1 \\ 4 & 0 & -1 \end{vmatrix}$$

$$= 2 \begin{vmatrix} -5 & 1 \\ 0 & -1 \end{vmatrix} - 5 \begin{vmatrix} 1 & 1 \\ 4 & -1 \end{vmatrix} + 0 \begin{vmatrix} 1 & -5 \\ 4 & 0 \end{vmatrix}$$

$$= 2(5 - 0) - 5(-1 - 4) + 0$$
$$= 10 + 25$$
$$= 35$$

$$D_x = \begin{vmatrix} 4 & 5 & 0 \\ -1 & -5 & 1 \\ 11 & 0 & -1 \end{vmatrix}$$

$$= 4\begin{vmatrix} -5 & 1 \\ 0 & -1 \end{vmatrix} - 5\begin{vmatrix} -1 & 1 \\ 11 & -1 \end{vmatrix} + 0\begin{vmatrix} -1 & -5 \\ 11 & 0 \end{vmatrix}$$

$$= 4(5-0) - 5(1-11) + 0$$

$$= 20 + 50$$

$$= 70$$

$$D_y = \begin{vmatrix} 2 & 4 & 0 \\ 1 & -1 & 1 \\ 4 & 11 & -1 \end{vmatrix}$$

$$= 2\begin{vmatrix} -1 & 1 \\ 11 & -1 \end{vmatrix} - 4\begin{vmatrix} 1 & 1 \\ 4 & -1 \end{vmatrix} + 0\begin{vmatrix} 1 & -1 \\ 4 & 11 \end{vmatrix}$$

$$= 2(1-11) - 4(-1-4) + 0$$

$$= -20 + 20$$

$$= 0$$

$$D_z = \begin{vmatrix} 2 & 5 & 4 \\ 1 & -5 & -1 \\ 4 & 0 & 11 \end{vmatrix}$$

$$= 4\begin{vmatrix} 5 & 4 \\ -5 & -1 \end{vmatrix} - 0\begin{vmatrix} 2 & 4 \\ 1 & -1 \end{vmatrix} + 11\begin{vmatrix} 2 & 5 \\ 1 & -5 \end{vmatrix}$$

$$= 4[-5-(-20)] - 0 + 11(-10-5)$$

$$= 4(15) + 11(-15)$$

$$= 60 - 165$$

$$= -105$$

$$x = \frac{D_x}{D} = \frac{70}{35} = 2, \quad y = \frac{D_y}{D} = \frac{0}{35} = 0,$$

$$z = \frac{D_z}{D} = \frac{-105}{35} = -3$$

The solution is (2, 0, –3).

51. $\begin{cases} x + 3y - z = 5 \\ 2x - y - 2z = 3 \\ x + 2y + 3z = 4 \end{cases}$

$$D = \begin{vmatrix} 1 & 3 & -1 \\ 2 & -1 & -2 \\ 1 & 2 & 3 \end{vmatrix}$$

$$= 1\begin{vmatrix} -1 & -2 \\ 2 & 3 \end{vmatrix} - 3\begin{vmatrix} 2 & -2 \\ 1 & 3 \end{vmatrix} + (-1)\begin{vmatrix} 2 & -1 \\ 1 & 2 \end{vmatrix}$$

$$= [-3-(-4)] - 3[6-(-2)] - [4-(-1)]$$

$$= 1 - 3(8) - 5$$

$$= -28$$

$$D_x = \begin{vmatrix} 5 & 3 & -1 \\ 3 & -1 & -2 \\ 4 & 2 & 3 \end{vmatrix}$$

$$= 5\begin{vmatrix} -1 & -2 \\ 2 & 3 \end{vmatrix} - 3\begin{vmatrix} 3 & -2 \\ 4 & 3 \end{vmatrix} + (-1)\begin{vmatrix} 3 & -1 \\ 4 & 2 \end{vmatrix}$$

$$= 5[-3-(-4)] - 3[9-(-8)] - [6-(-4)]$$

$$= 5(1) - 3(17) - 10$$

$$= 5 - 51 - 10$$

$$= -56$$

$$D_y = \begin{vmatrix} 1 & 5 & -1 \\ 2 & 3 & -2 \\ 1 & 4 & 3 \end{vmatrix}$$

$$= 1\begin{vmatrix} 3 & -2 \\ 4 & 3 \end{vmatrix} - 5\begin{vmatrix} 2 & -2 \\ 1 & 3 \end{vmatrix} + (-1)\begin{vmatrix} 2 & 3 \\ 1 & 4 \end{vmatrix}$$

$$= [9-(-8)] - 5[6-(-2)] - (8-3)$$

$$= 17 - 5(8) - 5$$

$$= 17 - 40 - 5$$

$$= -28$$

$$D_z = \begin{vmatrix} 1 & 3 & 5 \\ 2 & -1 & 3 \\ 1 & 2 & 4 \end{vmatrix}$$

$$= 1\begin{vmatrix} -1 & 3 \\ 2 & 4 \end{vmatrix} - 3\begin{vmatrix} 2 & 3 \\ 1 & 4 \end{vmatrix} + 5\begin{vmatrix} 2 & -1 \\ 1 & 2 \end{vmatrix}$$

$$= (-4-6) - 3(8-3) + 5[4-(-1)]$$

$$= -10 - 3(5) + 5(5)$$

$$= -10 - 15 + 25$$

$$= 0$$

$$x = \frac{D_x}{D} = \frac{-56}{-28} = 2, \quad y = \frac{D_y}{D} = \frac{-28}{-28} = 1,$$

$$z = \frac{D_z}{D} = \frac{0}{-28} = 0$$

The solution is (2, 1, 0).

52. $\begin{cases} 2x \quad\;\; -z = 1 \\ 3x - y + 2z = 3 \\ x + y + 3z = -2 \end{cases}$

$$D = \begin{vmatrix} 2 & 0 & -1 \\ 3 & -1 & 2 \\ 1 & 1 & 3 \end{vmatrix}$$

$= 2\begin{vmatrix} -1 & 2 \\ 1 & 3 \end{vmatrix} - 0\begin{vmatrix} 3 & 2 \\ 1 & 3 \end{vmatrix} + (-1)\begin{vmatrix} 3 & -1 \\ 1 & 1 \end{vmatrix}$

$= 2(-3 - 2) - 0 - [3 - (-1)]$

$= 2(-5) - 4$

$= -10 - 4$

$= -14$

$$D_x = \begin{vmatrix} 1 & 0 & -1 \\ 3 & -1 & 2 \\ -2 & 1 & 3 \end{vmatrix}$$

$= 1\begin{vmatrix} -1 & 2 \\ 1 & 3 \end{vmatrix} - 0\begin{vmatrix} 3 & 2 \\ -2 & 3 \end{vmatrix} + (-1)\begin{vmatrix} 3 & -1 \\ -2 & 1 \end{vmatrix}$

$= (-3 - 2) - 0 - (3 - 2)$

$= -5 - 1$

$= -6$

$$D_y = \begin{vmatrix} 2 & 1 & -1 \\ 3 & 3 & 2 \\ 1 & -2 & 3 \end{vmatrix}$$

$= 2\begin{vmatrix} 3 & 2 \\ -2 & 3 \end{vmatrix} - 1\begin{vmatrix} 3 & 2 \\ 1 & 3 \end{vmatrix} + (-1)\begin{vmatrix} 3 & 3 \\ 1 & -2 \end{vmatrix}$

$= 2[9 - (-4)] - (9 - 2) - (-6 - 3)$

$= 2(13) - 7 + 9$

$= 26 - 7 + 9$

$= 28$

$$D_z = \begin{vmatrix} 2 & 0 & 1 \\ 3 & -1 & 3 \\ 1 & 1 & -2 \end{vmatrix}$$

$= 2\begin{vmatrix} -1 & 3 \\ 1 & -2 \end{vmatrix} - 0\begin{vmatrix} 3 & 3 \\ 1 & -2 \end{vmatrix} + 1\begin{vmatrix} 3 & -1 \\ 1 & 1 \end{vmatrix}$

$= 2(2 - 3) - 0 + [3 - (-1)]$

$= 2(-1) + 4$

$= -2 + 4$

$= 2$

$x = \dfrac{D_x}{D} = \dfrac{-6}{-14} = \dfrac{3}{7}, \;\; y = \dfrac{D_y}{D} = \dfrac{28}{-14} = -2,$

$z = \dfrac{D_z}{D} = \dfrac{2}{-14} = -\dfrac{1}{7}$

The solution is $\left(\dfrac{3}{7}, -2, -\dfrac{1}{7}\right)$.

53. $\begin{cases} x + 2y + 3z = 14 \\ \quad\;\; y + 2z = 3 \\ 2x \quad\;\; - 2z = 10 \end{cases}$

$$D = \begin{vmatrix} 1 & 2 & 3 \\ 0 & 1 & 2 \\ 2 & 0 & -2 \end{vmatrix}$$

$= 1\begin{vmatrix} 1 & 2 \\ 0 & -2 \end{vmatrix} - 0\begin{vmatrix} 2 & 3 \\ 0 & -2 \end{vmatrix} + 2\begin{vmatrix} 2 & 3 \\ 1 & 2 \end{vmatrix}$

$= (-2 - 0) - 0 + 2(4 - 3)$

$= -2 + 2(1)$

$= 0$

This cannot be solved by Cramer's rule.

Solving E2 for y gives $y = -2z + 3$.

Solving E3 for x gives $2x = 2z + 10$

$\qquad\qquad\qquad\qquad\qquad x = z + 5.$

Replace x with $z + 5$ and y with $-2z + 3$ in E1.

$(z + 5) + 2(-2z + 3) + 3z = 14$

$\qquad z + 5 - 4z + 6 + 3z = 14$

$\qquad\qquad\qquad\qquad 11 = 14$ False

The system is inconsistent.

The solution is $\varnothing$.

54. $\begin{cases} 5x + 7y \quad\quad = 9 \\ \quad\;\; 14y - z = 28 \\ 4x \quad\quad + 2z = -4 \end{cases}$

$$D = \begin{vmatrix} 5 & 7 & 0 \\ 0 & 14 & -1 \\ 4 & 0 & 2 \end{vmatrix}$$

$= 5\begin{vmatrix} 14 & -1 \\ 0 & 2 \end{vmatrix} - 7\begin{vmatrix} 0 & -1 \\ 4 & 2 \end{vmatrix} + 0\begin{vmatrix} 0 & 14 \\ 4 & 0 \end{vmatrix}$

$= 5(28 - 0) - 7[0 - (-4)] + 0$

$= 5(28) - 7(4)$

$= 140 - 28 = 112$

$$D_x = \begin{vmatrix} 9 & 7 & 0 \\ 28 & 14 & -1 \\ -4 & 0 & 2 \end{vmatrix}$$

$$= 9\begin{vmatrix} 14 & -1 \\ 0 & 2 \end{vmatrix} - 7\begin{vmatrix} 28 & -1 \\ -4 & 2 \end{vmatrix} + 0\begin{vmatrix} 28 & 14 \\ -4 & 0 \end{vmatrix}$$

$$= 9(28 - 0) - 7(56 - 4) + 0$$

$$= 9(28) - 7(52)$$

$$= 252 - 364$$

$$= -112$$

$$D_y = \begin{vmatrix} 5 & 9 & 0 \\ 0 & 28 & -1 \\ 4 & -4 & 2 \end{vmatrix}$$

$$= 5\begin{vmatrix} 28 & -1 \\ -4 & 2 \end{vmatrix} - 9\begin{vmatrix} 0 & -1 \\ 4 & 2 \end{vmatrix} + 0\begin{vmatrix} 0 & 28 \\ 4 & -4 \end{vmatrix}$$

$$= 5(56 - 4) - 9[0 - (-4)] + 0$$

$$= 5(52) - 9(4)$$

$$= 260 - 36$$

$$= 224$$

$$D_z = \begin{vmatrix} 5 & 7 & 9 \\ 0 & 14 & 28 \\ 4 & 0 & -4 \end{vmatrix}$$

$$= 5\begin{vmatrix} 14 & 28 \\ 0 & -4 \end{vmatrix} - 0\begin{vmatrix} 7 & 9 \\ 0 & -4 \end{vmatrix} + 4\begin{vmatrix} 7 & 9 \\ 14 & 28 \end{vmatrix}$$

$$= 5(-56 - 0) - 0 + 4(196 - 126)$$

$$= -280 + 4(70)$$

$$= -280 + 280$$

$$= 0$$

$$x = \frac{D_x}{D} = \frac{-112}{112} = -1, \quad y = \frac{D_y}{D} = \frac{224}{112} = 2,$$

$$z = \frac{D_z}{D} = \frac{0}{112} = 0$$

The solution is $(-1, 2, 0)$.

Chapter 4 Test

1. $\begin{vmatrix} 4 & -7 \\ 2 & 5 \end{vmatrix} = 4(5) - (-7)(2) = 20 + 14 = 34$

2. $\begin{vmatrix} 4 & 0 & 2 \\ 1 & -3 & 5 \\ 0 & -1 & 2 \end{vmatrix}$

$$= 4\begin{vmatrix} -3 & 5 \\ -1 & 2 \end{vmatrix} - 0\begin{vmatrix} 1 & 5 \\ 0 & 2 \end{vmatrix} + 2\begin{vmatrix} 1 & -3 \\ 0 & -1 \end{vmatrix}$$

$$= 4[-6 - (-5)] - 0 + 2(-1 - 0)$$

$$= 4(-1) + 2(-1)$$

$$= -4 - 2$$

$$= -6$$

3. $\begin{cases} 2x - y = -1 & (1) \\ 5x + 4y = 17 & (2) \end{cases}$

By elimination:
Multiply E1 by 4 and add it to E2.

$$8x - 4y = -4$$
$$5x + 4y = 17$$
$$\overline{13x = 13}$$
$$x = 1$$

Replace x with 1 in E2.

$$5(1) + 4y = 17$$
$$4y = 12$$
$$y = 3$$

The solution is $(1, 3)$.

4. $\begin{cases} 7x - 14y = 5 & (1) \\ x = 2y & (2) \end{cases}$

By substitution:

Replace x with $2y$ in E1.

$7(2y) - 14y = 5$

$\quad 14y - 14y = 5$

$\qquad\qquad 0 = 5$ False

The system is inconsistent.

The solution is $\varnothing$.

5. $\begin{cases} 4x - 7y = 29 \\ 2x + 5y = -11 \end{cases}$

Multiply E2 by -2 and add to E1.

$-4x - 10y = 22$

$\underline{4x - 7y = 29}$

$\qquad -17y = 51$

$\qquad\qquad y = -3$

Replace y with -3 in E1.

$4x - 7(-3) = 29$

$\quad 4x + 21 = 29$

$\qquad\quad 4x = 8$

$\qquad\quad\; x = 2$

The solution is $(2, -3)$.

6. $\begin{cases} 15x + 6y = 15 \\ 10x + 4y = 10 \end{cases}$

Divide E1 by 3 and E2 by 2.

$\begin{cases} 5x + 2y = 5 \\ 5x + 2y = 5 \end{cases}$

The system is dependent.

The solution is $\{(x, y) \mid 10x + 4y = 10\}$.

7. $\begin{cases} 2x - 3y = 4 & (1) \\ 3y + 2z = 2 & (2) \\ x - z = -5 & (3) \end{cases}$

Add E1 and E2.

$2x + 2z = 6$ or $x + z = 3$ (4)

Add E3 and E4.

$x + z = 3$

$\underline{x - z = -5}$

$2x = -2$

$\quad x = -1$

Replace x with -1 in E3.

$-1 - z = -5$

$\quad -z = -4$ so $z = 4$

Replace x with -1 in E1.

$2(-1) - 3y = 4$

$\quad -2 - 3y = 4$

$\qquad\quad -3y = 6$

$\qquad\qquad y = -2$

The solution is $(-1, -2, 4)$.

8. $\begin{cases} 3x - 2y - z = -1 & (1) \\ 2x - 2y = 4 & (2) \\ 2x - 2z = -12 & (3) \end{cases}$

Multiply E2 by -1 and add to E1.

$3x - 2y - z = -1$

$\underline{-2x + 2y = -4}$

$x - z = -5$ (4)

Multiply E4 by -2 and add to E3.

$2x - 2z = -12$

$\underline{-2x + 2z = 10}$

$\qquad\quad 0 = -2$ False

The system is inconsistent.

The solution is $\varnothing$.

9. $\begin{cases} \dfrac{x}{2} + \dfrac{y}{4} = -\dfrac{3}{4} \\ x + \dfrac{3}{4}y = -4 \end{cases}$

Clear fractions by multiplying both equations by 4.

$\begin{cases} 2x + y = -3 & (1) \\ 4x + 3y = -16 & (2) \end{cases}$

Multiply E1 by -2 and add to E2.

$-4x - 2y = 6$

$\underline{4x + 3y = -16}$

$\qquad\quad y = -10$

Replace y with -10 in E1.

$2x + (-10) = -3$

$\qquad 2x = 7$ so $x = \dfrac{7}{2}$

The solution is $\left(\dfrac{7}{2}, -10\right)$.

10. $\begin{cases} 3x - y = 7 \\ 2x + 5y = -1 \end{cases}$.

$D = \begin{vmatrix} 3 & -1 \\ 2 & 5 \end{vmatrix} = 3(5) - (-1)(2) = 15 + 2 = 17$

$D_x = \begin{vmatrix} 7 & -1 \\ -1 & 5 \end{vmatrix}$

$= 7(5) - (-1)(-1)$

$= 35 - 1$

$= 34$

$D_y = \begin{vmatrix} 3 & 7 \\ 2 & -1 \end{vmatrix}$

$= -1(3) - 7(2)$

$= -3 - 14$

$= -17$

$x = \dfrac{D_x}{D} = \dfrac{34}{17} = 2$

$y = \dfrac{D_y}{D} = \dfrac{-17}{17} = -1$

The solution is $(2, -1)$.

11. $\begin{cases} x + y + z = 4 \\ 2x + 5y = 1 \\ x - y - 2z = 0 \end{cases}$

$D = \begin{vmatrix} 1 & 1 & 1 \\ 2 & 5 & 0 \\ 1 & -1 & -2 \end{vmatrix}$

$= 1\begin{vmatrix} 2 & 5 \\ 1 & -1 \end{vmatrix} - 0\begin{vmatrix} 1 & 1 \\ 1 & -1 \end{vmatrix} + (-2)\begin{vmatrix} 1 & 1 \\ 2 & 5 \end{vmatrix}$

$= (-2 - 5) - 0 - 2(5 - 2)$

$= -7 - 2(3)$

$= -13$

$D_x = \begin{vmatrix} 4 & 1 & 1 \\ 1 & 5 & 0 \\ 0 & -1 & -2 \end{vmatrix}$

$= 4\begin{vmatrix} 5 & 0 \\ -1 & -2 \end{vmatrix} - 1\begin{vmatrix} 1 & 1 \\ -1 & -2 \end{vmatrix} + 0\begin{vmatrix} 1 & 1 \\ 5 & 0 \end{vmatrix}$

$= 4(-10 - 0) - [-2 - (-1)] + 0$

$= 4(-10) - (-1) + 0$

$= -40 + 1$

$= -39$

$D_y = \begin{vmatrix} 1 & 4 & 1 \\ 2 & 1 & 0 \\ 1 & 0 & -2 \end{vmatrix}$

$= 1\begin{vmatrix} 2 & 1 \\ 1 & 0 \end{vmatrix} - 0\begin{vmatrix} 1 & 4 \\ 1 & 0 \end{vmatrix} + (-2)\begin{vmatrix} 1 & 4 \\ 2 & 1 \end{vmatrix}$

$= (0 - 1) - 0 - 2(1 - 8)$

$= -1 - 2(-7)$

$= -1 + 14$

$= 13$

$D_z = \begin{vmatrix} 1 & 1 & 4 \\ 2 & 5 & 1 \\ 1 & -1 & 0 \end{vmatrix}$

$= 1\begin{vmatrix} 1 & 4 \\ 5 & 1 \end{vmatrix} - (-1)\begin{vmatrix} 1 & 4 \\ 2 & 1 \end{vmatrix} + 0\begin{vmatrix} 1 & 1 \\ 2 & 5 \end{vmatrix}$

$= (1 - 20) + (1 - 8) + 0$

$= -19 - 7$

$= -26$

$x = \dfrac{D_x}{D} = \dfrac{-39}{-13} = 3, \quad y = \dfrac{D_y}{D} = \dfrac{13}{-13} = -1,$

$z = \dfrac{D_z}{D} = \dfrac{-26}{-13} = 2$

The solution is $(3, -1, 2)$.

12. $\begin{cases} x - y = -2 \\ 3x - 3y = -6 \end{cases}$

$\begin{bmatrix} 1 & -1 & | & -2 \\ 3 & -3 & | & -6 \end{bmatrix}$

Multiply R1 by -3 and add to R2.

$\begin{bmatrix} 1 & -1 & | & -2 \\ 0 & 0 & | & 0 \end{bmatrix}$

This corresponds to $\begin{cases} x - y = -2 \\ 0 = 0 \end{cases}$.

This is a dependent system.

The solution is $\{(x, y) \mid x - y = -2\}$.

13. $\begin{cases} x + 2y = -1 \\ 2x + 5y = -5 \end{cases}$

$\begin{bmatrix} 1 & 2 & | & -1 \\ 2 & 5 & | & -5 \end{bmatrix}$

Multiply R1 by -2 and add to R2.

$$\begin{bmatrix} 1 & 2 & | & -1 \\ 0 & 1 & | & -3 \end{bmatrix}$$

This corresponds to $\begin{cases} x+2y=-1 \\ y=-3. \end{cases}$

$$x+2(-3)=-1$$
$$x-6=-1$$
$$x=5$$

The solution is $(5, -3)$.

14. $\begin{cases} x-y-z=0 \\ 3x-y-5z=-2 \\ 2x+3y=-5 \end{cases}$

$$\begin{bmatrix} 1 & -1 & -1 & | & 0 \\ 3 & -1 & -5 & | & -2 \\ 2 & 3 & 0 & | & -5 \end{bmatrix}$$

Multiply R1 by -3 and add to R2.
Multiply R1 by -2 and add to R3.

$$\begin{bmatrix} 1 & -1 & -1 & | & 0 \\ 0 & 2 & -2 & | & -2 \\ 0 & 5 & 2 & | & -5 \end{bmatrix}$$

Divide R2 by 2.

$$\begin{bmatrix} 1 & -1 & -1 & | & 0 \\ 0 & 1 & -1 & | & -1 \\ 0 & 5 & 2 & | & -5 \end{bmatrix}$$

Multiply R2 by $-$ and add to R3.

$$\begin{bmatrix} 1 & -1 & -1 & | & 0 \\ 0 & 1 & -1 & | & -1 \\ 0 & 0 & 7 & | & 0 \end{bmatrix}$$

Divide R3 by 7.

$$\begin{bmatrix} 1 & -1 & -1 & | & 0 \\ 0 & 1 & -1 & | & -1 \\ 0 & 0 & 1 & | & 0 \end{bmatrix}$$

This corresponds to $\begin{cases} x-y-z=0 \\ y-z=-1 \\ z=0. \end{cases}$

$$y-0=-1$$
$$y=-1$$
$$x-(-1)-0=0$$
$$x+1=0$$
$$x=-1$$

The solution is $(-1, -1, 0)$.

15. Let x = double occupancy rooms and y = single occupancy rooms.

$$\begin{cases} x+y=80 & (1) \\ 90x+80y=6930 & (2) \end{cases}$$

Multiply E1 by -80 and add to E2.

$$-80x-80y=-6400$$
$$\underline{90x+80y=6930}$$
$$10x=530$$
$$x=53$$

Replace x with 53 in E1.

$$53+y=80$$
$$y=27$$

53 double-occupancy and 27 single-occupancy rooms are occupied.

16. Let x = gallons of 10% solution and y = gallons of 20% solution.

$$\begin{cases} x+y=20 & (1) \\ 0.10x+0.20y=0.175(20) & (2) \end{cases}$$

Multiply E1 by -0.10 add to E2.

$$-0.10x-0.10y=-2.0$$
$$\underline{0.10x+0.20y=3.5}$$
$$0.10y=1.5$$
$$y=15$$

Replace y with 15 in E1.

$$x+15=20$$
$$x=5$$

They should use 5 gallons of 10% fructose solution and 15 gallons of the 20% solution.

17. $R(x)=4x$ and $C(x)=1.5x+2000$
Break even occurs when $R(x)=C(x)$.

$4x = 1.5x + 2000$

$2.5x = 2000$

$x = 800$

The company must sell 800 packages to break even.

19. Let x = measure of the smallest angle. Then the largest angle has a measure of $5x - 3$, and the remaining angle has a measure of $2x - 1$. The sum of the three angles must add to $180°$ ::

$$a + b + c = 180$$

$$x + (5x - 3) + (2x - 1) = 180$$

$$x + 5x - 3 + 2x - 1 = 180$$

$$8x - 4 = 180$$

$$8x = 184$$

$$x = 23$$

$$5x - 3 = 5(23) - 3 = 115 - 3 = 112$$

$$2x - 1 = 2(23) - 1 = 46 - 1 = 45$$

The angle measures are $23°$, $45°$, and $112°$.

Chapter 4 Cumulative Review

1. **a.** True

   b. True

2. **a.** False

   b. True

3. **a.** $11 + 2 - 7 = 13$

   b. $-5 - 4 + 2 = -7$

4. **a.** $-7 - (-2) = -7 + 2 = -5$

   b. $14 - 38 = -24$

5. **a.** The opposite of 8 is -8.

   b. The opposite of $\frac{1}{5}$ is $-\frac{1}{5}$.

   c. The opposite of -9.6 is 9.6.

6. **a.** The reciprocal of 5 is $\frac{1}{5}$.

   b. The reciprocal of $-\frac{2}{3}$ is $-\frac{3}{2}$.

7. **a.** $3(2x + y) = 6x + 3y$

   b. $-(3x - 1) = -3x + 1$

   c. $0.7a(b - 2) = 0.7ab - 1.4a$

8. **a.** $7(3x - 2y + 4) = 21x - 14y + 28$

   b. $-(-2s - 3t) = 2s + 3t$

9. **a.** $3x - 5x + 4 = (3 - 5)x + 4 = -2x + 4$

   b. $7yz + yz = (7 + 1)yz = 8yz$

   c. $4z + 6.1 = 4z + 6.1$

10. **a.** $5y^2 - 1 + 2(y^2 + 2) = 5y^2 - 1 + 2y^2 + 4$
    $$= 7y^2 + 3$$

    b. $(7.8x - 1.2) - (5.6x - 2.4)$
    $$= 7.8x - 1.2 - 5.6x + 2.4$$
    $$= 14.4x + 1.2$$

11. $-6x - 1 + 5x = 3$
    $$-x - 1 = 3$$
    $$-x = 4$$
    $$x = -4$$
    The solution is -4.

12. $8y - 14 = 6y - 14$
    $$2y = 0$$
    $$y = 0$$
    The solution is 0.

13. $0.3x + 0.1 = 0.27x - 0.02$
    $$0.03x = -0.12$$
    $$x = -4$$
    The solution is -4.

14. $2(m - 6) - m = 4(m - 3) - 3m$
    $$2m - 12 - m = 4m - 12 - 3m$$
    $$m - 12 = m - 12$$
    $$0 = 0 \quad \text{Always True}$$

The solution is all real numbers.

15. Let x = length of the third side, then
$2x+12$ = length of the two equal sides.
$$x+(2x+12)+(2x+12)=149$$
$$5x+24=149$$
$$5x=125$$
$$x=25$$
$$2(25)+12=50+12=62$$
The sides are 25 cm, 62 cm, and 62 cm.

16. Let x = measure of the equal angles,
$x+10$ = measure of the third angle, and
$\frac{1}{2}x$ = measure of the fourth angle.

$$x+x+(x+10)+\frac{1}{2}x=360$$

$$\frac{7}{2}x+10=360$$

$$\frac{7}{2}x=350$$
$$7x=700$$
$$x=100$$
$$x+10=100+10=110$$
$$\frac{1}{2}x=\frac{1}{2}(100)=50$$

The measure of the angles are $100°$, $100°$, $110°$, and $50°$.

17. $3x+4\geq 2x-6$
$$x\geq -10$$

18. $5(2x-1)>-5$
$$10x-5>-5$$
$$10x>0$$
$$x>0$$
$(0,\infty)$

19. $2<4-x<7$
$$-2<-x<3$$
$$2>x>-3$$
$$-3<x<2$$
$(-3,2)$

20. $-1<\dfrac{-2x-1}{3}<1$
$$3(-1)<3\left[\dfrac{-2x-1}{3}\right]<3(1)$$
$$-3<-2x-1<3$$
$$-2<-2x<4$$
$$1>x>-2$$
$$-2<x<1$$
$(-2, 1)$

21. $|2x|+5=7$
$$|2x|=2$$
$$2x=2 \text{ or } 2x=-2$$
$$x=1 \text{ or } \quad x=-1$$
The solution is $\{-1, 1\}$.

22. $|x-5|=4$
$$x-5=4 \text{ or } x-5=-4$$
$$x=9 \text{ or } \quad x=1$$
The solution is $\{1, 9\}$.

23. $|m-6|<2$
$$-2<m-6<2$$
$$4<m<8$$
$(4, 8)$

24. $|2x+1|>5$
$$2x+1<-5 \text{ or } 2x+1>5$$
$$2x<-6 \text{ or } \quad 2x>4$$
$$x<-3 \text{ or } \quad\quad x>2$$
$(-\infty,-3)\cup(2,\infty)$

25.

   a. $(2, -1)$ is in Quadrant IV.

   b. $(0, 5)$ is on the y-axis.

   c. $(-3, 5)$ is in Quadrant II.

d. $(-2, 0)$ is on the x-axis.

e. $\left(-\dfrac{1}{2}, -4\right)$ is in Quadrant III.

f. $(1.5, 1.5)$ is in Quadrant I.

26. a. $(-1, -5)$ is in Quadrant III.

b. $(4, -2)$ is in Quadrant IV.

c. $(0, 2)$ is on the y-axis.

27. Yes. For each input, x, there is exactly one output, y.

28. $-2x + \dfrac{1}{2}y = -2$, or $y = 4x - 4$

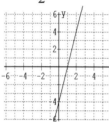

29. $f(x) = 7x^2 - 3x + 1$, $g(x) = 3x - 2$

　a. $f(1) = 7(1)^2 - 3(1) + 1 = 7 - 3 + 1 = 5$

　b. $g(1) = 3(1) - 2 = 3 - 2 = 1$

　c. $f(-2) = 7(-2)^2 - 3(-2) + 1$
　　　　$= 7(4) + 6 + 1$
　　　　$= 28 + 6 + 1$
　　　　$= 35$

　d. $g(0) = 3(0) - 2 = 0 - 2 = -2$

30. $f(x) = 3x^2$

　a. $f(5) = 3(5)^2 = 3(25) = 75$

　b. $f(-2) = 3(-2)^2 = 3(4) = 12$

31.

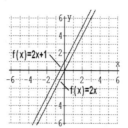

32. $m = \dfrac{9-6}{0-(-2)} = \dfrac{3}{2}$

33. $3x - 4y = 4$
　　　$-4y = -3x + 4$
　　　　$y = \dfrac{3}{4}x - 1$

　　$m = \dfrac{3}{4}$, y-intercept $= (0, -1)$

34. $y = 2$
　　$m = 0$, y-intercept $= (0, 2)$

35. a. $3x + 7y = 4$
　　　　$7y = -3x + 4$
　　　　　$y = -\dfrac{3}{7}x + \dfrac{4}{7}$

　　$m = -\dfrac{3}{7}$

　　　$6x + 14y = 7$
　　　　$14y = -6x + 7$
　　　　　$y = -\dfrac{3}{7}x + \dfrac{1}{2}$

　　$m = -\dfrac{3}{7}$

　　Parallel, since the slopes are equal.

　b. $-x + 3y = 2$
　　　　$3y = x + 2$
　　　　$y = \dfrac{1}{3}x + \dfrac{2}{3}$

　　$m = \dfrac{1}{3}$

$$2x + 6y = 5$$
$$6y = -2x + 5$$
$$y = -\frac{1}{3}x + \frac{5}{6}$$
$$m = -\frac{1}{3}$$

Neither, since the slopes are not equal and their product is not -1.

36. $y - (-9) = \frac{1}{5}(x - 0)$

$$y + 9 = \frac{1}{5}x$$

$$y = \frac{1}{5}x - 9$$

37. $m = \frac{-5 - 0}{-4 - 4} = \frac{-5}{-8} = \frac{5}{8}$

$$y - 0 = \frac{5}{8}(x - 4)$$

$$y = \frac{5}{8}x - \frac{5}{2}$$

$$f(x) = \frac{5}{8}x - \frac{5}{2}$$

38. $f(x) = \frac{1}{2}x - \frac{1}{3}$ or $y = \frac{1}{2}x - \frac{1}{3}$

$m = \frac{1}{2}$ so $m_\perp = -2$

$$y - 6 = -2[x - (-2)]$$
$$y - 6 = -2(x + 2)$$
$$y - 6 = -2x - 4$$
$$y = -2x + 2$$

39. $3x \geq y$, or $y \leq 3x$

Graph the boundary line $y = 3x$ with a solid line because the inequality symbol is $\leq$.

Test: $(0, 1)$

$$3x \geq y$$
$$3(0) \geq 1$$
$$0 \geq 1 \quad \text{False}$$

Shade the half-plane that does not

contain $(0, 1)$

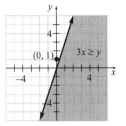

40. $x \geq 1$

Graph the boundary line $x = 1$ with a solid line because the inequality symbol is $\geq$.
Shade the half-plane that does not contain $(0, 0)$.

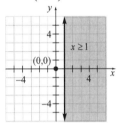

41. a. $\begin{cases} -x + y = 2 \\ 2x - y = -3 \end{cases}$

$\begin{array}{ll} -(-1) + 1 = 2 & 2(-1) - (1) = -3 \\ 1 + 1 = 2 & -2 - 1 = -3 \\ 2 = 2 \ \text{True} & -3 = -3 \ \text{True} \end{array}$

Yes, $(-1, 1)$ is a solution.

b. $\begin{cases} 5x + 3y = -1 \\ x - y = 1 \end{cases}$

$\begin{array}{ll} 5(-2) + 3(3) = -1 & -2 - 3 = -1 \\ -10 + 9 = -1 & -5 = -1 \ \text{False} \\ -1 = -1 \ \text{True} & \end{array}$

No, $(-2, 3)$ is not a solution.

42. $\begin{cases} 5x + y = -2 \quad (1) \\ 4x - 2y = -10 \quad (2) \end{cases}$

Multiply E1 by 2 and add to E2.
$$10x + 2y = -4$$
$$\underline{4x - 2y = -10}$$
$$14x \qquad = -14$$
$$x = -1$$
Replace x with -1 in E1.

$$5(-1) + y = -2$$
$$-5 + y = -2$$
$$y = 3$$

The solution is $(-1, 3)$.

43. $\begin{cases} 3x - y + z = -15 & (1) \\ x + 2y - z = 1 & (2) \\ 2x + 3y - 2z = 0 & (3) \end{cases}$

Add E1 and E2.
$$4x + y = -14 \quad (4)$$

Multiply E1 by 2 and add to E3.
$$6x - 2y + 2z = -30$$
$$\underline{2x + 3y - 2z = 0}$$
$$8x + y \quad\quad = -30 \quad (5)$$

Solve the new system:
$$\begin{cases} 4x + y = -14 & (4) \\ 8x + y = -30 & (5) \end{cases}$$

Multiply E4 by -1 and add to E5.
$$-4x - y = 14$$
$$\underline{8x + y = -30}$$
$$4x \quad\quad = -16$$
$$x = -4$$

Replace x with -4 in E4.
$$4(-4) + y = -14$$
$$-16 + y = -14$$
$$y = 2$$

Replace x with -4 and y with 2 in E1.
$$3(-4) - (2) + z = -15$$
$$-12 - 2 + z = -15$$
$$-14 + z = -15$$
$$z = -1$$

The solution is $(-4, 2, -1)$.

44. $\begin{cases} x - 2y + z = 0 & (1) \\ 3x - y - 2z = -15 & (2) \\ 2x - 3y + 3z = 7 & (3) \end{cases}$

Multiply E1 by 2 and add to E2.
$$2x - 4y + 2z = 0$$
$$\underline{3x - y - 2z = -15}$$
$$5x - 5y \quad\quad = -15 \text{ or } x - y = -3 \quad (4)$$

Multiply E1 by -3 and add to E3.

$$-3x + 6y - 3z = 0$$
$$\underline{2x - 3y + 3z = 7}$$
$$-x + 3y \quad\quad = 7 \quad (5)$$

Add E4 and E5.
$$2y = 4$$
$$y = 2$$

Replace y with 2 in E4.
$$x - 2 = -3$$
$$x = -1$$

Replace x with -1 and y with 2 in E1.
$$-1 - 2(2) + z = 0$$
$$-5 + z = 0$$
$$z = 5$$

The solution is $(-1, 2, 5)$.

45. $\begin{cases} x + 3y = 5 \\ 2x - y = -4 \end{cases}$

$$\begin{bmatrix} 1 & 3 & | & 5 \\ 2 & -1 & | & -4 \end{bmatrix}$$

Multiply R1 by -2 and add to R2.
$$\begin{bmatrix} 1 & 3 & | & 5 \\ 0 & -7 & | & -14 \end{bmatrix}$$

Divide R2 by -7.
$$\begin{bmatrix} 1 & 3 & | & 5 \\ 0 & 1 & | & 2 \end{bmatrix}$$

This corresponds to $\begin{cases} x + 3y = 5 \\ \quad\quad y = 2 \end{cases}$.

$$x + 3(2) = 5$$
$$x + 6 = 5$$
$$x = -1$$

The solution is $(-1, 2)$.

46. $\begin{cases} -6x + 8y = 0 & (1) \\ 9x - 12y = 2 & (2) \end{cases}$

Divide E1 by -2 and E2 by 3.
$$\begin{cases} 3x - 4y = 0 \\ 3x - 4y = \dfrac{2}{3} \end{cases}$$

This system is inconsistent.
The solution is $\varnothing$.

Chapter 5

Section 5.1

Calculator Explorations

1. $(3 \times 10^{11})(2 \times 10^{32}) = 6 \times 10^{43}$

3. $(5.2 \times 10^{23})(7.3 \times 10^4) = 3.796 \times 10^{28}$

Mental Math

1. $5x^{-1}y^{-2} = \dfrac{5}{xy^2}$

2. $7xy^{-4} = \dfrac{7x}{y^4}$

3. $a^2 b^{-1} c^{-5} = \dfrac{a^2}{bc^5}$

4. $a^{-4} b^2 c^{-6} = \dfrac{b^2}{a^4 c^6}$

5. $\dfrac{y^{-2}}{x^{-4}} = \dfrac{x^4}{y^2}$

6. $\dfrac{x^{-7}}{z^{-3}} = \dfrac{z^3}{x^7}$

Exercise Set 5.1

1. $4^2 \cdot 4^3 = 4^{2+3} = 4^5$

3. $x^5 \cdot x^3 = x^{5+3} = x^8$

5. $-7x^3 \cdot 20x^9 = -7 \cdot 20 x^{3+9} = -140 x^{12}$

7. $(4xy)(-5x) = -20x^{1+1}y = -20x^2 y$

9. $(-4x^3 p^2)(4y^3 x^3) = -16x^{3+3} y^3 p^2$
$$= -16x^6 y^3 p^2$$

11. $-8^0 = -(8^0) = -1$

13. $(4x+5)^0 = 1$

15. $(5x)^0 + 5x^0 = 1 + 5 \cdot 1 = 1 + 5 = 6$

17. Answers may vary.

19. $\dfrac{a^5}{a^2} = a^{5-2} = a^3$

21. $\dfrac{x^9 y^6}{x^8 y^6} = x^{9-8} y^{6-6} = x^1 y^0 = x$

23. $-\dfrac{26z^{11}}{2z^7} = -13z^{11-7} = -13z^4$

25. $\dfrac{-36a^5 b^7 c^{10}}{6ab^3 c^4} = -6a^{5-1} b^{7-3} c^{10-4}$
$$= -6a^4 b^4 c^6$$

27. $4^{-2} = \dfrac{1}{4^2} = \dfrac{1}{16}$

29. $\dfrac{x^7}{x^{15}} = x^{7-15} = x^{-8} = \dfrac{1}{x^8}$

31. $5a^{-4} = \dfrac{5}{a^4}$

33. $\dfrac{x^{-2}}{x^5} = x^{-2-5} = x^{-7} = \dfrac{1}{x^7}$

35. $\dfrac{8r^4}{2r^{-4}} = 4r^{4-(-4)} = 4r^8$

37. $\dfrac{x^{-9} x^4}{x^{-5}} = \dfrac{x^{-9+4}}{x^{-5}} = \dfrac{x^{-5}}{x^{-5}} = x^{-5-(-5)} = x^0 = 1$

39. $4^{-1} + 3^{-2} = \dfrac{1}{4^1} + \dfrac{1}{3^2}$
$$= \dfrac{1}{4} + \dfrac{1}{9}$$
$$= \dfrac{9}{36} + \dfrac{4}{36}$$
$$= \dfrac{13}{36}$$

41. $4x^0 + 5 = 4(1) + 5 = 4 + 5 = 9$

43. $x^7 \cdot x^8 \cdot x = x^{7+8+1} = x^{16}$

45. $2x^3 \cdot 5x^7 = 2 \cdot 5 x^{3+7} = 10x^{10}$

47. $\dfrac{z^{12}}{z^{15}} = z^{12-15} = z^{-3} = \dfrac{1}{z^3}$

49. $\dfrac{y^{-3}}{y^{-7}} = y^{-3-(-7)} = y^4$

51. $3x^{-1} = \dfrac{3}{x^1} = \dfrac{3}{x}$

53. $3^0 - 3t^0 = 1 - 3 \cdot 1 = 1 - 3 = -2$

55. $\dfrac{r^4}{r^{-4}} = r^{4-(-4)} = r^8$

57. $\dfrac{x^{-7}y^{-2}}{x^2 y^2} = x^{-7-2}y^{-2-2} = x^{-9}y^{-4} = \dfrac{1}{x^9 y^4}$

59. $\dfrac{2a^{-6}b^2}{18ab^{-5}} = \dfrac{a^{-6-1}b^{2-(-5)}}{9} = \dfrac{a^{-7}b^7}{9} = \dfrac{b^7}{9a^7}$

61. $\dfrac{(24x^8)(x)}{20x^{-7}} = \dfrac{6x^{8+1}}{5x^{-7}} = \dfrac{6x^{9-(-7)}}{5} = \dfrac{6x^{16}}{5}$

63. $31,250,000 = 3.125 \times 10^7$

65. $0.016 = 1.6 \times 10^{-2}$

67. $67,413 = 6.7413 \times 10^4$

69. $0.0125 = 1.25 \times 10^{-2}$

71. $0.000053 = 5.3 \times 10^{-5}$

73. $778,300,000 = 7.783 \times 10^8$

75. $737,000 = 7.37 \times 10^5$

77. $1,410,000,000 = 1.41 \times 10^9$

79. $0.001 = 1 \times 10^{-3}$

81. $3.6 \times 10^{-9} = 0.0000000036$

83. $9.3 \times 10^7 = 93,000,000$

85. $1.278 \times 10^6 = 1,278,000$

87. $7.35 \times 10^{12} = 7,350,000,000,000$

89. $4.03 \times 10^{-7} = 0.000000403$

91. $2.0 \times 10^8 = 200,000,000$

93. $4.9 \times 10^9 = 4,900,000,000$

95. $(5 \cdot 2)^2 = (10)^2 = 100$

97. $\left(\dfrac{3}{4}\right)^3 = \dfrac{3^3}{4^3} = \dfrac{27}{64}$

99. $(2^3)^2 = 8^2 = 64$

101. $(2^{-1})^4 = \left(\dfrac{1}{2}\right)^4 = \dfrac{1^4}{2^4} = \dfrac{1}{16}$

103. Answers may vary.

105. a. $x^a \cdot x^a = x^{a+a} = x^{2a}$

    b. $x^a + x^a = 2x^a$

    c. $\dfrac{x^a}{x^b} = x^{a-b}$

    d. $x^a \cdot x^b = x^{a+b}$

    e. $x^a + x^b = x^a + x^b$

107. 7^{13}

109. 7^{-11}

111. $x^5 \cdot x^{7a} = x^{5+7a}$

113. $\dfrac{x^{3t-1}}{x^t} = x^{(3t-1)-t} = x^{2t-1}$

115. $x^{4a} \cdot x^7 = x^{4a+7}$

117. $\dfrac{z^{6x}}{z^7} = z^{6x-7}$

119. $\dfrac{x^{3t} \cdot x^{4t-1}}{x^t} = \dfrac{x^{3t+(4t-1)}}{x^t} = x^{7t-1-t} = x^{6t-1}$

121. $x^{9+b} \cdot x^{3a-b} = x^{(9+b)+(3a-b)} = x^{3a+9}$

Section 5.2

Mental Math

1. $(x^4)^5 = x^{4(5)} = x^{20}$

2. $(5^6)^2 = 5^{6(2)} = 5^{12}$

3. $x^4 \cdot x^5 = x^{4+5} = x^9$

4. $x^7 \cdot x^8 = x^{7+8} = x^{15}$

5. $(y^6)^7 = y^{6(7)} = y^{42}$

6. $(x^3)^4 = x^{3(4)} = x^{12}$

7. $(z^4)^9 = z^{4(9)} = z^{36}$

8. $(z^3)^7 = z^{3(7)} = z^{21}$

9. $(z^{-6})^{-3} = z^{-6(-3)} = z^{18}$

10. $(y^{-4})^{-2} = y^{-4(-2)} = y^8$

Exercise Set 5.2

1. $(3^{-1})^2 = 3^{-1(2)} = 3^{-2} = \dfrac{1}{3^2} = \dfrac{1}{9}$

3. $(x^4)^{-9} = x^{4(-9)} = x^{-36} = \dfrac{1}{x^{36}}$

5. $(y)^{-5} = y^{-5} = \dfrac{1}{y^5}$

7. $(3x^2 y^3)^2 = 3^2 (x^2)^2 (y^3)^2$
$\qquad = 9x^{2(2)} y^{3(2)}$
$\qquad = 9x^4 y^6$

9. $\left(\dfrac{2x^5}{y^{-3}}\right)^4 = \dfrac{2^4 (x^5)^4}{(y^{-3})^4}$
$\qquad = \dfrac{16x^{5(4)}}{y^{-3(4)}}$
$\qquad = \dfrac{16x^{20}}{y^{-12}}$
$\qquad = 16x^{20} y^{12}$

11. $(a^2 bc^{-3})^{-6} = (a^2)^{-6} b^{-6} (c^{-3})^{-6}$
$\qquad = a^{2(-6)} b^{-6} c^{-3(-6)}$
$\qquad = a^{-12} b^{-6} c^{18}$
$\qquad = \dfrac{c^{18}}{a^{12} b^6}$

13. $\left(\dfrac{x^7 y^{-3}}{z^{-4}}\right)^{-5} = \dfrac{(x^7)^{-5} (y^{-3})^{-5}}{(z^{-4})^{-5}}$
$\qquad = \dfrac{x^{-35} y^{15}}{z^{20}}$
$\qquad = \dfrac{y^{15}}{x^{35} z^{20}}$

15. $(5^{-1})^3 = 5^{-1(3)} = 5^{-3} = \dfrac{1}{5^3} = \dfrac{1}{125}$

17. $\left(\dfrac{x^{-9}}{x^{-4}}\right)^{-3} = \dfrac{(x^{-9})^{-3}}{(x^{-4})^{-3}} = \dfrac{x^{27}}{x^{12}} = x^{27-12} = x^{15}$

19. $\left(\dfrac{5x^7 y^4}{10x^3 y^{-2}}\right)^{-3} = \left(\dfrac{x^4 y^6}{2}\right)^{-3}$
$\qquad = \dfrac{(x^4)^{-3} (y^6)^{-3}}{(2)^{-3}}$
$\qquad = (2)^3 x^{-12} y^{-18}$
$\qquad = \dfrac{8}{x^{12} y^{18}}$

21. $\dfrac{8^{-2} x^{-3} y^{11}}{x^2 y^{-5}} = \dfrac{x^{-3-2} y^{11-(-5)}}{8^2} = \dfrac{x^{-5} y^{16}}{64} = \dfrac{y^{16}}{64x^5}$

23. $\left(\dfrac{4p^6}{p^9}\right)^3 = \left(\dfrac{4}{p^3}\right)^3 = \dfrac{4^3}{(p^3)^3} = \dfrac{64}{p^9}$

25. $(-xy^0x^2a^3)^{-3} = (-x^3a^3)^{-3}$
$$= (-1)^3(x^3)^{-3}(a^3)^{-3}$$
$$= -1x^{-9}a^{-9}$$
$$= -\frac{1}{x^9a^9}$$

27. $\left(\dfrac{x^{-1}y^{-2}}{5^{-3}}\right)^{-5} = \dfrac{(x^{-1})^{-5}(y^{-2})^{-5}}{(5^{-3})^{-5}} = \dfrac{x^5y^{10}}{5^{15}}$

29. $(x^7)^{-9} = x^{7(-9)} = x^{-63} = \dfrac{1}{x^{63}}$

31. $\left(\dfrac{7}{8}\right)^3 = \dfrac{7^3}{8^3} = \dfrac{343}{512}$

33. $(4x^2)^2 = 4^2(x^2)^2 = 16x^4$

35. $(-2^{-2}y)^3 = (-2^{-2})^3y^3$
$$= -2^{-6}y^3$$
$$= -\frac{y^3}{2^6}$$
$$= -\frac{y^3}{64}$$

37. $\left(\dfrac{4^{-4}}{y^3x}\right)^{-2} = \dfrac{(4^{-4})^{-2}}{(y^3)^{-2}x^{-2}} = \dfrac{4^8}{y^{-6}x^{-2}} = 4^8x^2y^6$

39. $\left(\dfrac{1}{4}\right)^{-3} = \dfrac{1^{-3}}{4^{-3}} = \dfrac{4^3}{1^3} = \dfrac{64}{1} = 64$

41. $\left(\dfrac{3x^5}{6x^4}\right)^4 = \left(\dfrac{x^{5-4}}{2}\right)^4 = \left(\dfrac{x}{2}\right)^4 = \dfrac{x^4}{2^4} = \dfrac{x^4}{16}$

43. $\dfrac{(y^3)^{-4}}{y^3} = \dfrac{y^{-12}}{y^3} = y^{-12-3} = y^{-15} = \dfrac{1}{y^{15}}$

45. $\left(\dfrac{2x^{-3}}{y^{-1}}\right)^{-3} = \dfrac{2^{-3}(x^{-3})^{-3}}{(y^{-1})^{-3}} = \dfrac{x^9}{2^3y^3} = \dfrac{x^9}{8y^3}$

47. $\dfrac{3^{-2}a^{-5}b^6}{4^{-2}a^{-7}b^{-3}} = \dfrac{4^2a^{-5-(-7)}b^{6-(-3)}}{3^2} = \dfrac{16a^2b^9}{9}$

49. $(4x^6y^5)^{-2}(6x^4y^3)$
$$= 4^{-2}(x^6)^{-2}(y^5)^{-2}\cdot 6x^4y^3$$
$$= \frac{1}{4^2}x^{-12}y^{-10}\cdot 6x^4y^3$$
$$= \frac{6}{16}x^{-12+4}y^{-10+3}$$
$$= \frac{3x^{-8}y^{-7}}{8} = \frac{3}{8x^8y^7}$$

51. $x^6(x^6bc)^{-6} = x^6(x^6)^{-6}b^{-6}c^{-6}$
$$= \frac{x^6x^{-36}}{b^6c^6}$$
$$= \frac{x^{-30}}{b^6c^6}$$
$$= \frac{1}{x^{30}b^6c^6}$$

53. $\dfrac{2^{-3}x^2y^{-5}}{5^{-2}x^7y^{-1}} = \dfrac{5^2x^{2-7}y^{-5-(-1)}}{2^3}$
$$= \frac{25x^{-5}y^{-4}}{8}$$
$$= \frac{25}{8x^5y^4}$$

55. $\left(\dfrac{2x^2}{y^4}\right)^3\left(\dfrac{2x^5}{y}\right)^{-2} = \dfrac{2^3x^{2(3)}2^{-2}x^{5(-2)}}{y^{4(3)}y^{-2}}$
$$= \frac{8x^6x^{-10}}{2^2y^{12}y^{-2}}$$
$$= \frac{2x^{-4}}{y^{10}}$$
$$= \frac{2}{x^4y^{10}}$$

57. $(5\times10^{11})(2.9\times10^{-3}) = 5\times2.9\times10^{11+(-3)}$
$$= 14.5\times10^8$$
$$= 1.45\times10^1\times10^8$$
$$= 1.45\times10^9$$

59. $(2\times10^5)^3 = 2^3\times10^{5(3)} = 8\times10^{15}$

61. $\dfrac{3.6\times10^{-4}}{9\times10^{2}} = \dfrac{3.6}{9}\times10^{-4-2}$

$\qquad = 0.4\times10^{-6}$

$\qquad = 4\times10^{-1}\times10^{-6}$

$\qquad = 4\times10^{-7}$

63. $\dfrac{0.0069}{0.023} = \dfrac{6.9\times10^{-3}}{2.3\times10^{-2}}$

$\qquad = \dfrac{6.9}{2.3}\times10^{-3-(-2)}$

$\qquad = 3\times10^{-1}$

65. $\dfrac{18,200\times100}{91,000} = \dfrac{1.82\times10^{4}\times1\times10^{2}}{9.1\times10^{4}}$

$\qquad = \dfrac{1.82\times10^{6}}{9.1\times10^{4}}$

$\qquad = 0.2\times10^{6-4}$

$\qquad = 2\times10^{-1}\times10^{2}$

$\qquad = 2\times10^{-1+2}$

$\qquad = 2\times10^{1}$

67. $\dfrac{6000\times0.006}{0.009\times400} = \dfrac{6\times10^{3}\times6\times10^{-3}}{9\times10^{-3}\times4\times10^{2}}$

$\qquad = \dfrac{36\times10^{0}}{36\times10^{-1}}$

$\qquad = 1\times10^{0-(-1)}$

$\qquad = 1\times10^{1}$

69. $\dfrac{0.00064\times2000}{16,000} = \dfrac{6.4\times10^{-4}\times2\times10^{3}}{1.6\times10^{4}}$

$\qquad = \dfrac{12.8\times10^{-1}}{1.6\times10^{4}}$

$\qquad = 8\times10^{-1-4}$

$\qquad = 8\times10^{-5}$

71. $\dfrac{66,000\times0.001}{0.002\times0.003} = \dfrac{6.6\times10^{4}\times1\times10^{-3}}{2\times10^{-3}\times3\times10^{-3}}$

$\qquad = \dfrac{6.6\times10^{1}}{6\times10^{-6}}$

$\qquad = 1.1\times10^{1-(-6)}$

$\qquad = 1.1\times10^{7}$

73. $\dfrac{1.25\times10^{15}}{(2.2\times10^{-2})(6.4\times10^{-5})}$

$\qquad = \dfrac{1.25\times10^{15}}{14.08\times10^{-7}}$

$\qquad = 0.887784091\times10^{15-(-7)}$

$\qquad = 0.887784091\times10^{22}$

$\qquad = 8.877840909\times10^{-2}\times10^{22}$

$\qquad = 8.877840909\times10^{20}$

75. $200,000 = 2\times10^{-3}$ second

77. $(3.8\times10^{-6})(1.64\times10^{-5})$

$\qquad = 6.232\times10^{-11}$

The volume is 6.232×10^{-11} cubic meters.

79. $12m-14-15m-1 = -3m-15$

81. $-9y-(5-6y) = -9y-5+6y = -3y-5$

83. $5(x-3)-4(2x-5) = 5x-15-8x+20$

$\qquad\qquad\qquad\qquad\quad = -3x+5$

85. $(x^{2b+7})^{2} = x^{(2b+7)\cdot2} = x^{4b+14}$

87. $\dfrac{x^{-5y+2}x^{2y}}{x} = \dfrac{x^{-5y+2+2y}}{x}$

$\qquad = \dfrac{x^{-3y+2}}{x}$

$\qquad = x^{-3y+2-1}$

$\qquad = x^{-3y+1}$

89. $(c^{2a+3})^{3} = c^{(2a+3)\cdot3} = c^{6a+9}$

91. $\dfrac{(y^{4a})^{7}}{y^{2a-1}} = \dfrac{y^{28a}}{y^{2a-1}} = y^{28a-(2a-1)} = y^{26a+1}$

93. $\left(\dfrac{3y^{5a}}{y^{-a+1}}\right)^{2} = \dfrac{3^{2}y^{5a(2)}}{y^{(-a+1)\cdot2}}$

$\qquad = \dfrac{9y^{10a}}{y^{-2a+2}}$

$\qquad = 9y^{10a-(-2a+2)}$

$\qquad = 9y^{12a-2}$

95. $\dfrac{(y^{3-a})^b}{(y^{1-b})^a} = \dfrac{y^{(3-a)\cdot b}}{y^{(1-b)\cdot a}}$

$= \dfrac{y^{3b-ab}}{y^{a-ab}}$

$= y^{3b-ab-(a-ab)}$

$= y^{3b-a}$

97. $\dfrac{x^{-5-3a}\,y^{-2a-b}}{x^{-5+3b}\,y^{-2b-a}} = \left(\dfrac{x^{-5-3a}}{x^{-5+3b}}\right)\left(\dfrac{y^{-2a-b}}{y^{-2b-a}}\right)$

$= x^{-5-3a-(-5+3b)}\,y^{-2a-b-(-2b-a)}$

$= x^{-5-3a+5-3b}\,y^{-2a-b+2b+a}$

$= x^{-3a-3b}\,y^{b-a}$

99. $\left(\dfrac{3x^{-1}}{y^{-3}}\right)\left(5x^{-7}\right) = \dfrac{3\cdot 5 x^{-1+(-7)}}{y^{-3}}$

$= \dfrac{15x^{-8}}{y^{-3}}$

$= \dfrac{15y^{3}}{x^{8}}$

The area is $\dfrac{15y^{3}}{x^{8}}$ square feet.

101. $D = \dfrac{M}{V}$

$3.12\times10^{-2} = \dfrac{M}{4.269\times10^{14}}$

$(3.12\times10^{-2})(4.269\times10^{14}) = M$

$3.12\times4.269\times10^{-2+14} = M$

$13.31928\times10^{12} = M$

$1.331928\times10^{1}\times10^{12} = M$

$1.331928\times10^{13} = M$

The mass is 1.331928×10^{13} tons.

103. $a^{-2} = \dfrac{1}{a^{2}}$

Since both 1 and a^{2} are both positive, their quotient cannot be negative. Therefore, no, there is no such a.

105. $D = \dfrac{2.927\times10^{8}}{3.536\times10^{6}}$

$= \dfrac{2.927}{3.536}\times10^{8-6}$

$= 0.8277714932\times10^{2}$

$= 82.77714932$

≈ 83

The population density is about 83 people per square mile.

107. $\dfrac{1.27\times10^{8}}{3.22\times10^{7}} = \dfrac{1.27}{3.22}\times10^{8-7}$

$= 0.3944099379\times10^{1}$

$= 3.944099379$

Japan's population was about 3.9 times greater than Oceania's.

109. $\dfrac{3.16\times10^{9}}{7.01\times10^{8}} = \dfrac{3.16}{7.01}\times10^{9-8}$

$= 0.4507845934\times10^{1}$

$= 4.507845934$

Moscow's volume is about 4.5 times greater than São Paulo's.

Section 5.3

Graphing Calculator Explorations

1. $(2x^{2}+7x+6)+(x^{3}-6x^{2}-14)$
$= x^{3}-4x^{2}+7x-8$

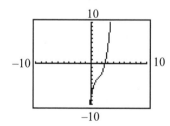

3. $(1.8x^2 - 6.8x - 1.7) - (3.9x^2 - 3.6x)$
$= -2.1x^2 - 3.2x - 1.7$

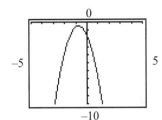

5. $(1.29x - 5.68) + (7.69x^2 - 2.55x + 10.98)$
$= 7.69x^2 - 1.26x + 5.3$

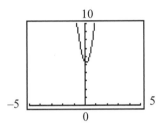

Exercise Set 5.3

1. 4 has degree 0.

3. $5x^2$ has degree 2.

5. $-3xy^2$ has degree $1 + 2 = 3$.

7. $6x + 3$ has degree 1 and is a binomial.

9. $3x^2 - 2x + 5$ has degree 2 and is a trinomial.

11. $-xyz$ has degree $1 + 1 + 1 = 3$ and is a monomial.

13. $x^2y - 4xy^2 + 5x + y$ has degree $2 + 1 = 3$ and is none of these.

15. Answers may vary.

17. $P(x) = x^2 + x + 1$
$P(7) = 7^2 + 7 + 1 = 49 + 7 + 1 = 57$

19. $Q(x) = 5x^2 - 1$
$Q(-10) = 5(-10)^2 - 1$
$= 5(100) - 1$
$= 500 - 1$
$= 499$

21. $P(x) = x^2 + x + 1$
$P(0) = 0^2 + 0 + 1 = 0 + 0 + 1 = 1$

23. $Q(x) = 5x^2 - 1$
$Q\left(\dfrac{1}{4}\right) = 5\left(\dfrac{1}{4}\right)^2 - 1$
$= 5\left(\dfrac{1}{16}\right) - 1$
$= \dfrac{5}{16} - \dfrac{16}{16}$
$= -\dfrac{11}{16}$

25. $P(t) = -16t^2 + 1053$
$P(2) = -16(2)^2 + 1053$
$= -16(4) + 1053$
$= -64 + 1053$
$= 989 \text{ ft}$

27. $P(t) = -16t^2 + 1053$
$P(6) = -16(6)^2 + 1053$
$= -16(36) + 1053$
$= -576 + 1053$
$= 477 \text{ ft}$

29. $5y + y = 6y$

31. $4x + 7x - 3 = 11x - 3$

33. $4xy + 2x - 3xy - 1 = xy + 2x - 1$

35. $7x^2 - 2xy + 5y^2 - x^2 + xy + 11y^2$
$= 6x^2 - xy + 16y^2$

37. $(9y^2 - 8) + (9y^2 - 9) = 18y^2 - 17$

39.
$$\begin{array}{r} x^2 + xy - y^2 \\ 2x^2 - 4xy + 7y^2 \\ \hline 3x^2 - 3xy + 6y^2 \end{array}$$

41.
$$\begin{array}{r} x^2 - 6x + 3 \\ +\quad (2x + 5) \\ \hline x^2 - 4x + 8 \end{array}$$

43. $(9y^2 - 7y + 5) - (8y^2 - 7y + 2)$
$= 9y^2 - 7y + 5 - 8y^2 + 7y - 2$
$= y^2 + 3$

45. $(4x^2 + 2x) - (6x^2 - 3x)$
$= 4x^2 + 2x - 6x^2 + 3x$
$= -2x^2 + 5x$

47.
$$\begin{array}{r} 3x^2 - 4x + 8 \\ -5x^2 \quad\ + 7 \\ \hline -2x^2 - 4x + 15 \end{array}$$

49. $(5x - 11) + (-x - 2) = 5x - 11 - x - 2$
$= 4x - 13$

51. $(7x^2 + x + 1) - (6x^2 + x - 1)$
$= 7x^2 + x + 1 - 6x^2 - x + 1$
$= x^2 + 2$

53. $(7x^3 - 4x + 8) + (5x^3 + 4x + 8x)$
$= 7x^3 - 4x + 8 + 5x^3 + 4x + 8x$
$= 12x^3 + 8x + 8$

55.
$$\begin{array}{r} 9x^3 - 2x^2 + 4x - 7 \\ -2x^3 + 6x^2 + 4x - 3 \\ \hline 7x^3 + 4x^2 + 8x - 10 \end{array}$$

57. $(y^2 + 4yx + 7) + (-19y^2 + 7yx + 7)$
$= y^2 + 4yx + 7 - 19y^2 + 7yx + 7$
$= -18y^2 + 11yx + 14$

59. $(3x^3 - b + 2a - 6) + (-4x^3 + b + 6a - 6)$
$= 3x^3 - b + 2a - 6 - 4x^3 + b + 6a - 6$
$= -x^3 + 8a - 12$

61. $(4x^2 - 6x + 2) - (-x^2 + 3x + 5)$
$= 4x^2 - 6x + 2 + x^2 - 3x - 5$
$= 5x^2 - 9x - 3$

63. $(-3x + 8) + (-3x^2 + 3x - 5)$
$= -3x + 8 - 3x^2 + 3x - 5$
$= -3x^2 + 3$

65. $(-3 + 4x^2 + 7xy^2) + (2x^3 - x^2 + xy^2)$
$= -3 + 4x^2 + 7xy^2 + 2x^3 - x^2 + xy^2$
$= 2x^3 + 3x^2 + 8xy^2 - 3$

67.
$$\begin{array}{r} 6y^2 - 6y + 4 \\ y^2 + 6y - 7 \\ \hline 7y^2 \quad\ - 3 \end{array}$$

69.
$$\begin{array}{r} 3x^2 + 15x + 8 \\ 2x^2 + 7x^2 + 8 \\ \hline 5x^2 + 22x + 16 \end{array}$$

71.
$$\begin{array}{r} \frac{1}{2}x^2 - \frac{1}{3}x^2 y \qquad\qquad + 2y^3 \\ \frac{1}{4}x^2 \qquad - \frac{8}{3}x^2 y^2 - \frac{1}{2}y^3 \\ \hline \frac{3}{4}x^2 - \frac{1}{3}x^2 y - \frac{8}{3}x^2 y^2 + \frac{3}{2}y^3 \end{array}$$

73. $(5q^4 - 2q^2 - 3q) + (-6q^4 + 3q^2 + 5)$
$= 5q^4 - 2q^2 - 3q - 6q^4 + 3q^2 + 5$
$= -q^4 + q^2 - 3q + 5$

75.
$$\begin{array}{r} 7x^2 + 4x + 9 \\ +\ 8x^2 + 7x - 8 \\ \hline 15x^2 + 11x + 1 \\ -\qquad\quad 3x + 7 \\ \hline 15x^2 + 8x - 6 \end{array}$$

77. $(4x^4 - 7x^2 + 3) + (2 - 3x^4)$
$= 4x^4 - 7x^2 + 3 + 2 - 3x^4$
$= x^4 - 7x^2 + 5$

79. $\left(\dfrac{2}{3}x^2 - \dfrac{1}{6}x + \dfrac{5}{6}\right) - \left(\dfrac{1}{3}x^2 + \dfrac{5}{6}x - \right)$

$= \dfrac{2}{3}x^2 - \dfrac{1}{6}x + \dfrac{5}{6} - \dfrac{1}{3}x^2 - \dfrac{5}{6}x + \dfrac{1}{6}$

$= \dfrac{1}{3}x^3 - x + 1$

81. If $L = 5$, $W = 4$, and $H = 9$, then
$2HL + 2LW + 2HW$
$= 2(9)(5) + 2(5)(4) + 2(9)(4)$
$= 90 + 40 + 72$
$= 202$
The surface area is 202 square inches.

83. $P(t) = -16t^2 + 300t$

   a. $P(1) = -16(1)^2 + 300(1) = 284$ feet

   b. $P(2) = -16(2)^2 + 300(2) = 536$ feet

   c. $P(1) = -16(3)^2 + 300(3) = 756$ feet

   d. $P(4) = -16(4)^2 + 300(4) = 944$ feet

   e. Answers may vary.

   f. $\quad 0 = -16t^2 + 300t$
$\quad\quad 0 = -4t(4t - 75)$
$\quad 4t - 75 = 0$
$\quad\quad 4t = 75$
$\quad\quad\quad t = \dfrac{75}{4} = 18.75$
$\quad$ 19 sec

85. $P(x) = 45x - 100,000$
$P(4000) = 45(4000) - 100,000$
$\quad\quad\quad\quad = 80,000$
The profit is \$80,000.

87. $R(x) = 2x$
$R(20,000) = 2(20,000)$
$\quad\quad\quad\quad\quad = 40,000$
The revenue is \$40,000.

89. A

91. D

93. $5(3x - 2) = 15x - 10$

95. $-2(x^2 - 5x + 6) = -2x^2 + 10x - 12$

97. $(4x^{2a} - 3x^a + 0.5) - (x^{2a} - 5x^a - 0.2)$
$= 4x^{2a} - 3x^a + 0.5 - x^{2a} + 5x^a + 0.2$
$= 3x^{2a} + 2x^a + 0.7$

99. $(8y^{2y} - 7x^y + 3) + (-4x^{2y} + 9x^y - 14)$
$= 8y^{2y} - 7x^y + 3 - 4x^{2y} + 9x^y - 14$
$= 4y^{2y} + 2x^y - 11$

101. $P = 2l + 2w$
$= 2(3x^2 - x + 2y) + 2(x + 5y)$
$= 6x^2 - 2x + 4y + 2x + 10y$
$= 6x^2 + 14y$
The perimeter is $P = (6x^2 + 14y)$ units.

103. $P(x) + Q(x) = (3x + 3) + (4x^2 - 6x + 3)$
$= 3x + 3 + 4x^2 - 6x + 3$
$= 4x^2 - 3x + 6$

105. $Q(x) - R(x) = (4x^2 - 6x + 3) - (5x^2 - 7)$
$= 4x^2 - 6x + 3 - 5x^2 + 7$
$= -x^2 - 6x + 10$

107. $2[Q(x)] - R(x)$
$= 2(4x^2 - 6x + 3) - (5x^2 - 7)$
$= 8x^2 - 12x + 6 - 5x^2 + 7$
$= 3x^2 - 12x + 13$

109. $3[R(x)] + 4[P(x)]$
$= 3(5x^2 - 7) + 4(3x + 3)$
$= 15x^2 - 21 + 12x + 12$
$= 15x^2 + 12x - 9$

111. $P(x) = 2x - 3$

   a. $P(a) = 2a - 3$

b. $P(-x) = 2(-x) - 3 = -2x - 3$

c. $P(x+h) = 2(x+h) - 3 = 2x + 2h - 3$

113. $P(x) = 4x$

    a. $P(a) = 4a$

    b. $P(-x) = 4(-x) = -4x$

    c. $P(x+h) = 4(x+h) = 4x + 4h$

115. $P(x) = 4x - 1$

    a. $P(a) = 4a - 1$

    b. $P(-x) = 4(-x) - 1 = -4x - 1$

    c. $P(x+h) = 4(x+h) - 1 = 4x + 4h - 1$

117. $f(x) = -246.7x^2 + 1887.9x + 1016.9$

    a. 1998 means that $x = 0$:
$$f(0) = -246.7(0)^2 + 1887.9(0) + 1016.9$$
$$= 1016.9$$
$$\approx 1017 \text{ stations}$$

    b. 2000 means that $x = 2$:
$$f(2) = -246.7(2)^2 + 1887.9(2) + 1016.9$$
$$= 3805.9$$
$$\approx 3806 \text{ stations}$$

    c. 2006 means $x = 8$:
$$f(8) = -246.7(8)^2 + 1887.9(8) + 1016.9$$
$$= 331.3$$
$$\approx 331 \text{ stations}$$

    d. Answers may vary.

119. $f(x) = 0.014x^2 + 0.12x + 0.85$

    a. 1999 means that $x = 9$:
$$f(9) = 0.014(9)^2 + 0.12(9) + 0.85$$
$$= 3.064$$
$$\approx 3.1 \text{ million SUV's}$$

    b. 2005 means that $x = 15$:
$$f(15) = 0.014(15)^2 + 0.12(15) + 0.85$$
$$= 5.8 \text{ million SUV's}$$

121. $f(x) = 1.4x^2 + 129.6x + 939$

    a. 1985 means that $x = 5$:
$$f(5) = 1.4(5)^2 + 129.6(5) + 939$$
$$= \$1622$$

    b. 1995 means that $x = 15$:
$$f(15) = 1.4(15)^2 + 129.6(15) + 939$$
$$= \$3198$$

    c. 2010 means $x = 30$:
$$f(30) = 1.4(30)^2 + 129.6(30) + 939$$
$$= \$6087$$

    d. No; $f(x)$ is not linear.

Section 5.4

Graphing Calculator Explorations

1. $(x+4)(x-4) = x^2 - 16$

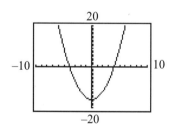

3. $(3x-7)^2 = 9x^2 - 42x + 49$

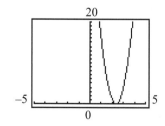

5. $(5x+1)(x^2-3x-2)$
$= 5x^3-14x^2-13x-2$

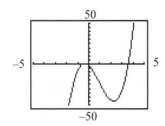

Exercise Set 5.4

1. $(-4x^3)(3x^2) = -4(3)x^5 = -12x^5$

3. $3x(4x+7) = 3x(4x)+3x(7)$
$= 12x^2+21x$

5. $-6xy(4x+y) = -6xy(4x)-6xy(y)$
$= -24x^2y-6xy^2$

7. $-4ab(xa^2+ya^2-3)$
$= -4ab(xa^2)-4ab(ya^2)-4ab(-3)$
$= -4a^3bx-4a^3by+12ab$

9.
$$\begin{array}{r} 2x+4 \\ \times\quad x-3 \\ \hline -6x-12 \\ 2x^2+4x\quad \\ \hline 2x^2-2x-12 \end{array}$$

11. $(2x+3)(x^3-x+2)$
$= 2x(x^3-x+2)+3(x^3-x+2)$
$= 2x^4-2x^2+4x+3x^3-3x+6$
$= 2x^4+3x^3-2x^2+x+6$

13.
$$\begin{array}{r} 3x-2 \\ \times\quad 5x+1 \\ \hline 3x-2 \\ 15x^2-10x\quad \\ \hline 15x^2-7x-2 \end{array}$$

15.
$$\begin{array}{r} 3m^2+2m-1 \\ \times\quad 5m+2 \\ \hline 6m^2+4m-2 \\ 15m^3+10m^2-5m\quad \\ \hline 15m^3+16m^2-m-2 \end{array}$$

17. Answers may vary.

19. $(x-3)(x+4) = x\cdot x+x(4)-3\cdot x-3(4)$
$= x^2+4x-3x-12$
$= x^2+x-12$

21. $(5x+8y)(2x-y)$
$= 5x\cdot 2x+5x(-y)+8y(2x)+8y(-y)$
$= 10x^2-5xy+16xy-8y^2$
$= 10x^2+11xy-8y^2$

23. $(3x-1)(x+3) = 3x\cdot x+3x\cdot 3-1\cdot x-1\cdot 3$
$= 3x^2+9x-x-3$
$= 3x^2+8x-3$

25. $\left(3x+\dfrac{1}{2}\right)\left(3x-\dfrac{1}{2}\right)$
$= 3x(3x)+3x\left(-\dfrac{1}{2}\right)+\dfrac{1}{2}(3x)+\dfrac{1}{2}\left(-\dfrac{1}{2}\right)$
$= 9x^2-\dfrac{3}{2}x+\dfrac{3}{2}x-\dfrac{1}{4}$
$= 9x^2-\dfrac{1}{4}$

27. $(x+4)^2 = x^2+2(x)(4)+4^2$
$= x^2+8x+16$

29. $(6y-1)(6y+1) = (6y)^2-1^2 = 36y^2-1$

31. $(3x-y)^2 = (3x)^2-2(3x)(y)+y^2$
$= 9x^2-6xy+y^2$

33. $(3b-6y)(3b+6y) = (3b)^2-(6y)^2$
$= 9b^2-36y^2$

35. $[3 + (4b + 1)]^2$
$= 3^2 + 2(3)(4b + 1) + (4b + 1)^2$
$= 9 + 6(4b + 1) + (4b)^2 + 2(4b)(1) + 1^2$
$= 9 + 24b + 6 + 16b^2 + 8b + 1$
$= 16b^2 + 32b + 16$

37. $[(2s - 3) - 1][(2s - 3) + 1]$
$= (2s - 3)^2 - 1^2$
$= (2s)^2 - 2(2s)(3) + 3^2 - 1$
$= 4s^2 - 12s + 9 - 1$
$= 4s^2 - 12s + 8$

39. $[(xy + 4) - 6]^2$
$= (xy + 4)^2 - 2(xy + 4)(6) + 6^2$
$= (xy)^2 + 2(xy)(4) + 4^2 - 12(xy + 4) + 36$
$= x^2 y^2 + 8xy + 16 - 12xy - 48 + 36$
$= x^2 y^2 - 4xy + 4$

41. Answers may vary.

43. $(x + y)(2x - 1)(x + 1)$
$= (x + y)(2x^2 + 2x - x - 1)$
$= (x + y)(2x^2 + x - 1)$
$= x(2x^2 + x - 1) + y(2x^2 + x - 1)$
$= 2x^3 + x^2 - x + 2x^2 y + xy - y$
$= 2x^3 + 2x^2 y + x^2 + xy - x - y$

45. $(x - 4)^4 = (x - 4)^2 (x - 4)^2$
$= (x^2 - 8x + 16)(x^2 - 8x + 16)$

$$
\begin{array}{r}
x^2 - 4x + 4 \\
\times \quad x^2 - 4x + 4 \\
\hline
4x^2 - 16x + 16 \\
-4x^3 + 16x^2 - 16x \\
x^4 - 4x^3 + 4x^2 \\
\hline
x^4 - 8x^3 + 24x^2 - 32x + 16
\end{array}
$$

47. $(x - 5)(x + 5)(x^2 + 25) = (x^2 - 25)(x^2 + 25)$
$= (x^2)^2 - 25^2$
$= x^4 - 625$

49. $(3x + 1)(3x + 5)$
$= (3x)^2 + 3x(5) + 1(3x) + 1(5)$
$= 9x^2 + 15x + 3x + 5$
$= 9x^2 + 18x + 5$

51. $(2x^3 + 5)(5x^2 + 4x + 1)$
$= 2x^3(5x^2 + 4x + 1) + 5(5x^2 + 4x + 1)$
$= 10x^5 + 8x^4 + 2x^3 + 25x^2 + 20x + 5$

53. $(7x - 3)(7x + 3) = (7x)^2 - 3^2 = 49x^2 - 9$

55.
$$
\begin{array}{r}
3x^2 + 4x - 4 \\
\times \quad 3x + 6 \\
\hline
18x^2 + 24x - 24 \\
9x^3 + 12x^2 - 12x \\
\hline
9x^3 + 30x^2 + 12x - 24
\end{array}
$$

57. $\left(4x + \dfrac{1}{3}\right)\left(4x - \dfrac{1}{2}\right)$
$= (4x)^2 + 4x\left(-\dfrac{1}{2}\right) + \dfrac{1}{3}(4x) + \dfrac{1}{3}\left(-\dfrac{1}{2}\right)$
$= 16x^2 - 2x + \dfrac{4}{3}x - \dfrac{1}{6}$
$= 16x^2 - \dfrac{2}{3}x - \dfrac{1}{6}$

59. $(6x + 1)^2 = (6x)^2 + 2(6x)(1) + 1^2$
$= 36x^2 + 12x + 1$

61. $(x^2 + 2y)(x^2 - 2y) = (x^2)^2 - (2y)^2$
$= x^4 - 4y^2$

63.
$$
\begin{array}{r}
5a^2 b^2 - 6a - 6b \\
\times \quad -6a^2 b^2 \\
\hline
-30a^4 b^4 + 36a^3 b^2 + 36a^2 b^3
\end{array}
$$

65. $(a - 4)(2a - 4) = 2a^2 - 4a - 8a + 16$
$= 2a^2 - 12a + 16$

67. $(7ab + 3c)(7ab - 3c) = (7ab)^2 - (3c)^2$
$= 49a^2 b^2 - 9c^2$

69. $(m - 4)^2 = m^2 - 2(m)(4) + 4^2$
$= m^2 - 8m + 16$

174

71. $(3x+1)^2 = (3x)^2 + 2(3x)(1) + 1^2$
$= 9x^2 + 6x + 1$

73. $(y-4)(y-3) = y^2 - 3y - 4y + 12$
$= y^2 - 7y + 12$

75. $(x+y)(2x-1)(x+1)$
$= (x+y)(2x^2 + 2x - x - 1)$
$= (x+y)(2x^2 + x - 1)$
$= x(2x^2 + x - 1) + y(2x^2 + x - 1)$
$= 2x^3 + x^2 - x + 2x^2 y + xy - y$
$= 2x^3 + 2x^2 y + x^2 + xy - x - y$

77.
$$\begin{array}{r} 3x^2 + 2x - 1 \\ \times \quad 3x^2 + 2x - 1 \\ \hline -3x^2 - 2x + 1 \\ 6x^3 + 4x^2 - 2x \\ 9x^4 + 6x^3 - 3x^2 \\ \hline 9x^4 + 12x^3 - 2x^2 - 4x + 1 \end{array}$$

79.
$$\begin{array}{r} 4x^2 - 2x + 5 \\ \times \quad 3x + 1 \\ \hline 4x^2 - 2x + 5 \\ 12x^3 - 6x^2 + 15x \\ \hline 12x^3 - 2x^2 + 13x + 5 \end{array}$$

81. $f(x) = x^2 - 3x$
$f(a) = a^2 - 3a$

83. $f(x) = x^2 - 3x$
$f(a+h) = (a+h)^2 - 3(a+h)$
$= a^2 + 2ah + h^2 - 3a - 3h$

85. $f(x) = x^2 - 3x$
$f(b-2) = (b-2)^2 - 3(b-2)$
$= b^2 - 4b + 4 - 3b + 6$
$= b^2 - 7b + 10$

87. $y = -2x + 7$
$m = -2$

89. $3x - 5y = 14$
$-5y = -3x + 14$
$y = \dfrac{3}{5}x - \dfrac{14}{5}$
$m = \dfrac{3}{5}$

91. Since any vertical line crosses the graph at most once, it is a function.

93. $F(x) = x^2 + 3x + 2$

   a. $F(a+h) = (a+h)^2 + 3(a+h) + 2$
$= a^2 + 2ah + h^2 + 3a + 3h + 2$

   b. $F(a) = a^2 + 3a + 2$

   c. $F(a+h) - F(a)$
$= a^2 + 2ah + h^2 + 3a + 3h + 2$
$\qquad\qquad - (a^2 + 3a + 2)$
$= 2ah + h^2 + 3h$

95. $5x^2 y^n (6y^{n+1} - 2)$
$= 5x^2 y^n (6y^{n+1}) + 5x^2 y^n (-2)$
$= 30x^2 y^{2n+1} - 10x^2 y^n$

97. $(x^a + 5)(x^{2a} - 3)$
$= x^a \cdot x^{2a} + x^a (-3) + 5(x^{2a}) + 5(-3)$
$= x^{3a} - 3x^a + 5x^{2a} - 15$

99. Area $= \Box r^2$
$= \Box(5x - 2)^2$
$= \Box(25x^2 - 20x + 4)$ square km

101. Area $= (3x - 2)^2 - x^2$
$= (9x^2 - 12x + 4) - x^2$
$= (8x^2 - 12x + 4)$ square inches

103. One operation is addition, the other is multiplication.

   a. $(3x + 5) + (3x + 7) = 6x + 12$

b. $(3x+5)(3x+7)$
$= 9x^2 + 21x + 15x + 35$
$= 9x^2 + 36x + 35$

105. $P(x) \cdot R(x) = (5x)(x+5)$
$= 5x \cdot x + 5x \cdot 5$
$= 5x^2 + 25x$

107. $[Q(x)]^2 = (x^2 - 2)^2$
$= (x^2)^2 - 2(x^2)(2) + 2^2$
$= x^4 - 4x^2 + 4$

109. $R(x) \cdot Q(x) = (x+5)(x^2-2)$
$= x^3 - 2x + 5x^2 - 10$
$= x^3 + 5x^2 - 2x - 10$

Section 5.5

Mental Math

1. $6 = 2 \cdot 3$
$12 = 2 \cdot 2 \cdot 3$
$GCF = 2 \cdot 3 = 6$

2. $9 = 3 \cdot 3$
$27 = 3 \cdot 3 \cdot 3$
$GCF = 3 \cdot 3 = 9$

3. $15x = 3 \cdot 5 \cdot x$
$10 = 2 \cdot 5$
$GCF = 5$

4. $9x = 3 \cdot 3 \cdot x$
$12 = 2 \cdot 2 \cdot 3$
$GCF = 3$

5. $13x = 13 \cdot x$
$2x = 2 \cdot x$
$GCF = x$

6. $4y = 4 \cdot y$
$5y = 5 \cdot y$
$GCF = y$

7. $7x = 7 \cdot x$
$14x = 2 \cdot 7 \cdot x$
$GCF = 7x$

8. $8z = 2 \cdot 2 \cdot 2 \cdot z$
$4z = 2 \cdot 2 \cdot z$
$GCF = 2 \cdot 2 \cdot z = 4z$

Exercise Set 5.5

1. $a^8, a^5,$ and a^3; GCF $= a^3$

3. $x^2 y^3 z^3, y^2 z^3,$ and $xy^2 z^2$; GCF $= y^2 z^2$

5. $6x^3 y, 9x^2 y^2,$ and $12x^2 y$; GCF $= 3x^2 y$

7. $10x^3 yz^3, 20x^2 z^5, 45xz^3$; GCF $= 5xz^3$

9. $18x - 12 = 6(3x - 2)$

11. $4y^2 - 16xy^3 = 4y^2(1 - 4xy)$

13. $6x^5 - 8x^4 + 2x^3 = 2x^3(3x^2 - 4x + 1)$

15. $8a^3 b^3 - 4a^2 b^2 + 4ab + 16ab^2$
$= 4ab(2a^2 b^2 - ab + 1 + 4b)$

17. $6(x+3) + 5a(x+3) = (x+3)(6+5a)$

19. $2x(z+7) + (z+7) = (z+7)(2x+1)$

21. $3x(x^2+5) - 2(x^2+5) = (x^2+5)(3x-2)$

23. Answers may vary.

25. $ab + 3a + 2b + 6 = a(b+3) + 2(b+3)$
$= (a+2)(b+3)$

27. $ac + 4a - 2c - 8 = a(c+4) - 2(c+4)$
$= (a-2)(c+4)$

29. $2xy - 3x - 4y + 6 = x(2y-3) - 2(2y-3)$
$= (x-2)(2y-3)$

31. $12xy - 8x - 3y + 2 = 4x(3y-2) - (3y-2)$
$= (4x-1)(3y-2)$

33. $6x^3 + 9 = 3(2x^3 + 3)$

35. $x^3 + 3x^2 = x^2(x+3)$

37. $8a^3 - 4a = 4a(2a^2 - 1)$

39. $-20x^2 y + 16xy^3 = -4xy(5x - 4y^2)$

41. $10a^2 b^3 + 5ab^2 - 15ab^3$
$= 5ab^2(2ab + 1 - 3b)$

43. $9abc^2 + 6a^2 bc - 6ab + 3bc$
$= 3b(3ac^2 + 2a^2 c - 2a + c)$

45. $4x(y-2) - 3(y-2) = (y-2)(4x-3)$

47. $6xy + 10x + 9y + 15$
$= 2x(3y+5) + 3(3y+5)$
$= (2x+3)(3y+5)$

49. $xy + 3y - 5x - 15 = y(x+3) - 5(x+3)$
$= (y-5)(x+3)$

51. $6ab - 2a - 9b + 3 = 2a(3b-1) - 3(3b-1)$
$= (2a-3)(3b-1)$

53. $12xy + 18x + 2y + 3$
$= 6x(2y+3) + 1(2y+3)$
$= (6x+1)(2y+3)$

55. $2m(n-8) - (n-8) = (2m-1)(n-8)$

57. $15x^3 y^2 - 18x^2 y^2 = 3x^2 y^2(5x-6)$

59. $2x^2 + 3xy + 4x + 6y$
$= x(2x+3y) + 2(2x+3y)$
$= (2x+3y)(x+2)$

61. $5x^2 + 5xy - 3x - 3y = 5x(x+y) - 3(x+y)$
$= (5x-3)(x+y)$

63. $x^3 + 3x^2 + 4x + 12 = x^2(x+3) + 4(x+3)$
$= (x^2+4)(x+3)$

65. $x^3 - x^2 - 2x + 2 = x^2(x-1) - 2(x-1)$
$= (x^2-2)(x-1)$

67. $(5x^2)(11x^5) = 5(11)x^2 x^5 = 55x^7$

69. $(5x^2)^3 = 5^3(x^2)^3 = 125x^6$

71. $(x+2)(x-5) = x^2 - 5x + 2x - 10$
$= x^2 - 3x - 10$

73. $(x+3)(x+2) = x^2 + 3x + 2x + 6$
$= x^2 + 5x + 6$

75. $(y-3)(y-1) = y^2 - 1y - 3y + 3$
$= y^2 - 4y + 3$

77. None

   a. $(2-x)(3-y) = 6 - 2y - 3x + xy$
$= xy - 3x - 2y + 6$

   b. $(-2+x)(-3+y) = 6 - 2y - 3x + xy$
$= xy - 3x - 2y + 6$

   c. $(x-2)(y-3) = xy - 3x - 2y + 6$

   d. $(-x+2)(-y+3) = xy - 3x - 2y + 6$

79. a is correct.

81. $I(R_1 + R_2) = E$

83. $x^2 + 4(10x) = x^2 + 40x = x(x+40)$ sq. in.

85. $h(t) = -16t^2 + 224$

   a. $h(t) = -16(t^2 - 14)$

   b. $h(2) = -16(2)^2 + 224$
$= -16(4) + 224$
$= -64 + 224$
$= 160$
$h(2) = -16(2^2 - 14)$
$= -16(4 - 14)$
$= -16(-10)$
$= 160$

c. Answers may vary.

87. $3y^n + 3y^{2n} + 5y^{8n} = y^n(3 + 3y^n + 5y^{7n})$

89. $3x^{5a} - 6x^{3a} + 9x^{2a} = 3x^{2a}(x^{3a} - 2x^a + 3)$

Section 5.6

Mental Math

1. $10 = 2 \cdot 5$
$7 = 2 + 5$
2 and 5

2. $12 = 2 \cdot 2 \cdot 3 = 2 \cdot 6$
$8 = 2 + 6$
2 and 6

3. $24 = 2 \cdot 2 \cdot 2 \cdot 3 = 8 \cdot 3$
$11 = 8 + 3$
8 and 3

4. $30 = 2 \cdot 3 \cdot 5 = 10 \cdot 3$
$13 = 10 + 3$
10 and 3

Exercise Set 5.6

1. $x^2 + 9x + 18 = (x + 6)(x + 3)$

3. $x^2 - 12x + 32 = (x - 4)(x - 8)$

5. $x^2 + 10x - 24 = (x + 12)(x - 2)$

7. $x^2 - 2x - 24 = (x - 6)(x + 4)$

9. Note that the GCF is 3, so that
$3x^2 - 18x + 24 = 3(x^2 - 6x + 8)$
$= 3(x - 2)(x - 4)$.

11. Note that the GCF is 4z, so that
$4x^2z + 28xz + 40z = 4z(x^2 + 7x + 10)$
$= 4z(x + 2)(x + 5)$.

13. Note that the GCF is 2, so that
$2x^2 + 30x - 108 = 2(x^2 + 15x - 54)$
$= 2(x + 18)(x - 3)$.

15. $x^2 + bx + 6$
$6 = 2 \cdot 3$ or $6 = (-2)(-3)$
$6 = 1 \cdot 6$ or $6 = (-1)(-6)$
$(x + 2)(x + 3) = x^2 + 5x + 6$
$(x - 2)(x - 3) = x^2 - 5x + 6$
$(x + 1)(x + 6) = x^2 + 7x + 6$
$(x - 1)(x - 6) = x^2 - 7x + 6$
$b = \pm 5$ and $b = \pm 7$

17. $5x^2 + 16x + 3 = (5x + 1)(x + 3)$

19. $2x^2 - 11x + 12 = (2x - 3)(x - 4)$

21. $2x^2 + 25x - 20$ is prime.

23. $4x^2 - 12x + 9 = (2x - 3)(2x - 3)$
$= (2x - 3)^2$

25. Note that the GCF is 2, so that
$12x^2 + 10x - 50 = 2(6x^2 + 5x - 25)$
$= 2(3x - 5)(2x + 5)$.

27. Note that the GCF is y^2, so that
$3y^4 - y^3 - 10y^2 = y^2(3y^2 - y - 10)$
$= y^2(3y + 5)(y - 2)$.

29. Note that the GCF is 2x, so that
$6x^3 + 8x^2 + 24x = 2x(3x^2 + 4x + 12)$.

31. $x^2 + 8xz + 7z^2 = (x + z)(x + 7z)$

33. $2x^2 - 5xy - 3y^2 = (2x + y)(x - 3y)$

35. $x^2 - x - 12$; $ac = -12$ so the two numbers are –4 and 3.
$x^2 - x - 12 = x^2 - 4x + 3x - 12$
$= x(x - 4) + 3(x - 4)$
$= (x + 3)(x - 4)$

37. Note that the GCF is 2, so that
$$28y^2 + 22y + 4 = 2(14y^2 + 11y + 2).$$
$ac = 28$; the two numbers are 4 and 7.
$$14y^2 + 11y + 2 = 14y^2 + 7y + 4y + 2$$
$$= 7y(2y+1) + 2(2y+1)$$
$$= (7y+2)(2y+1)$$
So, $28y^2 + 22y + 4 = 2(7y+2)(2y+1).$

39. $2x^2 + 15x - 27$; $ac = -54$ so the two numbers are 18 and -3.
$$2x^2 + 15x - 27 = 2x^2 + 18x - 3x - 27$$
$$= 2x(x+9) - 3(x+9)$$
$$= (2x-3)(x+9)$$

41. $3x^2 + bx + 5$
$3 = 1 \cdot 3$ or $3 = (-1)(-3)$
$5 = 1 \cdot 5$ or $5 = (-1)(-5)$

$(3x+1)(x+5) = 3x^2 + 16x + 5$
$(3x-1)(x-5) = 3x^2 - 16x + 5$
$(-3x+1)(-x+5) = 3x^2 - 16x + 5$
$(-3x-1)(-x-5) = 3x^2 + 16x + 5$
$(3x+5)(x+1) = 3x^2 + 8x + 5$
$(3x-5)(x-1) = 3x^2 - 8x + 5$
$(-3x+5)(-x+1) = 3x^2 - 8x + 5$
$(-3x-5)(-x-1) = 3x^2 + 8x + 5$
$b = \pm 8$ and $b = \pm 16$

43. Let $y = x^2$. Then we have
$$x^4 + x^2 - 6 = y^2 + y - 6 = (y+3)(y-2).$$
This yields $(x^2+3)(x^2-2).$

45. Let $y = 5x+1$. Then we have
$$(5x+1)^2 + 8(5x+1) + 7 = y^2 + 8y + 7$$
$$= (y+1)(y+7).$$
This yields
$$[(5x+1)+1][(5x+1)+7] = (5x+2)(5x+8).$$

47. Let $y = x^3$. Then we have
$$x^6 - 7x^3 + 12 = y^2 - 7y + 12$$
$$= (y-4)(y-3).$$
This yields $(x^3-4)(x^3-3).$

49. Let $y = a+5$. Then we have
$$(a+5)^2 - 5(a+5) - 24 = y^2 - 5y - 24$$
$$= (y-8)(y+3).$$
This yields
$$[(a+5)-8][(a+5)+3] = (a-3)(a+8).$$

51. Note that the GCF is x, so that
$$V(x) = 3x^3 - 2x^2 - 8x = x(3x^2 - 2x - 8)$$
$$= x(3x+4)(x-2).$$

53. $x^2 - 24x - 81 = (x-27)(x+3)$

55. $x^2 - 15x - 54 = (x-18)(x+3)$

57. $3x^2 - 6x + 3 = 3(x^2 - 2x + 1)$
$$= 3(x-1)(x-1)$$
$$= 3(x-1)^2$$

59. $3x^2 - 5x - 2 = (3x+1)(x-2)$

61. $8x^2 - 26x + 15 = (4x-3)(2x-5)$

63. $18x^4 + 21x^3 + 6x^2 = 3x^2(6x^2 + 7x + 2)$
$$= 3x^2(3x+2)(2x+1)$$

65. $3a^2 + 12ab + 12b^2 = 3(a^2 + 4ab + 4b^2)$
$$= 3(a+2b)(a+2b)$$
$$= 3(a+2b)^2$$

67. $x^2 + 4x + 5$ is prime.

69. Let $y = x+4$. Then
$$2(x+4)^2 + 3(x+4) - 5$$
$$= 2y^2 + 3y - 5$$
$$= (2y+5)(y-1)$$
$$= [2(x+4)+5][(x+4)-1]$$
$$= (2x+8+5)(x+3)$$
$$= (2x+13)(x+3)$$

71. $6x^2 - 49x + 30 = (3x - 2)(2x - 15)$

73. $x^4 - 5x^2 - 6 = (x^2 - 6)(x^2 + 1)$

75. $6x^3 - x^2 - x = x(6x^2 - x - 1)$
$$= x(3x + 1)(2x - 1)$$

77. $12a^2 - 29ab + 15b^2 = (4a - 3b)(3a - 5b)$

79. $9x^2 + 30x + 25 = (3x + 5)(3x + 5)$
$$= (3x + 5)^2$$

81. $3x^2 y - 11xy + 8y = y(3x^2 - 11x + 8)$
$$= y(3x - 8)(x - 1)$$

83. $2x^2 + 2x - 12 = 2(x^2 + x - 6)$
$$= 2(x + 3)(x - 2)$$

85. $(x - 4)^2 + 3(x - 4) - 18$
$$= [(x - 4) + 6][(x - 4) - 3]$$
$$= (x + 2)(x - 7)$$

87. $2x^6 + 3x^3 - 9 = (2x^3 - 3)(x^3 + 3)$

89. $72xy^4 - 24xy^2 z + 2xz^2$
$$= 2x(36y^4 - 12y^2 z + z^2)$$
$$= 2x(6y^2 - z)(6y^2 - z)$$
$$= 2x(6y^2 - z)^2$$

91. $(x - 3)(x + 3) = x^2 - 3^3 = x^2 - 9$

93. $(2x + 1)^2 = (2x)^2 + 2(2x)(1) + 1^2$
$$= 4x^2 + 4x + 1$$

95.
$$\begin{array}{r} x^2 + 2x + 4 \\ \hline x - 2 \\ \hline -2x^2 + 4x - 8 \\ x^3 + 2x^2 + 4x \\ \hline x^3 \qquad\qquad -8 \end{array}$$

97. $h(t) = -16t^2 + 80t + 576$

  a. $h(0) = -16(0)^2 + 80(0) + 576 = 576$ ft

$h(2) = -16(2)^2 + 80(2) + 576$
$$= -16(4) + 160 + 576$$
$$= -64 + 160 + 576$$
$$= 672 \text{ ft}$$
$h(4) = -16(4)^2 + 80(4) + 576$
$$= -16(16) + 320 + 576$$
$$= -256 + 320 + 576$$
$$= 640 \text{ ft}$$
$h(6) = -16(6)^2 + 80(6) + 576$
$$= -16(36) + 480 + 576$$
$$= -576 + 480 + 576$$
$$= 480 \text{ ft}$$

  b. Answers may vary.

  c. $h(t) = -16t^2 + 80t + 576$
$$= -16(t^2 - 5t - 36)$$
$$= -16(t - 9)(t + 4)$$

99. $x^{2n} + 10x^n + 16 = (x^n + 2)(x^n + 8)$

101. $x^{2n} - 3x^n - 18 = (x^n - 6)(x^n + 3)$

103. $2x^{2n} + 11x^n + 5 = (2x^n + 1)(x^n + 5)$

105. $4x^{2n} - 12x^n + 9 = (2x^n - 3)(2x^n - 3)$
$$= (2x^n - 3)^2$$

107. $x^4 + 6x^3 + 5x^2 = x^2(x^2 + 6x + 5)$
$$= x^2(x + 5)(x + 1)$$

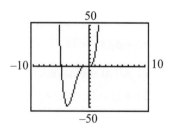

109. $30x^3 + 9x^2 - 3x = 3x(10x^2 + 3x - 1)$
$$= 3x(5x - 1)(2x + 1)$$

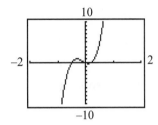

Exercise Set 5.7

1. $x^2 + 6x + 9 = (x + 3)^2$

3. $4x^2 - 12x + 9 = (2x - 3)^2$

5. $3x^2 - 24x + 48 = 3(x^2 - 8x + 16)$
$$= 3(x - 4)^2$$

7. $9y^2 x^2 + 12yx + 4x^2 = x^2(9y^2 + 12y + 4)$
$$= x^2(3y + 2)^2$$

9. $x^2 - 25 = (x + 5)(x - 5)$

11. $9 - 4z^2 = (3 + 2z)(3 - 2z)$

13. $(y + 2)^2 - 49 = [(y + 2) + 7][(y + 2) - 7]$
$$= (y + 9)(y - 5)$$

15. $64x^2 - 100 = 4(16x^2 - 25)$
$$= 4(4x + 5)(4x - 5)$$

17. $x^2 + 27 = x^3 + 3^3 = (x + 3)(x^2 - 3x + 9)$

19. $z^3 - 1 = z^3 - 1^3 = (z - 1)(z^2 + z + 1)$

21. $m^3 + n^3 = (m + n)(m^2 - mn + n^2)$

23. $x^3 y^2 - 27y^2 = y^2(x^3 - 27)$
$$= y^2(x^3 - 3^3)$$
$$= y^2(x - 3)(x^2 + 3x + 9)$$

25. $a^3 b + 8b^4 = b(a^3 + 8b^3)$
$$= b\left[a^3 + (2b)^3\right]$$
$$= b(a + 2b)(a^2 - 2ab + 4b^2)$$

27. $125y^3 - 8x^3$
$$= (5y)^3 - (2x)^3$$
$$= (5y - 2x)(25y^2 + 10xy + 4x^2)$$

29. $(x^2 + 6x + 9) - y^2 = (x + 3)^2 - y^2$
$$= (x + 3 + y)(x + 3 - y)$$

31. $(x^2 - 10x + 25) - y^2 = (x - 5)^2 - y^2$
$$= (x - 5 + y)(x - 5 - y)$$

33. $(4x^2 + 4x + 1) - z^2$
$$= (2x + 1)^2 - z^2$$
$$= (2x + 1 + z)(2x + 1 - z)$$

35. $9x^2 - 49 = (3x + 7)(3x - 7)$

37. $x^2 - 12x + 36 = (x - 6)^2$

39. $x^4 - 81 = (x^2 + 9)(x^2 - 9)$
$$= (x^2 + 9)(x + 3)(x - 3)$$

41. $(x^2 + 8x + 16) - 4y^2$
$$= (x + 4)^2 - (2y)^2$$
$$= (x + 4 + 2y)(x + 4 - 2y)$$

43. $(x + 2y)^2 - 9 = (x + 2y)^2 - 3^2$
$$= (x + 2y + 3)(x + 2y - 3)$$

45. $x^3 - 216 = x^3 - 6^3 = (x - 6)(x^2 + 6x + 36)$

47. $x^3 + 125 = x^3 + 5^3 = (x + 5)(x^2 - 5x + 25)$

49. $4x^2 + 25$ is prime.

51. $4a^2 + 12a + 9 = (2a + 3)(2a + 3)$
$$= (2a + 3)^2$$

53. $18x^2 y - 2y = 2y(9x^2 - 1)$
$$= 2y(3x + 1)(3x - 1)$$

55. $8x^3 + y^3 = (2x)^3 + y^3$
$$= (2x + y)(4x^2 - 2xy + y^2)$$

57. $x^6 - y^3 = (x^2)^3 - y^3$
$$= (x^2 - y)(x^4 + x^2 y + y^2)$$

59. $(x^2 + 16x + 64) - x^4$
$$= (x + 8)^2 - (x^2)^2$$
$$= (x + 8 + x^2)(x + 8 - x^2)$$

61. $3x^6 y^2 + 81y^2 = 3y^2(x^6 + 27)$
$$= 3y^2 \left[(x^2)^3 + 3^3 \right]$$
$$= 3y^2(x^2 + 3)(x^4 - 3x^2 + 9)$$

63. $(x + y)^3 + 125$
$$= (x + y)^3 + 5^3$$
$$= [(x + y) + 5][(x + y)^2 - 5(x + y) + 25]$$
$$= (x + y + 5)(x^2 + 2xy + y^2 - 5x - 5y + 25)$$

65. $(2x + 3)^3 - 64$
$$= (2x + 3)^3 - 4^3$$
$$= [(2x + 3) - 4][(2x + 3)^2 + 4(2x + 3) + 16]$$
$$= (2x - 1)(4x^2 + 12x + 9 + 8x + 12 + 16)$$
$$= (2x - 1)(4x^2 + 20x + 37)$$

67. $x - 5 = 0$
$$x = 5$$

69. $3x + 1 = 0$
$$3x = -1$$
$$x = -\frac{1}{3}$$

71. $-2x = 0$
$$x = 0$$

73. $-5x + 25 = 0$
$$-5x = -25$$
$$x = 5$$

75. Area $= \pi R^2 - \pi r^2 = \pi(R^2 - r^2)$ sq. units

77. Volume $= x^3 - y^2 x$
$$= x(x^2 - y^2)$$
$$= x(x + y)(x - y) \text{ cubic units}$$

79. $\frac{1}{2} \cdot b = \frac{1}{2} \cdot 6 = 3$ so $c = 3^2 = 9$

81. $\frac{1}{2} \cdot b = \frac{1}{2}(-14) = -7$ so $c = (-7)^2 = 49$

83. $\frac{1}{2} \cdot c = \frac{c}{2}$ so $\left(\frac{c}{2} \right)^2 = 16$
$$\frac{c^2}{4} = 16$$
$$c^2 = 64$$
$$c = \pm 8$$

85. $x^6 - 1$

a. $(x^3)^2 - 1^2$
$$= (x^3 + 1)(x^3 - 1)$$
$$= (x + 1)(x^2 - x + 1)(x - 1)(x^2 + x + 1)$$

b. $(x^2)^3 - 1^3$
$$= (x^2 - 1)(x^4 + x^2 + 1)$$
$$= (x + 1)(x - 1)(x^4 + x^2 + 1)$$

c. No; answers may vary.

87. $x^{2n} - 36 = (x^n)^2 - 6^2 = (x^n + 6)(x^n - 6)$

89. $25x^{2n} - 81 = (5x^n)^2 - 9^2$
$$= (5x^n + 9)(5x^n - 9)$$

91. $x^{4n} - 625 = (x^{2n})^2 - 25^2$
$$= (x^{2n} + 25)(x^{2n} - 25)$$
$$= (x^{2n} + 25)\left[(x^n)^2 - 5^2 \right]$$
$$= (x^{2n} + 25)(x^n + 5)(x^n - 5)$$

Integrated Review

1. $(-y^2+6y-1)+(3y^2-4y-10)$
$= -y^2+6y-1+3y^2-4y-10$
$= 2y^2+2y-11$

2. $(5z^4-6z^2+z+1)-(7z^4-2z+1)$
$= -2z^4-6z^2+3z$

3. $(x^2-6x+2)-(x-5)$
$= x^2-6x+2-x+5$
$= x^2-7x+7$

4. $(2x^2+6x-5)+(5x^2-10x)$
$= 7x^2-4x-5$

5. $(5x-3)^2 = (5x)^2-2(5x)(3)+3^2$
$= 25x^2-30x+9$

6. $(5x^2-14x-3)\div(5x+1) = x-3$

7. $(2x^4-3x^2+5x-2)\div(x+2)$
$= 2x^3-4x^2+5x-5+\dfrac{8}{x+2}$

8.
$$
\begin{array}{r}
x^2-3x-2 \\
\times \quad\quad 4x-1 \\
\hline
-x^2+3x+2 \\
4x^3-12x^2-8x \quad\quad \\
\hline
4x^3-13x^2-5x+2
\end{array}
$$

9. $x^2-8x+16-y^2 = (x-4)^2-y^2$
$= (x-4+y)(x-4-y)$

10. $12x^2-22x-20 = 2(6x^2-11x-10)$
$= 2(3x+2)(2x-5)$

11. $x^4-x = x(x^3-1) = x(x-1)(x^2+x+1)$

12. Let $y = 2x+1$. Then
$(2x+1)^2-3(2x+1)+2$
$= y^2-3y+2$
$= (y-2)(y-1)$
$= [(2x+1)-2][(2x+1)-1]$
$= (2x-1)(2x)$
$= 2x(2x-1)$

13. $14x^2y-2xy = 2xy(7x-1)$

14. $24ab^2-6ab = 6ab(4b-1)$

15. $4x^2-16 = 4(x^2-4) = 4(x+2)(x-2)$

16. $9x^2-81 = 9(x^2-9) = 9(x+3)(x-3)$

17. $3x^2-8x-11 = (3x-11)(x+1)$

18. $5x^2-2x-3 = (5x+3)(x-1)$

19. $4x^2+8x-12 = 4(x^2+2x-3)$
$= 4(x+3)(x-1)$

20. $6x^2-6x-12 = 6(x^2-x-2)$
$= 6(x-2)(x+1)$

21. $4x^2+36x+81 = (2x+9)(2x+9)$
$= (2x+9)^2$

22. $25x^2+40x+16 = (5x+4)(5x+4)$
$= (5x+4)^2$

23. $8x^3+125y^3 = (2x)^3+(5y)^3$
$= (2x+5y)(4x^2-10xy+25y^2)$

24. $27x^3-64y^3 = (3x)^3-(4y)^3$
$= (3x-4y)(9x^2+12xy+16y^2)$

25. $64x^2y^3-8x^2 = 8x^2(8y^3-1)$
$= 8x^2[(2y)^3-1^3]$
$= 8x^2(2y-1)(4y^2+2y+1)$

26. $27x^5y^4 - 216x^2y$
$= 27x^2y(x^3y^3 - 8)$
$= 27x^2y[(xy)^3 - 2^3]$
$= 27x^2y(xy - 2)(x^2y^2 + 2xy + 4)$

27. $(x+5)^3 + y^3$
$= [(x+5) + y][(x+5)^2 - (x+5)y + y^2]$
$= (x+y+5)(x^2 + 10x + 25 - xy - 5y + y^2)$
$= (x+y+5)(x^2 + 10x - xy - 5y + y^2 + 25]$

28. $(y-1)^3 + 27x^3$
$= (y-1)^3 + (3x)^3$
$= [(y-1) + 3x][(y-1)^2 - (y-1)(3x) + 9x^2]$
$= (y-1+3x)(y^2 - 2y + 1 - 3xy + 3x + 9x^2)$

29. Let $y = 5a - 3$. Then
$(5a-3)^2 - 6(5a-3) + 9$
$= y^2 - 6y + 9$
$= (y-3)(y-3)$
$= (y-3)^2$
$= [(5a-3) - 3]^2$
$= (5a-6)^2$

30. Let $y = 4r + 1$. Then
$(4r+1)^2 + 8(4r+1) + 16$
$= y^2 + 8y + 16$
$= (y+4)(y+4)$
$= (y+4)^2$
$= [(4r+1) + 4]^2$
$= (4r+5)^2$

31. $7x^2 - 63x = 7x(x-9)$

32. $20x^2 + 23x + 6 = (4x+3)(5x+2)$

33. $ab - 6a + 7b - 42 = a(b-6) + 7(b-6)$
$= (a+7)(b-6)$

34. $20x^2 - 220x + 600 = 20(x^2 - 11x + 30)$
$= 20(x-6)(x-5)$

35. $x^4 - 1 = (x^2)^2 - 1^2$
$= (x^2 + 1)(x^2 - 1)$
$= (x^2 + 1)(x+1)(x-1)$

36. $15x^2 - 20x = 5x(3x-4)$

37. $10x^2 - 7x - 33 = (5x-11)(2x+3)$

38. $45m^3n^3 - 27m^2n^2 = 9m^2n^2(5mn-3)$

39. $5a^3b^3 - 50a^3b = 5a^3b(b^2 - 10)$

40. $x^4 + x = x(x^3 + 1) = x(x+1)(x^2 - x + 1)$

41. $16x^2 + 25$ is a prime polynomial.

42. $20x^3 + 20y^3 = 20(x^3 + y^3)$
$= 20(x+y)(x^2 - xy + y^2)$

43. $10x^3 - 210x^2 + 1100x$
$= 10x(x^2 - 21x + 110)$
$= 10x(x-11)(x-10)$

44. $9y^2 - 42y + 49 = (3y-7)(3y-7)$
$= (3y-7)^2$

45. $64a^3b^4 - 27a^3b$
$= a^3b(64b^3 - 27)$
$= a^3b[(4b)^3 - 3^3]$
$= a^3b(4b-3)(16b^2 + 12b + 9)$

46. $y^4 - 16 = (y^2)^2 - 4^2$
$= (y^2 + 4)(y^2 - 4)$
$= (y^2 + 4)(y+2)(y-2)$

47. $2x^3 - 54 = 2(x^3 - 27)$
$= 2(x^3 - 3^3)$
$= 2(x-3)(x^2 + 3x + 9)$

48. $2sr + 10s - r - 5 = 2s(r+5) - 1(r+5)$
$= (2s-1)(r+5)$

49. $3y^5 - 5y^4 + 6y - 10 = y^4(3y-5) + 2(3y-5)$
$= (y^4 + 2)(3y-5)$

50. $64a^2 + b^2$ is a prime polynomial.

51. $100z^3 + 100 = 100(z^3 + 1)$
$$= 100(z+1)(z^2 - z + 1)$$

52. $250x^4 - 16x = 2x(125x^3 - 8)$
$$= 2x[(5x)^3 - 2^3]$$
$$= 2x(5x - 2)(25x^2 + 10x + 4)$$

53. $4b^2 - 36b + 81 = (2b - 9)(2b - 9)$
$$= (2b - 9)^2$$

54. $2a^5 - a^4 + 6a - 3 = a^4(2a - 1) + 3(2a - 1)$
$$= (a^4 + 3)(2a - 1)$$

55. Let $x = y - 6$. Then
$$(y - 6)^2 + 3(y - 6) + 2$$
$$= x^2 + 3x + 2$$
$$= (x + 2)(x + 1)$$
$$= [(y - 6) + 2][(y - 6) + 1]$$
$$= (y - 4)(y - 5)$$

56. Let $x = c + 2$. Then
$$(c + 2)^2 - 6(c + 2) + 5$$
$$= x^2 - 6x + 5$$
$$= (x - 5)(x - 1)$$
$$= [(c + 2) - 5][(c + 2) - 1]$$
$$= (c - 3)(c + 1)$$

57. Area $= 3^2 - 4x^2$
$$= 9 - 16x^2$$
$$= (3 + 2x)(3 - 2x)$$

Section 5.8

Graphing Calculator Explorations

1.

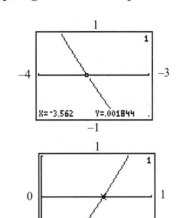

The intercepts are –3.562, 0.562.

3.

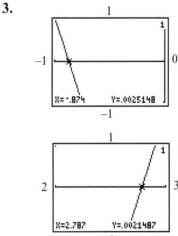

The intercepts are –0.874, 2.787.

5.

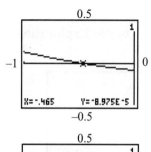

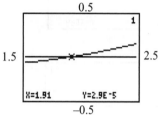

The intercepts are –0.465, 1.910.

Mental Math

1. $(x-3)(x+5) = 0$
$x-3 = 0$ or $x+5 = 0$
$x = 3$ or $x = -5$
The solutions are –5, 3.

2. $(y+5)(y+3) = 0$
$y+5 = 0$ or $y+3 = 0$
$y = -5$ or $y = -3$
The solutions are –5, –3.

3. $(z-3)(z+7) = 0$
$z-3 = 0$ or $z+7 = 0$
$z = 3$ or $z = -7$
The solutions are –7, 3.

4. $(c-2)(c-4) = 0$
$c-2 = 0$ or $c-4 = 0$
$c = 2$ or $c = 4$
The solutions are 2, 4.

5. $x(x-9) = 0$
$x = 0$ or $x-9 = 0$
$x = 9$
The solutions are 0, 9.

6. $w(w+7) = 0$
$w = 0$ or $w+7 = 0$
$w = -7$
The solutions are –7, 0.

Exercise Set 5.8

1. $(x+3)(3x-4) = 0$
$x+3 = 0$ or $3x-4 = 0$
$x = -3$ or $3x = 4$
$x = \dfrac{4}{3}$
The solutions are $-3, \dfrac{4}{3}$.

3. $3(2x-5)(4x+3) = 0$
$2x-5 = 0$ or $4x+3 = 0$
$2x = 5$ or $4x = -3$
$x = \dfrac{5}{2}$ or $x = -\dfrac{3}{4}$
The solutions are $-\dfrac{3}{4}, \dfrac{5}{2}$.

5. $x^2 +11x + 24 = 0$
$(x+8)(x+3) = 0$
$x+8 = 0$ or $x+3 = 0$
$x = -8$ or $x = -3$
The solutions are –8, –3.

7. $12x^2 + 5x - 2 = 0$
$(4x-1)(3x+2) = 0$
$4x-1 = 0$ or $3x+2 = 0$
$4x = 1$ or $3x = -2$
$x = \dfrac{1}{4}$ or $x = -\dfrac{2}{3}$
The solutions are $-\dfrac{2}{3}, \dfrac{1}{4}$.

9. $z^2 + 9 = 10z$
$z^2 -10z + 9 = 0$
$(z-9)(z-1) = 0$

$z - 9 = 0$ or $z - 1 = 0$
$z = 9$ or $z = 1$
The solutions are 1, 9.

11. $x(5x + 2) = 3$
$5x^2 + 2x - 3 = 0$
$(5x - 3)(x + 1) = 0$
$5x - 3 = 0$ or $x + 1 = 0$
$5x = 3$ or $x = -1$
$x = \dfrac{3}{5}$

The solutions are $-1, \dfrac{3}{5}$.

13. $x - 6x = x(8 + x)$
$x - 6x = 8x + x^2$
$-14x = 0$
$x = 0$
The solutions is 0.

15. $\dfrac{z^2}{6} - \dfrac{z}{2} - 3 = 0$
$z^2 - 3z - 18 = 0$
$(z - 6)(z + 3) = 0$
$z - 6 = 0$ or $z + 3 = 0$
$z = 6$ or $z = -3$
The solutions are –3, 6.

17. $\dfrac{x^2}{2} + \dfrac{x}{20} = \dfrac{1}{10}$
$10x^2 + x = 2$
$10x^2 + x - 2 = 0$
$(5x - 2)(2x + 1) = 0$
$5x - 2 = 0$ or $2x + 1 = 0$
$5x = 2$ or $2x = -1$
$x = \dfrac{2}{5}$ or $x = -\dfrac{1}{2}$

The solutions are $-\dfrac{1}{2}, \dfrac{2}{5}$.

19. $\dfrac{4t^2}{5} = \dfrac{t}{5} + \dfrac{3}{10}$
$8t^2 = 2t + 3$
$8t^2 - 2t - 3 = 0$

$(4t - 3)(2t + 1) = 0$
$4t - 3 = 0$ or $2t + 1 = 0$
$4t = 3$ or $2t = -1$
$t = \dfrac{3}{4}$ or $t = -\dfrac{1}{2}$

The solutions are $-\dfrac{1}{2}, \dfrac{3}{4}$.

21. $(x + 2)(x - 7)(3x - 8) = 0$
$x + 2 = 0$ or $x - 7 = 0$ or $3x - 8 = 0$
$x = -2$ or $x = 7$ or $3x = 8$
$x = \dfrac{8}{3}$

The solutions are $-2, 7, \dfrac{8}{3}$.

23. $y^3 = 9y$
$y^3 - 9y = 0$
$y(y^2 - 9) = 0$
$y(y + 1)(y - 9) = 0$
$y = 0$ or $y + 3 = 0$ or $y - 3 = 0$
$y = -3$ or $y = 3$
The solutions are –3, 0, 3.

25. $x^3 - x = 2x^2 - 2$
$x^3 - 2x^2 - x + 2 = 0$
$x^2(x - 2) - 1(x - 2) = 0$
$(x^2 - 1)(x - 2) = 0$
$(x + 1)(x - 1)(x - 2) = 0$
$x + 1 = 0$ or $x - 1 = 0$ or $x - 2 = 0$
$x = -1$ or $x = 1$ or $x = 2$
The solutions are –1, 1, 2.

27. Answers may vary.

29. $(2x + 7)(x - 10) = 0$
$2x + 7 = 0$ or $x - 10 = 0$
$2x = -7$ or $x = 10$
$x = -\dfrac{7}{2}$

The solutions are $-\dfrac{7}{2}, 10$.

31. $3x(x-5) = 0$
$3x = 0$ or $x - 5 = 0$
$x = 0$ or $x = 5$
The solutions are 0, 5.

33. $x^2 - 2x - 15 = 0$
$(x-5)(x+3) = 0$
$x - 5 = 0$ or $x + 3 = 0$
$x = 5$ or $x = -3$
The solutions are $-3, 5$.

35. $12x^2 + 2x - 2 = 0$
$2(6x^2 + x - 1) = 0$
$2(3x - 1)(2x + 1) = 0$
$3x - 1 = 0$ or $2x + 2 = 0$
$3x = 1$ or $2x = -2$
$x = \dfrac{1}{3}$ or $x = -\dfrac{1}{2}$
The solutions are $-\dfrac{1}{2}, \dfrac{1}{3}$.

37. $w^2 - 5w = 36$
$w^2 - 5w - 36 = 0$
$(w - 9)(w + 4) = 0$
$w - 9 = 0$ or $w + 4 = 0$
$w = 9$ or $w = -4$
The solutions are $-4, 9$.

39. $25x^2 - 40x + 16 = 0$
$(5x - 4)^2 = 0$
$5x - 4 = 0$
$5x = 4$
$x = \dfrac{4}{5}$
The solutions is $\dfrac{4}{5}$.

41. $2r^3 + 6r^2 = 20r$
$2r^3 + 6r^2 - 20r = 0$
$2r(r^2 + 3r - 10) = 0$
$2r(r + 5)(r - 2) = 0$

$2r = 0$ or $r + 5 = 0$ or $r - 2 = 0$
$r = 0$ or $r = -5$ or $r = 2$
The solutions are $-5, 0, 2$.

43. $z(5z - 4)(z + 3) = 0$
$z = 0$ or $5z - 4 = 0$ or $z + 3 = 0$
$5z = 4$ or $z = -3$
$z = \dfrac{4}{5}$
The solutions are $-3, 0, \dfrac{4}{5}$.

45. $2z(z + 6) = 2z^2 + 12z - 8$
$2z^2 + 12z = 2z^2 + 12z - 8$
$0 = -8$ False
No solution exist; $\varnothing$

47. $(x - 1)(x + 4) = 24$
$x^2 + 3x - 4 = 24$
$x^2 + 3x - 28 = 0$
$(x + 7)(x - 4) = 0$
$x + 7 = 0$ or $x - 4 = 0$
$x = -7$ or $x = 4$
The solutions are $-7, 4$.

49. $\dfrac{x^2}{4} - \dfrac{5}{2}x + 6 = 0$
$x^2 - 10x + 24 = 0$
$(x - 6)(x - 4) = 0$
$x - 6 = 0$ or $x - 4 = 0$
$x = 6$ or $x = 4$
The solutions are 4, 6.

51. $y^2 + \dfrac{1}{4} = -y$
$4y^2 + 1 = -4y$
$4y^2 + 4y + 1 = 0$
$(2y + 1)^2 = 0$
$2y + 1 = 0$
$2y = -1$
$y = -\dfrac{1}{2}$
The solution is $-\dfrac{1}{2}$.

53.
$$y^3 + 4y^2 = 9y + 36$$
$$y^3 + 4y^2 - 9y - 36 = 0$$
$$y^2(y+4) - 9(y+4) = 0$$
$$(y^2 - 9)(y+4) = 0$$
$$(y+3)(y-3)(y+4) = 0$$
$$y+3=0 \quad \text{or} \quad y-3=0 \quad \text{or} \quad y+4=0$$
$$y=-3 \text{ or} \qquad y=3 \text{ or} \qquad y=-4$$
The solutions are –4, –3, 3.

55.
$$2x^3 = 50x$$
$$2x^3 - 50x = 0$$
$$2x(x^2 - 25) = 0$$
$$2x(x+5)(x-5) = 0$$
$$2x=0 \text{ or } x+5=0 \quad \text{ or } x-5=0$$
$$x=0 \text{ or } \qquad x=-5 \text{ or } \qquad x=5$$
The solutions are –5, 0, 5.

57.
$$x^2 + (x+1)^2 = 61$$
$$x^2 + x^2 + 2x + 1 = 61$$
$$2x^2 + 2x - 60 = 0$$
$$2(x^2 + x - 30) = 0$$
$$2(x+6)(x-5) = 0$$
$$x+6=0 \quad \text{ or } x-5=0$$
$$x=-6 \text{ or } \qquad x=5$$
The solutions are –6, 5.

59.
$$m^2(3m-2) = m$$
$$3m^3 - 2m^2 = m$$
$$3m^3 - 2m^2 - m = 0$$
$$m(3m^2 - 2m - 1) = 0$$
$$m(3m+1)(m-1) = 0$$
$$m=0 \text{ or } 3m+1=0 \quad \text{ or } m-1=0$$
$$3m=-1 \text{ or } \qquad m=1$$
$$m = -\frac{1}{3}$$
The solutions are $-\frac{1}{3}, 0, 1$.

61.
$$3x^2 = -x$$
$$3x^2 + x = 0$$
$$x(3x+1) = 0$$

$$x=0 \text{ or } 3x+1=0$$
$$3x = -\frac{1}{3}$$
The solutions are $-\frac{1}{3}, 0$.

63.
$$x(x-3) = x^2 + 5x + 7$$
$$x^2 - 3x = x^2 + 5x + 7$$
$$-8x = 7$$
$$x = -\frac{7}{8}$$
The solution is $-\frac{7}{8}$.

65.
$$3(t-8) + 2t = 7 + t$$
$$3t - 24 + 2t = 7 + t$$
$$5t - 24 = 7 + t$$
$$4t = 31$$
$$t = \frac{31}{4}$$
The solution is $\frac{31}{4}$.

67.
$$-3(x-4) + x = 5(3-x)$$
$$-3x + 12 + x = 15 - 5x$$
$$-2x + 12 = 15 - 5x$$
$$3x = 3$$
$$x = 1$$
The solution is 1.

69. **a** and **d** are incorrect because the right side of the equation is not zero.

71. Let n = the one number and
$n + 5$ = the other number.
$$n(n+5) = 66$$
$$n^2 + 5n - 66 = 0$$
$$(n+11)(n-6) = 0$$
$$n+11=0 \quad \text{ or } n-6=0$$
$$n=-11 \text{ or } \qquad n=6$$
The two solutions are –11 and –6 and 6 and 11.

73. Let d = amount of cable needed. Then from the Pythagorean theorem,
$$d^2 = 45^2 + 60^2 = 5625$$
so $d = \sqrt{5625} = 75$ ft.

75. $C(x) = x^2 - 15x + 50$
$$9500 = x^2 - 15x + 50$$
$$0 = x^2 - 15x - 9450$$
$$0 = (x - 105)(x + 90)$$
$$x - 105 = 0 \quad \text{or} \quad x + 90 = 0$$
$$x = 105 \quad \text{or} \quad x = -90$$
Disregard the negative. 105 units.

77. Let x = one leg of a right triangle and $x - 3$ = the other leg of the right triangle.
$$15^2 = x^2 + (x - 3)^2$$
$$225 = x^2 + x^2 - 6x + 9$$
$$225 = 2x^2 - 6x + 9$$
$$0 = 2x^2 - 6x - 216$$
$$0 = 2(x^2 - 3x - 108)$$
$$0 = 2(x - 12)(x + 9)$$
$$x - 12 = 0 \quad \text{or} \quad x + 9 = 0$$
$$x = 12 \quad \text{or} \quad x = -9$$
Disregarding the extraneous solution –9, we find that one leg of the right triangle is 12 cm and the other leg is 9 cm.

79. Note that the outer rectangle has lengths of $2x + 12$ and $2x + 16$. Thus, the area of the border is $(2x + 12)(2x + 16) - 12 \cdot 16$. Set this equal to 128 and solve for x.
$$(2x + 12)(2x + 16) - 12 \cdot 16 = 128$$
$$4x^2 + 56x + 192 - 192 = 128$$
$$4x^2 + 56x = 128$$
$$4x^2 + 56x - 128 = 0$$
$$x^2 + 14x - 32 = 0$$
$$(x + 16)(x - 2) = 0$$
$$x + 16 = 0 \quad \text{or} \quad x - 2 = 0$$
$$x = -16 \quad \text{or} \quad x = 2$$
Since x must be positive, we see that $x = 2$ inches.

81. The sunglasses will hit the ground when $h(t)$ equals 0.
$$-16t^2 + 1600 = 0$$
$$-16(t^2 - 100) = 0$$
$$-16(t - 10)(t + 10) = 0$$
$$t - 10 = 0 \quad \text{or} \quad t + 10 = 0$$
$$t = 10 \quad \text{or} \quad t = -10$$
The sunglasses will hit the ground 10 seconds after being dropped.

83. Let the width of the floor = w. Then the length is $2w - 3$ and so the area is
$$(2w - 3)w = 90$$
$$2w^2 - 3w - 90 = 0$$
$$(2w - 15)(w + 6) = 0$$
$$2w - 15 = 0 \quad \text{or} \quad w + 6 = 0$$
$$2w = 15 \quad \text{or} \quad w = -6$$
$$w = \frac{15}{2} = 7.5$$
Disregard –6.
$$2w - 3 = 2(7.5) - 3 = 15 - 3 = 12$$
The width is 7.5 ft and the length is 12 ft.

85.
$$0.5x^2 = 50$$
$$0.5x^2 - 50 = 0$$
$$5x^2 - 500 = 0$$
$$x^2 - 100 = 0$$
$$(x + 10)(x - 10) = 0$$
$$x + 10 = 0 \quad \text{or} \quad x - 10 = 0$$
$$x = -10 \quad \text{or} \quad x = 10$$
Disregard the negative solution. A 10-inch square tier is needed, provided each person has one serving.

87. The object will hit the ground when $h(t)$ equals 0.
$$-16t^2 + 80t + 576 = 0$$
$$-16(t^2 - 5t - 36) = 0$$
$$-16(t - 9)(t + 4) = 0$$

$t - 9 = 0$ or $t + 4 = 0$
$\quad t = 9$ or $\quad\quad$ $t = -4$

The object will hit the ground 9 seconds after being dropped.

89. E

91. F

93. B

95. $(-3, 0)$, $(0, 2)$; function

97. $(-4, 0)$, $(0, 2)$, $(4, 0)$, $(0, -2)$; not a function

99. Answers may vary.

101. $(x^2 + x - 6)(3x^2 - 14x - 5) = 0$
$\quad x^2 + x - 6 = 0$ or $3x^2 - 14x - 5 = 0$
$(x + 3)(x - 2) = 0$ or $(3x + 1)(x - 5) = 0$
$x + 3 = 0,\ x - 2 = 0,\ 3x + 1 = 0$ or $x - 5 = 0$
$\quad x = -3, \quad x = 2, \quad 3x = -1, \quad\quad x = 5$
$$x = -\frac{1}{3}$$

The solutions are $-3, -\dfrac{1}{3}, 2, 5$.

103. No; answers may vary.

105. Answers may vary.
Ex.: $f(x) = (x - 6)(x - 7) = x^2 - 13x + 42$

107. Answers may vary.
Ex.: $f(x) = (x - 4)(x + 3) = x^2 - x - 12$

Chapter 5 Review

1. $(-2)^2 = (-2)(-2) = 4$

2. $(-3)^4 = (-3)(-3)(-3)(-3) = 81$

3. $-2^2 = -(2 \cdot 2) = -4$

4. $-3^4 = -(3 \cdot 3 \cdot 3 \cdot 3) = -81$

5. $8^0 = 1$

6. $-9^0 = -1$

7. $-4^{-2} = -\dfrac{1}{4^2} = -\dfrac{1}{16}$

8. $(-4)^2 = \dfrac{1}{(-4)^2} = \dfrac{1}{16}$

9. $-xy^2 \cdot y^3 \cdot xy^2 z = -x^{1+1} y^{2+3+2} z = -x^2 y^7 z$

10. $(-4xy)(-3xy^2 b) = 12x^2 y^3 b$

11. $a^{-14} a^5 = a^{-14+5} = a^{-9} = \dfrac{1}{a^9}$

12. $\dfrac{a^{16}}{a^{17}} = a^{16-17} = a^{-1} = \dfrac{1}{a}$

13. $\dfrac{x^{-7}}{x^4} = x^{-7-4} = x^{-11} = \dfrac{1}{x^{11}}$

14. $\dfrac{9a(a^{-3})}{18a^{15}} = \dfrac{a^{1-3-15}}{2} = \dfrac{a^{-17}}{2} = \dfrac{1}{2a^{17}}$

15. $\dfrac{y^{6p-3}}{y^{6p+2}} = y^{(6p-3)-(6p+2)}$
$$= y^{6p-3-6p-2} = y^{-5} = \dfrac{1}{y^5}$$

16. $36{,}890{,}000 = 3.689 \times 10^7$

17. $-0.000362 = -3.62 \times 10^{-4}$

18. $1.678 \times 10^{-6} = 0.000001678$

19. $4.1 \times 10^5 = 410{,}000$

20. $(8^5)^3 = 8^{5 \cdot 3} = 8^{15}$

21. $\left(\dfrac{a}{4}\right)^2 = \dfrac{a^2}{4^2} = \dfrac{a^2}{16}$

22. $(3x)^3 = 3^3 x^3 = 27x^3$

23. $(-4x)^{-2} = \dfrac{1}{(-4x)^2} = \dfrac{1}{(-4)^2 x^2} = \dfrac{1}{16x^2}$

24. $\left(\dfrac{6x}{5}\right)^2 = \dfrac{(6x)^2}{5^2} = \dfrac{36x^2}{25}$

25. $(8^6)^{-3} = 8^{6(-3)} = 8^{-18} = \dfrac{1}{8^{18}}$

26. $\left(\dfrac{4}{3}\right)^{-2} = \dfrac{4^{-2}}{3^{-2}} = \dfrac{3^2}{4^2} = \dfrac{9}{16}$

27. $(-2x^3)^{-3} = \dfrac{1}{(-2x^3)^3}$

$= \dfrac{1}{(-2)^3(x^3)^3}$

$= \dfrac{1}{-8x^9}$

$= -\dfrac{1}{8x^9}$

28. $\left(\dfrac{8p^6}{4p^4}\right)^{-2} = (2p^2)^{-2} = 2^{-2}p^{-4} = \dfrac{1}{4p^4}$

29. $(-3x^{-2}y^2)^3 = (-3)^3(x^{-2})^3(y^2)^3$

$= -27x^{-6}y^6$

$= -\dfrac{27y^6}{x^6}$

30. $\left(\dfrac{x^{-5}y^{-3}}{z^3}\right)^{-5} = \dfrac{x^{25}y^{15}}{z^{-15}} = x^{25}y^{15}z^{15}$

31. $\dfrac{4^{-1}x^3yz}{x^{-2}yx^4} = \dfrac{x^{3-(-2)-4}z}{4} = \dfrac{x^{3+2-4}z}{4} = \dfrac{xz}{4}$

32. $(5xyz)^{-4}(x^{-2})^{-3} = \dfrac{1}{(5xyz)^4}x^6$

$= \dfrac{x^6}{5^4x^4y^4z^4}$

$= \dfrac{x^2}{625y^4z^4}$

33. $\dfrac{2(3yz)^{-3}}{y^{-3}} = \dfrac{2(3)^{-3}y^{-3}z^{-3}}{y^{-3}} = \dfrac{2}{3^3z^3} = \dfrac{2}{27z^3}$

34. $x^{4a}(3x^{5a})^3 = x^{4a}(3^3x^{15a})$

$= 27x^{4a+15a}$

$= 27x^{19a}$

35. $\dfrac{4y^{3x-3}}{2y^{2x+4}} = 2y^{(3x-3)-(2x+4)}$

$= 2y^{3x-3-2x-4}$

$= 2y^{x-7}$

36. $\dfrac{(0.00012)(144,000)}{0.0003}$

$= \dfrac{(1.2\times10^{-4})(1.44\times10^5)}{3\times10^{-4}}$

$= 0.576\times10^5$

$= 5.76\times10^4$

37. $\dfrac{(-0.00017)0.00039)}{3000}$

$= \dfrac{(-1.7\times10^{-4})(3.9\times10^{-4})}{3\times10^3}$

$= -2.21\times10^{-4-4-3}$

$= -2.21\times10^{-11}$

38. $\dfrac{27x^{-5}y^5}{18x^{-6}y^2}\cdot\dfrac{x^4y^{-2}}{x^{-2}y^3} = \dfrac{3x^{-5+4}y^{5-2}}{2x^{-6-2}y^{2+3}}$

$= \dfrac{3x^{-1}y^3}{2x^{-8}y^5}$

$= \dfrac{3}{2}x^{-1-(-8)}y^{3-5}$

$= \dfrac{3}{2}x^7y^{-2}$

$= \dfrac{3x^7}{2y^2}$

39. $\dfrac{3x^5}{y^{-4}}\cdot\dfrac{(3xy^{-3})^{-2}}{(z^{-3})^{-4}} = \dfrac{3x^5\cdot3^{-2}x^{-2}y^6}{y^{-4}z^{12}}$

$= \dfrac{3^{1-2}x^{5-2}y^{6-(-4)}}{z^{12}}$

$= \dfrac{3^{-1}x^3y^{10}}{z^{12}}$

$= \dfrac{x^3y^{10}}{3z^{12}}$

40. $\dfrac{(x^w)^2}{(x^{w-4})^{-2}} = \dfrac{x^{2w}}{x^{-2(w-4)}}$

$= \dfrac{x^{2w}}{x^{-2w+8}} = x^{2w-(-2w+8)} = x^{4w-8}$

41. The degree of the polynomial
$x^2y - 3xy^3z + 5x + 7y$ is the degree of
the term $-3xy^3z$ which is 5.

42. $3x + 2$ has degree 1.

43. $4x + 8x - 6x^2 - 6x^2y$
$= (4+8)x - 6x^2 - 6x^2y$
$= 12x - 6x^2 - 6x^2y$

44. $-8xy^3 + 4xy^3 - 3x^3y$
$= (-8+4)xy^3 - 3x^3y$
$= -4xy^3 - 3x^3y$

45. $(3x + 7y) + (4x^2 - 3x + 7) + (y - 1)$
$= 3x + 7y + 4x^2 - 3x + 7 + y - 1$
$= 4x^2 + (3-3)x + (7+1)y + (7-1)$
$= 4x^2 + 8y + 6$

46. $(4x^2 - 6xy + 9y^2) - (8x^2 - 6xy - y^2)$
$= 4x^2 - 6xy + 9y^2 - 8x^2 + 6xy + y^2$
$= (4-8)x^2 + (9+1)y^2$
$= -4x^2 + 10y^2$

47. $(3x^2 - 4b + 28) + (9x^2 - 30)$
$\qquad\qquad\qquad - (4x^2 - 6b + 20)$
$= 3x^2 - 4b + 28 + 9x^2 - 30 - 4x^2 + 6b - 20$
$= (3+9-4)x^2 + (-4+6)b + (28-30-20)$
$= 8x^2 + 2b - 22$

48. $(9xy + 4x^2 + 18) + (7xy - 4x^3 - 9x)$
$= 9xy + 4x^2 + 18 + 7xy - 4x^3 - 9x$
$= -4x^3 + 4x^2 + (9+7)xy - 9x + 18$
$= -4x^3 + 4x^2 + 16xy - 9x + 18$

49. $(3x^2y - 7xy - 4) + (9x^2y + x) - (x - 7)$
$= 3x^2y - 7xy - 4 + 9x^2y + x - x + 7$
$= (3+9)x^2y - 7xy + (-4+7)$
$= 12x^2y - 7xy + 3$

50. $\begin{array}{r} x^2 - 5x + 7 \\ - \quad (x+4) \\ \hline x^2 - 6x + 3 \end{array}$

51. $\begin{array}{r} x^3 \quad + 2xy^2 - y \\ + \quad (x - 4xy^2 \quad -7) \\ \hline x^3 + x - 2xy^2 - y - 7 \end{array}$

52. $P(6) = 9(6)^2 - 7(6) + 8 = 290$

53. $P(-2) = 9(-2)^2 - 7(-2) + 8 = 58$

54. $P(-3) = 9(-3)^2 - 7(-3) + 8 = 110$

55. $P(x) + Q(x) = (2x - 1) + (x^2 + 2x - 5)$
$\qquad\qquad = 2x - 1 + x^2 + 2x - 5$
$\qquad\qquad = x^2 + 4x - 6$

56. $2[P(x)] - Q(x) = 2(2x - 1) - (x^2 + 2x - 5)$
$\qquad\qquad = 4x - 2 - x^2 - 2x + 5$
$\qquad\qquad = -x^2 + 2x + 3$

57. $2(2x^2y - 6x + 1) + 2(x^2y + 5)$
$= 4x^2y - 12x + 2 + 2x^2y + 10$
$= (6x^2y - 12x + 12)$ cm

58. $-6x(4x^2 - 6x + 1) = -24x^3 + 36x^2 - 6x$

59. $-4ab^2(3ab^3 + 7ab + 1)$
$= -4ab^2(3ab^3) - 4ab^2(7ab) - 4ab^2(1)$
$= -12a^2b^5 - 28a^2b^3 - 4ab^2$

60. $(x - 4)(2x + 9) = 2x^2 + 9x - 8x - 36$
$\qquad\qquad\qquad = 2x^2 + x - 36$

61. $(-3xa + 4b)^2$
$= (-3xa)^2 + 2(-3xa)(4b) + (4b)^2$
$= 9x^2a^2 - 24xab + 16b^2$

62.

$$\begin{array}{r} 9x^2 + 4x + 1 \\ 4x - 3 \\ \hline -27x^2 - 12x - 3 \\ 36x^3 + 16x^2 + 4x \\ \hline 36x^3 - 11x^2 - 8x - 3 \end{array}$$

63. $(5x - 9y)(3x + 9y)$
$= 15x^2 + 45xy - 27xy + 81y^2$
$= 15x^2 + 18xy - 81y^2$

64. $\left(x - \dfrac{1}{3}\right)\left(x + \dfrac{2}{3}\right) = x^2 + \dfrac{2}{3}x - \dfrac{1}{3}x - \dfrac{1}{3}\left(\dfrac{2}{3}\right)$
$\qquad\qquad\qquad = x^2 + \dfrac{1}{3}x - \dfrac{2}{9}$

65. $(x^2 + 9x + 1)^2$
$= (x^2 + 9x + 1)(x^2 + 9x + 1)$
$= x^2(x^2 + 9x + 1) + 9x(x^2 + 9x + 1)$
$\quad + 1(x^2 + 9x + 1)$
$= x^4 + 9x^3 + x^2 + 9x^3 + 81x^2 + 9x$
$\quad + x^2 + 9x + 1$
$= x^4 + 18x^3 + 83x^2 + 18x + 1$

66. $(3x - y)^2 = (3x)^2 - 2(3x)y + y^2$
$\qquad\qquad = 9x^2 - 6xy + y^2$

67. $(4x + 9)^2 = (4x)^2 + 2(4x)(9) + 9^2$
$\qquad\qquad = 16x^2 + 72x + 81$

68. $(x + 3y)(x - 3y) = x^2 - (3y)^2 = x^2 - 9y^2$

69. $[4 + (3a - b)][4 - (3a - b)]$
$= 4^2 - (3a - b)^2$
$= 16 - [(3a)^2 - 2(3a)b + b^2]$
$= 16 - (9a^2 - 6ab + b^2)$
$= 16 - 9a^2 + 6ab - b^2$

70. $P(x) \cdot Q(x)$
$= (2x - 1)(x^2 + 2x - 5)$
$= 2x(x^2 + 2x - 5) - 1(x^2 + 2x - 5)$
$= 2x^3 + 4x^2 - 10x - x^2 - 2x + 5$
$= 2x^3 + 3x^2 - 12x + 5$

71. $Area = lw$
$= (3y + 7z)(3y - 7z)$
$= (3y)^2 - (7z)^2$
$= (9y^2 - 49z^2)$ square units

72. $4a^b(3a^{b+2} - 7) = 4a^b(3a^{b+2}) + 4a^b(-7)$
$\qquad\qquad\qquad = 12a^{b+b+2} - 28a^b$
$\qquad\qquad\qquad = 12a^{2b+2} - 28a^b$

73. $(4xy^z - b)^2 = (4xy^z)^2 - 2(4xy^z)b + b^2$
$\qquad\qquad\quad = 4^2x^2(y^z)^2 - 8xy^zb + b^2$
$\qquad\qquad\quad = 16x^2y^{2z} - 8xy^zb + b^2$

74. $(3x^a - 4)(3x^a + 4) = (3x^a)^2 - 4^2$
$\qquad\qquad\qquad\qquad = 3^2(x^a)^2 - 16$
$\qquad\qquad\qquad\qquad = 9x^{2a} - 16$

75. $16x^3 - 24x^2 = 8x^2(2x - 3)$

76. $36y - 24y^2 = 12y(3 - 2y)$

77. $6ab^2 + 8ab - 4a^2b^2 = 2ab(3b + 4 - 2ab)$

78. $14a^2b^2 - 21ab^2 + 7ab$
$= 7ab(2ab - 3b + 1)$

79. $6a(a + 3b) - 5(a + 3b) = (a + 3b)(6a - 5)$

80. $4x(x - 2y) - 5(x - 2y) = (x - 2y)(4x - 5)$

81. $xy - 6y + 3x - 18 = y(x - 6) + 3(x - 6)$
$\qquad\qquad\qquad\quad = (x - 6)(y + 3)$

82. $ab - 8b + 4a - 32 = b(a - 8) + 4(a - 8)$
$\qquad\qquad\qquad\quad = (a - 8)(b + 4)$

83. $pq - 3p - 5q + 15 = p(q - 3) - 5(q - 3)$
$\qquad\qquad\qquad\quad = (q - 3)(p - 5)$

84. $x^3 - x^2 - 2x + 2 = x^2(x - 1) - 2(x - 1)$
$\qquad\qquad\qquad\quad = (x - 1)(x^2 - 2)$

85. Area $= 2xy - x^2 = x(2y - x)$ sq. units

86. $x^2 - 14x - 72 = (x - 18)(x + 4)$

87. $x^2 + 16x - 80 = (x-4)(x+20)$

88. $2x^2 - 18x + 28 = 2(x^2 - 9x + 14)$
$= 2(x-7)(x-2)$

89. $3x^2 + 33x + 54 = 3(x^2 + 11x + 18)$
$= 3(x+9)(x+2)$

90. $2x^3 - 7x^2 - 9x = x(2x^2 - 7x - 9)$
$= x(2x-9)(x+1)$

91. $3x^2 + 2x - 16 = (3x+8)(x-2)$

92. $6x^2 + 17x + 10 = (6x+5)(x+2)$

93. $15x^2 - 91x + 6 = (15x-1)(x-6)$

94. $4x^2 + 2x - 12 = 2(2x^2 + x - 6)$
$= 2(2x-3)(x+2)$

95. $9x^2 - 12x - 12 = 3(3x^2 - 4x - 4)$
$= 3(3x+2)(x-2)$

96. $y^2(x+6)^2 - 2y(x+6)^2 - 3(x+6)^2$
$= (x+6)^2(y^2 - 2y - 3)$
$= (x+6)^2(y-3)(y+1)$

97. Let $y = x + 5$. Then
$(x+5)^2 + 6(x+5) + 8$
$= y^2 + 6y + 8$
$= (y+4)(y+2)$
$= [(x+5)+4][(x+5)+2]$
$= (x+9)(x+7)$

98. $x^4 - 6x^2 - 16 = (x^2 - 8)(x^2 + 2)$

99. $x^4 + 8x^2 - 20 = (x^2 + 10)(x^2 - 2)$

100. $x^2 - 100 = (x+10)(x-10)$

101. $x^2 - 81 = (x+9)(x-9)$

102. $2x^2 - 32 = 2(x^2 - 16) = 2(x+4)(x-4)$

103. $6x^2 - 54 = 6(x^2 - 9) = 6(x+3)(x-3)$

104. $81 - x^4 = (9 + x^2)(9 - x^2)$
$= (9 + x^2)(3+x)(3-x)$

105. $16 - y^4 = (4 + y^2)(4 - y^2)$
$= (4 + y^2)(2+y)(2-y)$

106. $(y+2)^2 - 25 = [(y+2)+5][(y+2)-5]$
$= (y+7)(y-3)$

107. $(x-3)^2 - 16 = [(x-3)+4][(x-3)-4]$
$= (x+1)(x-7)$

108. $x^3 + 216 = x^3 + 6^3$
$= (x+6)(x^2 - 6 \cdot x + 6^2)$
$= (x+6)(x^2 - 6x + 36)$

109. $y^3 + 512 = y^3 + 8^3$
$= (y+8)(y^2 - 8 \cdot y + 8^2)$
$= (y+8)(y^2 - 8y + 64)$

110. $8 - 27y^3 = 2^3 - (3y)^3$
$= (2-3y)(4 + 2 \cdot 3y + (3y)^2)$
$= (2-3y)(4 + 6y + 9y^2)$

111. $1 - 64y^3 = 1^3 - (4y)^3$
$= (1-4y)(1^2 + 1 \cdot 4y + (4y)^2)$
$= (1-4y)(1 + 4y + 16y^2)$

112. $6x^4y + 48xy = 6xy(x^3 + 8)$
$= 6xy(x^3 + 2^3)$
$= 6xy(x+2)(x^2 - 2x + 2^2)$
$= 6xy(x+2)(x^2 - 2x + 4)$

113. $2x^5 + 16x^2y^3$
$= 2x^2(x^3 + 8y^3)$
$= 2x^2(x^3 + (2y)^3)$
$= 2x^2(x+2y)(x^2 - x \cdot 2y + (2y)^2)$
$= 2x^2(x+2y)(x^2 - 2xy + 4y^2)$

114. $x^2 - 2x + 1 - y^2$
$= (x^2 - 2x + 1) - y^2$
$= (x-1)^2 - y^2$
$= [(x-1) + y][(x-1) - y]$
$= (x - 1 + y)(x - 1 - y)$

115. $x^2 - 6x + 9 - 4y^2$
$= (x^2 - 6x + 9) - 4y^2$
$= (x-3)^2 - (2y)^2$
$= [(x-3) + 2y][(x-3) - 2y]$
$= (x - 3 + 2y)(x - 3 - 2y)$

116. $4x^2 + 12x + 9 = (2x + 3)(2x + 3)$
$= (2x + 3)^2$

117. $16a^2 - 40ab + 25b^2 = (4a - 5b)(4a - 5b)$
$= (4a - 5b)^2$

118. Volume $= \pi R^2 h - \pi r^2 h$
$= \pi h (R^2 - r^2)$
$= \pi h (R + r)(R - r)$ cubic units

119. $(3x - 1)(x + 7) = 0$
$3x - 1 = 0$ or $x + 7 = 0$
$x = \dfrac{1}{3}$ or $\quad x = -7$

The solutions are $-7, \dfrac{1}{3}$.

120. $3(x + 5)(8x - 3) = 0$
$x + 5 = 0$ or $8x - 3 = 0$
$x = -5$ or $\quad x = \dfrac{3}{8}$

The solutions are $-5, \dfrac{3}{8}$.

121. $5x(x - 4)(2x - 9) = 0$
$5x = 0$ or $x - 4 = 0$ or $2x - 9 = 0$
$x = 0$ or $\quad x = 4$ or $\quad x = \dfrac{9}{2}$

The solutions are $0, 4, \dfrac{9}{2}$.

122. $6(x + 3)(x - 4)(5x + 1) = 0$
$x + 3 = 0$ or $x - 4 = 0$ or $5x + 1 = 0$
$x = -3$ or $\quad x = 4$ or $\quad 5x = -1$
$x = -\dfrac{1}{5}$

The solutions are $-3, -\dfrac{1}{5}, 4..$

123. $\qquad 2x^2 = 12x$
$2x^2 - 12x = 0$
$2x(x - 6) = 0$
$2x = 0$ or $x - 6 = 0$
$x = 0$ or $\quad x = 6$

The solutions are 0, 6.

124. $\qquad 4x^3 - 36x = 0$
$4x(x^2 - 9) = 0$
$4x(x + 3)(x - 3) = 0$
$4x = 0$ or $x + 3 = 0$ or $x - 3 = 0$
$x = 0$ or $\quad x = -3$ or $\quad x = 3$

The solutions are $-3, 0, 3$.

125. $\qquad (1 - x)(3x + 2) = -4x$
$3x + 2 - 3x^2 - 2x = -4x$
$-3x^2 + x + 2 = -4x$
$-3x^2 + 5x + 2 = 0$
$3x^2 - 5x - 2 = 0$
$(3x + 1)(x - 2) = 0$
$3x + 1 = 0$ or $x - 2 = 0$
$3x = -1$ or $\quad x = 2$
$x = -\dfrac{1}{3}$

The solutions are $-\dfrac{1}{3}, 2$.

126. $\qquad 2x(x - 12) = -40$
$2x^2 - 24x = -40$
$2x^2 - 24x + 40 = 0$
$2(x^2 - 12x + 20) = 0$
$2(x - 10)(x - 2) = 0$

$x - 10 = 0$ or $x - 2 = 0$
$x = 10$ or $x = 2$
The solutions are 2, 10.

127. $3x^2 + 2x = 12 - 7x$
$3x^2 + 9x - 12 = 0$
$3(x^2 + 3x - 4) = 0$
$3(x + 4)(x - 1) = 0$
$x + 4 = 0$ or $x - 1 = 0$
$x = -4$ or $x = 1$
The solutions are –4, 1.

128. $2x^2 + 3x = 35$
$2x^2 + 3x - 35 = 0$
$(2x - 7)(x + 5) = 0$
$2x - 7 = 0$ or $x + 5 = 0$
$2x = 7$ or $x = -5$
$x = \dfrac{7}{2}$

The solutions are $-5, \dfrac{7}{2}$.

129. $x^3 - 18x = 3x^2$
$x^3 - 3x^2 - 18x = 0$
$x(x^2 - 3x - 18) = 0$
$x(x - 6)(x + 3) = 0$
$x = 0$ or $x - 6 = 0$ or $x + 3 = 0$
$x = 6$ or $x = -3$
The solutions are –3, 0, 6.

130. $19x^2 - 42x = -x^3$
$x^3 + 19x^2 - 42x = 0$
$x(x^2 + 19x - 42) = 0$
$x(x + 21)(x - 2) = 0$
$x = 0$ or $x + 21 = 0$ or $x - 2 = 0$
$x = -21$ or $x = 2$
The solutions are –21, 0, 2.

131. $12x = 6x^3 + 6x^2$
$-6x^3 - 6x^2 + 12x = 0$
$-6x(x^2 + x - 2) = 0$
$-6x(x + 2)(x - 1) = 0$

$-6x = 0$ or $x + 2 = 0$ or $x - 1 = 0$
$x = 0$ or $x = -2$ or $x = 1$
The solutions are –2, 0, 1.

132. $8x^3 + 10x^2 = 3x$
$8x^3 + 10x^2 - 3x = 0$
$x(8x^2 + 10x - 3) = 0$
$x(4x - 1)(2x + 3) = 0$
$x = 0$ or $4x - 1 = 0$ or $2x + 3 = 0$
$4x = 1$ or $2x = -3$
$x = \dfrac{1}{4}$ or $x = -\dfrac{3}{2}$

The solutions are $-\dfrac{3}{2}, 0, \dfrac{1}{4}$.

133. Let x = the number. Then
$x + 2x^2 = 105$
$2x^2 + x - 105 = 0$
$(2x + 15)(x - 7) = 0$
$2x + 15 = 0$ or $x - 7 = 0$
$2x = -15$ or $x = 7$
$x = -\dfrac{15}{2}$

The numbers are $-\dfrac{15}{2}$ and 7.

134. Let x = width; then $2x - 5$ = length.
$x(2x - 5) = 33$
$2x^2 - 5x = 33$
$2x^2 - 5x - 33 = 0$
$(2x - 11)(x + 3) = 0$
$2x - 11 = 0$ or $x + 3 = 0$
$2x = 11$ or $x = -3$
$x = \dfrac{11}{2}$

Disregard the negative.
Width = $\dfrac{11}{2} = 5\dfrac{1}{2}$ m

Length = $2\left(\dfrac{11}{2}\right) - 5 = 6$ m

135. $h(t) = -16t^2 + 400$

$0 = -16t^2 + 400$

$0 = -16(t^2 - 25)$

$0 = -16(t+5)(t-5)$

$t+5 = 0$ or $t-5 = 0$

$t = -5$ or $t = 5$

Disregard the negative. The stunt dummy will reach the ground after 5 seconds.

Chapter 5 Test

1. $(-9x)^{-2} = \dfrac{1}{(-9x)^2} = \dfrac{1}{81x^2}$

2. $-3xy^{-2}(4xy^2)z = -12x^{1+1}y^{-2+2}z$
$= -12x^2z$

3. $\dfrac{6^{-1}a^2b^{-3}}{3^{-2}a^{-5}b^2} = \dfrac{3^2 a^{2+5}}{6^1 b^{2+3}} = \dfrac{9a^7}{6b^5} = \dfrac{3a^7}{2b^5}$

4. $\left(\dfrac{-xy^{-5}z}{xy^3}\right)^{-5} = \dfrac{-x^5 y^{25} z^{-5}}{x^{-5} y^{-15}}$

$= \dfrac{x^{-5+5} y^{25-(-15)}}{z^5}$

$= -\dfrac{y^{40}}{z^5}$

5. $630,000,000 = 6.3 \times 10^8$

6. $0.01200 = 1.2 x 10^{-2}$

7. $5 \times 10^{-6} = 0.000005$

8. $\dfrac{(0.0024)(0.00012)}{0.00032}$

$= \dfrac{(2.4 \times 10^{-3})(1.2 \times 10^{-4})}{3.2 \times 10^{-4}}$

$= \dfrac{(2.4)(1.2)}{3.2} \times 10^{-3+(-4)-(-4)}$

$= 0.9 \times 10^{-3}$

$= 0.0009$

9. $(4x^3y - 3x - 4) - (9x^3y + 8x + 5)$
$= 4x^3y - 3x - 4 - 9x^3y - 8x - 5$
$= -5x^3y - 11x - 9$

10. $-3xy(4x + y) = -3xy(4x) - 3xy(y)$
$= -12x^2y - 3xy^2$

11. $(3x+4)(4x-7) = 12x^2 - 21x + 16x - 28$
$= 12x^2 - 5x - 28$

12. $(5a - 2b)(5a + 2b) = (5a)^2 - (2b)^2$
$= 25a^2 - 4b^2$

13. $(6m + n)^2 = (6m)^2 + 2(6m)n + n^2$
$= 36m^2 + 12mn + n^2$

14.
$$\begin{array}{r} x^2 - 6x + 4 \\ \times \quad 2x - 1 \\ \hline -x^2 + 6x - 4 \\ 2x^3 - 12x^2 + 8x \quad\quad \\ \hline 2x^3 - 13x^3 + 14x - 4 \end{array}$$

15. $16x^3y - 12x^2y^4 = 4x^2y(4x - 3y^3)$

16. $x^2 - 13x - 30 = (x-15)(x+2)$

17. $4y^2 + 20y + 25 = (2y+5)(2y+5)$
$= (2y+5)^2$

18. $6x^2 - 15x - 9 = 3(2x^2 - 5x - 3)$
$= 3(2x+1)(x-3)$

19. $4x^2 - 25 = (2x)^2 - 5^2 = (2x+5)(2x-5)$

20. $x^3 + 64 = x^3 + 4^3 = (x+4)(x^2 - 4x + 16)$

21. $3x^2y - 27y^3 = 3y(x^2 - 9y^2)$
$= 3y(x^2 - (3y)^2)$
$= 3y(x+3y)(x-3y)$

22. $6x^2 + 24 = 6(x^2 + 4)$

23. $16y^3 - 2 = 2(8y^3 - 1)$
$$= 2((2y)^3 - 1^3)$$
$$= 2(2y-1)(4y^2 + 2y + 1)$$

24. $x^2 y - 9y - 3x^2 + 27$
$$= y(x^2 - 9) - 3(x^2 - 9)$$
$$= (x^2 - 9)(y - 3)$$
$$= (x+3)(x-3)(y-3)$$

25. $3n(7n - 20) = 96$
$$21n^2 - 60n = 96$$
$$21n^2 - 60n - 96 = 0$$
$$3(7n^2 - 20n - 32) = 0$$
$$3(7n+8)(n-4) = 0$$
$$7n+8 = 0 \quad \text{or} \quad n-4 = 0$$
$$7n = -8 \ \text{or} \qquad n = 4$$
$$n = -\frac{8}{7}$$

The solutions are $-\frac{8}{7}, 4$.

26. $(x+2)(x-2) = 5(x+4)$
$$x^2 - 4 = 5x + 20$$
$$x^2 - 5x - 24 = 0$$
$$(x-8)(x+3) = 0$$
$$x-8 = 0 \quad \text{or} \quad x+3 = 0$$
$$x = 8 \ \text{or} \qquad x = -3$$
The solutions are $8, -3$.

27. $2x^3 + 5x^2 - 8x - 20 = 0$
$$x^2(2x+5) - 4(2x+5) = 0$$
$$(2x+5)(x^2 - 4) = 0$$
$$(2x+5)(x+2)(x-2) = 0$$
$$2x+5 = 0 \quad \text{or} \quad x+2 = 0 \quad \text{or} \quad x-2 = 0$$
$$2x = -5 \ \text{or} \qquad x = -2 \ \text{or} \qquad x = 2$$
$$x = -\frac{5}{2}$$

The solutions are $-\frac{5}{2}, -2, 2$.

28. Area $= x^2 - (2y)^2$
$$= (x+2y)(x-2y) \text{ square units}$$

29. $h(t) = -16t^2 + 96t + 880$

a. $-16(1)^2 + 96(1) + 880 = -16 + 96 + 880$
$$= 960 \text{ feet}$$

b. $-16(5.1)^2 + 96(5.1) + 880$
$$= -416.16 + 489.6 + 880$$
$$= 953.44 \text{ feet}$$

c. $\qquad 0 = -16t^2 + 96t + 880$
$$16t^2 - 96t - 880 = 0$$
$$16(t^2 - 6t - 55) = 0$$
$$(t-11)(t+5) = 0$$
$$t-11 = 0 \quad \text{or} \quad t+5 = 0$$
$$t = 11 \ \text{or} \qquad t = -5$$

Disregard the negative. The pebble will hit the ground in 11 seconds.

Chapter 5 Cumulative Review

1. a. $\sqrt[3]{27} = 3$

  b. $\sqrt[5]{1} = 1$

  c. $\sqrt[4]{16} = 2$

2. a. $\sqrt[3]{64} = 4$

  b. $\sqrt[4]{81} = 3$

  c. $\sqrt[5]{32} = 2$

3. $2(x-3) = 5x - 9$
$$2x - 6 = 5x - 9$$
$$-3x = -3$$
$$x = 1$$
The solution is 1.

4. $\qquad 0.3y + 2.4 = 0.1y + 4$
$$10(0.3y + 2.4) = 10(0.1y + 4)$$
$$3y + 24 = y + 40$$
$$2y = 16$$
$$y = 8$$
The solution is 8.

5. $A = 10,000\left(1 + \dfrac{0.05}{4}\right)^{4(3)}$

$= 10,000(1.0125)^{12}$

$= 10,000(1.160754518)$

$= 11,607.54518$

There will be $11,607.55 in the account.

6. The area of the room is
$2(14 \cdot 8) + 2(18 \cdot 8) = 512$ sq ft. Two
coats means $2 \cdot 512 = 1024$ sq ft of wall
needs paint.

$$\dfrac{1}{400} = \dfrac{x}{1024}$$

$$1024\left(\dfrac{1}{400}\right) = 1024\left(\dfrac{x}{1024}\right)$$

$$2.56 = x$$

$$x \approx 3$$

3 gallons of paint are needed.

7. a. $\dfrac{1}{4}x \le \dfrac{3}{8}$

$8\left(\dfrac{1}{4}x\right) \le 8\left(\dfrac{3}{8}\right)$

$2x \le 3$

$x \le \dfrac{3}{2}$

$\left\{x \,\middle|\, x \le \dfrac{3}{2}\right\}$

$-\dfrac{3}{2}$

b. $-2.3x < 6.9$

$10(-2.3x) < 10(6.9)$

$-23x < 69$

$\dfrac{-23x}{-23} > \dfrac{69}{-23}$

$x > -3$

$\{x \mid x > -3\}$

-3

8. $x + 2 \le \dfrac{1}{4}(x - 7)$

$4(x + 2) \le 4\left[\dfrac{1}{4}(x - 7)\right]$

$4x + 8 \le x - 7$

$3x \le -15$

$x \le -5$

$\{x \mid x \le -5\}$

-5

9. $-1 \le \dfrac{2x}{3} + 5 \le 2$

$3(-1) \le 3\left(\dfrac{2x}{3} + 5\right) \le 3(2)$

$-3 \le 2x + 15 \le 6$

$-18 \le 2x \le -9$

$-9 \le x \le -\dfrac{9}{2}$

$\left[-9, -\dfrac{9}{2}\right]$

10. $-\dfrac{1}{3} < \dfrac{3x + 1}{6} \le \dfrac{1}{3}$

$6\left(-\dfrac{1}{3}\right) < 6\left(\dfrac{3x + 1}{6}\right) \le 6\left(\dfrac{1}{3}\right)$

$-2 < 3x + 1 \le 2$

$-3 < 3x \le 1$

$-1 < x \le \dfrac{1}{3}$

$\left(-1, \dfrac{1}{3}\right]$

11. $|y| = 0$

$y = 0$

The solution is 0.

12. $8 + |4c| = 24$

$|4c| = 16$

$4c = 16$ or $4c = -16$

$c = 4$ or $c = -4$

The solutions are –4, 4.

13. $\left|2x-\dfrac{1}{10}\right| < -13$ is impossible.

$\varnothing$

14. $|5x-1|+9 > 5$

$\quad$ $|5x-1| > -4$ is always true.

$(-\infty, \infty)$

15. $y = \dfrac{1}{3}x$

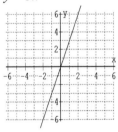

16. $y = 3x$

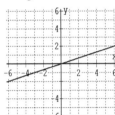

17. $x = y^2$ is not a function.

18. $f(x) = 3x^2 + 2x + 3$

$\quad f(-3) = 3(-3)^2 + 2(-3) + 3$

$\quad\quad\quad = 3(9) - 6 + 3$

$\quad\quad\quad = 27 - 6 + 3$

$\quad\quad\quad = 24$

19. $x = 2$

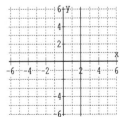

20. $y - 5 = 0$

$\quad y = 5$

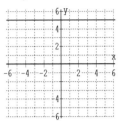

21. $y = 2$

$\quad m = 0$

22. $f(x) = -2x - 3$

$\quad m = -2$

23. $y = 3$

24. $x = -3$

25. $x + \dfrac{1}{2}y \ge -4 \qquad$ or $y \le -2$

$\quad\quad \dfrac{1}{2}y \ge -x - 4 \quad$ or $y \le -2$

$\quad\quad\quad y \ge -2x - 8 \;$ or $y \le -2$

Graph each inequality. The union of the two inequalities is both shaded regions, as shown by the shading in the graph below.

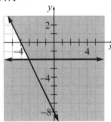

26. $y - 3 = 0[x - (-2)]$

$\quad y - 3 = 0$

$\quad\quad y = 3$

27. $\begin{cases} 2x + 4y = -6 & (1) \\ \quad\quad x = 2y - 5 & (2) \end{cases}$

Substitute $2y - 5$ for x in E1.

$$2(2y - 5) + 4y = -6$$
$$4y - 10 + 4y = -6$$
$$8y = 4$$
$$y = \frac{1}{2}$$

Replace $\frac{1}{2}$ for y in E2.

$$x = 2\left(\frac{1}{2}\right) - 5 = 1 - 5 = -4$$

The solution is $\left(-4, \frac{1}{2}\right)$.

28. $\begin{cases} 4x - 2y = 8 & (1) \\ y = 3x - 6 & (2) \end{cases}$

Substitute $3x - 6$ for y in E1.

$$4x - 2(3x - 6) = 8$$
$$4x - 6x + 12 = 8$$
$$-2x = -4$$
$$x = 2$$

Replace 2 for x in E2.

$$y = 3(2) - 6 = 6 - 6 = 0$$

The solution is $(2, 0)$.

29. $\begin{cases} 2x + 4y = 1 & (1) \\ 4x - 4z = -1 & (2) \\ y - 4z = -3 & (3) \end{cases}$

Multiply E2 by -1 and add to E3.

$$\begin{array}{r} -4x + 4z = 1 \\ y - 4z = -3 \\ \hline -4x + y = -2 \quad (4) \end{array}$$

Muliply E1 by 2 and add to E4.

$$\begin{array}{r} 4x + 8y = 2 \\ -4x + y = -2 \\ \hline 9y = 0 \\ y = 0 \end{array}$$

Replace y with 0 in E1.

$$2x + 4(0) = 1$$
$$2x = 1$$
$$x = \frac{1}{2}$$

Replace y with 0 in E3.

$$y - 4z = -3$$
$$-4z = -3$$
$$z = \frac{3}{4}$$

The solution is $\left(\frac{1}{2}, 0, \frac{3}{4}\right)$.

30. $\begin{cases} x + y - \dfrac{3}{2}z = \dfrac{1}{2} & (1) \\ -y - 2z = 14 & (2) \\ x - \dfrac{2}{3}y = -\dfrac{1}{3} & (3) \end{cases}$

Multiply E1 by 2 and E3 by 3 to clear fractions.

$$\begin{cases} 2x + 2y - 3z = 1 & (1) \\ -y - 2z = 14 & (2) \\ 3x - 2y = -1 & (3) \end{cases}$$

Add E1 and E3.

$$5x - 3z = 1 \quad (4)$$

Muliply E2 by 2 and add to E1.

$$\begin{array}{r} -2y - 4z = 28 \\ 2x + 2y - 3z = 1 \\ \hline 2x - 7z = 29 \quad (5) \end{array}$$

Solve the new system $\begin{cases} 5x - 3z = 0 & (4) \\ 2x - 7z = 29 & (5) \end{cases}$.

Multiply E4 by -2, multiply E5 by 5, and add.

$$\begin{array}{r} -10x + 6z = 0 \\ 10x - 35z = 145 \\ \hline -29z = 145 \end{array}$$

$$z = -5$$

Replace z with -5 in E4.

$$5x - 3(-5) = 0$$
$$5x + 15 = 0$$
$$5x = -15$$
$$x = -3$$

Replace z with 5 in E2.

$$-y - 2(-5) = 14$$
$$-y + 10 = 14$$
$$-y = 4$$
$$y = -4$$

The solution is $(-3, -4, -5)$.

31. Let x = the first number
and y = the second number.
$$\begin{cases} x = y - 4 & (1) \\ 4x = 2y + 6 & (2) \end{cases}$$
Multiply E1 by –4 and add to E2.
$$-4x = -4y + 16$$
$$\underline{4x = 2y + 6}$$
$$0 = -2y + 22$$
$$2y = 22$$
$$y = 11$$
Replace y with 11 in E1.
$$x = 11 - 4 = 7$$
The numbers are 7 and 11.

32. Let x = ounces of 20% solution
and y = ounces of 60% solution.
$$\begin{cases} x + y = 50 & (1) \\ 0.20x + 0.60y = 50(0.30) & (2) \end{cases}$$
Multiply E2 by 100 to clear decimals.
$$\begin{cases} x + y = 50 & (1) \\ 20x + 60y = 1500 & (2) \end{cases}$$
Multiply E1 by –20 and add to E2.
$$-20x - 20y = -1000$$
$$\underline{20x + 60y = 1500}$$
$$40y = 500$$
$$y = \frac{500}{40} = 12.5$$
Replace y with 12.5 in E1.
$$x + 12.5 = 50$$
$$x = 37.5$$
You should mix 37.5 ounces of the 20% solution and 12.5 ounces of the 60% solution.

33. $\begin{cases} 2x - y = 3 \\ 4x - 2y = 5 \end{cases}$
$$\begin{bmatrix} 2 & -1 & | & 3 \\ 4 & -2 & | & 5 \end{bmatrix}$$
Divide R1 by 2.
$$\begin{bmatrix} 1 & -\frac{1}{2} & | & \frac{3}{2} \\ 4 & -2 & | & 5 \end{bmatrix}$$

Multiply R1 by –4 and add to R2.
$$\begin{bmatrix} 1 & -\frac{1}{2} & | & \frac{3}{2} \\ 0 & 0 & | & 11 \end{bmatrix}$$
This corresponds to $\begin{cases} x - \frac{1}{2}y = \frac{3}{2} \\ 0 = 11. \end{cases}$
The last equation is impossible.
The system is inconsistent.
The solution is $\varnothing$.

34. $\begin{cases} 4y = 8 \\ x + y = 7 \end{cases}$
$$\begin{bmatrix} 0 & 4 & | & 8 \\ 1 & 1 & | & 7 \end{bmatrix}$$
Interchange R1 and R2.
$$\begin{bmatrix} 1 & 1 & | & 7 \\ 0 & 4 & | & 8 \end{bmatrix}$$
Divide R2 by 4.
$$\begin{bmatrix} 1 & 1 & | & 7 \\ 0 & 1 & | & 2 \end{bmatrix}$$
This corresponds to $\begin{cases} x + y = 7 \\ y = 2. \end{cases}$
Replace y with 2 in the equation
$x + y = 7$.
$$x + 2 = 7$$
$$x = 5$$
The solution is (5, 2).

35. $\begin{cases} x - 2y + z = 4 \\ 3x + y - 2z = 3 \\ 5x + 5y + 3z = -8 \end{cases}$
$$D = \begin{vmatrix} 1 & -2 & 1 \\ 3 & 1 & -2 \\ 5 & 5 & 3 \end{vmatrix}$$
$$= 1\begin{vmatrix} 1 & -2 \\ 5 & 3 \end{vmatrix} - (-2)\begin{vmatrix} 3 & -2 \\ 5 & 3 \end{vmatrix} + 1\begin{vmatrix} 3 & 1 \\ 5 & 5 \end{vmatrix}$$
$$= [3 - (-10)] + 2[9 - (-10)] + (15 - 5)$$
$$= 13 + 2(19) + 10$$
$$= 61$$

$$D_x = \begin{vmatrix} 4 & -2 & 1 \\ 3 & 1 & -2 \\ -8 & 5 & 3 \end{vmatrix}$$

$$= 4\begin{vmatrix} 1 & -2 \\ 5 & 3 \end{vmatrix} - (-2)\begin{vmatrix} 3 & -2 \\ -8 & 3 \end{vmatrix} + 1\begin{vmatrix} 3 & 1 \\ -8 & 5 \end{vmatrix}$$

$$= 4[3 - (-10)] + 2(9 - 16) + [15 - (-8)]$$

$$= 4(13) + 2(-7) + 23$$

$$= 52 - 14 + 23$$

$$= 61$$

$$D_y = \begin{vmatrix} 1 & 4 & 1 \\ 3 & 3 & -2 \\ 5 & -8 & 3 \end{vmatrix}$$

$$= 1\begin{vmatrix} 3 & -2 \\ -8 & 3 \end{vmatrix} - 4\begin{vmatrix} 3 & -2 \\ 5 & 3 \end{vmatrix} + 1\begin{vmatrix} 3 & 3 \\ 5 & -8 \end{vmatrix}$$

$$= (9 - 16) - 4[9 - (-10)] + (-24 - 15)$$

$$= -7 - 4(19) + (-39)$$

$$= -7 - 76 - 39$$

$$= -122$$

$$D_z = \begin{vmatrix} 1 & -2 & 4 \\ 3 & 1 & 3 \\ 5 & 5 & -8 \end{vmatrix}$$

$$= 1\begin{vmatrix} 1 & 3 \\ 5 & -8 \end{vmatrix} - (-2)\begin{vmatrix} 3 & 3 \\ 5 & -8 \end{vmatrix} + 4\begin{vmatrix} 3 & 1 \\ 5 & 5 \end{vmatrix}$$

$$= (-8 - 15) + 2(-24 - 15) + 4(15 - 5)$$

$$= -23 + 2(-39) + 4(10)$$

$$= -23 - 78 + 40$$

$$= -61$$

$$x = \frac{D_x}{D} = \frac{61}{61} = 1, \quad y = \frac{D_y}{D} = \frac{-122}{61} = -2,$$

$$z = \frac{D_z}{D} = \frac{-61}{61} = -1$$

The solution is $(1, -2, -1)$.

36. $\begin{cases} x + y + z = 0 \\ 2x - 3y + z = 5 \\ 2x + y + 2z = 2 \end{cases}$

$$D = \begin{vmatrix} 1 & 1 & 1 \\ 2 & -3 & 1 \\ 2 & 1 & 2 \end{vmatrix}$$

$$= 1\begin{vmatrix} -3 & 1 \\ 1 & 2 \end{vmatrix} - 1\begin{vmatrix} 2 & 1 \\ 2 & 2 \end{vmatrix} + 1\begin{vmatrix} 2 & -3 \\ 2 & 1 \end{vmatrix}$$

$$= (-6 - 1) - (4 - 2) + [2 - (-6)]$$

$$= -7 - 2 + 8$$

$$= -1$$

$$D_x = \begin{vmatrix} 0 & 1 & 1 \\ 5 & -3 & 1 \\ 2 & 1 & 2 \end{vmatrix}$$

$$= 0\begin{vmatrix} -3 & 1 \\ 1 & 2 \end{vmatrix} - 1\begin{vmatrix} 5 & 1 \\ 2 & 2 \end{vmatrix} + 1\begin{vmatrix} 5 & -3 \\ 2 & 1 \end{vmatrix}$$

$$= -(10 - 2) + [5 - (-6)]$$

$$= -8 + 11$$

$$= 3$$

$$D_y = \begin{vmatrix} 1 & 0 & 1 \\ 2 & 5 & 1 \\ 2 & 2 & 2 \end{vmatrix}$$

$$= 1\begin{vmatrix} 5 & 1 \\ 2 & 2 \end{vmatrix} - 0\begin{vmatrix} 2 & 1 \\ 2 & 2 \end{vmatrix} + 1\begin{vmatrix} 2 & 5 \\ 2 & 2 \end{vmatrix}$$

$$= (10 - 2) + (4 - 10)$$

$$= 8 + (-6)$$

$$= 2$$

$$D_z = \begin{vmatrix} 1 & 1 & 0 \\ 2 & -3 & 5 \\ 2 & 1 & 2 \end{vmatrix}$$

$$= 1\begin{vmatrix} -3 & 5 \\ 1 & 2 \end{vmatrix} - 1\begin{vmatrix} 2 & 5 \\ 2 & 2 \end{vmatrix} + 0\begin{vmatrix} 2 & -3 \\ 2 & 1 \end{vmatrix}$$

$$= (-6 - 5) - (4 - 10)$$

$$= -11 - (-6)$$

$$= -11 + 6$$

$$= -5$$

$$x = \frac{D_x}{D} = \frac{3}{-1} = -3, \quad y = \frac{D_y}{D} = \frac{2}{-1} = -2,$$

$$z = \frac{D_z}{D} = \frac{-5}{-1} = 5$$

The solution is $(-3, -2, 5)$.

37. a. $730,000 = 7.3 \times 10^5$

b. $0.00000104 = 1.04 \times 10^{-6}$

38. a. $8,250,000 = 8.25 \times 10^6$

b. $0.0000346 = 3.46 \times 10^{-5}$

39. a. $(2x^0 y^{-3})^{-2} = 2^{-2}(1)^2 (y^{-3})^{-2} = \dfrac{y^6}{2^2} = \dfrac{y^6}{4}$

b. $\left(\dfrac{x^{-5}}{x^{-2}}\right)^{-3} = \dfrac{(x^{-5})^{-3}}{(x^{-2})^{-3}} = \dfrac{x^{15}}{x^6} = x^{15-6} = x^9$

c. $\left(\dfrac{2}{7}\right)^{-2} = \dfrac{2^{-2}}{7^{-2}} = \dfrac{7^2}{2^2} = \dfrac{49}{4}$

d. $\dfrac{5^{-2} x^{-3} y^{11}}{x^2 y^{-5}} = \dfrac{x^{-3-2} y^{11-(-5)}}{5^2}$
$= \dfrac{x^{-5} y^{16}}{25}$
$= \dfrac{y^{16}}{25x^5}$

40. a. $(4a^{-1}b^0)^{-3} = 4^{-3}(a^{-1})^{-3}(1)^{-3} = \dfrac{a^3}{4^3} = \dfrac{a^3}{64}$

b. $\left(\dfrac{a^{-6}}{a^{-8}}\right)^{-2} = \dfrac{(a^{-6})^{-2}}{(a^{-8})^{-2}}$
$= \dfrac{a^{12}}{a^{16}} = a^{12-16} = a^{-4} = \dfrac{1}{a^4}$

c. $\left(\dfrac{2}{3}\right)^{-3} = \dfrac{2^{-3}}{3^{-3}} = \dfrac{3^3}{2^3} = \dfrac{27}{8}$

d. $\dfrac{3^{-2} a^{-2} b^{12}}{a^4 b^{-5}} = \dfrac{a^{-2-4} b^{12-(-5)}}{3^2}$
$= \dfrac{a^{-6} b^{17}}{9}$
$= \dfrac{b^{17}}{9a^6}$

41. The degree is the degree of the term $x^2 y^2$, which is $2 + 2 = 4$.

42. $(3x^2 - 2x) - (5x^2 + 3x)$
$= 3x^2 - 2x - 5x^2 - 3x$
$= -2x^2 - 5x$

43. a. $(2x^3)(5x^6) = 2(5)x^{3+6} = 10x^9$

b. $(7y^4 z^4)(-xy^{11}z^5) = -7xy^{4+11}z^{4+5}$
$= -7xy^{15}z^9$

44. a. $(3y^6)(4y^2) = 3(4)y^{6+2} = 12y^8$

b. $(6a^3 b^2)(-a^2 bc^4) = -6a^{3+2}b^{2+1}c^4$
$= -6a^5 b^3 c^4$

45. $17x^3 y^2 - 34x^4 y^2 = 17x^3 y^2 (1 - 2x)$

46. $12x^3 y - 3xy^3 = 3xy(4x^2 - y^2)$
$= 3xy((2x)^2 - y^2)$
$= 3xy(2x + y)(2x - y)$

47. $x^2 + 10x + 16 = (x + 8)(x + 2)$

48. $5a^2 + 14a - 3 = (5a - 1)(a + 3)$

49. $\quad 2x^2 + 9x - 5 = 0$
$\quad (2x - 1)(x + 5) = 0$
$\quad 2x - 1 = 0 \ \text{ or } \ x + 5 = 0$
$\qquad 2x = 1 \ \text{ or } \qquad x = -5$
$\qquad\quad x = \dfrac{1}{2}$

The solution is $-5, \dfrac{1}{2}$.

50. $\quad 3x^2 - 10x - 8 = 0$
$\quad (3x + 2)(x - 4) = 0$
$\quad 3x + 2 = 0 \ \text{ or } \ x - 4 = 0$
$\qquad 3x = -2 \ \text{ or } \qquad x = 4$
$\qquad\quad x = -\dfrac{2}{3}$

The solution is $-\dfrac{2}{3}, 4$.

Chapter 6

Section 6.1

Calculator Explorations

1. $f(x) = \dfrac{x+1}{x^2-4} = \dfrac{x+1}{(x+2)(x-2)}$

Domain: $\{x | x$ is a real number and $x \neq -2, x \neq 2\}$

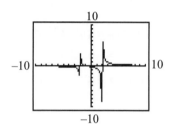

3. $h(x) = \dfrac{x^2}{2x^2+7x-4} = \dfrac{x^2}{(2x-1)(x+4)}$

Domain: $\left\{x | x \text{ is a real number and } x \neq -4, x \neq \dfrac{1}{2}\right\}$

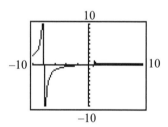

Exercise Set 6.1

1. $f(x) = \dfrac{x+8}{2x-1}$

$f(2) = \dfrac{2+8}{2(2)-1} = \dfrac{10}{4-1} = \dfrac{10}{3}$

$f(0) = \dfrac{0+8}{2(0)-1} = \dfrac{8}{-1} = -8$

$f(-1) = \dfrac{-1+8}{2(-1)-1} = \dfrac{7}{-3} = -\dfrac{7}{3}$

3. $g(x) = \dfrac{x^2+8}{x^3-25x}$

$g(3) = \dfrac{(3)^2+8}{(3)^3-25(3)}$

$= \dfrac{9+8}{27-75}$

$= \dfrac{17}{-48}$

$= -\dfrac{17}{48}$

$g(-2) = \dfrac{(-2)^2+8}{(-2)^3-25(-2)}$

$= \dfrac{4+8}{-8+50}$

$= \dfrac{12}{42}$

$= \dfrac{2}{7}$

$g(1) = \dfrac{(1)^2+8}{(1)^3-25(1)} = \dfrac{1+8}{1-25} = \dfrac{9}{-24} = -\dfrac{3}{8}$

5. $f(x) = \dfrac{5x-7}{4}$

Domain: $\{x \,|\, x$ is a real number$\}$

7. $s(t) = \dfrac{t^2+1}{2t}$

Undefined values when

$2t = 0$

$t = \dfrac{0}{2} = 0$

Domain: $\{t | t$ is a real number and $t \neq 0\}$

9. $f(x) = \dfrac{3x}{7-x}$

Undefined values when

$7 - x = 0$

$7 = x$

Domain: $\{x | x$ is a real number and $x \neq 7\}$

11. $f(x) = \dfrac{x}{3x-1}$

Undefined values when

$3x - 1 = 0$

$3x = 1$

$x = \dfrac{1}{3}$

Domain: $\left\{ x \middle| x \text{ is a real number and } x \neq \dfrac{1}{3} \right\}$

13. $C(x) = \dfrac{3+2x}{x^3 + x^2 - 2x}$

Undefined values when

$x^3 + x^2 - 2x = 0$

$x(x^2 + x - 2) = 0$

$x(x+2)(x-1) = 0$

$x = 0 \text{ or } x + 2 = 0 \quad \text{ or } x - 1 = 0$

$\qquad\qquad x = -2 \text{ or } \qquad x = 1$

Domain: $\{x | x \text{ is a real number}$

and $x \neq -2, x \neq 0, x \neq 1\}$

15. $C(x) = \dfrac{x+3}{x^2 - 4}$

Undefined values when

$x^2 - 4 = 0$

$(x+2)(x-2) = 0$

$x + 2 = 0 \quad \text{ or } x - 2 = 0$

$\quad x = -2 \text{ or } \qquad x = 2$

Domain: $\{x | x \text{ is a real number}$

and $x \neq -2, x \neq 2\}$

17. Answers may vary.

19. $\dfrac{4x-8}{3x-6} = \dfrac{4(x-2)}{3(x-2)} = \dfrac{4}{3}$

21. $\dfrac{2x-14}{7-x} = \dfrac{2(x-7)}{-1(x-7)} = \dfrac{2}{-1} = -2$

23. $\dfrac{x^2 - 2x - 3}{x^2 - 6x + 9} = \dfrac{(x-3)(x+1)}{(x-3)^2} = \dfrac{x+1}{x-3}$

25. $\dfrac{2x^2 + 12x + 18}{x^2 - 9} = \dfrac{2(x^2 + 6x + 9)}{x^2 - 3^2}$

$\qquad = \dfrac{2(x+3)^2}{(x+3)(x-3)}$

$\qquad = \dfrac{2(x+3)}{x-3}$

27. $\dfrac{3x+6}{x^2 + 2x} = \dfrac{3(x+2)}{x(x+2)} = \dfrac{3}{x}$

29. $\dfrac{2x^2 - x - 3}{2x^3 - 3x^2 + 2x - 3} = \dfrac{(2x-3)(x+1)}{x^2(2x-3) + 1(2x-3)}$

$\qquad = \dfrac{(2x-3)(x+1)}{(2x-3)(x^2+1)}$

$\qquad = \dfrac{x+1}{x^2 + 1}$

31. $\dfrac{8q^2}{16q^3 - 16q^2} = \dfrac{8q^2}{16q^2(q-1)} = \dfrac{1}{2(q-1)}$

33. $\dfrac{x^2 + 6x - 40}{10 + x} = \dfrac{(x+10)(x-4)}{x+10} = x - 4$

35. $\dfrac{x^3 - 125}{5 - x} = \dfrac{(x-5)(x^2 + 5x + 25)}{-1(x-5)}$

$\qquad = \dfrac{x^2 + 5x + 25}{-1}$

$\qquad = -x^2 - 5x - 25$

37. $\dfrac{8x^3 - 27}{4x - 6} = \dfrac{(2x)^3 - 3^3}{2(2x-3)}$

$\qquad = \dfrac{(2x-3)(4x^2 + 6x + 9)}{2(2x-3)}$

$\qquad = \dfrac{4x^2 + 6x + 9}{2}$

39. D

41. $\dfrac{3xy^3}{4x^3 y^2} \cdot \dfrac{-8x^3 y^4}{9x^4 y^7} = \dfrac{-24x^4 y^7}{36x^7 y^9} = -\dfrac{2}{3x^3 y^2}$

43.
$$\frac{8a}{3a^4b^2} \div \frac{4b^5}{6a^2b} = \frac{8a}{3a^4b^2} \cdot \frac{6a^2b}{4b^5}$$
$$= \frac{48a^3b}{12a^4b^7}$$
$$= \frac{4}{ab^6}$$

45.
$$\frac{a^2b}{a^2-b^2} \cdot \frac{a+b}{4a^3b} = \frac{a^2b(a+b)}{4a^3b(a+b)(a-b)}$$
$$= \frac{1}{4a(a-b)}$$

47.
$$\frac{x^2-9}{4} \div \frac{x^2-6x+9}{x^2-x-6}$$
$$= \frac{x^2-9}{4} \cdot \frac{x^2-x-6}{x^2-6x+9}$$
$$= \frac{(x+3)(x-3)}{4} \cdot \frac{(x-3)(x+2)}{(x-3)^2}$$
$$= \frac{(x+3)(x+2)}{4}$$

49.
$$\frac{9x+9}{4x+8} \cdot \frac{2x+4}{3x^2-3} = \frac{9(x+1)\cdot 2(x+2)}{4(x+2)\cdot 3(x^2-1)}$$
$$= \frac{18(x+1)(x+2)}{12(x+2)(x+1)(x-1)}$$
$$= \frac{3}{2(x-1)}$$

51.
$$\frac{a+b}{ab} \div \frac{a^2-b^2}{4a^3b} = \frac{a+b}{ab} \cdot \frac{4a^3b}{a^2-b^2}$$
$$= \frac{4a^3b(a+b)}{ab(a+b)(a-b)}$$
$$= \frac{4a^2}{a-b}$$

53.
$$\frac{2x^2-4x-30}{5x^2-40x-75} \div \frac{x^2-8x+15}{x^2-6x+9}$$
$$= \frac{2(x^2-2x-15)}{5(x^2-8x-15)} \cdot \frac{(x-3)^2}{(x-5)(x-3)}$$
$$= \frac{2(x-5)(x+3)(x-3)}{5(x^2-8x-15)(x-5)}$$
$$= \frac{2(x+3)(x-3)}{5(x^2-8x-15)}$$

55.
$$\frac{2x^3-16}{6x^2+6x-36} \cdot \frac{9x+18}{3x^2+6x+12}$$
$$= \frac{2(x^3-8)\cdot 9(x+2)}{6(x^2+x-6)\cdot 3(x^2+2x+4)}$$
$$= \frac{18(x-2)(x^2+2x+4)(x+2)}{18(x+3)(x-2)(x^2+2x+4)}$$
$$= \frac{x+2}{x+3}$$

57.
$$\frac{15b-3a}{b^2-a^2} \div \frac{a-5b}{ab+b^2}$$
$$= \frac{3(5b-a)}{(b+a)(b-a)} \cdot \frac{b(a+b)}{-(5b-a)}$$
$$= \frac{3b(5b-a)(a+b)}{-(b+a)(b-a)(5b-a)}$$
$$= \frac{3b}{-(b-a)}$$
$$= \frac{3b}{a-b}$$

59.
$$\frac{a^3+a^2b+a+b}{a^3+a} \cdot \frac{6a^2}{2a^2-2b^2}$$
$$= \frac{a^2(a+b)+1(a+b)}{a(a^2+1)} \cdot \frac{6a^2}{2(a^2-b^2)}$$
$$= \frac{6a^2(a+b)(a^2+1)}{2a(a^2+1)(a+b)(a-b)}$$
$$= \frac{3a}{a-b}$$

61.
$$\frac{5a}{12} \cdot \frac{2}{25a^2} \cdot \frac{15a}{2} = \frac{1}{6} \cdot \frac{1}{5a} \cdot \frac{15a}{2}$$
$$= \frac{1}{6} \cdot \frac{1}{1} \cdot \frac{3}{2} = \frac{3}{12} = \frac{1}{4}$$

63.
$$\frac{3x-x^2}{x^3-27} \div \frac{x}{x^2+3x+9}$$
$$= \frac{x(3-x)}{(x-3)(x^2+3x+9)} \cdot \frac{x^2+3x+9}{x}$$
$$= \frac{3-x}{x-3}$$
$$= \frac{-1(x-3)}{x-3}$$
$$= -1$$

65. $\dfrac{4a}{7} \div \left(\dfrac{a^2}{14} \cdot \dfrac{3}{a} \right) = \dfrac{4a}{7} \div \dfrac{3a}{14}$

$ = \dfrac{4a}{7} \cdot \dfrac{14}{3a}$

$ = \dfrac{4}{1} \cdot \dfrac{2}{3}$

$ = \dfrac{8}{3}$

67. $\dfrac{8b+24}{3a+6} \div \dfrac{ab-2b+3a-6}{a^2-4a+4}$

$= \dfrac{8(b+3)}{3(a+2)} \cdot \dfrac{(a-2)^2}{b(a-2)+3(a-2)}$

$= \dfrac{8(b+3)}{3(a+2)} \cdot \dfrac{(a-2)^2}{(a-2)(b+3)}$

$= \dfrac{8(a-2)}{3(a+2)}$

69. $\dfrac{4}{x} \div \dfrac{3xy}{x^2} \cdot \dfrac{6x^2}{x^4} = \dfrac{4}{x} \cdot \dfrac{x^2}{3xy} \cdot \dfrac{6x^2}{x^4}$

$ = \dfrac{24x^4}{3x^6 y}$

$ = \dfrac{8}{x^2 y}$

71. $\dfrac{3x^2-5x-2}{y^2+y-2} \cdot \dfrac{y^2+4y-5}{12x^2+7x+1} \div \dfrac{5x^2-9x-2}{8x^2-2x-1}$

$= \dfrac{(3x+1)(x-2)(y+5)(y+1)(4x+1)(2x-1)}{(y+2)(y-1)(4x+1)(3x+1)(5x+1)(x-2)}$

$= \dfrac{(y+5)(2x-1)}{(y+2)(5x+1)}$

73. $\dfrac{5a^2-20}{3a^2-12a} \div \dfrac{a^3+2a^2}{2a^2-8a} \cdot \dfrac{9a^3+6a^2}{2a^2-4a}$

$= \dfrac{5(a^2-4)}{3a(a-4)} \cdot \dfrac{2a(a-4)}{a^2(a+2)} \cdot \dfrac{3a^2(3a+2)}{2a(a-2)}$

$= \dfrac{5(a+2)(a-2)(3a+2)}{a(a+2)(a-2)}$

$= \dfrac{5(3a+2)}{a}$

75. $\dfrac{5x^4+3x^2-2}{x-1} \cdot \dfrac{x+1}{x^4-1}$

$= \dfrac{(5x^2-2)(x^2+1)}{x-1} \cdot \dfrac{x+1}{(x^2+1)(x^2-1)}$

$= \dfrac{(5x^2-2)(x^2+1)}{x-1} \cdot \dfrac{x+1}{(x^2+1)(x+1)(x-1)}$

$= \dfrac{5x^2-2}{(x-1)^2}$

77. $\dfrac{4}{5} + \dfrac{3}{5} = \dfrac{4+3}{5} = \dfrac{7}{5}$

79. $\dfrac{5}{28} - \dfrac{2}{21}$ The LCD = 84.

$\dfrac{5}{28} \cdot \dfrac{3}{3} - \dfrac{2}{21} \cdot \dfrac{4}{4} = \dfrac{15}{84} - \dfrac{8}{84}$

$ = \dfrac{7}{84}$

$ = \dfrac{7 \cdot 1}{7 \cdot 12}$

$ = \dfrac{1}{12}$

81. $\dfrac{3}{8} + \dfrac{1}{2} - \dfrac{3}{16}$

The LCD = 16.

$\dfrac{3}{8} \cdot \dfrac{2}{2} + \dfrac{1}{2} \cdot \dfrac{8}{8} - \dfrac{3}{16} = \dfrac{6}{16} + \dfrac{8}{16} - \dfrac{3}{16}$

$ = \dfrac{6+8-3}{16}$

$ = \dfrac{11}{16}$

83. $A = l \cdot w$

$ = \dfrac{x+2}{x} \cdot \dfrac{5x}{x^2-4}$

$ = \dfrac{x+2}{x} \cdot \dfrac{5x}{(x+2)(x-2)}$

$ = \dfrac{5}{x-2}$ square meters

85. $f(x) = \dfrac{100,000x}{100-x}$

     a. $\{x \mid 0 \le x < 100\}$

b. $f(30) = \dfrac{100,000(30)}{100-30} = \$42,857.14$

c. $f(60) = \dfrac{100,000(60)}{100-60} = \$150,000$

$f(80) = \dfrac{100,000(80)}{100-80} = \$400,000$

d. $f(90) = \dfrac{100,000(90)}{100-90} = \$900,000$

$f(95) = \dfrac{100,000(95)}{100-95} = \$1,900,000$

$f(99) = \dfrac{100,000(99)}{100-99} = \$9,900,000$

87. Answers may vary.

89. Since $A = b \cdot h$, we have $b = \dfrac{A}{h}$. Now

$$b = \dfrac{\dfrac{x^2+x-2}{x^3}}{\dfrac{x^2}{x-1}} = \dfrac{(x+2)(x-1)}{x^3} \cdot \dfrac{x-1}{x^2}$$

$$= \dfrac{(x+2)(x-1)^2}{x^5} \text{ feet}$$

91. $f(x) = \dfrac{20x}{100-x}$

x	0	10	30	50	70	90	95	99
y	0	$\dfrac{20}{9}$	$\dfrac{60}{7}$	20	$\dfrac{140}{3}$	180	380	1980

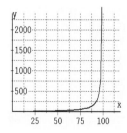

93. $\dfrac{x^{2n}-4}{7x} \cdot \dfrac{14x^3}{x^n-2} = \dfrac{(x^n+2)(x^n-2) \cdot 2x^2}{x^n-2}$
$= 2x^2(x^n+2)$

95. $\dfrac{y^{2n}+9}{10y} \cdot \dfrac{y^n-3}{y^{4n}-81}$

$= \dfrac{y^{2n}+9}{10y} \cdot \dfrac{y^n-3}{(y^{2n}+9)(y^{2n}-9)}$

$= \dfrac{y^n-3}{10y(y^n+3)(y^n-3)}$

$= \dfrac{1}{10y(y^n+3)}$

97. $\dfrac{y^{2n}-y^n-2}{2y^n-4} \div \dfrac{y^{2n}-1}{1+y^n}$

$= \dfrac{(y^n-2)(y^n+1)}{2(y^n-2)} \cdot \dfrac{1+y^n}{(y^n+1)(y^n-1)}$

$= \dfrac{1+y^n}{2(y^n-1)}$

Section 6.2

Graphing Calculator Explorations

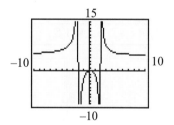

Mental Math 6.2

1. A, B

2. D

3. C

4. A, C

5. $\dfrac{5}{y} + \dfrac{7}{y} = \dfrac{5+7}{y} = \dfrac{12}{y}$

6. $\dfrac{5}{y} - \dfrac{7}{y} = \dfrac{5-7}{y} = -\dfrac{2}{y}$

7. $\dfrac{5}{y} \cdot \dfrac{7}{y} = \dfrac{5 \cdot 7}{y \cdot y} = \dfrac{35}{y^2}$

8. $\dfrac{5}{y} \div \dfrac{7}{y} = \dfrac{5}{y} \cdot \dfrac{y}{7} = \dfrac{5}{7}$

Exercise Set 6.2

1. $\dfrac{2}{x} - \dfrac{5}{x} = \dfrac{2-5}{x} = -\dfrac{3}{x}$

3. $\dfrac{2}{x-2} + \dfrac{x}{x-2} = \dfrac{2+x}{x-2} = \dfrac{x+2}{x-2}$

5. $\dfrac{x^2}{x+2} - \dfrac{4}{x+2} = \dfrac{x^2-4}{x+2}$

$\qquad = \dfrac{(x+2)(x-2)}{x+2}$

$\qquad = x-2$

7. $\dfrac{2x-6}{x^2+x-6} + \dfrac{3-3x}{x^2+x-6} = \dfrac{-x-3}{x^2+x-6}$

$\qquad = \dfrac{-(x+3)}{(x+3)(x-2)}$

$\qquad = \dfrac{-1}{x-2}$

$\qquad = \dfrac{1}{2-x}$

9. $P = 4 \cdot s = 4\left(\dfrac{x}{x+5}\right) = \dfrac{4x}{x+5}$ feet

$A = s^2 = \left(\dfrac{x}{x+5}\right)^2$

$\qquad = \dfrac{x^2}{(x+5)^2} = \dfrac{x^2}{x^2+10x+25}$ sq. feet

11. $7 = 7$

$5x = 5 \cdot x$

$\text{LCD} = 7 \cdot 5 \cdot x = 35x$

13. $x = x$

$x+1 = x+1$

$\text{LCD} = x(x+1)$

15. $x+7 = x+7$

$x-7 = x-7$

$\text{LCD} = (x+7)(x-7)$

17. $3x+6 = 3(x+2)$

$2x-4 = 2(x-2)$

$\text{LCD} = 2 \cdot 3 \cdot (x+2)(x-2)$

$\qquad = 6(x+2)(x-2)$

19. $(3x-1)(x+2) = (3x-1) \cdot (x+2)$

$3x-1 = 3x-1$

$\text{LCD} = (3x-1)(x+2)$

21. $a^2 - b^2 = (a+b)(a-b)$

$a^2 - 2ab + b^2 = (a-b)^2$

$\text{LCD} = (a+b)(a-b)^2$

23. $x^2 - 9 = (x+3)(x-3)$

$x = x$

$12 - 4x = 4(3-x) = -4(x-3)$

$\text{LCD} = -4x(x+3)(x-3)$

25. Answers may vary.

27. $\dfrac{4}{3x} + \dfrac{3}{2x} = \dfrac{4 \cdot 2}{3x \cdot 2} + \dfrac{3 \cdot 3}{2x \cdot 3} = \dfrac{8+9}{6x} = \dfrac{17}{6x}$

29. $\dfrac{5}{2y^2} - \dfrac{2}{7y} = \dfrac{5 \cdot 7}{2y^2 \cdot 7} - \dfrac{2 \cdot 2y}{7y \cdot 2y} = \dfrac{35-4y}{14y^2}$

31. $\dfrac{x-3}{x+4} - \dfrac{x+2}{x-4}$

$\qquad = \dfrac{(x-3)(x-4)}{(x+4)(x-4)} - \dfrac{(x+2)(x+4)}{(x-4)(x+4)}$

$\qquad = \dfrac{(x^2-7x+12)-(x^2+6x+8)}{(x+4)(x-4)}$

$\qquad = \dfrac{-13x+4}{(x+4)(x-4)}$

33. $\dfrac{1}{x-5}+\dfrac{x}{x^2-x-20}$

$=\dfrac{1}{x-5}+\dfrac{x}{(x-5)(x+4)}$

$=\dfrac{1(x+4)}{(x-5)(x+4)}+\dfrac{x}{(x-5)(x+4)}$

$=\dfrac{x+4+x}{(x-5)(x+4)}$

$=\dfrac{2x+4}{(x-5)(x+4)}$

35. $\dfrac{1}{a-b}+\dfrac{1}{b-a}=\dfrac{1}{a-b}+\dfrac{1}{-(a-b)}$

$\qquad\qquad\qquad =\dfrac{1}{a-b}-\dfrac{1}{a-b}$

$\qquad\qquad\qquad =0$

37. $\dfrac{x+1}{1-x}+\dfrac{1}{x-1}=\dfrac{x+1}{-(x-1)}+\dfrac{1}{x-1}$

$\qquad\qquad\qquad =\dfrac{-(x+1)+1}{x-1}$

$\qquad\qquad\qquad =\dfrac{-x}{x-1}$

$\qquad\qquad\qquad =-\dfrac{x}{x-1}$

39. $\dfrac{5}{x-2}+\dfrac{x+4}{2-x}=\dfrac{5}{x-2}+\dfrac{x+4}{-(x-2)}$

$\qquad\qquad\qquad =\dfrac{5}{x-2}-\dfrac{x+4}{x-2}$

$\qquad\qquad\qquad =\dfrac{5-(x+4)}{x-2}$

$\qquad\qquad\qquad =\dfrac{5-x-4}{x-2}$

$\qquad\qquad\qquad =\dfrac{-x+1}{x-2}$

41. $\dfrac{y+1}{y^2-6y+8}-\dfrac{3}{y^2-16}$

$=\dfrac{y+1}{(y-4)(y-2)}-\dfrac{3}{(y+4)(y-4)}$

$=\dfrac{(y+1)(y+4)}{(y-4)(y-2)(y+4)}$

$\qquad -\dfrac{3(y-2)}{(y+4)(y-4)(y-2)}$

$=\dfrac{(y^2+5y+4)-(3y-6)}{(y-4)(y+4)(y-2)}$

$=\dfrac{y^2+5y+4-3y+6}{(y-4)(y+4)(y-2)}$

$=\dfrac{y^2+2y+10}{(y-4)(y+4)(y-2)}$

43. $\dfrac{x+4}{3x^2+11x+6}+\dfrac{x}{2x^2+x-15}$

$=\dfrac{x+4}{(3x+2)(x+3)}+\dfrac{x}{(2x-5)(x+3)}$

$=\dfrac{(x+4)(2x-5)}{(3x+2)(x+3)(2x-5)}$

$\qquad +\dfrac{x(3x+2)}{(2x-5)(x+3)(3x+2)}$

$=\dfrac{2x^2+3x-20+(3x^2+2x)}{(3x+2)(x+3)(2x-5)}$

$=\dfrac{5x^2+5x-20}{(3x+2)(x+3)(2x-5)}$

$=\dfrac{5(x^2+x-4)}{(3x+2)(x+3)(2x-5)}$

45. $\dfrac{7}{x^2-x-2}+\dfrac{x}{x^2+4x+3}$

$=\dfrac{7}{(x-2)(x+1)}+\dfrac{x}{(x+3)(x+1)}$

$=\dfrac{7(x+3)+x(x-2)}{(x-2)(x+1)(x+3)}$

$=\dfrac{7x+21+x^2-2x}{(x-2)(x+1)(x+3)}$

$=\dfrac{x^2+5x+21}{(x-2)(x+1)(x+3)}$

47. $\dfrac{2}{x+1} - \dfrac{3x}{3x+3} + \dfrac{1}{2x+2}$

$= \dfrac{2}{x+1} - \dfrac{3x}{3(x+1)} + \dfrac{1}{2(x+1)}$

$= \dfrac{2}{x+1} - \dfrac{x}{x+1} + \dfrac{1}{2(x+1)}$

$= \dfrac{2 \cdot 2}{(x+1) \cdot 2} - \dfrac{x \cdot 2}{(x+1) \cdot 2} + \dfrac{1}{2(x+1)}$

$= \dfrac{4 - 2x + 1}{2(x+1)}$

$= \dfrac{-2x + 5}{2(x+1)}$

49. $\dfrac{3}{x+3} + \dfrac{5}{x^2 + 6x + 9} - \dfrac{x}{x^2 - 9}$

$= \dfrac{3}{x+3} + \dfrac{5}{(x+3)^2} - \dfrac{x}{(x+3)(x-3)}$

$= \dfrac{3(x+3)(x-3)}{(x+3)^2(x-3)} + \dfrac{5(x-3)}{(x+3)^2(x-3)}$

$\qquad\qquad - \dfrac{x(x+3)}{(x+3)^2(x-3)}$

$= \dfrac{3(x^2 - 9) + 5x - 15 - (x^2 + 3x)}{(x+3)^2(x-3)}$

$= \dfrac{3x^2 - 27 + 5x - 15 - x^2 - 3x}{(x+3)^2(x-3)}$

$= \dfrac{2x^2 + 2x - 42}{(x+3)^2(x-3)}$

$= \dfrac{2(x^2 + x - 21)}{(x+3)^2(x-3)}$

51. $\dfrac{4}{3x^2y^3} + \dfrac{5}{3x^2y^3} = \dfrac{4+5}{3x^2y^3} = \dfrac{9}{3x^2y^3} = \dfrac{3}{x^2y^3}$

53. $\dfrac{x-5}{2x} - \dfrac{x+5}{2x} = \dfrac{x-5-(x+5)}{2x}$

$\qquad\qquad = \dfrac{x-5-x-5}{2x}$

$\qquad\qquad = \dfrac{-10}{2x}$

$\qquad\qquad = -\dfrac{5}{x}$

55. $\dfrac{3}{2x+10} + \dfrac{8}{3x+15}$

$= \dfrac{3}{2(x+5)} + \dfrac{8}{3(x+5)}$

$= \dfrac{3 \cdot 3}{2(x+5) \cdot 3} + \dfrac{8 \cdot 2}{3(x+5) \cdot 2}$

$= \dfrac{9+16}{6(x+5)}$

$= \dfrac{25}{6(x+5)}$

57. $\dfrac{-2}{x^2 - 3x} - \dfrac{1}{x^3 - 3x^2}$

$= \dfrac{-2}{x(x-3)} - \dfrac{1}{x^2(x-3)}$

$= \dfrac{-2 \cdot x}{x(x-3) \cdot x} - \dfrac{1}{x^2(x-3)}$

$= \dfrac{-2x - 1}{x^2(x-3)}$

59. $\dfrac{ab}{a^2 - b^2} + \dfrac{b}{a+b}$

$= \dfrac{ab}{(a+b)(a-b)} + \dfrac{b}{a+b}$

$= \dfrac{ab}{(a+b)(a-b)} + \dfrac{b(a-b)}{(a+b)(a-b)}$

$= \dfrac{ab + ba - b^2}{(a+b)(a-b)}$

$= \dfrac{2ab - b^2}{(a+b)(a-b)}$

61. $\dfrac{5}{x^2 - 4} - \dfrac{3}{x^2 + 4x + 4}$

$= \dfrac{5}{(x+2)(x-2)} - \dfrac{3}{(x+2)^2}$

$= \dfrac{5(x+2)}{(x+2)^2(x-2)} - \dfrac{3(x-2)}{(x+2)^2(x-2)}$

$= \dfrac{5x + 10 - 3x + 6}{(x+2)^2(x-2)}$

$= \dfrac{2x + 16}{(x+2)^2(x-2)}$

63. $\dfrac{2}{a^2+2a+1}+\dfrac{3}{a^2-1}$

$=\dfrac{2}{(a+1)^2}+\dfrac{3}{(a+1)(a-1)}$

$=\dfrac{2(a-1)}{(a+1)^2(a-1)}+\dfrac{3(a+1)}{(a+1)^2(a-1)}$

$=\dfrac{2a-2+3a+3}{(a+1)^2(a-1)}$

$=\dfrac{5a+1}{(a+1)^2(a-1)}$

65. Answers may vary.

67. Answers may vary.

69. $\left(\dfrac{2}{3}-\dfrac{1}{x}\right)\cdot\left(\dfrac{3}{x}+\dfrac{1}{2}\right)=\left(\dfrac{2x}{3x}-\dfrac{3}{3x}\right)\cdot\left(\dfrac{6}{2x}+\dfrac{x}{2x}\right)$

$=\dfrac{2x-3}{3x}\cdot\dfrac{x+6}{2x}$

$=\dfrac{2x^2+9x-18}{6x^2}$

71. $\left(\dfrac{1}{x}+\dfrac{2}{3}\right)-\left(\dfrac{1}{x}-\dfrac{2}{3}\right)=\dfrac{1}{x}+\dfrac{2}{3}-\dfrac{1}{x}+\dfrac{2}{3}=\dfrac{4}{3}$

73. $\left(\dfrac{2a}{3}\right)^2\div\left(\dfrac{a^2}{a+1}-\dfrac{1}{a+1}\right)$

$=\dfrac{4a^2}{9}\div\dfrac{a^2-1}{a+1}$

$=\dfrac{4a^2}{9}\cdot\dfrac{a+1}{(a+1)(a-1)}$

$=\dfrac{4a^2}{9(a-1)}$

75. $\left(\dfrac{2x}{3}\right)^2\div\left(\dfrac{x}{3}\right)^2=\dfrac{4x^2}{9}\div\dfrac{x^2}{9}=\dfrac{4x^2}{9}\cdot\dfrac{9}{x^2}=4$

77. $\dfrac{x}{x^2-9}+\dfrac{3}{x^2-6x+9}-\dfrac{1}{x+3}$

$=\dfrac{x}{(x+3)(x-3)}+\dfrac{3}{(x-3)^2}-\dfrac{1}{x+3}$

$=\dfrac{x(x-3)+3(x+3)-1(x-3)^2}{(x+3)(x-3)^2}$

$=\dfrac{x^2-3x+3x+9-(x^2-6x+9)}{(x+3)(x-3)^2}$

$=\dfrac{6x}{(x+3)(x-3)^2}$

79. $\left(\dfrac{x}{x+1}-\dfrac{x}{x-1}\right)\div\dfrac{x}{2x+2}$

$=\dfrac{x(x-1)-x(x+1)}{(x+1)(x-1)}\cdot\dfrac{2(x+1)}{x}$

$=\dfrac{x^2-x-x^2-x}{x-1}\cdot\dfrac{2}{x}$

$=\dfrac{-2x}{x-1}\cdot\dfrac{2}{x}$

$=-\dfrac{4}{x-1}$

81. $\dfrac{4}{x}\cdot\left(\dfrac{2}{x+2}-\dfrac{2}{x-2}\right)=\dfrac{4}{x}\cdot\dfrac{2(x-2)-2(x+2)}{(x+2)(x-2)}$

$=\dfrac{4}{x}\cdot\dfrac{2x-4-2x-4}{(x+2)(x-2)}$

$=\dfrac{4}{x}\cdot\dfrac{-8}{(x+2)(x-2)}$

$=-\dfrac{32}{x(x+2)(x-2)}$

83. $12\left(\dfrac{2}{3}+\dfrac{1}{6}\right)=\dfrac{24}{3}+\dfrac{12}{6}=8+2=10$

85. $x^2\left(\dfrac{4}{x^2}+1\right)=x^2\cdot\dfrac{4}{x^2}+x^2\cdot1=4+x^2$

87. $\sqrt{100}=10$

89. $\sqrt[3]{8}=2$

91. $\sqrt[4]{81}=3$

93. $a^2+b^2=c^2$

$3^2+4^2=c^2$

$9+16=c^2$

$25=c^2$

$c=5$ meters

95. $x^{-1} + (2x)^{-1} = \dfrac{1}{x} + \dfrac{1}{2x} = \dfrac{2}{2x} + \dfrac{1}{2x} = \dfrac{3}{2x}$

97. $4x^{-2} - 3x^{-1} = \dfrac{4}{x^2} - \dfrac{3}{x} = \dfrac{4}{x^2} - \dfrac{3x}{x^2} = \dfrac{4-3x}{x^2}$

99. $x^{-3}(2x+1) - 5x^{-2} = \dfrac{2x+1}{x^3} - \dfrac{5}{x^2}$

$$= \dfrac{2x+1}{x^3} - \dfrac{5x}{x^3}$$

$$= \dfrac{2x+1-5x}{x^3}$$

$$= \dfrac{1-3x}{x^3}$$

101. $\dfrac{2}{x-2} + \dfrac{x}{x-2} = \dfrac{x+2}{x-2}$

Let $y_1 = \dfrac{2}{x-2} + \dfrac{x}{x-2}$ and $y_2 = \dfrac{x+2}{x-2}$.
Graph both equations and trace to see that the graphs coincide.

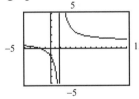

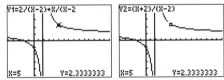

103. $\dfrac{x-3}{x+4} - \dfrac{x+2}{x-4} = \dfrac{-13x+4}{(x+4)(x-4)}$

Let $y_1 = \dfrac{x-3}{x+4} - \dfrac{x+2}{x-4}$ and

$y_2 = \dfrac{-13x+4}{(x+4)(x-4)}$.

Graph both equations using a standard viewing window and trace to see that

the graphs coincide.

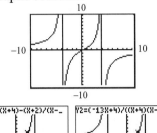

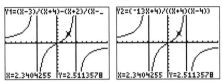

Exercise Set 6.3

1. $\dfrac{\frac{10}{3x}}{\frac{5}{6x}} = \dfrac{10}{3x} \cdot \dfrac{6x}{5} = \dfrac{2}{1} \cdot \dfrac{2}{1} = 4$

3. $\dfrac{1+\frac{2}{5}}{2+\frac{3}{5}} = \dfrac{\left(1+\frac{2}{5}\right)5}{\left(2+\frac{3}{5}\right)5} = \dfrac{5+2}{10+3} = \dfrac{7}{13}$

5. $\dfrac{\frac{4}{x-1}}{\frac{x}{x-1}} = \dfrac{4}{x-1} \cdot \dfrac{x-1}{x} = \dfrac{4}{x}$

7. $\dfrac{1-\frac{2}{x}}{x+\frac{4}{9x}} = \dfrac{\left(1-\frac{2}{x}\right)9x}{\left(x+\frac{4}{9x}\right)9x} = \dfrac{9x-18}{9x^2+4} = \dfrac{9(x-2)}{9x^2+4}$

9. $\dfrac{\frac{4x^2-y^2}{xy}}{\frac{2}{y}-\frac{1}{x}} = \dfrac{\left(\frac{4x^2-y^2}{xy}\right)xy}{\left(\frac{2}{y}-\frac{1}{x}\right)xy}$

$$= \dfrac{4x^2-y^2}{2x-y}$$

$$= \dfrac{(2x+y)(2x-y)}{2x-y}$$

$$= 2x+y$$

11. $\dfrac{\dfrac{x+1}{3}}{\dfrac{2x-1}{6}} = \dfrac{x+1}{3} \cdot \dfrac{6}{2x-1}$

$\qquad = \dfrac{x+1}{1} \cdot \dfrac{2}{2x-1}$

$\qquad = \dfrac{2(x+1)}{2x-1}$

13. $\dfrac{\dfrac{2}{x}+\dfrac{3}{x^2}}{\dfrac{4}{x^2}-\dfrac{9}{x}} = \dfrac{\left(\dfrac{2}{x}+\dfrac{3}{x^2}\right)x^2}{\left(\dfrac{4}{x^2}-\dfrac{9}{x}\right)x^2} = \dfrac{2x+3}{4-9x}$

15. $\dfrac{\dfrac{1}{x}+\dfrac{2}{x^2}}{x+\dfrac{8}{x^2}} = \dfrac{\left(\dfrac{1}{x}+\dfrac{2}{x^2}\right)x^2}{\left(x+\dfrac{8}{x^2}\right)x^2}$

$\qquad = \dfrac{x+2}{x^3+8}$

$\qquad = \dfrac{x+2}{(x+2)(x^2-2x+4)}$

$\qquad = \dfrac{1}{x^2-2x+4}$

17. $\dfrac{\dfrac{4}{5-x}+\dfrac{5}{x-5}}{\dfrac{2}{x}+\dfrac{3}{x-5}} = \dfrac{-\dfrac{4}{x-5}+\dfrac{5}{x-5}}{\dfrac{2(x-5)+3x}{x(x-5)}}$

$\qquad = \dfrac{\dfrac{1}{x-5}}{\dfrac{2x-10+3x}{x(x-5)}}$

$\qquad = \dfrac{1}{x-5} \cdot \dfrac{x(x-5)}{5x-10}$

$\qquad = \dfrac{x}{5x-10}$

19. $\dfrac{\dfrac{x+2}{x}-\dfrac{2}{x-1}}{\dfrac{x+1}{x}+\dfrac{x+1}{x-1}} = \dfrac{\left(\dfrac{x+2}{x}-\dfrac{2}{x-1}\right)x(x-1)}{\left(\dfrac{x+1}{x}+\dfrac{x+1}{x-1}\right)x(x-1)}$

$\qquad = \dfrac{(x+2)(x-1)-2x}{(x+1)(x-1)+x(x+1)}$

$\qquad = \dfrac{x^2+x-2-2x}{x^2-1+x^2+x}$

$\qquad = \dfrac{x^2-x-2}{2x^2+x-1}$

$\qquad = \dfrac{(x-2)(x+1)}{(2x-1)(x+1)} = \dfrac{x-2}{2x-1}$

21. $\dfrac{\dfrac{2}{x}+3}{\dfrac{4}{x^2}-9} = \dfrac{\left(\dfrac{2}{x}+3\right)x^2}{\left(\dfrac{4}{x^2}-9\right)x^2}$

$\qquad = \dfrac{2x+3x^2}{4-9x^2}$

$\qquad = \dfrac{x(2+3x)}{(2+3x)(2-3x)}$

$\qquad = \dfrac{x}{2-3x}$

23. $\dfrac{1-\dfrac{x}{y}}{\dfrac{x^2}{y^2}-1} = \dfrac{\left(1-\dfrac{x}{y}\right)y^2}{\left(\dfrac{x^2}{y^2}-1\right)y^2}$

$\qquad = \dfrac{y^2-xy}{x^2-y^2}$

$\qquad = \dfrac{y(y-x)}{(x+y)(x-y)}$

$\qquad = \dfrac{-y(x-y)}{(x+y)(x-y)} = -\dfrac{y}{x+y}$

25. $\dfrac{\dfrac{-2x}{x-y}}{\dfrac{y}{x^2}} = \dfrac{-2x}{x-y} \cdot \dfrac{x^2}{y} = -\dfrac{2x^3}{y(x-y)}$

27. $\dfrac{\dfrac{2}{x}+\dfrac{1}{x^2}}{\dfrac{y}{x^2}} = \dfrac{\left(\dfrac{2}{x}+\dfrac{1}{x^2}\right)x^2}{\left(\dfrac{y}{x^2}\right)x^2} = \dfrac{2x+1}{y}$

29. $\dfrac{\dfrac{x}{9}-\dfrac{1}{x}}{1+\dfrac{3}{x}} = \dfrac{\left(\dfrac{x}{9}-\dfrac{1}{x}\right)9x}{\left(1+\dfrac{3}{x}\right)9x}$

$= \dfrac{x^2-9}{9x+27}$

$= \dfrac{(x+3)(x-3)}{9(x+3)}$

$= \dfrac{x-3}{9}$

31. $\dfrac{\dfrac{x-1}{x^2-4}}{1+\dfrac{1}{x-2}} = \dfrac{\dfrac{x-1}{x^2-4}}{\dfrac{(x-2)+1}{x-2}}$

$= \dfrac{x-1}{(x+2)(x-2)} \cdot \dfrac{x-2}{x-1}$

$= \dfrac{1}{x+2}$

33. $\dfrac{x^{-1}}{x^{-2}+y^{-2}} = \dfrac{\left(\dfrac{1}{x}\right)x^2y^2}{\left(\dfrac{1}{x^2}+\dfrac{1}{y^2}\right)x^2y^2} = \dfrac{xy^2}{y^2+x^2}$

35. $\dfrac{2a^{-1}+3b^{-2}}{a^{-1}-b^{-1}} = \dfrac{\left(\dfrac{2}{a}+\dfrac{3}{b^2}\right)ab^2}{\left(\dfrac{1}{a}-\dfrac{1}{b}\right)ab^2}$

$= \dfrac{2b^2+3a}{b^2-ab}$

$= \dfrac{2b^2+3a}{b(b-a)}$

37. $\dfrac{1}{x-x^{-1}} = \dfrac{1\cdot x}{\left(x-\dfrac{1}{x}\right)x}$

$= \dfrac{x}{x^2-1}$

$= \dfrac{x}{(x+1)(x-1)}$

39. $\dfrac{a^{-1}+1}{a^{-1}-1} = \dfrac{\left(\dfrac{1}{a}+1\right)a}{\left(\dfrac{1}{a}-1\right)a} = \dfrac{1+a}{1-a}$

41. $\dfrac{3x^{-1}+(2y)^{-1}}{x^{-2}} = \dfrac{\left(\dfrac{3}{x}+\dfrac{1}{2y}\right)2x^2y}{\left(\dfrac{1}{x^2}\right)2x^2y}$

$= \dfrac{6xy+x^2}{2y}$

$= \dfrac{x(6y+x)}{2y}$

43. $\dfrac{2a^{-1}+(2a)^{-1}}{a^{-1}+2a^{-2}} = \dfrac{\left(\dfrac{2}{a}+\dfrac{1}{2a}\right)2a^2}{\left(\dfrac{1}{a}+\dfrac{2}{a^2}\right)2a^2}$

$= \dfrac{4a+a}{2a+4}$

$= \dfrac{5a}{2a+4}$

45. $\dfrac{5x^{-1}+2y^{-1}}{x^{-2}y^{-2}} = \dfrac{\dfrac{5}{x}+\dfrac{2}{y}}{\dfrac{1}{x^2y^2}}$

$= \dfrac{\dfrac{5y+2x}{xy}}{\dfrac{1}{x^2y^2}}$

$= \dfrac{5y+2x}{xy} \cdot \dfrac{x^2y^2}{1}$

$= xy(5y+2x) \text{ or } 5xy^2+2x^2y$

47. $\dfrac{5x^{-1}-2y^{-1}}{25x^{-2}-4y^{-2}}=\dfrac{\left(\dfrac{5}{x}-\dfrac{2}{y}\right)x^2y^2}{\left(\dfrac{25}{x^2}-\dfrac{4}{y^2}\right)x^2y^2}$

$=\dfrac{5xy^2-2x^2y}{25y^2-4x^2}$

$=\dfrac{xy(5y-2x)}{(5y+2x)(5y-2x)}$

$=\dfrac{xy}{2x+5y}$

49. $\dfrac{3x^3y^2}{12x}=\dfrac{x^{3-1}y^2}{4}=\dfrac{x^2y^2}{4}$

51. $\dfrac{144x^5y^5}{-16x^2y}=-9x^{5-2}y^{5-1}=-9x^3y^4$

53. $|x-5|=9$

$x-5=9$ or $x-5=-9$

$x=14$ or $x=-4$

The solutions are –4, 14.

55. $\dfrac{a}{1-\dfrac{s}{770}}=\dfrac{a}{\dfrac{770}{770}-\dfrac{s}{770}}$

$=\dfrac{a}{\dfrac{770-s}{770}}$

$=a\cdot\dfrac{770}{770-s}$

$=\dfrac{770a}{770-s}$

57. $\dfrac{\dfrac{1}{x}}{\dfrac{3}{y}}$; a and b

59. $\dfrac{1}{1+(1+x)^{-1}}=\dfrac{1(1+x)}{\left(1+\dfrac{1}{1+x}\right)(1+x)}$

$=\dfrac{1+x}{(1+x)+1}$

$=\dfrac{1+x}{2+x}$

61. $\dfrac{x}{1-\dfrac{1}{1+\dfrac{1}{x}}}=\dfrac{x}{1-\dfrac{1}{\left(\dfrac{x+1}{x}\right)}}=\dfrac{x}{1-\dfrac{x}{x+1}}$

$=\dfrac{x(x+1)}{\left(1-\dfrac{x}{x+1}\right)(x+1)}$

$=\dfrac{x(x+1)}{x+1-x}$

$=x(x+1)=x^2+x$

63. $\dfrac{\dfrac{2}{y^2}-\dfrac{5}{xy}-\dfrac{3}{x^2}}{\dfrac{2}{y^2}+\dfrac{7}{xy}+\dfrac{3}{x^2}}=\dfrac{\left(\dfrac{2}{y^2}-\dfrac{5}{xy}-\dfrac{3}{x^2}\right)x^2y^2}{\left(\dfrac{2}{y^2}+\dfrac{7}{xy}+\dfrac{3}{x^2}\right)x^2y^2}$

$=\dfrac{2x^2-5xy-3y^2}{2x^2+7xy+3y^2}$

$=\dfrac{(2x+y)(x-3y)}{(2x+y)(x+3y)}$

$=\dfrac{x-3y}{x+3y}$

65. $\dfrac{3(a+1)^{-1}+4a^{-2}}{(a^3+a^2)^{-1}}=\dfrac{\dfrac{3}{a+1}+\dfrac{4}{a^2}}{\dfrac{1}{a^3+a^2}}$

$=\dfrac{\dfrac{3a^2+4(a+1)}{a^2(a+1)}}{\dfrac{1}{a^2(a+1)}}$

$=\dfrac{3a^2+4a+4}{1}$

$=3a^2+4a+4$

67. $f(x)=\dfrac{1}{x}$

   a. $f(a+h)=\dfrac{1}{a+h}$

   b. $f(a)=\dfrac{1}{a}$

c. $\dfrac{f(a+h)-f(a)}{h} = \dfrac{\dfrac{1}{a+h}-\dfrac{1}{a}}{h}$

d. $\dfrac{f(a+h)-f(a)}{h} = \dfrac{\left(\dfrac{1}{a+h}-\dfrac{1}{a}\right)a(a+h)}{h\cdot a(a+h)}$

$= \dfrac{a-(a+h)}{ha(a+h)}$

$= \dfrac{-h}{ha(a+h)}$

$= \dfrac{-1}{a(a+h)}$

69. $f(x) = \dfrac{3}{x+1}$

a. $f(a+h) = \dfrac{3}{a+h+1}$

b. $f(a) = \dfrac{3}{a+1}$

c. $\dfrac{f(a+h)-f(a)}{h} = \dfrac{\dfrac{3}{a+h+1}-\dfrac{3}{a+1}}{h}$

d. $\dfrac{f(a+h)-f(a)}{h}$

$= \dfrac{\left(\dfrac{3}{a+h+1}-\dfrac{3}{a+1}\right)(a+h+1)(a+1)}{h\cdot(a+h+1)(a+1)}$

$= \dfrac{3(a+1)-3(a+h+1)}{h(a+h+1)(a+1)}$

$= \dfrac{3a+3-3a-3h-3}{h(a+h+1)(a+1)}$

$= \dfrac{-3h}{h(a+h+1)(a+1)}$

$= \dfrac{-3}{(a+h+1)(a+1)}$

3. $\dfrac{12a^5b^2+16a^4b}{4a^4b} = \dfrac{12a^5b^2}{4a^4b} + \dfrac{16a^4b}{4a^4b}$

$= 3ab+4$

5. $\dfrac{4x^2y^2+6xy^2-4y^3}{2x^2y}$

$= \dfrac{4x^2y^2}{2x^2y} + \dfrac{6xy^2}{2x^2y} - \dfrac{4y^3}{2x^2y}$

$= 2y + \dfrac{3y}{x} - \dfrac{2y}{x^2}$

7. $\dfrac{4x^2+8x+4}{4} = \dfrac{4x^2}{4} + \dfrac{8x}{4} + \dfrac{4}{4}$

$= x^2+2x+1$

9. $\dfrac{3x^4+6x^2-18}{3} = \dfrac{3x^4}{3} + \dfrac{6x^2}{3} - \dfrac{18}{3}$

$= (x^4+2x^2-6)$ meters

11. $\begin{array}{r} x+1 \\ x+2\overline{\smash{)}x^2+3x+2} \\ \underline{x^2+2x} \\ x+2 \\ \underline{x+2} \\ 0 \end{array}$

Answer: $x+1$

13. $\begin{array}{r} 2x-8 \\ x+1\overline{\smash{)}2x^2-6x-7} \\ \underline{2x^2+2x} \\ -8x-7 \\ \underline{-8x-8} \\ 1 \end{array}$

Answer: $2x-8+\dfrac{1}{x+1}$

Exercise Set 6.4

1. $\dfrac{4a^2+8a}{2a} = \dfrac{4a^2}{2a} + \dfrac{8a}{2a} = 2a+4$

15.
$$\begin{array}{r} x - \frac{1}{2} \\ 2x+4\overline{\smash{\big)}\,2x^2+3x-2} \\ \underline{2x^2+4x} \\ -x-2 \\ \underline{-x-2} \\ 0 \end{array}$$

Answer: $x - \dfrac{1}{2}$

17.
$$\begin{array}{r} 2x^2 - \frac{1}{2}x+5 \\ 2x+4\overline{\smash{\big)}\,4x^3+7x^2+8x+20} \\ \underline{4x^3+8x^2} \\ -x^2+8x \\ \underline{-x^2-2x} \\ 10x+20 \\ \underline{10x+20} \\ 0 \end{array}$$

Answer: $2x^2 - \dfrac{1}{2}x+5$

19. $A = l \cdot w$ so
$$w = \frac{A}{l} = \frac{15x^2-29x-14}{5x+2}.$$

$$\begin{array}{r} 3x-7 \\ 5x+2\overline{\smash{\big)}\,15x^2-29x-14} \\ \underline{15x^2+6x} \\ -35x-14 \\ \underline{-35x-14} \\ 0 \end{array}$$

The width is $(3x-7)$ in.

21. $\dfrac{25a^2b^{12}}{10a^5b^7} = \dfrac{5a^{2-5}b^{12-7}}{2} = \dfrac{5a^{-3}b^5}{2} = \dfrac{5b^5}{2a^3}$

23. $\dfrac{x^6y^6-x^3y^3}{x^3y^3} = \dfrac{x^6y^6}{x^3y^3} - \dfrac{x^3y^3}{x^3y^3} = x^3y^3 - 1$

25.
$$\begin{array}{r} a+3 \\ a+1\overline{\smash{\big)}\,a^2+4a+3} \\ \underline{a^2+a} \\ 3a+3 \\ \underline{3a+3} \\ 0 \end{array}$$

Answer: $a+3$

27.
$$\begin{array}{r} 2x+5 \\ x-2\overline{\smash{\big)}\,2x^2+x-10} \\ \underline{2x^2-4x} \\ 5x-10 \\ \underline{5x-10} \\ 0 \end{array}$$

Answer: $2x+5$

29. $\dfrac{-16y^3+24y^4}{-4y^2} = \dfrac{-16y^3}{-4y^2} + \dfrac{24y^4}{-4y^2}$
$$= 4y - 6y^2$$

31.
$$\begin{array}{r} 2x+23 \\ x-5\overline{\smash{\big)}\,2x^2+13x+15} \\ \underline{2x^2-10x} \\ 23x+15 \\ \underline{23x-115} \\ 130 \end{array}$$

Answer: $2x+23+\dfrac{130}{x-5}$

33. $\dfrac{20x^2y^3+6xy^4-12x^3y^5}{2xy^3}$
$$= \frac{20x^2y^3}{2xy^3} + \frac{6xy^4}{2xy^3} - \frac{12x^3y^5}{2xy^3}$$
$$= 10x+3y-6x^2y^2$$

35.
$$\begin{array}{r} 2x+4 \\ 3x+2\overline{\smash{\big)}\,6x^2+16x+8} \\ \underline{6x^2+4x} \\ 12x+8 \\ \underline{12x+8} \\ 0 \end{array}$$

Answer: $2x+4$

37.

$$
\begin{array}{r}
y+5 \\
2y-3\overline{)2y^2+7y-15} \\
\underline{2y^2-3y} \\
10y-15 \\
\underline{10y-15} \\
0
\end{array}
$$

Answer: $y+5$

39.

$$
\begin{array}{r}
2x+3 \\
2x-3\overline{)4x^2+0x-9} \\
\underline{4x^2-6x} \\
6x-9 \\
\underline{6x-9} \\
0
\end{array}
$$

Answer: $2x+3$

41.

$$
\begin{array}{r}
2x^2-8x+38 \\
x+4\overline{)2x^3+0x^2+6x\ -4} \\
\underline{2x^3+8x^2} \\
-8x^2+\ 6x \\
\underline{-8x^2-32x} \\
38x-\ 4 \\
\underline{38x+152} \\
-156
\end{array}
$$

Answer: $2x^2-8x+38-\dfrac{156}{x+4}$

43.

$$
\begin{array}{r}
3x+3 \\
x-1\overline{)3x^2+0x-4} \\
\underline{3x^2-3x} \\
3x-4 \\
\underline{3x-3} \\
-1
\end{array}
$$

Answer: $3x+3-\dfrac{1}{x-1}$

45.

$$
\begin{array}{r}
-2x^3+3x^2-x+4 \\
-x+5\overline{)2x^4-13x^3+16x^2-9x+20} \\
\underline{2x^4-10x^3} \\
-3x^3+16x^2 \\
\underline{-3x^3+15x^2} \\
x^2-9x \\
\underline{x^2-5x} \\
-4x+20 \\
\underline{-4x+20} \\
0
\end{array}
$$

Answer: $-2x^3+3x^2-x+4$

47.

$$
\begin{array}{r}
3x^3\ \ \ \ \ \ \ \ +5x+4 \\
x^2-2\overline{)3x^5+0x^4-x^3+4x^2-12x-8} \\
\underline{3x^5\ \ \ \ \ \ -6x^3} \\
5x^3+4x^2-12x \\
\underline{5x^3+0x^2-10x} \\
4x^2-2x-8 \\
\underline{4x^2+0x-8} \\
-2x
\end{array}
$$

Answer: $3x^3+5x+4-\dfrac{2x}{x^2-2}$

49. $\dfrac{3x^3-5}{3x^2}=\dfrac{3x^3}{3x^2}-\dfrac{5}{3x^2}=x-\dfrac{5}{3x^2}$

51. $3^2=(-3)^2$

53. $-2^3=(-2)^3$

55. $|x+5|<4$
$-4<x+5<4$
$-9<x<-1$
$(-9,-1)$

57. $|2x+7|\geq 9$
$2x+7\leq -9$ or $2x+7\geq 9$
$2x\leq -16$ or $2x\geq 2$
$x\leq -8$ or $x\geq 1$
$(-\infty,-8]\cup[1,\infty)$

59. $P(x) = 3x^3 + 2x^2 - 4x + 3$

$P(1) = 3(1)^3 + 2(1)^2 - 4(1) + 3$

$\quad = 3 + 2 - 4 + 3$

$\quad = 4$

$$
\begin{array}{r}
2x^2 + x - 2 \\
x-1 \overline{\smash)3x^3 + 2x^2 - 4x + 3} \\
\underline{3x^3 - 3x^2} \\
5x^2 - 4x \\
\underline{5x^2 - 5x} \\
x + 3 \\
\underline{x - 1} \\
4
\end{array}
$$

Remainder: 4

61. $P(x) = 5x^4 - 2x^2 + 3x - 6$

$P(-3) = 5(-3)^4 - 2(-3)^2 + 3(-3) - 6$

$\quad = 5(81) - 2(9) - 9 - 6$

$\quad = 405 - 18 - 6 - 6$

$\quad = 372$

$$
\begin{array}{r}
5x^3 - 15x^2 + 43x - 126 \\
x+2 \overline{\smash)5x^4 + 0x^3 - 2x^2 + 3x - 6} \\
\underline{5x^4 + 15x^3} \\
-15x^3 - 2x^2 \\
\underline{-15x^3 - 45x^2} \\
43x^2 + 3x \\
\underline{43x^2 + 129x} \\
-126x - 6 \\
\underline{-126x - 378} \\
372
\end{array}
$$

Remainder: 372

63. Answers may vary.

65.

$$
\begin{array}{r}
x^3 + \frac{5}{3}x^2 + \frac{5}{3}x + \frac{8}{3} \\
x-1 \overline{\smash)x^4 + \frac{2}{3}x^3 + 0x^2 + x + 0} \\
\underline{x^4 - x^3} \\
\frac{5}{3}x^3 - 0x^2 \\
\underline{\frac{5}{3}x^3 - \frac{5}{3}x^2} \\
\frac{5}{3}x^2 + x \\
\underline{\frac{5}{3}x^2 - \frac{5}{3}x} \\
\frac{8}{3}x + 0 \\
\underline{\frac{8}{3}x - \frac{8}{3}} \\
\frac{8}{3}
\end{array}
$$

Answer: $x^3 + \dfrac{5}{3}x^2 + \dfrac{5}{3}x + \dfrac{8}{3} + \dfrac{8}{3(x-1)}$

67.

$$
\begin{array}{r}
\frac{3}{2}x^3 + \frac{1}{4}x^2 + \frac{1}{8}x - \frac{7}{16} \\
2x-1 \overline{\smash)3x^4 + x^3 + 0x^2 - x + \frac{1}{2}} \\
\underline{3x^4 - \frac{3}{2}x^3} \\
\frac{1}{2}x^3 - 0x^2 \\
\underline{\frac{1}{2}x^3 - \frac{1}{4}x^2} \\
\frac{1}{4}x^2 - x \\
\underline{\frac{1}{4}x^2 - \frac{1}{8}x} \\
-\frac{7}{8}x + \frac{1}{2} \\
\underline{-\frac{7}{8}x + \frac{7}{16}} \\
\frac{1}{16}
\end{array}
$$

Answer:

$$
\frac{3}{2}x^3 + \frac{1}{4}x^2 + \frac{1}{8}x - \frac{7}{16} + \frac{1}{16(2x-1)}
$$

69.

$$\begin{array}{r} x^3 \qquad -\dfrac{2}{5}x \\[4pt] 5x+10{\overline{\smash{\big)}\,5x^4 + 10x^3 - 2x^2 - 4x + \;\; 0}} \\[-2pt] \underline{5x^4 + 10x^3} \\[2pt] -2x^2 - 4x \\[-2pt] \underline{-2x^2 - 4x} \\[2pt] 0 \end{array}$$

Answer: $x^3 - \dfrac{2}{5}x$

71. $\dfrac{f(x)}{g(x)} = \dfrac{25x^2 - 5x + 30}{5x}$

$= \dfrac{25x^2}{5x} - \dfrac{5x}{5x} + \dfrac{30}{5x}$

$= 5x - 1 + \dfrac{6}{x}$

Setting the denominator equal to 0 we get $5x = 0$

$x = 0$.

Thus, $x = 0$ is not in the domain of $\dfrac{f(x)}{g(x)}$.

73. $\dfrac{f(x)}{g(x)} = \dfrac{7x^4 - 3x^2 + 2}{x - 2}$

$$\begin{array}{r} 7x^3 + 14x^2 + \;\;25x + 50 \\[2pt] x-2{\overline{\smash{\big)}\,7x^4 + 0x^3 - \;\;3x^2 + \;\;0x + \;\;2}} \\[-2pt] \underline{7x^4 - 14x^3} \\[2pt] 14x^3 - \;\;3x^2 \\[-2pt] \underline{14x^3 - 28x^2} \\[2pt] 25x^2 + \;\;0x \\[-2pt] \underline{25x^2 - 50x} \\[2pt] 50x + \;\;2 \\[-2pt] \underline{50x - 100} \\[2pt] 102 \end{array}$$

Therefore,

$\dfrac{f(x)}{g(x)} = \dfrac{7x^4 - 3x^2 + 2}{x - 2}$

$= 7x^3 + 14x^2 + 25x + 50 + \dfrac{102}{x-2}$

Setting the denominator equal to 0 we get $x - 2 = 0$, or $x = 2$.

Thus, $x = 2$ is not in the domain of $\dfrac{f(x)}{g(x)}$.

75. Answers may vary.

Exercise Set 6.5

1. $\begin{array}{r} 5\,\underline{\rvert}\;\;1 \quad\;\; 3 \quad -40 \\ \quad\;\; 5 \quad\;\; 40 \\ \hline 1 \quad\;\; 8 \quad\;\;\; 0 \end{array}$

$x + 8$

3. $\begin{array}{r} -6\,\underline{\rvert}\;\;1 \quad\;\;\; 5 \quad -6 \\ \quad\;\; -6 \quad\;\; 6 \\ \hline 1 \quad -1 \quad\;\;\; 0 \end{array}$

$x - 1$

5. $\begin{array}{r} 2\,\underline{\rvert}\;\;1 \quad -7 \quad -13 \quad\;\; 5 \\ \quad\;\; 2 \quad -10 \quad -46 \\ \hline 1 \quad -5 \quad -23 \quad -41 \end{array}$

$x^2 - 5x - 23 - \dfrac{41}{x - 2}$

7. $\begin{array}{r} 2\,\underline{\rvert}\;\;4 \quad\;\; 0 \quad -9 \\ \quad\;\; 8 \quad\;\; 16 \\ \hline 4 \quad\;\; 8 \quad\;\; 7 \end{array}$

$4x + 8 + \dfrac{7}{x - 2}$

9. a. $P(2) = 3(2)^2 - 4(2) - 1 = 12 - 8 - 1 = 3$

    b. $\begin{array}{r} 2\,\underline{\rvert}\;\;3 \quad -4 \quad -1 \\ \quad\;\; 6 \quad\;\; 4 \\ \hline 3 \quad\;\; 2 \quad\;\; 3 \end{array}$

    $P(2) = 3$

11. a. $P(-2) = 4(-2)^4 + 7(-2)^2 + 9(-2) - 1$

          $= 64 + 28 - 18 - 1$

          $= 73$

    b. $\begin{array}{r} -2\,\underline{\rvert}\;\;4 \quad\;\; 0 \quad\;\; 7 \quad\;\; 9 \quad -1 \\ \quad\;\; -8 \quad\;\; 16 \quad -46 \quad\;\; 74 \\ \hline 4 \quad -8 \quad\;\; 23 \quad -37 \quad\;\; 73 \end{array}$

    $P(-2) = 73$

13. a. $P(-1) = (-1)^5 + 3(-1)^4 + 3(-1) - 7$
$$= -1 + 3 - 3 - 7$$
$$= -8$$

b. $\underline{-1}|\ 1\quad 3\quad 0\quad 0\quad 3\quad -7$
$$\qquad\qquad\ -1\ -2\quad 2\ -2\ -1$$
$$\overline{\qquad 1\quad 2\ -2\quad 2\quad 1\ -8}$$
$$P(-1) = -8$$

15. $\underline{3}|\ 1\quad -3\quad 0\quad 2$
$$\qquad\qquad\quad 3\quad 0\quad 0$$
$$\overline{\qquad 1\quad 0\quad 0\quad 2}$$
$$x^2 + \frac{2}{x-3}$$

17. $\underline{-1}|\ 6\quad 13\quad 8$
$$\qquad\qquad\ -6\quad -7$$
$$\overline{\qquad 6\quad 7\quad 1}$$
$$6x + 7 + \frac{1}{x+1}$$

19. $\underline{5}|\ 2\quad -13\quad 16\quad -9\quad 20$
$$\qquad\qquad\ 10\ -15\quad 5\ -20$$
$$\overline{\qquad 2\quad -3\quad 1\ -4\quad 0}$$
$$2x^3 - 3x^2 + x - 4$$

21. $\underline{-3}|\ 3\quad 0\quad -15$
$$\qquad\qquad\ -9\quad 27$$
$$\overline{\qquad 3\quad -9\quad 12}$$
$$3x - 9 + \frac{12}{x+3}$$

23. $\dfrac{1}{2}\bigg|\ 3\quad -6\quad 4\quad 5$
$$\qquad\qquad\quad \frac{3}{2}\ -\frac{9}{4}\quad \frac{7}{8}$$
$$\overline{\qquad 3\quad -\frac{9}{2}\quad \frac{7}{4}\quad \frac{47}{8}}$$
$$3x^2 - \frac{9}{2}x + \frac{7}{4} + \frac{47}{8\left(x - \frac{1}{2}\right)}$$

25. $\dfrac{1}{3}\bigg|\ 3\quad 2\quad -4\quad 1$
$$\qquad\qquad\quad 1\quad 1\quad -1$$
$$\overline{\qquad 3\quad 3\quad -3\quad 0}$$
$$3x^2 + 3x - 3$$

27. $\underline{-1}|\ 3\quad 7\quad -4\quad 12$
$$\qquad\qquad\ -3\quad -4\quad 8$$
$$\overline{\qquad 3\quad 4\quad -8\quad 20}$$
$$3x^2 + 4x - 8 + \frac{20}{x+1}$$

29. $\underline{1}|\ 1\quad 0\quad 0\quad -1$
$$\qquad\qquad\ 1\quad 1\quad 1$$
$$\overline{\qquad 1\quad 1\quad 1\quad 0}$$
$$x^2 + x + 1$$

31. $\underline{-6}|\ 1\quad 0\quad -36$
$$\qquad\qquad\ -6\quad 36$$
$$\overline{\qquad 1\quad -6\quad 0}$$
$$x - 6$$

33. $\underline{1}|\ 1\quad 3\quad -7\quad 4$
$$\qquad\qquad\ 1\quad 4\quad -3$$
$$\overline{\qquad 1\quad 4\quad -3\quad 1}$$
Thus, $P(1) = 1$.

35. $\underline{-3}|\ 3\quad -7\quad -2\quad 5$
$$\qquad\qquad\ -9\quad 48\quad -138$$
$$\overline{\qquad 3\quad -16\quad 46\quad -133}$$
Thus, $P(-3) = -133$.

37. $\underline{-1}|\ 4\quad 0\quad 1\quad 0\quad -2$
$$\qquad\qquad\ -4\quad 4\quad -5\quad 5$$
$$\overline{\qquad 4\quad -4\quad 5\quad -5\quad 3}$$
Thus, $P(-1) = 3$.

39.

$$\frac{1}{3}\begin{array}{|rrrrr} 2 & 0 & -3 & 0 & -2 \\ & \frac{2}{3} & \frac{2}{9} & -\frac{25}{27} & -\frac{25}{81} \\ \hline 2 & \frac{2}{3} & -\frac{25}{9} & -\frac{25}{27} & -\frac{187}{81} \end{array}$$

Thus, $P\left(\dfrac{1}{3}\right) = -\dfrac{187}{81}$.

41.

$$\frac{1}{2}\begin{array}{|rrrrrr} 1 & 1 & -1 & 0 & 0 & 3 \\ & \frac{1}{2} & \frac{3}{4} & -\frac{1}{8} & -\frac{1}{16} & -\frac{1}{32} \\ \hline 1 & \frac{3}{2} & -\frac{1}{4} & -\frac{1}{8} & -\frac{1}{16} & \frac{95}{32} \end{array}$$

Thus, $P\left(\dfrac{1}{2}\right) = \dfrac{95}{32}$.

43. Answers may vary.

45. $7x + 2 = x - 3$
$$7x - x = -3 - 2$$
$$6x = -5$$
$$x = -\frac{5}{6}$$
The solution is $-\dfrac{5}{6}$.

47.
$$x^2 = 4x - 4$$
$$x^2 - 4x + 4 = 0$$
$$(x-2)^2 = 0$$
$$x - 2 = 0$$
$$x = 2$$
The solution is 2.

49.
$$\frac{x}{3} - 5 = 13$$
$$3\left(\frac{x}{3} - 5\right) = (13) \cdot 3$$
$$x - 15 = 39$$
$$x = 54$$
The solution is 54.

51. $x^3 - 1 = x^3 - 1^3 = (x-1)(x^2 + x + 1)$

53. $125z^3 + 8 = (5z)^3 + 2^3$
$$= (5z+2)(25z^2 - 10z + 4)$$

55. $xy + 2x + 3y + 6 = (xy + 2x) + (3y + 6)$
$$= x(y+2) + 3(y+2)$$
$$= (y+2)(x+3)$$

57. $x^3 - 9x = x(x^2 - 9) = x(x+3)(x-3)$

59.

$$-3\begin{array}{|rrrr} 1 & 3 & 4 & 12 \\ & -3 & 0 & -12 \\ \hline 1 & 0 & 4 & 0 \end{array}$$

Remainder $= 0$ and
$$(x+3)(x^2+4) = x^3 + 3x^2 + 4x + 12$$

61. $P(x)$ is equal to the remainder when $P(x)$ is divided by $x - c$. Therefore, $P(c) = 0$.

63. Multiply $(x^2 - x + 10)$ by $(x+3)$ and add the remainder, -2.
$$(x^2 - x + 10)(x+3) - 2$$
$$= (x^3 + 3x^2 - x^2 - 3x + 10x + 30) - 2$$
$$= x^3 + 2x^2 + 7x + 28$$

65. $V = lwh$ so $w = \dfrac{V}{lh}$
$$= \frac{x^4 + 6x^3 - 7x^2}{x^2(x+7)}$$
$$= \frac{x^4 + 6x^3 - 7x^2}{x^3 + 7x^2}$$

$$\begin{array}{r} x - 1 \\ x^3 + 7x^2 \overline{)x^4 + 6x^3 - 7x^2} \\ \underline{x^4 + 7x^3} \\ -x^3 - 7x^2 \\ \underline{-x^3 - 7x^2} \\ 0 \end{array}$$

The width is $(x-1)$ meters.

Exercise Set 6.6

1. $\dfrac{x}{2} - \dfrac{x}{3} = 12$

$6\left(\dfrac{x}{2} - \dfrac{x}{3}\right) = 6(12)$

$3x - 2x = 72$

$x = 72$

3. $\dfrac{x}{3} = \dfrac{1}{6} + \dfrac{x}{4}$

$12\left(\dfrac{x}{3}\right) = 12\left(\dfrac{1}{6} + \dfrac{x}{4}\right)$

$4x = 2 + 3x$

$x = 2$

5. $\dfrac{2}{x} + \dfrac{1}{2} = \dfrac{5}{x}$

$2x\left(\dfrac{2}{x} + \dfrac{1}{2}\right) = 2x\left(\dfrac{5}{x}\right)$

$4 + x = 10$

$x = 6$

7. $\dfrac{x^2 + 1}{x} = \dfrac{5}{x}$

$x\left(\dfrac{x^2 + 1}{x}\right) = x\left(\dfrac{5}{x}\right)$

$x^2 + 1 = 5$

$x^2 - 4 = 0$

$x + 2 = 0 \quad \text{or } x - 2 = 0$

$x = -2 \ \text{or} \qquad x = 2$

9. $\dfrac{x+5}{x+3} = \dfrac{2}{x+3}$

$(x+3) \cdot \dfrac{x+5}{x+3} = (x+3) \cdot \dfrac{2}{x+3}$

$x + 5 = 2$

$x = -3$

which we disregard as extraneous.
No solution, or $\varnothing$.

11. $\dfrac{5}{x-2} - \dfrac{2}{x+4} = -\dfrac{4}{x^2 + 2x - 8}$

$\dfrac{5}{x-2} - \dfrac{2}{x+4} = -\dfrac{4}{(x+4)(x-2)}$

$5(x+4) - 2(x-2) = -4$

$5x + 20 - 2x + 4 = -4$

$3x + 24 = -4$

$3x = -28$

$x = -\dfrac{28}{3}$

13. $\dfrac{1}{x-1} = \dfrac{2}{x+1}$

$(x+1)(x-1) \cdot \dfrac{1}{x-1} = (x+1)(x-1) \cdot \dfrac{2}{x+1}$

$1(x+1) = 2(x-1)$

$x + 1 = 2x - 2$

$-x = -3$

$x = 3$

15. $\dfrac{x^2 - 23}{2x^2 - 5x - 3} + \dfrac{2}{x-3} = \dfrac{-1}{2x+1}$

$\dfrac{x^2 - 23}{(2x+1)(x-3)} + \dfrac{2}{x-3} = \dfrac{-1}{2x+1}$

$(x^2 - 23) + 2(2x+1) = -1(x-3)$

$x^2 - 23 + 4x + 2 = -x + 3$

$x^2 + 5x - 24 = 0$

$(x+8)(x-3) = 0$

$x + 8 = 0 \quad \text{or } x - 3 = 0$

$x = -8 \ \text{or} \qquad x = 3$

We discard 3 as extraneous.
$x = -8$.

17. $\dfrac{1}{x-4} - \dfrac{3x}{x^2 - 16} = \dfrac{2}{x+4}$

$\dfrac{1}{x-4} - \dfrac{3x}{(x+4)(x-4)} = \dfrac{2}{x+4}$

$1(x+4) - 3x = 2(x-4)$

$x + 4 - 3x = 2x - 8$

$-2x + 4 = 2x - 8$

$-4x = -12$

$x = 3$

19.
$$\frac{1}{x-4} = \frac{8}{x^2-16}$$
$$\frac{1}{x-4} = \frac{8}{(x+4)(x-4)}$$
$$1(x+4) = 8$$
$$x+4 = 8$$
$$x = -4$$

which we discard as extraneous.
No solution, or $\varnothing$.

21.
$$\frac{1}{x-2} - \frac{2}{x^2-2x} = 1$$
$$\frac{1}{x-2} - \frac{2}{x(x-2)} = 1$$
$$x(x-2)\left[\frac{1}{x-2} - \frac{2}{x(x-2)}\right] = x(x-2)\cdot 1$$
$$x-2 = x(x-2)$$
$$x-2 = x^2-2x$$
$$0 = x^2-3x+2$$
$$0 = (x-2)(x-1)$$
$$x-2 = 0 \text{ or } x-1 = 1$$
$$x = 2 \text{ or } \qquad x = 1$$

We discard 2 as extraneous.
$$x = 1$$

23.
$$\frac{5}{x} = \frac{20}{12}$$
$$12x\left(\frac{5}{x}\right) = 12x\left(\frac{20}{12}\right)$$
$$60 = 20x$$
$$3 = x$$

25.
$$1 - \frac{4}{a} = 5$$
$$a\left(1 - \frac{4}{a}\right) = a(5)$$
$$a - 4 = 5a$$
$$-4 = 4a$$
$$-1 = a$$

27.
$$\frac{x^2+5}{x} - 1 = \frac{5(x+1)}{x}$$
$$x\left(\frac{x^2+5}{x} - 1\right) = x\left[\frac{5(x+1)}{x}\right]$$
$$x^2+5-x = 5x+5$$
$$x^2-6x = 0$$
$$x(x-6) = 0$$
$$x = 0 \text{ or } x-6 = 0$$
$$x = 6$$

We discard 0 as extraneous.
$$x = 6$$

29.
$$\frac{1}{2x} - \frac{1}{x+1} = \frac{1}{3x^2+3x}$$
$$\frac{1}{2x} - \frac{1}{x+1} = \frac{1}{3x(x+1)}$$
$$1\cdot 3(x+1) - 1\cdot 6x = 1\cdot 2$$
$$3x+3-6x = 2$$
$$-3x+3 = 2$$
$$-3x = -1$$
$$x = \frac{1}{3}$$

31.
$$\frac{1}{x} - \frac{x}{25} = 0$$
$$25x\left(\frac{1}{x} - \frac{x}{25}\right) = 25x(0)$$
$$25 - x^2 = 0$$
$$-(x^2-25) = 0$$
$$-(x+5)(x-5) = 0$$
$$x+5 = 0 \text{ or } x-5 = 0$$
$$x = -5 \text{ or } \qquad x = 5$$

33.
$$5 - \frac{2}{2y-5} = \frac{3}{2y-5}$$
$$(2y-5)\left(5 - \frac{2}{2y-5}\right) = (2y-5)\cdot\frac{3}{2y-5}$$
$$5(2y-5) - 2 = 3$$
$$10y-25-2 = 3$$
$$10y-27 = 3$$
$$10y = 30$$
$$y = 3$$

35. $\dfrac{x-1}{x+2} = \dfrac{2}{3}$
$$3(x-1) = 2(x+2)$$
$$3x-3 = 2x+4$$
$$x = 7$$

37. $\dfrac{x+3}{x+2} = \dfrac{1}{x+2}$
$$x+3 = 1$$
$$x = -2$$
which we discard as extraneous.
No solution, or $\varnothing$.

39. $\dfrac{1}{a-3} + \dfrac{2}{a+3} = \dfrac{1}{a^2-9}$
$$\dfrac{1}{a-3} + \dfrac{2}{a+3} = \dfrac{1}{(a+3)(a-3)}$$
$$1(a+3) + 2(a-3) = 1$$
$$a+3+2a-6 = 1$$
$$3a-3 = 1$$
$$3a = 4$$
$$a = \dfrac{4}{3}$$

41. $\dfrac{64}{x^2-16} + 1 = \dfrac{2x}{x-4}$
$$\dfrac{64}{(x+4)(x-4)} + 1 = \dfrac{2x}{x-4}$$
$$64 + 1(x+4)(x-4) = 2x(x+4)$$
$$64 + (x^2-16) = 2x^2 + 8x$$
$$x^2 + 48 = 2x^2 + 8x$$
$$0 = x^2 + 8x - 48$$
$$0 = (x+12)(x-4)$$
$$x+12 = 0 \quad \text{or } x-4 = 0$$
$$x = -12 \quad \text{or} \qquad x = 4$$
We discard 4 as extraneous.
$x = -12$

43. $\dfrac{-15}{4y+1} + 4 = y$
$$(4y+1)\left(\dfrac{-15}{4y+1} + 4\right) = (4y+1)y$$
$$-15 + 4(4y+1) = 4y^2 + y$$
$$-15 + 16y + 4 = 4y^2 + y$$
$$-11 + 16y = 4y^2 + y$$
$$0 = 4y^2 - 15y + 11$$
$$0 = (4y-11)(y-1)$$
$$4y-11 = 0 \quad \text{or } y-1 = 0$$
$$4y = 11 \quad \text{or} \qquad y = 1$$
$$y = \dfrac{11}{4}$$

45. $\dfrac{28}{x^2-9} + \dfrac{2x}{x-3} + \dfrac{6}{x+3} = 0$
$$\dfrac{28}{(x+3)(x-3)} + \dfrac{2x}{x-3} + \dfrac{6}{x+3} = 0$$
$$28 + 2x(x+3) + 6(x-3) = 0$$
$$28 + 2x^2 + 6x + 6x - 18 = 0$$
$$2x^2 + 12x + 10 = 0$$
$$2(x^2 + 6x + 5) = 0$$
$$2(x+5)(x+1) = 0$$
$$x+5 = 0 \quad \text{or } x+1 = 0$$
$$x = -5 \quad \text{or} \qquad x = -1$$

47. $\dfrac{x+2}{x^2+7x+10} = \dfrac{1}{3x+6} - \dfrac{1}{x+5}$
$$\dfrac{x+2}{(x+5)(x+2)} = \dfrac{1}{3(x+2)} - \dfrac{1}{x+5}$$
$$3(x+2) = 1(x+5) - 1 \cdot 3(x+2)$$
$$3x+6 = x+5 - 3x - 6$$
$$3x+6 = -2x - 1$$
$$5x = -7$$
$$x = -\dfrac{7}{5}$$

49. Let x = the number.
$$3x + 4 = 19$$
$$3x = 15$$
$$x = 5$$
The number is 5.

51. Let w = width. Then
$w + 5 =$ length.
$$2l + 2w = 50$$
$$2(w + 5) + 2w = 50$$
$$2w + 10 + 2w = 50$$
$$4w + 10 = 50$$
$$4w = 40$$
$$w = 10;$$
$$w + 5 = 10 + 5 = 15$$
The length is 15 inches and the width is 10 inches.

53. 10% (reading from the graph)

55. The tallest bars are for categories 25-29 years and 30-34 years.

57. 19% of 35,710
$0.19(35,710) = 6784.9$
We would expect approximately 6785 inmates 25-29 years old in 2001.

59. $f(x) = 20 + \dfrac{4000}{x}$
$$25 = 20 + \dfrac{4000}{x}$$
$$5 = \dfrac{4000}{x}$$
$$5x = 4000$$
$$x = 800$$
800 pencil sharpeners

61. $x^{-2} - 5x^{-1} - 36 = 0$
$$\dfrac{1}{x^2} - \dfrac{5}{x} - 36 = 0$$
$$1 - 5x - 36x^2 = 0$$
$$36x^2 - 5x - 1 = 0$$
$$(9x - 1)(4x + 1) = 0$$
$9x - 1 = 0$ or $4x + 1 = 0$
$9x = 1$ or $4x = -1$
$x = \dfrac{1}{9}$ or $x = -\dfrac{1}{4}$

63. $6p^{-2} - 5p^{-1} + 1 = 0$
$$\dfrac{6}{p^2} - \dfrac{5}{p} + 1 = 0$$
$$6 - 5p + p^2 = 0$$
$$p^2 - 5p + 6 = 0$$
$$(p - 2)(p - 3) = 0$$
$p - 2 = 0$ or $p - 3 = 0$
$p = 2$ or $p = 3$

65. $\dfrac{-8.5}{x + 1.9} = \dfrac{5.7}{x - 3.6}$
$$-8.5(x - 3.6) = 5.7(x + 1.9)$$
$$-8.5x + 30.6 = 5.7x + 10.83$$
$$-14.2x = -19.77$$
$$x = 1.39$$

67. $\dfrac{12.2}{x} + 17.3 = \dfrac{9.6}{x} - 14.7$
$$x\left(\dfrac{12.2}{x} + 17.3\right) = x\left(\dfrac{9.6}{x} - 14.7\right)$$
$$12.2 + 17.3x = 9.6 - 14.7x$$
$$32x = -2.6$$
$$x = -0.08$$

69. $(4 - x)^2 - 5(4 - x) + 6 = 0$
Let $u = 4 - x$. The $u^2 = (4 - x)^2$ and
$$u^2 - 5u + 6 = 0$$
$$(u - 3)(u - 2) = 0$$
$u = 3$ or $u = 2$
$4 - x = 3$ or $4 - x = 2$
$-x = -1$ or $-x = -2$
$x = 1$ or $x = 2$

71. $\left(\dfrac{5}{2 + x}\right)^2 + \left(\dfrac{5}{2 + x}\right) - 20 = 0$
Let $u = \dfrac{5}{2 + x}$. Then $u^2 = \left(\dfrac{5}{2 + x}\right)^2$ and
$$u^2 + u - 20 = 0$$
$$(u + 5)(u - 4) = 0$$

$$u = -5 \quad \text{or} \quad u = 4$$
$$\frac{5}{2+x} = -5 \quad \text{or} \quad \frac{5}{2+x} = 4$$
$$5 = -5(2+x) \quad \text{or} \quad 5 = 4(2+x)$$
$$5 = -10 - 5x \quad \text{or} \quad 5 = 8 + 4x$$
$$5x = -15 \quad \text{or} \quad -3 = 4x$$
$$x = -3 \quad \text{or} \quad -\frac{3}{4} = x$$

73. $\dfrac{2}{x} = \dfrac{10}{5}$

Let $y_1 = \dfrac{2}{x}$ and $y_2 = \dfrac{10}{5}$. Graph both equations and find the point of intersection. The x-coordinate of the intersection point is the solution to the equation.

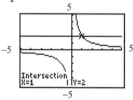

75. $\dfrac{6x+7}{2x+9} = \dfrac{5}{3}$

Let $y_1 = \dfrac{6x+7}{2x+9}$ and $y_2 = \dfrac{5}{3}$. Graph both equations and find the point of intersection. The x-coordinate of the intersection point is the solution to the equation.

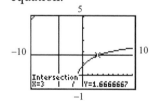

Integrated Review

1. $\dfrac{x}{2} = \dfrac{1}{8} + \dfrac{x}{4}$
$$8\left(\frac{x}{2}\right) = 8\left(\frac{1}{8} + \frac{x}{4}\right)$$
$$4x = 1 + 2x$$
$$2x = 1$$
$$x = \frac{1}{2}$$

2. $\dfrac{x}{4} = \dfrac{3}{2} + \dfrac{x}{10}$
$$20\left(\frac{x}{4}\right) = 20\left(\frac{3}{2} + \frac{x}{10}\right)$$
$$5x = 30 + 2x$$
$$3x = 30$$
$$x = 10$$

3. $\dfrac{1}{8} + \dfrac{x}{4} = \dfrac{1}{8} + \dfrac{x \cdot 2}{4 \cdot 2} = \dfrac{1}{8} + \dfrac{2x}{8} = \dfrac{1+2x}{8}$

4. $\dfrac{3}{2} + \dfrac{x}{10} = \dfrac{3 \cdot 5}{2 \cdot 5} + \dfrac{x}{10} = \dfrac{15}{10} + \dfrac{x}{10} = \dfrac{15+x}{10}$

5. $\dfrac{4}{x+2} - \dfrac{2}{x-1}$
$$= \frac{4(x-1)}{(x+2)(x-1)} - \frac{2(x+2)}{(x-1)(x+2)}$$
$$= \frac{4x-4-2x-4}{(x+2)(x-1)}$$
$$= \frac{2x-8}{(x+2)(x-1)}$$
$$= \frac{2(x-4)}{(x+2)(x-1)}$$

6. $\dfrac{5}{x-2} - \dfrac{10}{x+4}$
$$= \frac{5(x+4)}{(x-2)(x+4)} - \frac{10(x-2)}{(x-2)(x+4)}$$
$$= \frac{5x+20-10x+20}{(x-2)(x+4)}$$
$$= \frac{-5x+40}{(x-2)(x+4)}$$
$$= \frac{-5(x-8)}{(x-2)(x+4)}$$

7. $\dfrac{4}{x+2} = \dfrac{2}{x-1}$

$4(x-1) = 2(x+2)$

$4x - 4 = 2x + 4$

$2x = 8$

$x = 4$

8. $\dfrac{5}{x-2} = \dfrac{10}{x+4}$

$5(x+4) = 10(x-2)$

$5x + 20 = 10x - 20$

$-5x = -40$

$x = 8$

9. $\dfrac{2}{x^2-4} = \dfrac{1}{x+2} - \dfrac{3}{x-2}$

$\dfrac{2}{(x+2)(x-2)} = \dfrac{1}{x+2} - \dfrac{3}{x-2}$

$2 = 1(x-2) - 3(x+2)$

$2 = x - 2 - 3x - 6$

$2 = -2x - 8$

$2x = -10$

$x = -5$

10. $\dfrac{3}{x^2-25} = \dfrac{1}{x+5} + \dfrac{2}{x-5}$

$\dfrac{3}{(x+5)(x-5)} = \dfrac{1}{x+5} + \dfrac{2}{x-5}$

$3 = 1(x-5) + 2(x+5)$

$3 = x - 5 + 2x + 10$

$3 = 3x + 5$

$-2 = 3x$

$-\dfrac{2}{3} = x$

11. $\dfrac{5}{x^2-3x} + \dfrac{4}{2x-6} = \dfrac{5}{x(x-3)} + \dfrac{4}{2(x-3)}$

$= \dfrac{5}{x(x-3)} + \dfrac{2}{(x-3)}$

$= \dfrac{5}{x(x-3)} + \dfrac{2 \cdot x}{(x-3) \cdot x}$

$= \dfrac{5 + 2x}{x(x-3)}$

12. $\dfrac{5}{x^2-3x} \div \dfrac{4}{2x-6} = \dfrac{5}{x^2-3x} \cdot \dfrac{2x-6}{4}$

$= \dfrac{5}{x(x-3)} \cdot \dfrac{2(x-3)}{4}$

$= \dfrac{5}{2x}$

13. $\dfrac{x-1}{x+1} + \dfrac{x+7}{x-1} = \dfrac{4}{x^2-1}$

$\dfrac{x-1}{x+1} + \dfrac{x+7}{x-1} = \dfrac{4}{(x+1)(x-1)}$

$(x-1)^2 + (x+7)(x+1) = 4$

$x^2 - 2x + 1 + x^2 + 8x + 7 = 4$

$2x^2 + 6x + 8 = 4$

$2x^2 + 6x + 4 = 0$

$2(x^2 + 3x + 2) = 0$

$2(x+2)(x+1) = 0$

$x + 2 = 0 \quad \text{or} \quad x + 1 = 0$

$x = -2 \quad \text{or} \quad x = -1$

We discard -1 as extraneous.

$x = -2$

14. $\left(1 - \dfrac{y}{x}\right) \div \left(1 - \dfrac{x}{y}\right) = \dfrac{x-y}{x} \div \dfrac{y-x}{y}$

$= \dfrac{x-y}{x} \cdot \dfrac{y}{y-x}$

$= \dfrac{x-y}{x} \cdot \dfrac{y}{-(x-y)}$

$= -\dfrac{y}{x}$

15. $\dfrac{a^2-9}{a-6} \cdot \dfrac{a^2-5a-6}{a^2-a-6}$

$= \dfrac{(a+3)(a-3)}{a-6} \cdot \dfrac{(a-6)(a+1)}{(a-3)(a+2)}$

$= \dfrac{(a+3)(a+1)}{a+2}$

16. $\dfrac{2}{a-6}+\dfrac{3a}{a^2-5a-6}-\dfrac{a}{5a+5}$

$=\dfrac{2}{a-6}+\dfrac{3a}{(a-6)(a+1)}-\dfrac{a}{5(a+1)}$

$=\dfrac{2\cdot 5(a+1)+3a\cdot 5-a(a-6)}{5(a-6)(a+1)}$

$=\dfrac{10a+10+15a-a^2+6a}{5(a-6)(a+1)}$

$=\dfrac{-a^2+31a+10}{5(a-6)(a+1)}$

17. $\dfrac{2x+3}{3x-2}=\dfrac{4x+1}{6x+1}$

$(2x+3)(6x+1)=(4x+1)(3x-2)$

$12x^2+20x+3=12x^2-5x-2$

$20x+3=-5x-2$

$25x=-5$

$x=-\dfrac{1}{5}$

18. $\dfrac{5x-3}{2x}=\dfrac{10x+3}{4x+1}$

$(5x-3)(4x+1)=2x(10x+3)$

$20x^2-7x-3=20x^2+6x$

$-7x-3=6x$

$-13x=3$

$x=-\dfrac{3}{13}$

19. $\dfrac{a}{9a^2-1}+\dfrac{2}{6a-2}$

$=\dfrac{a}{(3a+1)(3a-1)}+\dfrac{2}{2(3a-1)}$

$=\dfrac{a}{(3a+1)(3a-1)}+\dfrac{1}{3a-1}$

$=\dfrac{a+(3a+1)}{(3a+1)(3a-1)}$

$=\dfrac{4a+1}{(3a+1)(3a-1)}$

20. $\dfrac{3}{4a-8}-\dfrac{a+2}{a^2-2a}$

$=\dfrac{3}{4(a-2)}-\dfrac{a+2}{a(a-2)}$

$=\dfrac{3a-4(a+2)}{4a(a-2)}$

$=\dfrac{3a-4a-8}{4a(a-2)}$

$=\dfrac{-a-8}{4a(a-2)}$ or $-\dfrac{a+8}{4a(a-2)}$

21. $-\dfrac{3}{x^2}-\dfrac{1}{x}+2=0$

$x^2\left(-\dfrac{3}{x^2}-\dfrac{1}{x}+2\right)=x^2(0)$

$-3-x+2x^2=0$

$2x^2-x-3=0$

$(2x-3)(x+1)=0$

$2x-3=0$ or $x+1=0$

$2x=3$ or $x=-1$

$x=\dfrac{3}{2}$

22. $\dfrac{x}{2x+6}+\dfrac{5}{x^2-9}=\dfrac{x}{2(x+3)}+\dfrac{5}{(x+3)(x-3)}$

$=\dfrac{x(x-3)+5\cdot 2}{2(x+3)(x-3)}$

$=\dfrac{x^2-3x+10}{2(x+3)(x-3)}$

23. $\dfrac{x-8}{x^2-x-2}+\dfrac{2}{x-2}$

$=\dfrac{x-8}{(x-2)(x+1)}+\dfrac{2}{x-2}$

$=\dfrac{x-8+2(x+1)}{(x-2)(x+1)}$

$=\dfrac{x-8+2x+2}{(x-2)(x+1)}$

$=\dfrac{3x-6}{(x-2)(x+1)}=\dfrac{3(x-2)}{(x-2)(x+1)}=\dfrac{3}{x+1}$

24.

$$\frac{x-8}{x^2-x-2}+\frac{2}{x-2}=\frac{3}{x+1}$$

$$\frac{x-8}{(x-2)(x+1)}+\frac{2}{x-2}=\frac{3}{x+1}$$

$$x-8+2(x+1)=3(x-2)$$

$$x-8+2x+2=3x-6$$

$$3x-6=3x-6$$

$$-6=-6$$

which is true for any real number. Therefore, the solution is $\{x|x$ is a real number and $x\neq 2,\ x\neq -1\}$.

25.

$$\frac{3}{a}-5=\frac{7}{a}-1$$

$$a\left(\frac{3}{a}-5\right)=a\left(\frac{7}{a}-1\right)$$

$$3-5a=7-a$$

$$-4a=4$$

$$a=-1$$

26.

$$\frac{7}{3z-9}+\frac{5}{z}=\frac{7}{3(z-3)}+\frac{5}{z}$$

$$=\frac{7z}{3(z-3)\cdot z}+\frac{5\cdot 3(z-3)}{z\cdot 3(z-3)}$$

$$=\frac{7z+15z-45}{3z(z-3)}$$

$$=\frac{22z-45}{3z(z-3)}$$

27. a. $\dfrac{x}{5}-\dfrac{x}{4}+\dfrac{1}{10}$ is an expression.

   b. Write each rational term so that the denominator is the LCD, 20.

c.

$$\frac{x}{5}-\frac{x}{4}+\frac{1}{10}=\frac{x\cdot 4}{5\cdot 4}-\frac{x\cdot 5}{4\cdot 5}+\frac{1\cdot 2}{10\cdot 2}$$

$$=\frac{4x-5x+2}{20}$$

$$=\frac{-x+2}{20}$$

28. a. $\dfrac{x}{5}-\dfrac{x}{4}=\dfrac{1}{10}$ is an equation.

   b. Clear the equation of fractions by multiplying each term by the LCD, 20.

c.

$$\frac{x}{5}-\frac{x}{4}=\frac{1}{10}$$

$$20\left(\frac{x}{5}-\frac{x}{4}\right)=20\left(\frac{1}{10}\right)$$

$$4x-5x=2$$

$$-x=2$$

$$x=-2$$

29. b

30. d

31. d

32. a

33. d

Exercise Set 6.7

1.

$$F=\frac{9}{5}C+32$$

$$F-32=\frac{9}{5}C$$

$$C=\frac{5}{9}(F-32)$$

3.

$$Q=\frac{A-I}{L}$$

$$QL=L\left(\frac{A-I}{L}\right)$$

$$QL=A-I$$

$$I=A-QL$$

5.
$$\frac{1}{R} = \frac{1}{R_1} + \frac{1}{R_2}$$
$$RR_1R_2 \cdot \frac{1}{R} = RR_1R_2\left(\frac{1}{R_1} + \frac{1}{R_2}\right)$$
$$R_1R_2 = RR_2 + RR_1$$
$$R_1R_2 = R(R_2 + R_1)$$
$$R = \frac{R_1R_2}{R_1 + R_2}$$

7. $S = \frac{n(a+L)}{2}$
$$2S = n(a+L)$$
$$n = \frac{2S}{a+L}$$

9.
$$A = \frac{h(a+b)}{2}$$
$$2A = h(a+b)$$
$$2A = ah + bh$$
$$2A - ah = bh$$
$$b = \frac{2A - ah}{h}$$

11.
$$\frac{P_1V_1}{T_1} = \frac{P_2V_2}{T_2}$$
$$T_1T_2 \cdot \frac{P_1V_1}{T_1} = T_1T_2 \cdot \frac{P_2V_2}{T_2}$$
$$P_1V_1T_2 = P_2V_2T_1$$
$$T_2 = \frac{P_2V_2T_1}{P_1V_1}$$

13.
$$f = \frac{f_1f_2}{f_1 + f_2}$$
$$(f_1 + f_2)f = f_1f_2$$
$$f_1f + f_2f = f_1f_2$$
$$f_1f = f_1f_2 - f_2f$$
$$f_1f = f_2(f_1 - f)$$
$$\frac{f_1f}{(f_1 - f)} = f_2$$

15. $\lambda = \frac{2L}{n}$
$$n\lambda = 2L$$
$$\frac{n\lambda}{2} = L$$

17.
$$\frac{\theta}{\omega} = \frac{2L}{c}$$
$$c\omega \cdot \frac{\theta}{\omega} = c\omega \cdot \frac{2L}{c}$$
$$c\theta = 2L\omega$$
$$c = \frac{2L\omega}{\theta}$$

19. Let n = the number. Then
$$\frac{1}{n} = \text{the reciprocal of the number.}$$
$$n + 5\left(\frac{1}{n}\right) = 6$$
$$n + \frac{5}{n} = 6$$
$$n\left(n + \frac{5}{n}\right) = 6n$$
$$n^2 + 5 = 6n$$
$$n^2 - 6n + 5 = 0$$
$$(n-5)(n-1) = 0$$
$$n - 5 = 0 \text{ or } n - 1 = 0$$
$$n = 5 \text{ or } \quad n = 1$$
The numbers are 1 and 5.

21. Let x = the number.
$$\frac{12 + x}{41 + 2x} = \frac{1}{3}$$
$$3(12 + x) = 1 \cdot (41 + 2x)$$
$$36 + 3x = 41 + 2x$$
$$x = 5$$
The number is 5.

23. Let a = amount of water in 3 minutes.
$$\frac{15}{10} = \frac{a}{3}$$
$$10a = 15(3)$$
$$10a = 45$$
$$a = 4.5$$
The camel can drink 4.5 gallons.

25. Let w = the number of women.
$$\frac{10.2}{100} = \frac{w}{35,712}$$
$$100w = 10.2(35,712)$$
$$10 \cdot 100w = 10 \cdot 10.2(35,712)$$
$$1000w = 102(35,712)$$
$$1000w = 3,642,624$$
$$w = 3642.624$$
There are about 3643 women.

27. Let x = number of hours needed working together.
$$\frac{1}{26} + \frac{1}{39} = \frac{1}{x}$$
$$78x\left(\frac{1}{26} + \frac{1}{39}\right) = 78x\left(\frac{1}{x}\right)$$
$$3x + 2x = 78$$
$$5x = 78$$
$$x = \frac{78}{5} = 15.6$$
The roofers together would take 15.6 hours.

29. Let x = time to sort the stack working together.
$$\frac{1}{20} + \frac{1}{30} + \frac{1}{60} = \frac{1}{x}$$
$$3x + 2x + x = 60$$
$$6x = 60$$
$$x = 10$$
It takes them 10 minutes to sort the mail when all three work together.

31. Let r = speed of the car. Then $r + 150$ = speed of the plane.
$$t_{\text{plane}} = t_{\text{car}}$$
$$\frac{600}{r+150} = \frac{150}{r}$$
$$600r = 150(r+150)$$
$$600r = 150r + 22,500$$
$$450r = 22,500$$
$$r = \frac{22,500}{450} = 50$$
$$r + 150 = 50 + 150 = 200$$
The speed of the plane was 200 mph.

33. Let r = speed of the boat in still water.
$$t_{\text{downstream}} = t_{\text{upstream}}$$
$$\frac{20}{r+5} = \frac{10}{r-5}$$
$$20(r-5) = 10(r+5)$$
$$20r - 100 = 10r + 50$$
$$10r = 150$$
$$r = 15$$
The speed of the boat in still water is 15 mph.

35. Let x = the first integer. Then $x + 1$ = the next integer.
$$\frac{1}{x} + \frac{1}{x+1} = -\frac{15}{56}$$
$$56(x+1) + 56x = -15x(x+1)$$
$$56x + 56 + 56x = -15x^2 - 15x$$
$$112x + 56 = -15x^2 - 15x$$
$$15x^2 + 127x + 56 = 0$$
$$(15x + 7)(x + 8) = 0$$
$$15x + 7 = 0 \quad \text{or} \quad x + 8 = 0$$
$$15x = -7 \quad \text{or} \qquad x = -8$$
$$x = -\frac{7}{15} \qquad x + 1 = -8 + 1 = -7$$
The integers are –8 and –7.

37. Let t = time for 2nd hose to fill the pond.
$$\frac{1}{45} + \frac{1}{t} = \frac{1}{20}$$
$$180t\left(\frac{1}{45} + \frac{1}{t}\right) = 180t \cdot \frac{1}{20}$$
$$4t + 180 = 9t$$
$$180 = 5t$$
$$36 = t$$
The second hose will take 36 minutes to fill the pond alone.

39. Let r = the speed of the first train. Then $r + 15$ = the speed of the 2nd train.
$$d_{\text{train 1}} + d_{\text{train 2}} = 630$$
$$6r + 6(r+15) = 630$$
$$6r + 6r + 90 = 630$$
$$12r = 540$$
$$r = 45$$

$r + 15 = 45 + 15 = 60$
The speed of the trains were 45 mph and 60 mph.

41. Let $t =$ time to travel 1 mile.
$$\frac{0.17}{1} = \frac{1}{t}$$
$$0.17t = 1$$
$$100(0.17t) = 100(1)$$
$$17t = 100$$
$$t = \frac{100}{17} = 5.882352941$$
It would take 5.9 hours.

43. Let $t =$ time to fill quota working together.
$$\frac{1}{5} + \frac{1}{6} + \frac{1}{7.5} = \frac{1}{t}$$
$$225t\left(\frac{1}{5} + \frac{1}{6} + \frac{1}{7.5}\right) = 225t \cdot \frac{1}{t}$$
$$45t + 37.5t + 30t = 225$$
$$112.5t = 225$$
$$t = 2$$
It would take 2 hours using all three machines.

45. Let $r =$ the speed of plane in still air.
$$t_{\text{with}} = t_{\text{against}}$$
$$\frac{465}{r + 20} = \frac{345}{r - 20}$$
$$465(r - 20) = 345(r + 20)$$
$$465r - 9300 = 345r + 6900$$
$$120r = 16,200$$
$$r = 135$$
The planes speed in still air is 135 mph.

47. Let $d =$ the distance of the run.
$$t_{\text{jogger 2}} = \frac{1}{2} + t_{\text{jogger 1}}$$
$$\frac{d}{6} = \frac{1}{2} + \frac{d}{8}$$
$$24\left(\frac{d}{6}\right) = 24\left(\frac{1}{2} + \frac{d}{8}\right)$$
$$4d = 12 + 3d$$
$$d = 12$$
The run was 12 miles.

49. Let $t =$ time to complete the job working together.
$$\frac{1}{4} + \frac{1}{5} = \frac{1}{t}$$
$$20t\left(\frac{1}{4} + \frac{1}{5}\right) = 20t \cdot \frac{1}{t}$$
$$5t + 4t = 20$$
$$9t = 20$$
$$t = \frac{20}{9} = 2\frac{2}{9}$$
It would take them $2\frac{2}{9}$ hours.

51. Let $n =$ numerator. Then $n + 1 =$ the denominator, and $\dfrac{n}{n+1} =$ the fraction.
Now
$$\frac{n - 3}{(n + 1) - 3} = \frac{4}{5}$$
$$\frac{n - 3}{n - 2} = \frac{4}{5}$$
$$5(n - 3) = 4(n - 2)$$
$$5n - 15 = 4n - 8$$
$$n = 7$$
Thus, $\dfrac{n}{n+1} = \dfrac{7}{7+1} = \dfrac{7}{8}$ is the fraction.

53. Let $t =$ time to move the cans working together.
$$\frac{1}{2} + \frac{1}{6} = \frac{1}{t}$$
$$6t\left(\frac{1}{2} + \frac{1}{6}\right) = 6t \cdot \frac{1}{t}$$
$$3t + t = 6$$
$$4t = 6$$
$$t = \frac{6}{4} = 1\frac{1}{2}$$
It would take them $1\frac{1}{2}$ minutes.

55. Let $r =$ the speed in still air.
$$t_{\text{into wind}} = t_{\text{wind behind}}$$
$$\frac{10}{r - 3} = \frac{11}{r + 3}$$

$10(r+3) = 11(r-3)$
$10r+30 = 11r-33$
$63 = r$
The speed in still air is 63 mph.

57. Let r = the speed in still water.

$t_{\text{downstream}} = t_{\text{upstream}}$

$$\frac{9}{r+6} = \frac{3}{r-6}$$
$$9(r-6) = 3(r+6)$$
$$9r-54 = 3r+18$$
$$6r = 72$$
$$r = 12$$

Thus, $t = t_{\text{downstream}} + t_{\text{upstream}}$

$$= \frac{9}{12+6} + \frac{3}{12-6}$$
$$= \frac{1}{2} + \frac{1}{2}$$
$$= 1$$

It takes him 1 hour to cover the 12 miles.

59. Let t = time if they worked together.

$$\frac{1}{3} + \frac{1}{6} = \frac{1}{t}$$
$$6t\left(\frac{1}{3} + \frac{1}{6}\right) = 6t \cdot \frac{1}{t}$$
$$2t+t = 6$$
$$3t = 6$$
$$t = 2$$

It would take them 2 hours.

61. $\dfrac{x}{4} = \dfrac{x+3}{6}$

$$6x = 4(x+3)$$
$$6x = 4x+12$$
$$2x = 12$$
$$x = 6$$

63. $\dfrac{x-6}{4} = \dfrac{x-2}{5}$

$$5(x-6) = 4(x-2)$$
$$5x-30 = 4x-8$$
$$x = 22$$

65. Answers may vary.

$$\frac{705w}{h^2} = 47$$
$$\frac{705(240)}{h^2} = 47$$
$$169,200 = 47h^2$$
$$3600 = h^2$$
$$0 = 3600 - h^2$$
$$0 = (60+h)(60-h)$$

$60+h = 0 \quad$ or $\quad 60-h = 0$
$\quad h = -60 \quad$ or $\quad\quad 60 = h$

Discard the –60.
The patient is 60 inches, or 5 feet tall.

67. $\dfrac{1}{R} = \dfrac{1}{R_1} + \dfrac{1}{R_2}$

$$\frac{1}{2} = \frac{1}{3} + \frac{1}{R_2}$$
$$6R_2\left(\frac{1}{2}\right) = 6R_2\left(\frac{1}{3} + \frac{1}{R_2}\right)$$
$$3R_2 = 2R_2 + 6$$
$$R_2 = 6$$

6 ohms

69. $\dfrac{1}{R} = \dfrac{1}{R_1} + \dfrac{1}{R_2} + \dfrac{1}{R_3}$

$$\frac{1}{R} = \frac{1}{5} + \frac{1}{6} + \frac{1}{2}$$
$$30R\left(\frac{1}{R}\right) = 30R\left(\frac{1}{5} + \frac{1}{6} + \frac{1}{2}\right)$$
$$30 = 6R + 5R + 15R$$
$$30 = 26R$$
$$R = \frac{30}{26} = \frac{15}{13}$$

$\dfrac{15}{13}$ ohms

Section 6.8

Mental Math

1. $y = 5x$ represents direct variation

2. $y = \dfrac{300}{x}$ represents inverse variation

3. $y = 5xz$ represents joint variation

4. $y = \dfrac{1}{2}abc$ represents joint variation

5. $y = \dfrac{9.1}{x}$ represents inverse variation

6. $y = 2.3x$ represents direct variation

7. $y = \dfrac{2}{3}x$ represents direct variation

8. $y = 3.1st$ represents joint variation

Exercise Set 6.8

1. $y = kx$

3. $a = \dfrac{k}{b}$

5. $y = kxz$

7. $y = \dfrac{k}{x^3}$

9. $y = \dfrac{kx}{p^2}$

11. $y = kx$
$4 = k(20)$
$k = \dfrac{1}{5}$
$y = \dfrac{1}{5}x$

13. $y = kx$
$6 = k(4)$
$k = \dfrac{3}{2}$
$y = \dfrac{3}{2}x$

15. $y = kx$
$7 = k\left(\dfrac{1}{2}\right)$
$k = 14$
$y = 14x$

17. $y = kx$
$0.2 = k(0.8)$
$k = 0.25$
$y = 0.25x$

19. $W = kr^3$
$1.2 = k(2)^3$
$1.2 = 8k$
$k = \dfrac{1.2}{8} = 0.15$
$W = 0.15r^3$
$\quad = 0.15(3)^3$
$\quad = 0.15(27)$
$\quad = 4.05 \text{ lb}$

21. $\quad P = kN$
$260,000 = k(450,000)$
$\quad k = \dfrac{260,000}{450,000} = \dfrac{26}{45}$
$P = \dfrac{26}{45}N = \dfrac{26}{45}(980,000)$
$\quad = 566,222 \text{ tons}$

23. $y = \dfrac{k}{x}$
$6 = \dfrac{k}{5}$
$k = 30$
$y = \dfrac{30}{x}$

25. $y = \dfrac{k}{x}$

$100 = \dfrac{k}{7}$

$k = 700$

$y = \dfrac{700}{x}$

27. $y = \dfrac{k}{x}$

$\dfrac{1}{8} = \dfrac{k}{16}$

$k = 2$

$y = \dfrac{2}{x}$

29. $y = \dfrac{k}{x}$

$0.2 = \dfrac{k}{0.7}$

$k = 0.14$

$y = \dfrac{0.14}{x}$

31. $R = \dfrac{k}{T}$

$45 = \dfrac{k}{6}$

$k = 45(6) = 270$

$R = \dfrac{270}{T} = \dfrac{270}{5} = 54$ mph

33. $I = \dfrac{k}{R}$

$40 = \dfrac{k}{270}$

$k = 40(270) = 10,800$

$I = \dfrac{10,800}{R} = \dfrac{10,800}{150} = 72$ amps

35. $I_1 = \dfrac{k}{d^2}$

Replace d with $2d$.

$I_2 = \dfrac{k}{(2d)^2} = \dfrac{k}{4d^2} = \dfrac{1}{4} \cdot \dfrac{k}{d^2} = \dfrac{1}{4} I_1$

Thus, the intensity is divided by 4.

37. $x = kyz$

39. $r = kst^3$

41. $y = kx^3$

$9 = k(3)^3$

$9 = 27k$

$k = \dfrac{9}{27} = \dfrac{1}{3}$

$y = \dfrac{1}{3} x^3$

43. $y = k\sqrt{x}$

$0.4 = k\sqrt{4}$

$0.4 = 2k$

$k = \dfrac{0.4}{2} = 0.2$

$y = 0.2\sqrt{x}$

45. $y = \dfrac{k}{x^2}$

$0.052 = \dfrac{k}{(5)^2}$

$0.052 = \dfrac{k}{25}$

$k = 0.052(25) = 1.3$

$y = \dfrac{1.3}{x^2}$

47. $y = kxz^3$

$120 = k(5)(2)^3$

$120 = 40k$

$k = \dfrac{120}{40} = 3$

$y = 3xz^3$

49. $W = \dfrac{kwh^2}{l}$

$12 = \dfrac{k\left(\dfrac{1}{2}\right)\left(\dfrac{1}{3}\right)^2}{10}$

$120 = \dfrac{1}{18} k$

$k = 120(18) = 2160$

$$W = \frac{2160wh^2}{l}$$

$$= \frac{2160\left(\frac{2}{3}\right)\left(\frac{1}{2}\right)^2}{16}$$

$$= \frac{360}{16}$$

$$= \frac{45}{2} \text{ tons or } 22.5 \text{ tons}$$

51. $V = kr^2h$

$$32\pi = k(4)^2(6)$$

$$32\pi = 96k$$

$$k = \frac{32\pi}{96} = \frac{\pi}{3}$$

$$V = \frac{\pi}{3}r^2h = \frac{\pi}{3}(3)^2(5)$$

$$= \frac{45\pi}{3}$$

$$= 15\pi \text{ cu. in.}$$

53. $H = ksd^3$

$$40 = k(120)(3)^3$$

$$40 = 1080k$$

$$k = \frac{40}{1080} = \frac{1}{24}$$

$$H = \frac{1}{24}sd^3 = \frac{1}{24}(80)(3)^3 = 90 \text{ hp}$$

55. $y = \dfrac{k}{x}$

$$400 = \frac{k}{8}$$

$$k = 400(8) = 3200$$

$$y = \frac{3200}{x} = \frac{3200}{4} = 800 \text{ millibars}$$

57. $r = 6$ cm

$$C = 2\pi r = 2\pi(6) = 12\pi \text{ cm}$$

$$A = \pi r^2 = \pi(6)^2 = 36\pi \text{ sq. cm}$$

59. $r = 7$ m

$$C = 2\pi r = 2\pi(7) = 14\pi \text{ m}$$

$$A = \pi r^2 = \pi(7)^2 = 49\pi \text{ sq. m}$$

61. $\sqrt{36} = 6$

63. $\sqrt{4} = 2$

65. $\sqrt{\dfrac{1}{25}} = \dfrac{1}{5}$

67. $\sqrt{\dfrac{25}{121}} = \dfrac{5}{11}$

69. $V_1 = khr^2$

$$V_2 = k\left(\frac{1}{2}h\right)(2r)^2$$

$$= k\left(\frac{1}{2}h\right)(4r^2) = 2(khr^2) = 2V_1$$

It is multiplied by 2.

71. $y_1 = kx^2$

$$y_2 = k(2x)^2 = k(4x^2) = 4(kx^2) = 4y_1$$

It is multiplied by 4.

73.

x	$\frac{1}{4}$	$\frac{1}{2}$	1	2	4
$y = \frac{3}{x}$	12	6	3	$\frac{3}{2}$	$\frac{3}{4}$

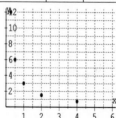

75.

x	$\frac{1}{4}$	$\frac{1}{2}$	1	2	4
$y = \frac{1}{2x}$	2	1	$\frac{1}{2}$	$\frac{1}{4}$	$\frac{1}{8}$

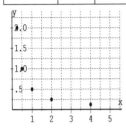

Chapter 6 Review

1. $f(x) = \dfrac{3-5x}{7}$

Domain $\{x \mid x \text{ is a real number}\}$

2. $g(x) = \dfrac{2x+4}{11}$

Domain $\{x \mid x \text{ is a real number}\}$

3. $F(x) = \dfrac{-3x^2}{x-5}$

Undefined values when

$x - 5 = 0$

$x = 5$

Domain $\{x \mid x \text{ is a real number and } x \neq 5\}$

4. $h(x) = \dfrac{4x}{3x-12}$

Undefined values when

$3x - 12 = 0$

$3x = 12$

$x = 4$

Domain $\{x \mid x \text{ is a real number and } x \neq 4\}$

5. $f(x) = \dfrac{x^3+2}{x^2+8x}$

Undefined values when

$x^2 + 8x = 0$

$x(x+8) = 0$

$x = 0 \ \text{ or } \ x + 8 = 0$

$\qquad\qquad x = -8$

Domain

$\{x \mid x \text{ is a real number and } x \neq 0, \ x \neq -8\}$

6. $G(x) = \dfrac{20}{3x^2-48}$

Undefined values when

$3x^2 - 48 = 0$

$3(x^2 - 16) = 0$

$3(x+4)(x-4) = 0$

$x + 4 = 0 \quad \text{ or } \ x - 4 = 0$

$x = -4 \ \text{ or } \qquad x = 4$

Domain

$\{x \mid x \text{ is a real number and } x \neq -4, \ x \neq 4\}$

7. $\dfrac{15x^4}{45x^2} = \dfrac{15x^{4-2}}{45} = \dfrac{x^2}{3}$

8. $\dfrac{x+2}{2+x} = \dfrac{x+2}{x+2} = 1$

9. $\dfrac{18m^6 p^2}{10m^4 p} = \dfrac{2 \cdot 9 m^{6-4} p^{2-1}}{2 \cdot 5} = \dfrac{9m^2 p}{5}$

10. $\dfrac{x-12}{12-x} = \dfrac{x-12}{-1(x-12)} = \dfrac{1}{-1} = -1$

11. $\dfrac{5x-15}{25x-75} = \dfrac{5(x-3)}{5 \cdot 5(x-3)} = \dfrac{1}{5}$

12. $\dfrac{22x+8}{11x+4} = \dfrac{2(11x+4)}{11x+4} = 2$

13. $\dfrac{2x}{2x^2-2x} = \dfrac{2x}{2x(x-1)} = \dfrac{1}{x-1}$

14. $\dfrac{x+7}{x^2-49} = \dfrac{x+7}{(x+7)(x-7)} = \dfrac{1}{x-7}$

15.
$$\frac{2x^2+4x-30}{x^2+x-20} = \frac{2(x^2+2x-15)}{(x+5)(x-4)}$$
$$= \frac{2(x+5)(x-3)}{(x+5)(x-4)}$$
$$= \frac{2(x-3)}{x-4}$$

16.
$$\frac{xy-3x+2y-6}{x^2+4x+4} = \frac{x(y-3)+2(y-3)}{(x+2)^2}$$
$$= \frac{(y-3)(x+2)}{(x+2)^2} = \frac{y-3}{x+2}$$

17. $C(x) = \dfrac{35x+4200}{x}$

   a. $C(50) = \dfrac{35(50)+4200}{50} = \$119$

   b. $C(100) = \dfrac{35(100)+4200}{100} = \$77$

   c. Decrease

18. $\dfrac{5}{x^3} \cdot \dfrac{x^2}{15} = \dfrac{1}{3x}$

19.
$$\frac{3x^4yz^3}{15x^2y^2} \cdot \frac{10xy}{z^6} = \frac{30x^{4+1}y^{1+1}z^3}{15x^2y^2z^6}$$
$$= \frac{2x^5y^2z^3}{x^2y^2z^6}$$
$$= 2x^{5-2}z^{3-6}$$
$$= \frac{2x^3}{z^3}$$

20. $\dfrac{4-x}{5} \cdot \dfrac{15}{2x-8} = \dfrac{-1(x-4)}{5} \cdot \dfrac{15}{2(x-4)} = -\dfrac{3}{2}$

21.
$$\frac{x^2-6x+9}{2x^2-18} \cdot \frac{4x+12}{5x-15}$$
$$= \frac{(x-3)^2}{2(x^2-9)} \cdot \frac{4(x+3)}{5(x-3)}$$
$$= \frac{x-3}{(x+3)(x-3)} \cdot \frac{2(x+3)}{5} = \frac{2}{5}$$

22.
$$\frac{a-4b}{a^2+ab} \cdot \frac{b^2-a^2}{8b-2a}$$
$$= \frac{a-4b}{a(a+b)} \cdot \frac{(b-a)(b+a)}{2(4b-a)}$$
$$= \frac{-1(4b-a)}{a} \cdot \frac{b-a}{2(4b-a)}$$
$$= \frac{-(b-a)}{2a}$$
$$= \frac{a-b}{2a}$$

23.
$$\frac{x^2-x-12}{2x^2-32} \cdot \frac{x^2+8x+16}{3x^2+21x+36}$$
$$= \frac{(x-4)(x+3)}{2(x^2-16)} \cdot \frac{(x+4)^2}{3(x^2+7x+12)}$$
$$= \frac{(x-4)(x+3)}{2(x+4)(x-4)} \cdot \frac{(x+4)^2}{3(x+4)(x+3)} = \frac{1}{6}$$

24.
$$\frac{2x^3+54}{5x^2+5x-30} \cdot \frac{6x+12}{3x^2-9x+27}$$
$$= \frac{2(x^3+27)}{5(x^2+x-6)} \cdot \frac{6(x+2)}{3(x^2-3x+9)}$$
$$= \frac{2(x+3)(x^2-3x+9)}{5(x+3)(x-2)} \cdot \frac{2(x+2)}{x^2-3x+9}$$
$$= \frac{4(x+2)}{5(x-2)}$$

25. $\dfrac{3}{4x} \div \dfrac{8}{2x^2} = \dfrac{3}{4x} \cdot \dfrac{2x^2}{8} = \dfrac{3x}{16}$

26.
$$\frac{4x+8y}{3} \div \frac{5x+10y}{9} = \frac{4x+8y}{3} \cdot \frac{9}{5x+10y}$$
$$= \frac{4(x+2y)\cdot 9}{3\cdot 5(x+2y)}$$
$$= \frac{12}{5}$$

27. $\dfrac{5ab}{14c^3} \div \dfrac{10a^4b^2}{6ac^5} = \dfrac{5ab}{14c^3} \cdot \dfrac{6ac^5}{10a^4b^2}$

$= \dfrac{30a^2bc^5}{140a^4b^2c^3}$

$= \dfrac{3a^{2-4}b^{1-2}c^{5-3}}{14}$

$= \dfrac{3a^{-2}b^{-1}c^2}{14}$

$= \dfrac{3c^2}{14a^2b}$

28. $\dfrac{2}{5x} \div \dfrac{4-18x}{6-27x} = \dfrac{2}{5x} \cdot \dfrac{6-27x}{4-18x}$

$= \dfrac{2 \cdot 3(2-9x)}{5x \cdot 2(2-9x)}$

$= \dfrac{3}{5x}$

29. $\dfrac{x^2-25}{3} \div \dfrac{x^2-10x+25}{x^2-x-20}$

$= \dfrac{x^2-25}{3} \cdot \dfrac{x^2-x-20}{x^2-10x+25}$

$= \dfrac{(x+5)(x-5)}{3} \cdot \dfrac{(x-5)(x+4)}{(x-5)^2}$

$= \dfrac{(x+5)(x+4)}{3}$

30. $\dfrac{a-4b}{a^2+ab} \div \dfrac{20b-5a}{b^2-a^2}$

$= \dfrac{a-4b}{a(a+b)} \cdot \dfrac{(b+a)(b-a)}{5(4b-a)}$

$= \dfrac{-1(4b-a)(b-a)}{5a(4b-a)}$

$= \dfrac{a-b}{5a}$

31. $\dfrac{7x+28}{2x+4} \div \dfrac{x^2+2x-8}{x^2-2x-8}$

$= \dfrac{7(x+4)}{2(x+2)} \cdot \dfrac{(x-4)(x+2)}{(x+4)(x-2)}$

$= \dfrac{7(x-4)}{2(x-2)}$

32. $\dfrac{3x+3}{x-1} \div \dfrac{x^2-6x-7}{x^2-1}$

$= \dfrac{3(x+1)}{x-1} \cdot \dfrac{(x+1)(x-1)}{(x-7)(x+1)}$

$= \dfrac{3(x+1)}{x-7}$

33. $\dfrac{2x-x^2}{x^3-8} \div \dfrac{x^2}{x^2+2x+4}$

$= \dfrac{x(2-x)}{(x-2)(x^2+2x+4)} \cdot \dfrac{x^2+2x+4}{x^2}$

$= \dfrac{-(x-2)}{x(x-2)}$

$= -\dfrac{1}{x}$

34. $\dfrac{5a^2-20}{a^3+2a^2+a+2} \div \dfrac{7a}{a^3+a}$

$= \dfrac{5(a^2-4)}{a^2(a+2)+1(a+2)} \cdot \dfrac{a(a^2+1)}{7a}$

$= \dfrac{5(a+2)(a-2)(a^2+1)}{7(a+2)(a^2+1)} = \dfrac{5(a-2)}{7}$

35. $\dfrac{2a}{21} \div \dfrac{3a^2}{7} \cdot \dfrac{4}{a} = \dfrac{2a}{21} \cdot \dfrac{7}{3a^2} \cdot \dfrac{4}{a} = \dfrac{8}{9a^2}$

36. $\dfrac{5x-15}{3-x} \cdot \dfrac{x+2}{10x+20} \cdot \dfrac{x^2-9}{x^2-x-6}$

$= \dfrac{5(x-3)}{-1(x-3)} \cdot \dfrac{x+2}{10(x+2)} \cdot \dfrac{(x+3)(x-3)}{(x-3)(x+2)}$

$= -\dfrac{x+3}{2(x+2)}$

37. $\dfrac{4a+8}{5a^2-20} \cdot \dfrac{3a^2-6a}{a+3} \div \dfrac{2a^2}{5a+15}$

$= \dfrac{4(a+2)}{5(a^2-4)} \cdot \dfrac{3a(a-2)}{a+3} \cdot \dfrac{5(a+3)}{2a^2}$

$= \dfrac{2(a+2)}{(a+2)(a-2)} \cdot \dfrac{3(a-2)}{1} \cdot \dfrac{1}{a} = \dfrac{6}{a}$

38. LCD $= (9)(2) = 18.$

39. LCD $= 60x^2y^5$

40. LCD $= 2x(x-2)$

41. LCD $= 5x(x-5)$

42. The first denominator $= 5x^3$.
The second denominator $= (x-4)(x+7)$
The third denominator $= 10x(x-3)$.
The LCD $= 10x^3(x-4)(x+7)(x-3)$.

43. $\dfrac{2}{15}+\dfrac{4}{15}=\dfrac{2+4}{15}=\dfrac{6}{15}=\dfrac{2}{5}$

44. $\dfrac{4}{x-4}+\dfrac{x}{x-4}=\dfrac{4+x}{x-4}$

45. $\dfrac{4}{3x^2}+\dfrac{2}{3x^2}=\dfrac{4+2}{3x^2}=\dfrac{6}{3x^2}=\dfrac{2}{x^2}$

46. $\dfrac{1}{x-2}-\dfrac{1}{4-2x}=\dfrac{1}{x-2}-\dfrac{1}{-2(x-2)}$
$=\dfrac{1\cdot 2}{(x-2)\cdot 2}+\dfrac{1}{2(x-2)}$
$=\dfrac{2+1}{2(x-2)}$
$=\dfrac{3}{2(x-2)}$

47. $\dfrac{2x+1}{x^2+x-6}+\dfrac{2-x}{x^2+x-6}=\dfrac{2x+1+(2-x)}{x^2+x-6}$
$=\dfrac{x+3}{(x+3)(x-2)}$
$=\dfrac{1}{x-2}$

48. $\dfrac{7}{2x}+\dfrac{5}{6x}=\dfrac{7\cdot 3}{2x\cdot 3}+\dfrac{5}{6x}$
$=\dfrac{21}{6x}+\dfrac{5}{6x}=\dfrac{26}{6x}=\dfrac{13}{3x}$

49. $\dfrac{1}{3x^2y^3}-\dfrac{1}{5x^4y}=\dfrac{5x^2}{15x^4y^3}-\dfrac{3y^2}{15x^4y^3}$
$=\dfrac{5x^2-3y^2}{15x^4y^3}$

50. $\dfrac{1}{10-x}+\dfrac{x-1}{x-10}=\dfrac{-1}{x-10}+\dfrac{x-1}{x-10}$
$=\dfrac{-1+x-1}{x-10}=\dfrac{x-2}{x-10}$

51. $\dfrac{x-2}{x+1}-\dfrac{x-3}{x-1}$
$=\dfrac{(x-2)(x-1)-(x-3)(x+1)}{(x+1)(x-1)}$
$=\dfrac{x^2-3x+2-(x^2-2x-3)}{(x+1)(x-1)}$
$=\dfrac{x^2-3x+2-x^2+2x+3}{(x+1)(x-1)}$
$=\dfrac{-x+5}{(x+1)(x-1)}$

52. $\dfrac{x}{9-x^2}-\dfrac{2}{5x-15}$
$=\dfrac{x}{(3+x)(3-x)}-\dfrac{2}{5(x-3)}$
$=\dfrac{-5x}{5(3+x)(x-3)}-\dfrac{2(3+x)}{5(3+x)(x-3)}$
$=\dfrac{-5x-6-2x}{5(x+3)(x-3)}$
$=\dfrac{-7x-6}{5(x+3)(x-3)}$

53. $2x+1-\dfrac{1}{x-3}=\dfrac{(2x+1)(x-3)-1}{x-3}$
$=\dfrac{2x^2-5x-3-1}{x-3}$
$=\dfrac{2x^2-5x-4}{x-3}$

54. $\dfrac{2}{a^2-2a+1}+\dfrac{3}{a^2-1}$
$=\dfrac{2}{(a-1)^2}+\dfrac{3}{(a+1)(a-1)}$
$=\dfrac{2(a+1)+3(a-1)}{(a-1)^2(a+1)}$
$=\dfrac{2a+2+3a-3}{(a+1)^2(a-1)}$
$=\dfrac{5a-1}{(a-1)^2(a+1)}$

55. $\dfrac{x}{9x^2+12x+16}-\dfrac{3x+4}{27x^3-64}$

$=\dfrac{x}{9x^2+12x+16}-\dfrac{3x+4}{(3x-4)(9x^2+12x+16)}$

$=\dfrac{x(3x-4)-(3x+4)}{(3x-4)(9x^2+12x+16)}$

$=\dfrac{3x^2-4x-3x-4}{(3x-4)(9x^2+12x+16)}$

$=\dfrac{3x^2-7x-4}{(3x-4)(9x^2+12x+16)}$

56. $\dfrac{2}{x-1}-\dfrac{3x}{3x-3}+\dfrac{1}{2(x-1)}$

$=\dfrac{2}{x-1}-\dfrac{3x}{3(x-1)}+\dfrac{1}{2(x-1)}$

$=\dfrac{2\cdot6-3x\cdot2+1\cdot3}{6(x-1)}$

$=\dfrac{15-6x}{6(x-1)}$

$=\dfrac{3(5-2x)}{6(x-1)}=\dfrac{5-2x}{2(x-1)}$

57. $\dfrac{3}{2x}\cdot\left(\dfrac{2}{x+1}-\dfrac{2}{x-3}\right)$

$=\dfrac{3}{2x}\cdot\dfrac{2(x-3)-2(x+1)}{(x+1)(x-3)}$

$=\dfrac{3}{2x}\cdot\dfrac{-8}{(x+1)(x-3)}$

$=-\dfrac{12}{x(x+1)(x-3)}$

58. $\left(\dfrac{2}{x}-\dfrac{1}{5}\right)\cdot\left(\dfrac{2}{x}+\dfrac{1}{3}\right)=\left(\dfrac{10-x}{5x}\right)\left(\dfrac{6+x}{3x}\right)$

$=\dfrac{(10-x)(6+x)}{15x^2}$

59. $\dfrac{2}{x^2-16}-\dfrac{3x}{x^2+8x+16}+\dfrac{3}{x+4}$

$=\dfrac{2}{(x+4)(x-4)}-\dfrac{3x}{(x+4)^2}+\dfrac{3}{x+4}$

$=\dfrac{2(x+4)}{(x+4)^2(x-4)}-\dfrac{3x(x-4)}{(x+4)^2(x-4)}$

$\quad+\dfrac{3(x+4)(x-4)}{(x+4)^2(x-4)}$

$=\dfrac{2x+8-3x^2+12x+3x^2-48}{(x+4)^2(x-4)}$

$=\dfrac{14x-40}{(x+4)^2(x-4)}$

60. $P=\dfrac{1}{x}+\dfrac{1}{x}+\dfrac{1}{x}+\dfrac{2}{x}+\dfrac{5}{2x}+\dfrac{2}{x}+\dfrac{3}{2x}$

$=\dfrac{1+1+1+2+2}{x}+\dfrac{5+3}{2x}$

$=\dfrac{7}{x}+\dfrac{8}{2x}$

$=\dfrac{7\cdot2}{x\cdot2}+\dfrac{8}{2x}$

$=\dfrac{14}{2x}+\dfrac{8}{2x}=\dfrac{22}{2x}=\dfrac{11}{x}$

61. $\dfrac{\frac{2}{5}}{\frac{3}{5}}=\dfrac{2}{5}\cdot\dfrac{5}{3}=\dfrac{2}{3}$

62. $\dfrac{1-\frac{3}{4}}{2+\frac{1}{4}}=\dfrac{\left(1-\frac{3}{4}\right)4}{\left(2+\frac{1}{4}\right)4}=\dfrac{4-3}{8+1}=\dfrac{1}{9}$

63. $\dfrac{\frac{1}{x}-\frac{2}{3x}}{\frac{5}{2x}-\frac{1}{3}}=\dfrac{\left(\frac{1}{x}-\frac{2}{3x}\right)6x}{\left(\frac{5}{2x}-\frac{1}{3}\right)6x}=\dfrac{6-4}{15-2x}=\dfrac{2}{15-2x}$

64. $\dfrac{\frac{x^2}{15}}{\frac{x+1}{5x}}=\dfrac{x^2}{15}\cdot\dfrac{5x}{x+1}=\dfrac{x^3}{3(x+1)}$

65. $\dfrac{\dfrac{3}{y^2}}{\dfrac{6}{y^3}} = \dfrac{3}{y^2} \cdot \dfrac{y^3}{6} = \dfrac{y}{2}$

66. $\dfrac{\dfrac{x+2}{3}}{\dfrac{5}{x-2}} = \dfrac{x+2}{3} \cdot \dfrac{x-2}{5} = \dfrac{(x+2)(x-2)}{15}$

67. $\dfrac{2-\dfrac{3}{2x}}{x-\dfrac{2}{5x}} = \dfrac{\left(2-\dfrac{3}{2x}\right)10x}{\left(x-\dfrac{2}{5x}\right)10x} = \dfrac{20x-15}{10x^2-4}$

68. $\dfrac{1+\dfrac{x}{y}}{\dfrac{x^2}{y^2}-1} = \dfrac{\left(1+\dfrac{x}{y}\right)y^2}{\left(\dfrac{x^2}{y^2}-1\right)y^2}$

$= \dfrac{y^2+xy}{x^2-y^2}$

$= \dfrac{y(y+x)}{(x+y)(x-y)}$

$= \dfrac{y}{x-y}$

69. $\dfrac{\dfrac{5}{x}+\dfrac{1}{xy}}{\dfrac{3}{x^2}} = \dfrac{\dfrac{5y}{xy}+\dfrac{1}{xy}}{\dfrac{3}{x^2}}$

$= \dfrac{\dfrac{5y+1}{xy}}{\dfrac{3}{x^2}}$

$= \dfrac{5y+1}{xy} \cdot \dfrac{x^2}{3}$

$= \dfrac{x(5y+1)}{3y}$ or $\dfrac{5xy+x}{3y}$

70. $\dfrac{\dfrac{x}{3}-\dfrac{3}{x}}{1+\dfrac{3}{x}} = \dfrac{\left(\dfrac{x}{3}-\dfrac{3}{x}\right)3x}{\left(1+\dfrac{3}{x}\right)3x}$

$= \dfrac{x^2-9}{3x+9}$

$= \dfrac{(x+3)(x-3)}{3(x+3)}$

$= \dfrac{x-3}{3}$

71. $\dfrac{\dfrac{1}{x-1}+1}{\dfrac{1}{x+1}-1} = \dfrac{\left(\dfrac{1}{x-1}+1\right)(x+1)(x-1)}{\left(\dfrac{1}{x+1}-1\right)(x+1)(x-1)}$

$= \dfrac{(x+1)+(x+1)(x-1)}{(x-1)-(x+1)(x-1)}$

$= \dfrac{x+1+x^2-1}{x-1-(x^2-1)}$

$= \dfrac{x^2+x}{-x^2+x}$

$= \dfrac{x(x+1)}{-x(x-1)}$

$= -\dfrac{x+1}{x-1}$

72. $\dfrac{2}{1-\dfrac{2}{x}} = \dfrac{2\cdot x}{\left(1-\dfrac{2}{x}\right)x} = \dfrac{2x}{x-2}$

73. $\dfrac{1}{1+\dfrac{2}{1-\dfrac{1}{x}}} = \dfrac{1}{1+\dfrac{2}{\dfrac{x-1}{x}}} = \dfrac{1}{1+\dfrac{2x}{x-1}}$

$= \dfrac{1(x-1)}{\left(1+\dfrac{2x}{x-1}\right)(x-1)}$

$= \dfrac{x-1}{x-1+2x}$

$= \dfrac{x-1}{3x-1}$

74. $\dfrac{\dfrac{x^2+5x-6}{4x+3}}{\dfrac{(x+6)^2}{8x+6}} = \dfrac{x^2+5x-6}{4x+3} \cdot \dfrac{8x+6}{(x+6)^2}$

$\qquad = \dfrac{(x+6)(x-1)}{4x+3} \cdot \dfrac{2(4x+3)}{(x+6)^2}$

$\qquad = \dfrac{2(x-1)}{x+6}$

75. $\dfrac{\dfrac{x-3}{x+3}+\dfrac{x+3}{x-3}}{\dfrac{x-3}{x+3}-\dfrac{x+3}{x-3}} = \dfrac{\left(\dfrac{x-3}{x+3}+\dfrac{x+3}{x-3}\right)(x+3)(x-3)}{\left(\dfrac{x-3}{x+3}-\dfrac{x+3}{x-3}\right)(x+3)(x-3)}$

$\qquad = \dfrac{(x-3)^2+(x+3)^2}{(x-3)^2-(x+3)^2}$

$\qquad = \dfrac{x^2-6x+9+x^2+6x+9}{x^2-6x+9-x^2-6x-9}$

$\qquad = \dfrac{2x^2+18}{-12x}$

$\qquad = -\dfrac{2(x^2+9)}{12x}$

$\qquad = -\dfrac{x^2+9}{6x}$

76. $\dfrac{\dfrac{3}{x-1}-\dfrac{2}{1-x}}{\dfrac{2}{x-1}-\dfrac{2}{x}} = \dfrac{\dfrac{3}{x-1}+\dfrac{2}{x-1}}{\dfrac{2}{x-1}-\dfrac{2}{x}}$

$\qquad = \dfrac{\dfrac{5}{x-1}}{\dfrac{2x-2(x-1)}{x(x-1)}}$

$\qquad = \dfrac{5}{x-1} \cdot \dfrac{x(x-1)}{2}$

$\qquad = \dfrac{5x}{2}$

77. $f(x) = \dfrac{3}{x}$

    a. $f(a+h) = \dfrac{3}{a+h}$

    b. $f(a) = \dfrac{3}{a}$

    c. $\dfrac{f(a+h)-f(a)}{h} = \dfrac{\dfrac{3}{a+h}-\dfrac{3}{a}}{h}$

    d. $\dfrac{f(a+h)-f(a)}{h} = \dfrac{\dfrac{3}{a+h}-\dfrac{3}{a}}{h}$

$\qquad = \dfrac{\left(\dfrac{3}{a+h}-\dfrac{3}{a}\right)a(a+h)}{h \cdot a(a+h)}$

$\qquad = \dfrac{3a-3(a+h)}{ha(a+h)}$

$\qquad = \dfrac{-3h}{ha(a+h)} = \dfrac{-3}{a(a+h)}$

78. $\dfrac{3x^5yb^9}{9xy^7} = \dfrac{x^{5-1}y^{1-7}b^9}{3} = \dfrac{x^4y^{-6}b^9}{3} = \dfrac{x^4b^9}{3y^6}$

79. $\dfrac{-9xb^4z^3}{-4axb^2} = \dfrac{9b^{4-2}z^3}{4a} = \dfrac{9b^2z^3}{4a}$

80. $\dfrac{4xy+2x^2-9}{4xy} = \dfrac{4xy}{4xy} + \dfrac{2x^2}{4xy} - \dfrac{9}{4xy}$

$\qquad = 1 + \dfrac{x}{2y} - \dfrac{9}{4xy}$

81. $\dfrac{12xb^2+16xb^4}{4xb^3} = \dfrac{12xb^2}{4xb^3} + \dfrac{16xb^4}{4xb^3}$

$\qquad = \dfrac{3}{b} + 4b$

82.

$$\begin{array}{r}
3x^3+9x^2+2x+6 \\
x-3\overline{\smash{)}\,3x^4+0x^3-25x^2+0x-20} \\
\underline{3x^4-9x^3}\\
9x^3-25x^2 \\
\underline{9x^3-27x^2}\\
2x^2+0x \\
\underline{2x^2-6x}\\
6x-20 \\
\underline{6x-18}\\
-2
\end{array}$$

Answer: $3x^3+9x^2+2x+6 - \dfrac{2}{x-3}$

83.

$$
\begin{array}{r}
2x^3-4x^2+7x-9 \\
x+2\overline{)2x^4+0x^3-x^2+\ 5x-12} \\
\underline{2x^4+4x^3} \\
-4x^3-x^2 \\
\underline{-4x^3-8x^2} \\
7x^2+\ 5x \\
\underline{7x^2+14x} \\
-9x-12 \\
\underline{-9x-18} \\
6
\end{array}
$$

Answer: $2x^3-4x^2+7x-9+\dfrac{6}{x+2}$

84.

$$
\begin{array}{r}
x^3+x-1 \\
2x-1\overline{)2x^4-x^3+2x^2-3x+1} \\
\underline{2x^4-x^3} \\
2x^2-3x \\
\underline{2x^2-\ x} \\
-2x+1 \\
\underline{-2x+1} \\
0
\end{array}
$$

Answer: x^3+x-1

85.

$$
\begin{array}{r}
x^2-1 \\
2x+3\overline{)2x^3+3x^2-2x+2} \\
\underline{2x^3+3x^2} \\
-2x+2 \\
\underline{-2x-3} \\
5
\end{array}
$$

Answer: $x^2-1+\dfrac{5}{2x+3}$

86.

$$
\begin{array}{r}
3x^2+2x-1 \\
x^2+x+2\overline{)3x^4+5x^3+7x^2+3x-2} \\
\underline{3x^4+3x^3+6x^2} \\
2x^3+\ x^2+3x \\
\underline{2x^3+2x^2+4x} \\
-x^2-\ x-2 \\
\underline{-x^2-\ x-2} \\
0
\end{array}
$$

Answer: $3x^2+2x-1$

87.

$$
\begin{array}{r}
3x^2\qquad\ +6 \\
3x^2-2x-5\overline{)9x^4-6x^3+\ 3x^2-12x-30} \\
\underline{9x^4-6x^3-15x^2} \\
18x^2-12x-30 \\
\underline{18x^2-12x-30} \\
0
\end{array}
$$

Answer: $3x^2+6$

88.

$$
\begin{array}{r|rrrr}
2 & 3 & 0 & 12 & -4 \\
& & 6 & 12 & 48 \\
\hline
& 3 & 6 & 24 & 44
\end{array}
$$

Answer: $3x^2+6x+24+\dfrac{44}{x-2}$

89.

$$
\begin{array}{r|rrrr}
-\frac{3}{2} & 3 & 2 & -4 & -1 \\
& & -\frac{9}{2} & \frac{15}{4} & \frac{3}{8} \\
\hline
& 3 & -\frac{5}{2} & -\frac{1}{4} & -\frac{5}{8}
\end{array}
$$

Answer: $3x^2-\dfrac{5}{2}x-\dfrac{1}{4}-\dfrac{5}{8\left(x+\frac{3}{2}\right)}$

90.

$$
\begin{array}{r|rrrrrr}
-1 & 1 & 0 & 0 & 0 & 0 & -1 \\
& & -1 & 1 & -1 & 1 & -1 \\
\hline
& 1 & -1 & 1 & -1 & 1 & -2
\end{array}
$$

Answer: $x^4-x^3+x^2-x+1-\dfrac{2}{x+1}$

91.

$$
\begin{array}{r|rrrr}
3 & 1 & 0 & 0 & -81 \\
& & 3 & 9 & 27 \\
\hline
& 1 & 3 & 9 & -54
\end{array}
$$

Answer: $x^2+3x+9-\dfrac{54}{x-3}$

92.

$$
\begin{array}{r|rrrrr}
4 & 3 & 1 & -1 & 0 & -2 \\
& & 12 & 52 & 204 & 816 \\
\hline
& 3 & 13 & 51 & 204 & 814
\end{array}
$$

Answer: $3x^3+13x^2+51x+204+\dfrac{814}{x-4}$

93.

$$
\begin{array}{r|rrrrr}
-2 & 3 & 0 & -2 & 0 & 10 \\
& & -6 & 12 & -20 & 40 \\
\hline
& 3 & -6 & 10 & -20 & 50
\end{array}
$$

Answer: $3x^3 - 6x^2 + 10x - 20 + \dfrac{50}{x+2}$

94.

$$
\begin{array}{r|rrrrrr}
4 & 3 & 0 & 0 & 0 & -9 & 7 \\
& & 12 & 48 & 192 & 768 & 3036 \\
\hline
& 3 & 12 & 48 & 192 & 759 & 3043
\end{array}
$$

Thus, $P(4) = 3043$.

95.

$$
\begin{array}{r|rrrrrr}
-5 & 3 & 0 & 0 & 0 & -9 & 7 \\
& & -15 & 75 & -375 & 1875 & -9330 \\
\hline
& 3 & -15 & 75 & -375 & 1866 & -9323
\end{array}
$$

Thus, $P(-5) = -9323$.

96.

$$
\begin{array}{r|rrrrrr}
\frac{2}{3} & 3 & 0 & 0 & 0 & -9 & 7 \\
& & 2 & \frac{4}{3} & \frac{8}{9} & \frac{16}{27} & -\frac{454}{81} \\
\hline
& 3 & -15 & \frac{4}{3} & \frac{8}{9} & -\frac{227}{27} & \frac{113}{81}
\end{array}
$$

Thus, $P\left(\dfrac{2}{3}\right) = \dfrac{113}{81}$.

97.

$$
\begin{array}{r|rrrrrr}
-\frac{1}{2} & 3 & 0 & 0 & 0 & -9 & 7 \\
& & -\frac{3}{2} & \frac{3}{4} & -\frac{3}{8} & \frac{3}{16} & \frac{141}{32} \\
\hline
& 3 & -\frac{3}{2} & \frac{3}{4} & -\frac{3}{8} & -\frac{141}{16} & \frac{365}{32}
\end{array}
$$

Thus, $P\left(-\dfrac{1}{2}\right) = \dfrac{365}{32}$.

98.

$$
\begin{array}{r|rrrrr}
3 & 1 & -1 & -6 & -6 & 18 \\
& & 3 & 6 & 0 & -18 \\
\hline
& 1 & 2 & 0 & -6 & 0
\end{array}
$$

length $= (x^3 + 2x^2 - 6)$ miles

99. $\dfrac{2}{5} = \dfrac{x}{15}$

$5x = 2(15)$

$5x = 30$

$x = 6$

100. $\dfrac{3}{x} + \dfrac{1}{3} = \dfrac{5}{x}$

$3x\left(\dfrac{3}{x} + \dfrac{1}{3}\right) = 3x \cdot \dfrac{5}{x}$

$9 + x = 15$

$x = 6$

101. $4 + \dfrac{8}{x} = 8$

$x\left(4 + \dfrac{8}{x}\right) = x \cdot 8$

$4x + 8 = 8x$

$8 = 4x$

$2 = x$

102. $\dfrac{2x+3}{5x-9} = \dfrac{3}{2}$

$3(5x - 9) = 2(2x + 3)$

$15x - 27 = 4x + 6$

$11x = 33$

$x = 3$

103. $\dfrac{1}{x-2} - \dfrac{3x}{x^2-4} = \dfrac{2}{x+2}$

$\dfrac{1}{x-2} - \dfrac{3x}{(x+2)(x-2)} = \dfrac{2}{x+2}$

$(x + 2) - 3x = 2(x - 2)$

$-2x + 2 = 2x - 4$

$-4x = -6$

$x = \dfrac{-6}{-4} = \dfrac{3}{2}$

104. $\dfrac{7}{x} - \dfrac{x}{7} = 0$

$\dfrac{7}{x} = \dfrac{x}{7}$

$x^2 = 49$

$x^2 - 49 = 0$

$(x + 7)(x - 7) = 0$

$x + 7 = 0$ or $x - 7 = 0$

$x = -7$ or $x = 7$

105. $\dfrac{x-2}{x^2-7x+10}=\dfrac{1}{5x-10}-\dfrac{1}{x-5}$

$\dfrac{x-2}{(x-5)(x-2)}=\dfrac{1}{5(x-2)}-\dfrac{1}{x-5}$

$5(x-2)=1(x-5)-1\cdot 5(x-2)$

$5x-10=x-5-5x+10$

$5x-10=-4x+5$

$9x=15$

$x=\dfrac{15}{9}=\dfrac{5}{3}$

106. $\dfrac{5}{x^2-7x}+\dfrac{4}{2x-14}=\dfrac{5}{x(x-7)}+\dfrac{2}{x-7}$

$=\dfrac{5+2\cdot x}{x(x-7)}$

$=\dfrac{2x+5}{x(x-7)}$

107. $3-\dfrac{5}{x}-\dfrac{2}{x^2}=0$

$3x^2-5x-2=0$

$(3x+1)(x-2)=0$

$3x+1=0\quad\text{or}\quad x-2=0$

$3x=-1\quad\text{or}\quad x=2$

$x=-\dfrac{1}{3}$

108. $\dfrac{4}{3-x}-\dfrac{7}{2x-6}+\dfrac{5}{x}$

$=\dfrac{-4}{x-3}-\dfrac{7}{2(x-3)}+\dfrac{5}{x}$

$=\dfrac{-4(2x)-7x+5\cdot 2(x-3)}{2x(x-3)}$

$=\dfrac{-8x-7x+10x-30}{2x(x-3)}$

$=\dfrac{-5x-30}{2x(x-3)}$

109. $A=\dfrac{h(a+b)}{2}$

$2A=h(a+b)$

$2A=ah+bh$

$2A-bh=ah$

$\dfrac{2A-bh}{h}=a$

110. $\dfrac{1}{R}=\dfrac{1}{R_1}+\dfrac{1}{R_2}$

$RR_1R_2\left(\dfrac{1}{R}\right)=RR_1R_2\left(\dfrac{1}{R_1}+\dfrac{1}{R_2}\right)$

$R_1R_2=RR_2+RR_1$

$R_1R_2-RR_2=RR_1$

$R_2(R_1-R)=RR_1$

$R_2=\dfrac{RR_1}{R_1-R}$

111. $I=\dfrac{E}{R+r}$

$I(R+r)=E$

$IR+Ir=E$

$IR=E-Ir$

$R=\dfrac{E-Ir}{I}$

112. $A=P+Prt$

$A-P=Prt$

$\dfrac{A-P}{Pt}=r$

113. $H=\dfrac{kA(T_1-T_2)}{L}$

$HL=kA(T_1-T_2)$

$\dfrac{HL}{k(T_1-T_2)}=A$

114. Let $x=$ the number.

$x+2\left(\dfrac{1}{x}\right)=3$

$x\left(x+\dfrac{2}{x}\right)=3x$

$x^2+2=3x$

$x^2-3x+2=0$

$(x-2)(x-1)=0$

$x-2=0\quad\text{or}\quad x-1=0$

$x=2\quad\text{or}\quad\quad x=1$

The number is either 1 or 2.

115. Let x = the number added to the numerator. Then $2x$ = the number added to the denominator. Thus,

$$\frac{3+x}{7+2x} = \frac{10}{21}$$
$$21(3+x) = 10(7+2x)$$
$$63 + 21x = 70 + 20x$$
$$x = 7$$

The number is 7.

116. Let x = the numerator. Then $x + 2$ = the denominator.

$$\frac{x-3}{(x+2)+5} = \frac{2}{3}$$
$$\frac{x-3}{x+7} = \frac{2}{3}$$
$$3(x-3) = 2(x+7)$$
$$3x - 9 = 2x + 14$$
$$x = 23; \; x + 2 = 23 + 2 = 25$$

The original fraction was $\frac{23}{25}$.

117. Let n = the first even integer. Then $n + 2$ = the next even integer.

$$\frac{1}{n} + \frac{1}{n+2} = -\frac{9}{40}$$
$$40n(n+2)\left(\frac{1}{n} + \frac{1}{n+2}\right) = 40n(n+2)\left(-\frac{9}{40}\right)$$
$$40(n+2) + 40n = -9n(n+2)$$
$$40n + 80 + 40n = -9n^2 - 18n$$
$$9n^2 + 98n + 80 = 0$$
$$(9n+8)(n+10) = 0$$
$$9n + 8 = 0 \quad \text{or} \quad n + 10 = 0$$
$$9n = -8 \quad \text{or} \qquad n = -10$$
$$n \ne -\frac{8}{9} \qquad n + 2 = -10 + 2 = -8$$

Discard the non-integer solution $-\frac{8}{9}$.

The two integers are –10 and –8.

118. Let t = time for all three to paint room.

$$\frac{1}{4} + \frac{1}{5} + \frac{1}{6} = \frac{1}{t}$$
$$120t\left(\frac{1}{4} + \frac{1}{5} + \frac{1}{6}\right) = 120t\left(\frac{1}{t}\right)$$
$$30t + 24t + 20t = 120$$
$$74t = 120$$
$$t = \frac{120}{74} = 1\frac{23}{37} \text{ hours}$$

119. Let t = time for Tom to type labels alone.

$$\frac{1}{6} + \frac{1}{t} = \frac{1}{4}$$
$$12t\left(\frac{1}{6} + \frac{1}{t}\right) = 12t\left(\frac{1}{4}\right)$$
$$2t + 12 = 3t$$
$$12 = t$$

Tom can type the mailing labels in 12 hours.

120. Let t = time to empty a full tank.

$$\frac{1}{2} - \frac{1}{2.5} = \frac{1}{t}$$
$$10t\left(\frac{1}{2} - \frac{1}{2.5}\right) = 10t\left(\frac{1}{t}\right)$$
$$5t - 4t = 10$$
$$t = 10 \text{ hours}$$

121. Let r = the speed of the car. Then $r + 430$ = the speed of the jet.

$$t_{driving} = t_{flying}$$
$$\frac{210}{r} = \frac{1715}{r+430}$$
$$1715r = 210(r+430)$$
$$1715r = 210r + 90,300$$
$$1505r = 90,300$$
$$r = 60;$$
$$r + 430 = 60 + 430 = 490$$

The speed of the jet is 490 mph.

122.

$$\frac{1}{R} = \frac{1}{R_1} + \frac{1}{R_2}$$

$$\frac{1}{\frac{30}{11}} = \frac{1}{5} + \frac{1}{R_2}$$

$$30R_2\left(\frac{11}{30}\right) = 30R_2\left(\frac{1}{5} + \frac{1}{R_2}\right)$$

$$11R_2 = 6R_2 + 30$$

$$5R_2 = 30$$

$$R_2 = 6 \text{ ohms}$$

123. Let r = the speed of the current.

$$t_{\text{upstream}} = t_{\text{downstream}}$$

$$\frac{72}{32-t} = \frac{120}{32+r}$$

$$72(32+r) = 120(32-t)$$

$$2304 + 72r = 3840 - 120t$$

$$192r = 1536$$

$$r = 8$$

The speed of the current is 8 mph.

124. Let r = the speed of the wind.

$$t_{\text{with wind}} = t_{\text{against wind}}$$

$$\frac{445}{400+r} = \frac{355}{400-r}$$

$$445(400-r) = 355(400+r)$$

$$178,000 - 445r = 142,000 + 355r$$

$$36,000 = 800x$$

$$45 = x$$

The speed of the wind is 45 mph.

125. Let r = the speed of the walker. Then $r+3$ = the speed of the jogger.

$$t_{\text{jogging}} = t_{\text{walking}}$$

$$\frac{14}{r+3} = \frac{8}{r}$$

$$14r = 8(r+3)$$

$$14r = 8r + 24$$

$$6r = 24$$

$$r = 4$$

The speed of the walker is 4 mph.

126. Let r = speed of the faster train. Then $r-18$ = speed of the slower train.

$$t_{\text{faster train}} = t_{\text{slower train}}$$

$$\frac{378}{r} = \frac{270}{r-18}$$

$$378(r-18) = 270r$$

$$378r - 6804 = 270r$$

$$108r = 6804$$

$$r = 63;$$

$$r - 18 = 63 - 18 = 45$$

The speed of the trains is 63 mph and 45 mph.

127. $A = kB$

$$6 = k(14)$$

$$k = \frac{6}{14} = \frac{3}{7}$$

$$A = \frac{3}{7}B = \frac{3}{7}(21) = 3(3) = 9$$

128. $C = \dfrac{k}{D}$

$$12 = \frac{k}{8}$$

$$96 = k$$

$$C = \frac{96}{D}$$

$$C = \frac{96}{24} = 4$$

129. $P = \dfrac{k}{V}$

$$1250 = \frac{k}{2}$$

$$k = 1250(2) = 2500$$

$$P = \frac{2500}{V}$$

$$800 = \frac{2500}{V}$$

$$V = \frac{2500}{800} = 3.125 \text{ cu. ft.}$$

130.
$$A = kr^2$$
$$36\pi = k(3)^2$$
$$36\pi = 9k$$
$$4\pi = k$$
$$A = 4\pi r^2$$
$$a = 4\pi(4)^2 = 64\pi \text{ sq. in.}$$

Chapter 6 Test

1. $f(x) = \dfrac{5x^2}{1-x}$

Undefined values when

$$1 - x = 0$$
$$1 = x$$

Domain $\{x | x$ is a real number and $x \neq 1\}$

2. $f(x) = \dfrac{9x^2 - 9}{x^2 + 4x + 3}$

Undefined values when

$$x^2 + 4x + 3 = 0$$
$$(x+3)(x+1) = 0$$
$$x + 3 = 0 \quad \text{or} \quad x + 1 = 0$$
$$x = -3 \quad \text{or} \quad x = -1$$

Domain

$\{x | x$ is a real number and $x \neq -3, x \neq -1\}$

3. $\dfrac{7x - 21}{24 - 8x} = \dfrac{7(x-3)}{8(3-x)} = \dfrac{7(x-3)}{-8(x-3)} = -\dfrac{7}{8}$

4. $\dfrac{x^2 - 4x}{x^2 + 5x - 36} = \dfrac{x(x-4)}{(x+9)(x-4)} = \dfrac{x}{x+9}$

5.
$$\dfrac{2x^3 + 16}{6x^2 + 12x} \cdot \dfrac{5}{x^2 - 2x + 4}$$
$$= \dfrac{2(x^3 + 8)}{6x(x+2)} \cdot \dfrac{5}{x^2 - 2x + 4}$$
$$= \dfrac{2(x+2)(x^2 - 2x + 4) \cdot 5}{6x(x+2)(x^2 - 2x + 4)} = \dfrac{5}{3x}$$

6.
$$\dfrac{3x^2 - 12}{x^2 + 2x - 8} \div \dfrac{6x + 18}{x + 4}$$
$$= \dfrac{3(x^2 - 4)}{(x+4)(x-2)} \cdot \dfrac{x+4}{6(x+3)}$$
$$= \dfrac{(x+2)(x-2)}{2(x-2)(x+3)}$$
$$= \dfrac{x+2}{2(x+3)}$$

7.
$$\dfrac{4x - 12}{2x - 9} \div \dfrac{3 - x}{4x^2 - 81} \cdot \dfrac{x+3}{5x + 15}$$
$$= \dfrac{4(x-3)}{2x - 9} \cdot \dfrac{(2x+9)(2x-9)}{-1(x-3)} \cdot \dfrac{x+3}{5(x+3)}$$
$$= -\dfrac{4(2x+9)}{5}$$

8.
$$\dfrac{3 + 2x}{10 - x} + \dfrac{13 + x}{x - 10} = \dfrac{3 + 2x}{-1(x - 10)} + \dfrac{13 + x}{x - 10}$$
$$= \dfrac{-3 - 2x + 13 + x}{x - 10}$$
$$= \dfrac{10 - x}{x - 10}$$
$$= \dfrac{-1(x - 10)}{x - 10} = -1$$

9.
$$\dfrac{3}{x^2 - x - 6} + \dfrac{2}{x^2 - 5x + 6}$$
$$= \dfrac{3}{(x-3)(x+2)} + \dfrac{2}{(x-3)(x-2)}$$
$$= \dfrac{3(x-2) + 2(x+2)}{(x-3)(x+2)(x-2)}$$
$$= \dfrac{3x - 6 + 2x + 4}{(x-3)(x+2)(x-2)}$$
$$= \dfrac{5x - 2}{(x-3)(x+2)(x-2)}$$

10. $\dfrac{5}{x-7} - \dfrac{2x}{3x-21} + \dfrac{x}{2x-14}$

$= \dfrac{5}{x-7} - \dfrac{2x}{3(x-7)} + \dfrac{x}{2(x-7)}$

$= \dfrac{5(6) - 2x(2) + 3x}{6(x-7)}$

$= \dfrac{30 - 4x + 3x}{6(x-7)} = \dfrac{30 - x}{6(x-7)}$

11. $\dfrac{3x}{5} \cdot \left(\dfrac{5}{x} - \dfrac{5}{2x} \right) = \dfrac{3x}{5} \cdot \left(\dfrac{5(2) - 5}{2x} \right)$

$= \dfrac{3x}{5} \cdot \left(\dfrac{5}{2x} \right)$

$= \dfrac{3x \cdot 5}{5 \cdot 2x}$

$= \dfrac{3}{2}$

12. $\dfrac{\dfrac{4x}{13}}{\dfrac{20x}{13}} = \dfrac{4x}{13} \cdot \dfrac{13}{20x} = \dfrac{1}{5}$

13. $\dfrac{\dfrac{5}{x} - \dfrac{7}{3x}}{\dfrac{9}{8x} - \dfrac{1}{x}} = \dfrac{\left(\dfrac{5}{x} - \dfrac{7}{3x} \right) 24x}{\left(\dfrac{9}{8x} - \dfrac{1}{x} \right) 24x} = \dfrac{120 - 56}{27 - 24} = \dfrac{64}{3}$

14. $\dfrac{4x^2 y + 9x + 3xz}{3xz} = \dfrac{4x^2 y}{3xz} + \dfrac{9x}{3xz} + \dfrac{3xz}{3xz}$

$= \dfrac{4xy}{3z} + \dfrac{3}{z} + 1$

15.
$$
\begin{array}{r}
2x^2 - x - 2 \\
2x+1 \overline{\smash{)}\; 4x^3 + 0x^2 - 5x + 0} \\
\underline{4x^3 + 2x^2} \\
-2x^2 - 5x \\
\underline{-2x^2 - x} \\
-4x + 0 \\
\underline{-4x - 2} \\
2
\end{array}
$$

Answer: $2x^2 - x - 2 + \dfrac{2}{2x+1}$

16.
$$
\begin{array}{r|rrrrr}
-3 & 4 & -3 & 0 & -1 & -1 \\
   &   & -12 & 45 & -135 & 408 \\
\hline
   & 4 & -15 & 45 & -136 & 407
\end{array}
$$

Answer: $4x^3 - 15x^2 + 45x - 136 + \dfrac{407}{x+3}$

17.
$$
\begin{array}{r|rrrrr}
-2 & 4 & 0 & 7 & -2 & -5 \\
   &   & -8 & 16 & -46 & 96 \\
\hline
   & 4 & -8 & 23 & -48 & 91
\end{array}
$$

Thus, $P(-2) = 91$.

18. $\dfrac{3}{x+2} - \dfrac{1}{5x} = \dfrac{2}{5x^2 + 10x}$

$\dfrac{3}{x+2} - \dfrac{1}{5x} = \dfrac{2}{5x(x+2)}$

$3(5x) - 1(x+2) = 2$

$15x - x - 2 = 2$

$14x = 4$

$x = \dfrac{4}{14} = \dfrac{2}{7}$

19. $\dfrac{x^2 + 8}{x} - 1 = \dfrac{2(x+4)}{x}$

$x\left(\dfrac{x^2 + 8}{x} - 1 \right) = x\left(\dfrac{2(x+4)}{x} \right)$

$(x^2 + 8) - x = 2(x+4)$

$x^2 - x + 8 = 2x + 8$

$x^2 - 3x = 0$

$x(x-3) = 0$

$x = 0 \text{ or } x - 3 = 0$

$x = 3$

Discard the answer 0 as extraneous.

$x = 3$

20.
$$\frac{x+b}{a} = \frac{4x-7a}{b}$$
$$ab\left(\frac{x+b}{a}\right) = ab\left(\frac{4x-7a}{b}\right)$$
$$b(x+b) = a(4x-7a)$$
$$xb+b^2 = 4ax-7a^2$$
$$b^2+7a^2 = 4ax-xb$$
$$b^2+7a^2 = x(4a-b)$$
$$x = \frac{b^2+7a^2}{4a-b}$$

21. Let x = the number.
$$(x+1)\cdot\frac{2}{x} = \frac{12}{5}$$
$$\frac{2(x+1)}{x} = \frac{12}{5}$$
$$\frac{2x+2}{x} = \frac{12}{5}$$
$$5(2x+2) = 12x$$
$$10x+10 = 12x$$
$$10 = 2x$$
$$5 = x$$
The number is 5.

22. Let t = time to weed garden together.

Note that 1 hr and 30 min = $\frac{3}{2}$ hours.

$$\frac{1}{2}+\frac{1}{3} = \frac{1}{t}$$
$$\frac{1}{2}+\frac{2}{3} = \frac{1}{t}$$
$$6t\left(\frac{1}{2}+\frac{2}{3}\right) = 6t\left(\frac{1}{t}\right)$$
$$3t+4t = 6$$
$$7t = 6$$
$$t = \frac{6}{7}$$

It takes them $\frac{6}{7}$ hour.

23.
$$W = \frac{k}{V}$$
$$20 = \frac{k}{12}$$
$$k = 20(12) = 240$$
$$W = \frac{240}{V} = \frac{240}{15} = 16$$

24.
$$Q = kRS^2$$
$$24 = k(3)(4)^2$$
$$24 = 48k$$
$$k = \frac{24}{48} = \frac{1}{2}$$
$$Q = \frac{1}{2}RS^2 = \frac{1}{2}(2)(3)^2 = 9$$

25.
$$S = k\sqrt{d}$$
$$160 = k\sqrt{400}$$
$$160 = 20k$$
$$k = \frac{160}{20} = 8$$
$$S = 8\sqrt{d}$$
$$128 = 8\sqrt{d}$$
$$\sqrt{d} = \frac{128}{8}$$
$$\sqrt{d} = 16$$
$$d = 256$$
The height of the cliff is 256 feet.

Chapter 6 Cumulative Review

1. a. $8x$

   b. $8x+3$

   c. $x \div (-7)$ or $\frac{x}{-7}$

   d. $2x - 1\frac{6}{10} = 2x - 1.6$

2. a. $x - \dfrac{1}{3}$

   b. $5x - 6$

   c. $8x + 3$

   d. $\dfrac{7}{2-x}$

3.
$$\dfrac{y}{3} - \dfrac{y}{4} = \dfrac{1}{6}$$
$$12\left(\dfrac{y}{3} - \dfrac{y}{4}\right) = 12\left(\dfrac{1}{6}\right)$$
$$4y - 3y = 2$$
$$y = 2$$

4.
$$\dfrac{x}{7} + \dfrac{x}{5} = \dfrac{12}{5}$$
$$35\left(\dfrac{x}{7} + \dfrac{x}{5}\right) = 35\left(\dfrac{12}{5}\right)$$
$$5x + 7x = 84$$
$$12x = 84$$
$$x = 7$$

5.
$$c < 200$$
$$-14.25t + 598.69 < 200$$
$$-14.25t < -398.69$$
$$t > 27.98 \approx 28$$
$$1985 + 28 = 2013$$
The consumption of cigarettes will be less than 200 billion per year in 2013 and after.

6. Let x = score on final exam.
$$\dfrac{78 + 65 + 82 + 79 + 2x}{6} \geq 78$$
$$\dfrac{304 + 2x}{6} \geq 78$$
$$6\left(\dfrac{304 + 2x}{6}\right) \geq 6(78)$$
$$304 + 2x \geq 468$$
$$2x \geq 164$$
$$x \geq 82$$
The minimum score she can make on her final is 82.

7. $\left|\dfrac{3x+1}{2}\right| = -2$ is impossible.

The solution is $\varnothing$.

8. $\left|\dfrac{2x-1}{3}\right| + 6 = 3$

$\left|\dfrac{2x-1}{3}\right| = -3$, which is impossible.

The solution is $\varnothing$.

9. $\left|\dfrac{2(x+1)}{3}\right| \leq 0$
$$\dfrac{2(x+1)}{3} = 0$$
$$2(x+1) = 0$$
$$2x + 2 = 0$$
$$2x = -2$$
$$x = -1$$
The solution is -1.

10. $\left|\dfrac{3(x-1)}{4}\right| \geq 2$
$$\dfrac{3(x-1)}{4} \leq -2 \ \text{ or } \ \dfrac{3(x-1)}{4} \geq 2$$
$$\dfrac{3x-3}{4} \leq -2 \ \text{ or } \ \dfrac{3x-3}{4} \geq 2$$
$$3x - 3 \leq -8 \ \text{ or } \ 3x - 3 \geq 8$$
$$3x \leq -5 \ \text{ or } \ 3x \geq 11$$
$$x \leq -\dfrac{5}{3} \ \text{ or } \ x \geq \dfrac{11}{3}$$
$$\left(-\infty, -\dfrac{5}{3}\right] \cup \left[\dfrac{11}{3}, \infty\right)$$

11. $y = -2x + 3$

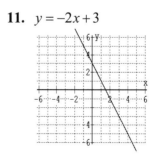

12. $y = -x + 3$

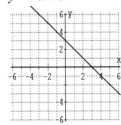

13. a. Function

b. Not a function

c. Function

14. $f(x) = -x^2 + 3x - 2$

a. $f(0) = -(0)^2 + 3(0) - 2 = -2$

b. $f(-3) = -(-3)^2 + 3(-3) - 2$
$= -9 - 9 - 2$
$= -20$

c. $f\left(\dfrac{1}{3}\right) = -\left(\dfrac{1}{3}\right)^2 + 3\left(\dfrac{1}{3}\right) - 2$
$= -\dfrac{1}{9} + 1 - 2$
$= -\dfrac{1}{9} + \dfrac{9}{9} - \dfrac{18}{9}$
$= \dfrac{-1 + 9 - 10}{9}$
$= -\dfrac{10}{9}$

15. $x - 3y = 6$
$-3y = -x + 6$
$y = \dfrac{1}{3}x - 2$

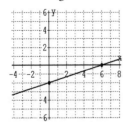

16. $3x - y = 6$
$-y = -3x + 6$
$y = 3x - 6$

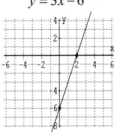

17. $y - (-5) = -3(x - 1)$
$y + 5 = -3x + 3$
$y = -3x - 2$

18. $y - 3 = \dfrac{1}{2}\left(x - (-1)\right)$
$y - 3 = \dfrac{1}{2}(x + 1)$
$y - 3 = \dfrac{1}{2}x + \dfrac{1}{2}$
$y = \dfrac{1}{2}x + \dfrac{1}{2} + 3$
$y = \dfrac{1}{2}x + \dfrac{7}{2}$
$f(x) = \dfrac{1}{2}x + \dfrac{7}{2}$

19. $x \geq 1$ and $y \geq 2x - 1$

Graph each inequality. The intersection of the two inequalities is the overlap of the two shaded regions, as shown by the graph below.

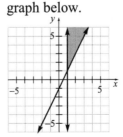

20. $2x + y \leq 4$ or $y > 2$

Graph each inequality. The union of the two inequalities is both shaded regions, as shown by the graph below.

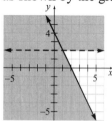

21. $\begin{cases} 3x - 2y = 10 & (1) \\ 4x - 3y = 15 & (2) \end{cases}$

Multiply E1 by -4 and E2 by 3, and add.
$$-12x + 8y = -40$$
$$\underline{12x - 9y = 45}$$
$$-y = 5$$
$$y = -5$$

Replace y with -5 in E1.
$$3x - 2(-5) = 10$$
$$3x + 10 = 10$$
$$3x = 0$$
$$x = 0$$
The solution is $(0, -5)$.

22. $\begin{cases} -2x + 3y = 6 & (1) \\ 3x - y = 5 & (2) \end{cases}$

Solve E2 for y: $y = 3x - 5$.
Replace y with $3x - 5$ in E1.
$$-2x + 3(3x - 5) = 6$$
$$-2x + 9x - 15 = 6$$
$$7x = 21$$
$$x = 3$$
Replace x with 3 in the equation
$y = 3x - 5$.
$$y = 3(3) - 5 = 9 - 5 = 4$$
The solution is $(3, 4)$.

23. $\begin{cases} 2x - 4y + 8z = 2 & (1) \\ -x - 3y + z = 11 & (2) \\ x - 2y + 4z = 0 & (3) \end{cases}$

Add E2 and E3.
$$-5y + 5z = 11 \ (4)$$
Multiply E2 by 2 and add to E1.
$$-2x - 6y + 2z = 22$$
$$\underline{2x - 4y + 8z = 2}$$
$$-10y + 10z = 24 \ (5)$$
Solve the new system
$$\begin{cases} -5y + 5z = 11 & (4) \\ -10y + 10z = 24 & (5). \end{cases}$$
Multiply E4 by -2 and add to E5.
$$10y - 10z = -22$$
$$\underline{-10y + 10z = 24}$$
$$0 = 2, \text{ which is impossible.}$$
The solution is $\varnothing$.

24. $\begin{cases} 2x - 2y + 4z = 6 & (1) \\ -4x - y + z = -8 & (2) \\ 3x - y + z = 6 & (3) \end{cases}$

Multiply E2 by -1 and add to E3.
$$4x + y - z = 8$$
$$\underline{3x - y + z = 6}$$
$$7x = 14$$
$$x = 2$$
Multiply E2 by -2 and add to E1.
$$8x + 2y - 2z = 16$$
$$\underline{2x - 2y + 4z = 6}$$
$$10x + 2z = 22 \text{ or } 5x + z = 11$$
Replace x with 2 in the equation
$5x + z = 11$.
$$5(2) + z = 11$$
$$10 + z = 11$$
$$z = 1$$
Replace x with 2 and z with 1 in E3.
$$3(2) - y + 1 = 6$$
$$7 - y = 6$$
$$-y = -1$$
$$y = 1$$
The solution is $(2, 1, 1)$.

25. Let x = measure of the smallest angle, y = measure of the largest angle and, z = measure of the remaining angle.

$$\begin{cases} x + y + z = 180 & (1) \\ \quad\quad y = x + 80 & (2) \\ \quad\quad\quad z = x + 10 & (3) \end{cases}$$

Substitute $x + 80$ for y and $x + 10$ for z in E1.

$x + (x + 80) + (x + 10) = 180$

$\quad\quad\quad\quad 3x + 90 = 180$

$\quad\quad\quad\quad\quad 3x = 90$

$\quad\quad\quad\quad\quad x = 30$

Replace x with 30 in E2 and E3.

$y = 30 + 80 = 110$

$z = 30 + 10 = 40$

The angles measure 30°, 110°, and 40°.

26. Let x = the price of a ream of paper and y = the price of a box of manila folders.

$$\begin{cases} 3x + 2y = 21.90 & (1) \\ 5x + \quad y = 24.25 & (2) \end{cases}$$

Multiply E2 by -2 and add to E1.

$-10x - 2y = -48.50$

$\underline{\quad 3x + 2y = 21.90\quad}$

$\;-7x \quad\quad = -26.60$

$\quad\quad x = 3.80$

Replace x with 3.80 in E2.

$5(3.80) + y = 24.25$

$\quad 19 + y = 24.25$

$\quad\quad\quad y = 5.25$

A ream of paper cost \$3.80 and a box of manila folders cost \$5.25.

27. $\begin{cases} x + 2y + \quad z = 2 \\ -2x - \quad y + 2z = 5 \\ \quad x + 3y - 2z = -8 \end{cases}$

$$\left[\begin{array}{ccc|c} 1 & 2 & 1 & 2 \\ -2 & -1 & 2 & 5 \\ 1 & 3 & -2 & -8 \end{array}\right]$$

Multiply R1 by 2 and add to R2.
Multiply R1 by -1 and add to R3.

$$\left[\begin{array}{ccc|c} 1 & 2 & 1 & 2 \\ 0 & 3 & 4 & 9 \\ 0 & 1 & -3 & -10 \end{array}\right]$$

Interchange R2 and R3.

$$\left[\begin{array}{ccc|c} 1 & 2 & 1 & 2 \\ 0 & 1 & -3 & -10 \\ 0 & 3 & 4 & 9 \end{array}\right]$$

Mutiply R2 by -3 and add to R3.

$$\left[\begin{array}{ccc|c} 1 & 2 & 1 & 2 \\ 0 & 1 & -3 & -10 \\ 0 & 0 & 13 & 39 \end{array}\right]$$

Divide R3 by 13.

$$\left[\begin{array}{ccc|c} 1 & 2 & 1 & 2 \\ 0 & 1 & -3 & -10 \\ 0 & 0 & 1 & 3 \end{array}\right]$$

This corresponds to $\begin{cases} x + 2y + \quad z = 2 \\ \quad\quad y - 3z = -10 \\ \quad\quad\quad z = 3. \end{cases}$

$\begin{aligned} y - 3z &= -10 \\ y - 3(3) &= -10 \\ y - 9 &= -10 \\ y &= -1 \end{aligned}$ and so $\begin{aligned} x + 2(-1) + (3) &= 2 \\ x - 2 + 3 &= 2 \\ x + 1 &= 2 \\ x &= 1 \end{aligned}$

The soltuon is $(1, -1, 3)$.

28. $\begin{cases} \quad x + \quad y + z = 9 \\ 2x - 2y + 3z = 2 \\ -3x + \quad y - \quad z = 1 \end{cases}$

$$\left[\begin{array}{ccc|c} 1 & 1 & 1 & 9 \\ 2 & -2 & 3 & 2 \\ -3 & 1 & -1 & 1 \end{array}\right]$$

Multiply R1 by -2 and add to R2.
Multiply R1 by 3 and add to R3.

$$\left[\begin{array}{ccc|c} 1 & 1 & 1 & 9 \\ 0 & -4 & 1 & -16 \\ 0 & 4 & 2 & 28 \end{array}\right]$$

Divide R2 by -4.

$$\left[\begin{array}{ccc|c} 1 & 1 & 1 & 9 \\ 0 & 1 & -\dfrac{1}{4} & 4 \\ 0 & 4 & 2 & 28 \end{array}\right]$$

Multiply R2 by -4 and add to R3.

$$\begin{bmatrix} 1 & 1 & 1 & | & 9 \\ 0 & 1 & -\frac{1}{4} & | & 4 \\ 0 & 0 & 3 & | & 12 \end{bmatrix}$$

Divide R3 by 3.

$$\begin{bmatrix} 1 & 1 & 1 & | & 9 \\ 0 & 1 & -\frac{1}{4} & | & 4 \\ 0 & 0 & 1 & | & 4 \end{bmatrix}$$

This corresponds to $\begin{cases} x + y + z = 9 \\ y - \frac{1}{4}z = 4 \\ z = 4. \end{cases}$

$$y - \frac{1}{4}(4) = 4 \qquad x + 5 + 4 = 9$$
$$y - 1 = 4 \text{ and so} \qquad x + 9 = 9$$
$$y = 5 \qquad x = 0$$

The solution is $(0, 5, 4)$.

29. a. $7^0 = 1$

   b. $-7^0 = -1 \cdot 7^0 = -1 \cdot 1 = -1$

   c. $(2x + 5)^0 = 1$

   d. $2x^0 = 2 \cdot x^0 = 2 \cdot 1 = 2$

30. a. $2^{-2} + 3^{-1} = \frac{1}{2^2} + \frac{1}{3}$
$$= \frac{1}{4} + \frac{1}{3}$$
$$= \frac{3}{12} + \frac{4}{12}$$
$$= \frac{7}{12}$$

   b. $-6a^0 = -6 \cdot a^0 = -6 \cdot 1 = -6$

   c. $\dfrac{x^{-5}}{x^{-2}} = x^{-5-(-2)} = x^{-3} = \dfrac{1}{x^3}$

31. a. $x^{-b}(2x^b)^2 = \dfrac{2^2(x^b)^2}{x^b}$
$$= \frac{4x^{2b}}{x^b}$$
$$= 4x^{2b-b}$$
$$= 4x^b$$

   b. $\dfrac{(y^{3a})^2}{y^{a-6}} = \dfrac{y^{6a}}{y^{a-6}} = y^{6a-(a-6)} = y^{5a+6}$

32. a. $3x^{4a}(4x^{-a})^2 = 3x^{4a} \cdot 16x^{-2a}$
$$= 48x^{4a+(-2a)}$$
$$= 48x^{2a}$$

   b. $\dfrac{(y^{4b})^3}{y^{2b-3}} = \dfrac{y^{12b}}{y^{2b-3}}$
$$= y^{12b-(2b-3)}$$
$$= y^{12b-2b+3}$$
$$= y^{10b+3}$$

33. a. $3x^2$ has degree $= 2$

   b. $-2^3 x^5 = -8x^5$ has degree $= 5$

   c. y has degree $= 1$

   d. $12x^2yz^3$ has degree $= 2 + 1 + 3 = 6$

   e. 5 has degree $= 0$

34. $(2x^2 + 8x - 3) - (2x - 7)$
$$= 2x^2 + 8x - 3 - 2x + 7$$
$$= 2x^2 + 6x + 4$$

35. $[3 + (2a + b)]^2$
$$= 3^2 + 2(3)(2a + b) + (2a + b)^2$$
$$= 9 + 6(2a + b) + (4a^2 + 2(2a)b + b^2)$$
$$= 9 + 12a + 6b + 4a^2 + 4ab + b^2$$

36. $[4 + (3x - y)]^2$
$$= 4^2 + 2(4)(3x - y) + (3x - y)^2$$
$$= 16 + 8(3x - y) + (9x^2 - 2(3x)y + y^2)$$
$$= 16 + 24x - 8y + 9x^2 - 6xy + y^2$$

37. $ab - 6a + 2b - 12 = a(b - 6) + 2(b - 6)$
$$= (b - 6)(a + 2)$$

38. $xy + 2x - 5y - 10 = x(y + 2) - 5(y + 2)$
$$= (x - 5)(y + 2)$$

39. $2n^2 - 38n + 80 = 2(n^2 - 19n + 40)$

40. $6x^2 - x - 35 = (2x - 5)(3x + 7)$

41. $x^2 + 4x + 4 - y^2$
$$= (x^2 + 4x + 4) - y^2$$
$$= (x + 2)^2 - y^2$$
$$= [(x + 2) + y][(x + 2) - y]$$
$$= (x + 2 + y)(x + 2 - y)$$

42. $4x^2 - 4x + 1 - 9y^2$
$$= (4x^2 - 4x + 1) - 9y^2$$
$$= (2x - 1)^2 - (3y)^2$$
$$= [(2x - 1) + 3y][(2x - 1) - 3y]$$
$$= (2x - 1 + 3y)(2x - 1 - 3y)$$

43. $(x + 2)(x - 6) = 0$
$x + 2 = 0$ or $x - 6 = 0$
$x = -2$ or $x = 6$
The solutions are -2 and 6.

44. $2x(3x + 1)(x - 3) = 0$
$x = 0$ or $3x + 1 = 0$ or $x - 3 = 0$
            $3x = -1$ or $x = 3$
              $x = -\dfrac{1}{3}$
The solutions are $-\dfrac{1}{3}, 0, 3$.

45. $\dfrac{2x^2}{10x^3 - 2x^2} = \dfrac{2x^2}{2x^2(5x - 1)} = \dfrac{1}{5x - 1}$

46. a. Domain: $(-\infty, \infty)$; Range: $[-4, \infty)$

    b. x-intercepts: $(-2, 0)$, $(2, 0)$
        y-intercept: $(0, -4)$

    c. There is no such point.

    d. The point with the least y-value is
        $(0, -4)$.

    e. -2 and 2

    f. between $x = -2$ and $x = 2$

g. The solutions are -2 and 2.

47. $\dfrac{5k}{k^2 - 4} - \dfrac{2}{k^2 + k - 2}$
$$= \dfrac{5k}{(k + 2)(k - 2)} - \dfrac{2}{(k + 2)(k - 1)}$$
$$= \dfrac{5k(k - 1) - 2(k - 2)}{(k + 2)(k - 2)(k - 1)}$$
$$= \dfrac{5k^2 - 5k - 2k + 4}{(k + 2)(k - 2)(k - 1)}$$
$$= \dfrac{5k^2 - 7k + 4}{(k + 2)(k - 2)(k - 1)}$$

48. $\dfrac{5a}{a^2 - 4} - \dfrac{3}{2 - a}$
$$= \dfrac{5a}{(a + 2)(a - 2)} + \dfrac{3}{a - 2}$$
$$= \dfrac{5a + 3(a + 2)}{(a + 2)(a - 2)}$$
$$= \dfrac{5a + 3a + 6}{(a + 2)(a - 2)}$$
$$= \dfrac{8a + 6}{(a + 2)(a - 2)}$$

49. $\dfrac{3}{x} - \dfrac{x + 21}{3x} = \dfrac{5}{3}$
$$3x\left(\dfrac{3}{x} - \dfrac{x + 21}{3x}\right) = 3x\left(\dfrac{5}{3}\right)$$
$$9 - (x + 21) = 5x$$
$$9 - x - 21 = 5x$$
$$-x - 12 = 5x$$
$$-12 = 6x$$
$$-2 = x$$
The solution is -2.

50. $\dfrac{3x - 4}{2x} = -\dfrac{8}{x}$
$$x(3x - 4) = -8(2x)$$
$$3x^2 - 4x = -16x$$
$$3x^2 + 12x = 0$$
$$3x(x + 4) = 0$$
$$3x = 0 \text{ or } x + 4 = 0$$
$$x = 0 \text{ or } \quad x = -4$$
Discard the answer 0 as extraneous.
The solution is -4.

Chapter 7

Section 7.1

Mental Math

1. D

2. A, C

3. D

4. C

Exercise Set 7.1

1. $\sqrt{100} = 10$ because $10^2 = 100$.

3. $\sqrt{\frac{1}{4}} = \frac{1}{2}$ because $\left(\frac{1}{2}\right)^2 = \frac{1}{4}$.

5. $\sqrt{0.0001} = 0.01$ because $(0.01)^2 = 0.0001$.

7. $-\sqrt{36} = -1 \cdot \sqrt{36} = -1 \cdot 6 = -6$ because $6^2 = 36$.

9. $\sqrt{x^{10}} = x^5$ because $(x^5)^2 = x^{10}$.

11. $\sqrt{16y^6} = 4y^3$ because $(4y^3)^2 = 16y^6$.

13. $\sqrt{7} \approx 2.646$
Since $4 < 7 < 9$, then $\sqrt{4} < \sqrt{7} < \sqrt{9}$, or $2 < \sqrt{7} < 3$. The approximation is between 2 and 3 and thus is reasonable.

15. $\sqrt{38} \approx 6.164$
Since $36 < 38 < 49$, then $\sqrt{36} < \sqrt{38} < \sqrt{49}$, or $6 < \sqrt{38} < 7$. The approximation is between 6 and 7 and thus is reasonable.

17. $\sqrt{200} \approx 14.142$
Since $196 < 200 < 225$, then $\sqrt{196} < \sqrt{200} < \sqrt{225}$, or $14 < \sqrt{200} < 15$. The approximation is between 14 and 15 and thus is reasonable.

19. $\sqrt[3]{64} = 4$ because $4^3 = 64$.

21. $\sqrt[3]{\frac{1}{8}} = \frac{1}{2}$ because $\left(\frac{1}{2}\right)^3 = \frac{1}{8}$.

23. $\sqrt[3]{-1} = -1$ because $(-1)^3 = -1$.

25. $\sqrt[3]{x^{12}} = x^4$ because $(x^4)^3 = x^{12}$.

27. $\sqrt[3]{-27x^9} = -3x^3$ because $(-3x^3)^3 = -27x^9$.

29. $-\sqrt[4]{16} = -2$ because $2^4 = 16$.

31. $\sqrt[4]{-16}$ is not a real number. There is no real number that, when raised to the fourth power, is -16.

33. $\sqrt[5]{-32} = -2$ because $(-2)^5 = -32$.

35. $\sqrt[5]{x^{20}} = x^4$ because $(x^4)^5 = x^{20}$.

37. $\sqrt[6]{64x^{12}} = 2x^2$ because $(2x^2)^6 = 64x^{12}$.

39. $\sqrt{81x^4} = 9x^2$ because $(9x^2)^2 = 81x^4$.

41. $\sqrt[4]{256x^8} = 4x^2$ because $(4x^2)^4 = 256x^8$.

43. $\sqrt{(-8)^2} = |-8| = 8$

45. $\sqrt[3]{(-8)^3} = -8$

47. $\sqrt{4x^2} = 2|x|$

49. $\sqrt[3]{x^3} = x$

51. $\sqrt{(x-5)^2} = |x-5|$

53. $\sqrt{x^2 + 4x + 4} = \sqrt{(x+2)^2} = |x+2|$

55. $-\sqrt{121} = -11$

57. $\sqrt[3]{8x^3} = 2x$

59. $\sqrt{y^{12}} = y^6$

61. $\sqrt{25a^2b^{20}} = 5ab^{10}$

63. $\sqrt[3]{-27x^{12}y^9} = -3x^4y^3$

65. $\sqrt[4]{a^{16}b^4} = a^4b$

67. $\sqrt[5]{-32x^{10}y^5} = -2x^2y$

69. $\sqrt{\dfrac{25}{49}} = \dfrac{5}{7}$

71. $\sqrt{\dfrac{x^2}{4y^2}} = \dfrac{x}{2y}$

73. $-\sqrt[3]{\dfrac{z^{21}}{27x^3}} = -\dfrac{z^7}{3x}$

75. $\sqrt[4]{\dfrac{x^4}{16}} = \dfrac{x}{2}$

77. $f(x) = \sqrt{2x+3}$
$f(0) = \sqrt{2(0)+3} = \sqrt{3}$

79. $g(x) = \sqrt[3]{x-8}$
$g(7) = \sqrt[3]{7-8} = \sqrt[3]{-1} = -1$

81. $g(x) = \sqrt[3]{x-8}$
$g(-19) = \sqrt[3]{-19-8} = \sqrt[3]{-27} = -3$

83. $f(x) = \sqrt{2x+3}$
$f(2) = \sqrt{2(2)+3} = \sqrt{7}$

85. $f(x) = \sqrt{x} + 2$
Domain: $[0, \infty)$

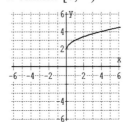

87. $f(x) = \sqrt{x-3}$
Domain: $[3, \infty)$

x	$f(x) = \sqrt{x-3}$
3	$\sqrt{3-3} = \sqrt{0} = 0$
4	$\sqrt{4-3} = \sqrt{1} = 1$
7	$\sqrt{7-3} = \sqrt{4} = 2$
12	$\sqrt{12-3} = \sqrt{9} = 3$

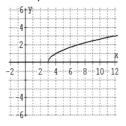

89. $f(x) = \sqrt[3]{x} + 1$
Domain: $(-\infty, \infty)$

91. $g(x) = \sqrt[3]{x} - 1$

Domain: $(-\infty, \infty)$

x	$g(x) = \sqrt[3]{x} - 1$
1	$\sqrt[3]{1} - 1 = \sqrt[3]{0} = 0$
2	$\sqrt[3]{2} - 1 = \sqrt[3]{1} = 1$
0	$\sqrt[3]{0} - 1 = \sqrt[3]{-1} = -1$
9	$\sqrt[3]{9} - 1 = \sqrt[3]{8} = 2$
-7	$\sqrt[3]{-7} - 1 = \sqrt[3]{8} - 8 = -2$

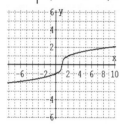

93. $(-2x^3 y^2)^5 = (-2)^5 x^{3 \cdot 5} y^{2 \cdot 5} = -32x^{15} y^{10}$

95. $(-3x^2 y^3 z^5)(20x^5 y^7) = -3(20)x^{2+5} y^{3+7} z^5$
$$= -60x^7 y^{10} z^5$$

97. $\dfrac{7x^{-1} y}{14(x^5 y^2)^{-2}} = \dfrac{7x^{-1} y}{14x^{-10} y^{-4}} = \dfrac{x^9 y^5}{2}$

99. Answers may vary.

101. $144 < 160 < 169$ so
$\sqrt{144} < \sqrt{160} < \sqrt{169}$, or
$12 < \sqrt{160} < 13$. Thus $\sqrt{160}$ is between
12 and 13. Therefore, the answer is **b.**

103. $\sqrt{30} \approx 5, \sqrt{10} \approx 3$, and $\sqrt{90} \approx 10$ so
$P = \sqrt{30} + \sqrt{10} + \sqrt{90} \approx 5 + 3 + 10 = 18$.
Therefore, the answer is **b.**

105. $B = \sqrt{\dfrac{hw}{3131}} = \sqrt{\dfrac{66 \cdot 135}{3131}}$
$$= \sqrt{\dfrac{8910}{3131}} \approx 1.69 \text{ sq meters}$$

107. Answers may vary.

109. $f(x) = \sqrt{x} + 2$

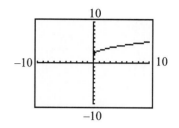

Domain: $[0, \infty)$

111. $f(x) = \sqrt[3]{x} + 1$

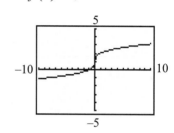

Domain: $(-\infty, \infty)$

Section 7.2

Mental Math

1. A

2. B

3. C

4. A

5. B

6. B

7. B

8. C

9. B

10. B

Exercise Set 7.2

1. $49^{1/2} = \sqrt{49} = 7$

3. $27^{1/3} = \sqrt[3]{27} = 3$

5. $\left(\dfrac{1}{16}\right)^{1/4} = \sqrt[4]{\dfrac{1}{16}} = \dfrac{1}{2}$

7. $169^{1/2} = \sqrt{169} = 13$

9. $2m^{1/3} = 2\sqrt[3]{m}$

11. $(9x^4)^{1/2} = \sqrt{9x^4} = 3x^2$

13. $(-27)^{1/3} = \sqrt[3]{-27} = -3$

15. $-16^{1/4} = -\sqrt[4]{16} = -2$

17. $16^{3/4} = \left(\sqrt[4]{16}\right)^3 = 2^3 = 8$

19. $(-64)^{2/3} = \left(\sqrt[3]{-64}\right)^2 = (-4)^2 = 16$

21. $(-16)^{3/4} = \left(\sqrt[4]{-16}\right)^3$ is not a real number.

23. $(2x)^{3/5} = \sqrt[5]{(2x)^3}$ or $\left(\sqrt[5]{2x}\right)^3$

25. $(7x+2)^{2/3} = \sqrt[3]{(7x+2)^2}$ or $\left(\sqrt[3]{7x+2}\right)^2$

27. $\left(\dfrac{16}{9}\right)^{3/2} = \left(\sqrt{\dfrac{16}{9}}\right)^3 = \left(\dfrac{4}{3}\right)^3 = \dfrac{64}{27}$

29. $8^{-4/3} = \dfrac{1}{8^{4/3}} = \dfrac{1}{\left(\sqrt[3]{8}\right)^4} = \dfrac{1}{2^4} = \dfrac{1}{16}$

31. $(-64)^{-2/3} = \dfrac{1}{(-64)^{2/3}}$

$\qquad = \dfrac{1}{\left(\sqrt[3]{-64}\right)^2} = \dfrac{1}{(-4)^2} = \dfrac{1}{16}$

33. $(-4)^{-3/2} = \dfrac{1}{(-4)^{3/2}} = \dfrac{1}{\left(\sqrt{-4}\right)^3}$ is not a real number.

35. $x^{-1/4} = \dfrac{1}{x^{1/4}}$

37. $\dfrac{1}{a^{-2/3}} = a^{2/3}$

39. $\dfrac{5}{7x^{-3/4}} = \dfrac{5x^{3/4}}{7}$

41. Answers may vary.

43. $a^{2/3}a^{5/3} = a^{2/3+5/3} = a^{7/3}$

45. $x^{-2/5} \cdot x^{7/5} = x^{-\frac{2}{5}+\frac{7}{5}} = x^{5/5} = x$

47. $3^{1/4} \cdot 3^{3/8} = 3^{\frac{1}{4}+\frac{3}{8}} = 3^{\frac{2}{8}+\frac{3}{8}} = 3^{5/8}$

49. $\dfrac{y^{1/3}}{y^{1/6}} = y^{\frac{1}{3}-\frac{1}{6}} = y^{\frac{2}{6}-\frac{1}{6}} = y^{1/6}$

51. $(4u^2)^{3/2} = 4^{3/2}u^{2(3/2)}$

$\qquad = \left(\sqrt{4}\right)^3 u^3$

$\qquad = 2^3 u^3$

$\qquad = 8u^3$

53. $\dfrac{b^{1/2}b^{3/4}}{-b^{1/4}} = -b^{\frac{1}{2}+\frac{3}{4}-\frac{1}{4}} = -b^{\frac{2}{4}+\frac{3}{4}-\frac{1}{4}} = -b^1 = -b$

55. $\dfrac{(3x^{1/4})^3}{x^{1/12}} = \dfrac{3^3 x^{3/4}}{x^{1/12}}$

$\qquad = 27x^{\frac{3}{4}-\frac{1}{12}}$

$\qquad = 27x^{\frac{9}{12}-\frac{1}{12}}$

$\qquad = 27x^{8/12}$

$\qquad = 27x^{2/3}$

57. $y^{1/2}(y^{1/2} - y^{2/3}) = y^{1/2}y^{1/2} - y^{1/2}y^{2/3}$
$$= y^{1/2+1/2} - y^{1/2+2/3}$$
$$= y^1 - y^{7/6}$$
$$= y - y^{7/6}$$

59. $x^{2/3}(2x - 2) = 2xx^{2/3} - 2x^{2/3}$
$$= 2x^{1+2/3} - 2x^{2/3}$$
$$= 2x^{5/3} - 2x^{2/3}$$

61. $(2x^{1/3} + 3)(2x^{1/3} - 3) = (2x^{1/3})^2 - 3^2$
$$= 2^2(x^{1/3})^2 - 9$$
$$= 4x^{2/3} - 9$$

63. $x^{8/3} + x^{10/3} = x^{8/3}(1) + x^{8/3}(x^{2/3})$
$$= x^{8/3}(1 + x^{2/3})$$

65. $x^{2/5} - 3x^{1/5} = x^{1/5}(x^{1/5}) - x^{1/5}(3)$
$$= x^{1/5}(x^{1/5} - 3)$$

67. $5x^{-1/3} + x^{2/3} = x^{-1/3}(5) + x^{-1/3}(x^{3/3})$
$$= x^{-1/3}(5 + x)$$

69. $\sqrt[6]{x^3} = x^{3/6} = x^{1/2} = \sqrt{x}$

71. $\sqrt[6]{4} = 4^{1/6} = (2^2)^{1/6} = 2^{1/3} = \sqrt[3]{2}$

73. $\sqrt[4]{16x^2} = (16x^2)^{1/4}$
$$= 16^{1/4}x^{2/4} = 2x^{1/2} = 2\sqrt{x}$$

75. $\sqrt[8]{x^4 y^4} = (x^4 y^4)^{1/8}$
$$= x^{4/8}y^{4/8}$$
$$= x^{1/2}y^{1/2}$$
$$= (xy)^{1/2}$$
$$= \sqrt{xy}$$

77. $\sqrt[3]{y} \cdot \sqrt[5]{y^2} = y^{1/3} \cdot y^{2/5}$
$$= y^{\frac{5}{15}+\frac{6}{15}}$$
$$= y^{11/15}$$
$$= \sqrt[15]{y^{11}}$$

79. $\dfrac{\sqrt[3]{b^2}}{\sqrt[4]{b}} = \dfrac{b^{2/3}}{b^{1/4}} = b^{\frac{2}{3} - \frac{1}{4}} = b^{\frac{8}{12} - \frac{3}{12}} = b^{5/12} = \sqrt[12]{b^5}$

81. $\dfrac{\sqrt[3]{a^2}}{\sqrt[6]{a}} = \dfrac{a^{2/3}}{a^{1/6}}$
$$= a^{\frac{2}{3} - \frac{1}{6}} = a^{\frac{4}{6} - \frac{1}{6}} = a^{3/6} = a^{1/2} = \sqrt{a}$$

83. $\sqrt{3} \cdot \sqrt[3]{4} = 3^{1/2} \cdot 4^{1/3}$
$$= 3^{3/6} \cdot 4^{2/6}$$
$$= (3^3 \cdot 4^2)^{1/6}$$
$$= (432)^{1/6}$$
$$= \sqrt[6]{432}$$

85. $\sqrt[5]{7} \cdot \sqrt[3]{y} = 7^{1/5} \cdot y^{1/3}$
$$= 7^{3/15} \cdot y^{5/15}$$
$$= (7^3 \cdot y^5)^{1/15}$$
$$= (343y^5)^{1/15}$$
$$= \sqrt[15]{343y^5}$$

87. $75 = 25 \cdot 3$

89. $48 = 4 \cdot 12$ or $16 \cdot 3$

91. $16 = 8 \cdot 2$

93. $54 = 27 \cdot 2$

95. $B(w) = 70w^{3/4}$
$$B(60) = 70(60)^{3/4}$$
$$\approx 1509 \text{ calories}$$

97. $f(x) = 1.54x^{9/5}$
$$f(10) = 1.54(10)^{9/5}$$
$$\approx 97.2 \text{ million subscriptions}$$

99. $\square \cdot a^{2/3} = a^{3/3}$
$$\square = \dfrac{a^{3/3}}{a^{2/3}}$$
$$\square = a^{3/3 - 2/3}$$
$$\square = a^{1/3}$$

101.
$$\frac{\Box}{x^{-2/5}} = x^{3/5}$$
$$x^{-2/5}\left(\frac{\Box}{x^{-2/5}}\right) = x^{3/5} \cdot x^{-2/5}$$
$$\Box = x^{3/5-2/5}$$
$$\Box = x^{1/5}$$

103. $8^{1/4} \approx 1.6818$

105. $18^{3/5} \approx 5.6645$

107. $\dfrac{\sqrt{t}}{\sqrt{u}} = \dfrac{t^{1/2}}{u^{1/2}}$

Exercise Set 7.3

1. $\sqrt{7} \cdot \sqrt{2} = \sqrt{7 \cdot 2} = \sqrt{14}$

3. $\sqrt[4]{8} \cdot \sqrt[4]{2} = \sqrt[4]{8 \cdot 2} = \sqrt[4]{16} = 2$

5. $\sqrt[3]{4} \cdot \sqrt[3]{9} = \sqrt[3]{4 \cdot 9} = \sqrt[3]{36}$

7. $\sqrt{2} \cdot \sqrt{3x} = \sqrt{2 \cdot 3x} = \sqrt{6x}$

9. $\sqrt{\dfrac{7}{x}} \cdot \sqrt{\dfrac{2}{y}} = \sqrt{\dfrac{7}{x} \cdot \dfrac{2}{y}} = \sqrt{\dfrac{14}{xy}}$

11. $\sqrt[4]{4x^3} \cdot \sqrt[4]{5} = \sqrt[4]{4x^3 \cdot 5} = \sqrt[4]{20x^3}$

13. $\sqrt{\dfrac{6}{49}} = \dfrac{\sqrt{6}}{\sqrt{49}} = \dfrac{\sqrt{6}}{7}$

15. $\sqrt{\dfrac{2}{49}} = \dfrac{\sqrt{2}}{\sqrt{49}} = \dfrac{\sqrt{2}}{7}$

17. $\sqrt[4]{\dfrac{x^3}{16}} = \dfrac{\sqrt[4]{x^3}}{\sqrt[4]{16}} = \dfrac{\sqrt[4]{x^3}}{2}$

19. $\sqrt[3]{\dfrac{4}{27}} = \dfrac{\sqrt[3]{4}}{\sqrt[3]{27}} = \dfrac{\sqrt[3]{4}}{3}$

21. $\sqrt[4]{\dfrac{8}{x^8}} = \dfrac{\sqrt[4]{8}}{\sqrt[4]{x^8}} = \dfrac{\sqrt[4]{8}}{x^2}$

23. $\sqrt[3]{\dfrac{2x}{81y^{12}}} = \dfrac{\sqrt[3]{2x}}{\sqrt[3]{81y^{12}}}$
$$= \dfrac{\sqrt[3]{2x}}{\sqrt[3]{27y^{12}} \cdot \sqrt[3]{3}}$$
$$= \dfrac{\sqrt[3]{2x}}{3y^4\sqrt[3]{3}}$$

25. $\sqrt{\dfrac{x^2 y}{100}} = \dfrac{\sqrt{x^2 y}}{\sqrt{100}} = \dfrac{\sqrt{x^2}\sqrt{y}}{10} = \dfrac{x\sqrt{y}}{10}$

27. $\sqrt{\dfrac{5x^2}{4y^2}} = \dfrac{\sqrt{5x^2}}{\sqrt{4y^2}} = \dfrac{\sqrt{5}\sqrt{x^2}}{2y} = \dfrac{\sqrt{5}x}{2y}$

29. $-\sqrt[3]{\dfrac{z^7}{27x^3}} = -\dfrac{\sqrt[3]{z^7}}{\sqrt[3]{27x^3}} = -\dfrac{\sqrt[3]{z^6 z}}{3x} = -\dfrac{z^2\sqrt[3]{z}}{3x}$

31. $\sqrt{32} = \sqrt{16 \cdot 2} = \sqrt{16} \cdot \sqrt{2} = 4\sqrt{2}$

33. $\sqrt[3]{192} = \sqrt[3]{64 \cdot 3} = \sqrt[3]{64} \cdot \sqrt[3]{3} = 4\sqrt[3]{3}$

35. $5\sqrt{75} = 5\sqrt{25 \cdot 3}$
$$= 5\sqrt{25} \cdot \sqrt{3} = 5(5)\sqrt{3} = 25\sqrt{3}$$

37. $\sqrt{24} = \sqrt{4 \cdot 6} = \sqrt{4} \cdot \sqrt{6} = 2\sqrt{6}$

39. $\sqrt{100x^5} = \sqrt{100x^4 \cdot x}$
$$= \sqrt{100x^4} \cdot \sqrt{x} = 10x^2\sqrt{x}$$

41. $\sqrt[3]{16y^7} = \sqrt[3]{8y^6 \cdot 2y}$
$$= \sqrt[3]{8y^6} \cdot \sqrt[3]{2y} = 2y^2\sqrt[3]{2y}$$

43. $\sqrt[4]{a^8 b^7} = \sqrt[4]{a^8 b^4 \cdot b^3}$
$$= \sqrt[4]{a^8 b^4} \cdot \sqrt[4]{b^3} = a^2 b\sqrt[4]{b^3}$$

45. $\sqrt{y^5} = \sqrt{y^4 \cdot y} = \sqrt{y^4} \cdot \sqrt{y} = y^2\sqrt{y}$

47. $\sqrt{25a^2 b^3} = \sqrt{25a^2 b^2 \cdot b}$
$$= \sqrt{25a^2 b^2} \cdot \sqrt{b} = 5ab\sqrt{b}$$

49. $\sqrt[5]{-32x^{10}y} = \sqrt[5]{-32x^{10} \cdot y}$
$$= \sqrt[5]{-32x^{10}} \cdot \sqrt[5]{y} = -2x^2 \sqrt[5]{y}$$

51. $\sqrt[3]{50x^{14}} = \sqrt[3]{x^{12} \cdot 50x^2}$
$$= \sqrt[3]{x^{12}} \cdot \sqrt[3]{50x^2} = x^4 \sqrt[3]{50x^2}$$

53. $-\sqrt{32a^8b^7} = -\sqrt{16a^8b^6 \cdot 2b}$
$$= -\sqrt{16a^8b^6} \cdot \sqrt{2b}$$
$$= -4a^4b^3\sqrt{2b}$$

55. $\sqrt{9x^7y^9} = \sqrt{9x^6y^8 \cdot xy}$
$$= \sqrt{9x^6y^8} \cdot \sqrt{xy} = 3x^3y^4\sqrt{xy}$$

57. $\sqrt[3]{125r^9s^{12}} = 5r^3s^4$

59. $\dfrac{\sqrt{14}}{\sqrt{7}} = \sqrt{\dfrac{14}{7}} = \sqrt{2}$

61. $\dfrac{\sqrt[3]{24}}{\sqrt[3]{3}} = \sqrt[3]{\dfrac{24}{3}} = \sqrt[3]{8} = 2$

63. $\dfrac{5\sqrt[4]{48}}{\sqrt[4]{3}} = 5\sqrt[4]{\dfrac{48}{3}} = 5\sqrt[4]{16} = 5(2) = 10$

65. $\dfrac{\sqrt{x^5y^3}}{\sqrt{xy}} = \sqrt{\dfrac{x^5y^3}{xy}} = \sqrt{x^4y^2} = x^2y$

67. $\dfrac{8\sqrt[3]{54m^7}}{\sqrt[3]{2m}} = 8\sqrt[3]{\dfrac{54m^7}{2m}}$
$$= 8\sqrt[3]{27m^6} = 8(3m^2) = 24m^2$$

69. $\dfrac{3\sqrt{100x^2}}{2\sqrt{2x^{-1}}} = \dfrac{3}{2}\sqrt{\dfrac{100x^2}{2x^{-1}}}$
$$= \dfrac{3}{2}\sqrt{50x^3}$$
$$= \dfrac{3}{2}\sqrt{25x^2 \cdot 2}$$
$$= \dfrac{3}{2}(5x)\sqrt{2}$$
$$= \dfrac{15x}{2}\sqrt{2}$$

71. $\dfrac{\sqrt[4]{96a^{10}b^3}}{\sqrt[4]{3a^2b^3}} = \sqrt[4]{\dfrac{96a^{10}b^3}{3a^2b^3}}$
$$= \sqrt[4]{32a^8}$$
$$= \sqrt[4]{16a^8 \cdot 2}$$
$$= 2a^2\sqrt[4]{2}$$

73. $6x + 8x = 14x$

75. $(2x+3)(x-5) = 2x^2 - 10x + 3x - 15$
$$= 2x^2 - 7x - 15$$

77. $9y^2 - 8y^2 = 1y^2 = y^2$

79. $-3(x+5) = -3x - 3(5) = -3x - 15$

81. $(x-4)^2 = x^2 - 2(x)(4) + 4^2$
$$= x^2 - 8x + 16$$

83. $A = \pi r \sqrt{r^2 + h^2}$

   a. $A = \pi(4)\sqrt{4^2 + 3^2}$
$$= 4\pi\sqrt{16+9}$$
$$= 4\pi\sqrt{25}$$
$$= 4\pi(5)$$
$$= 20\pi \text{ sq. centimeters}$$

   b. $A = \pi(6.8)\sqrt{(6.8)^2 + (7.2)^2}$
$$= 6.8\pi\sqrt{46.24 + 51.84}$$
$$= 6.8\pi\sqrt{98.08}$$
$$\approx 211.57 \text{ sq. centimeters}$$

85. $F(x) = 0.6\sqrt{49 - x^2}$

   a. $F(3) = 0.6\sqrt{49 - 3^2}$
$$= 0.6\sqrt{49 - 9}$$
$$= 0.6\sqrt{40} \approx 3.8 \text{ times}$$

   b. $F(5) = 0.6\sqrt{49 - 5^2}$
$$= 0.6\sqrt{49 - 25}$$
$$= 0.6\sqrt{24} \approx 2.9 \text{ times}$$

   c. Answers may vary.

Section 7.4

Mental Math

1. $2\sqrt{3} + 4\sqrt{3} = 6\sqrt{3}$

3. $8\sqrt{x} - 5\sqrt{x} = 3\sqrt{x}$

5. $7\sqrt[3]{x} + 5\sqrt[3]{x} = 12\sqrt[3]{x}$

7. $\sqrt{11} + \sqrt{11} = 2\sqrt{11}$

9. $9\sqrt{13} - \sqrt{13} = 8\sqrt{13}$

11. $8\sqrt[3]{2x} + 3\sqrt[3]{2x} - \sqrt[3]{2x} = 10\sqrt[3]{2x}$

Exercise Set 7.4

1. $\sqrt{8} - \sqrt{32} = \sqrt{4 \cdot 2} - \sqrt{16 \cdot 2}$
$= \sqrt{4} \cdot \sqrt{2} - \sqrt{16} \cdot \sqrt{2}$
$= 2\sqrt{2} - 4\sqrt{2}$
$= -2\sqrt{2}$

3. $2\sqrt{2x^3} + 4x\sqrt{8x}$
$= 2\sqrt{x^2 \cdot 2x} + 4x\sqrt{4 \cdot 2x}$
$= 2\sqrt{x^2} \cdot \sqrt{2x} + 4x\sqrt{4} \cdot \sqrt{2x}$
$= 2x\sqrt{2x} + 4x(2)\sqrt{2x}$
$= 2x\sqrt{2x} + 8x\sqrt{2x}$
$= 10x\sqrt{2x}$

5. $2\sqrt{50} - 3\sqrt{125} + \sqrt{98}$
$= 2\sqrt{25 \cdot 2} - 3\sqrt{25 \cdot 5} + \sqrt{49 \cdot 2}$
$= 2\sqrt{25} \cdot \sqrt{2} - 3\sqrt{25} \cdot \sqrt{5} + \sqrt{49} \cdot \sqrt{2}$
$= 2(5)\sqrt{2} - 3(5)\sqrt{5} + 7\sqrt{2}$
$= 10\sqrt{2} - 15\sqrt{5} + 7\sqrt{2}$
$= 17\sqrt{2} - 15\sqrt{5}$

7. $\sqrt[3]{16x} - \sqrt[3]{54x} = \sqrt[3]{8 \cdot 2x} - \sqrt[3]{27 \cdot 2x}$
$= \sqrt[3]{8} \cdot \sqrt[3]{2x} - \sqrt[3]{27} \cdot \sqrt[3]{2x}$
$= 2\sqrt[3]{2x} - 3\sqrt[3]{2x}$
$= -\sqrt[3]{2x}$

9. $\sqrt{9b^3} - \sqrt{25b^3} + \sqrt{49b^3}$
$= \sqrt{9b^2 \cdot b} - \sqrt{25b^2 \cdot b} + \sqrt{49b^2 \cdot b}$
$= \sqrt{9b^2} \cdot \sqrt{b} - \sqrt{25b^2} \cdot \sqrt{b} + \sqrt{49b^2} \cdot \sqrt{b}$
$= 3b\sqrt{b} - 5b\sqrt{b} + 7b\sqrt{b}$
$= 5b\sqrt{b}$

11. $\dfrac{5\sqrt{2}}{3} + \dfrac{2\sqrt{2}}{5} = \dfrac{5\left(5\sqrt{2}\right) + 3\left(2\sqrt{2}\right)}{3(5)}$
$= \dfrac{25\sqrt{2} + 6\sqrt{2}}{15}$
$= \dfrac{31\sqrt{2}}{15}$

13. $\sqrt[3]{\dfrac{11}{8}} - \dfrac{\sqrt[3]{11}}{6} = \dfrac{\sqrt[3]{11}}{\sqrt[3]{8}} - \dfrac{\sqrt[3]{11}}{6}$
$= \dfrac{\sqrt[3]{11}}{2} - \dfrac{\sqrt[3]{11}}{6}$
$= \dfrac{3\sqrt[3]{11} - \sqrt[3]{11}}{6}$
$= \dfrac{2\sqrt[3]{11}}{6}$
$= \dfrac{\sqrt[3]{11}}{3}$

15. $\dfrac{\sqrt{20x}}{9} + \sqrt{\dfrac{5x}{9}} = \dfrac{\sqrt{4 \cdot 5x}}{9} + \dfrac{\sqrt{5x}}{\sqrt{9}}$
$= \dfrac{2\sqrt{5x}}{9} + \dfrac{\sqrt{5x}}{3}$
$= \dfrac{2\sqrt{5x} + 3\sqrt{5x}}{9}$
$= \dfrac{5\sqrt{5x}}{9}$

17. $7\sqrt{9} - 7 + \sqrt{3} = 7(3) - 7 + \sqrt{3}$
$= 21 - 7 + \sqrt{3}$
$= 14 + \sqrt{3}$

19. $2 + 3\sqrt{y^2} - 6\sqrt{y^2} + 5 = 2 + 3y - 6y + 5$
$= 7 - 3y$

269

21. $3\sqrt{108} - 2\sqrt{18} - 3\sqrt{48}$
$= 3\sqrt{36 \cdot 3} - 2\sqrt{9 \cdot 2} - 3\sqrt{16 \cdot 3}$
$= 3\sqrt{36} \cdot \sqrt{3} - 2\sqrt{9} \cdot \sqrt{2} - 3\sqrt{16} \cdot \sqrt{3}$
$= 3(6)\sqrt{3} - 2(3)\sqrt{2} - 3(4)\sqrt{3}$
$= 18\sqrt{3} - 6\sqrt{2} - 12\sqrt{3}$
$= 6\sqrt{3} - 6\sqrt{2}$

23. $-5\sqrt[3]{625} + \sqrt[3]{40} = -5\sqrt[3]{125 \cdot 5} + \sqrt[3]{8 \cdot 5}$
$= -5(5)\sqrt[3]{5} + 2\sqrt[3]{5}$
$= -25\sqrt[3]{5} + 2\sqrt[3]{5}$
$= -23\sqrt[3]{5}$

25. $\sqrt{9b^3} - \sqrt{25b^3} + \sqrt{16b^3}$
$= \sqrt{9b^2 \cdot b} - \sqrt{25b^2 \cdot b} + \sqrt{16b^2 \cdot b}$
$= 3b\sqrt{b} - 5b\sqrt{b} + 4b\sqrt{b}$
$= 2b\sqrt{b}$

27. $5y\sqrt{8y} + 2\sqrt{50y^3}$
$= 5y\sqrt{4 \cdot 2y} + 2\sqrt{25y^2 \cdot 2y}$
$= 5y(2)\sqrt{2y} + 2(5y)\sqrt{2y}$
$= 10y\sqrt{2y} + 10y\sqrt{2y}$
$= 20y\sqrt{2y}$

29. $\sqrt[3]{54xy^3} - 5\sqrt[3]{2xy^3} + y\sqrt[3]{128x}$
$= \sqrt[3]{27y^3 \cdot 2x} - 5\sqrt[3]{y^3 \cdot 2x} + y\sqrt[3]{64 \cdot 2x}$
$= 3y\sqrt[3]{2x} - 5y\sqrt[3]{2x} + 4y\sqrt[3]{2x}$
$= 2y\sqrt[3]{2x}$

31. $6\sqrt[3]{11} + 8\sqrt{11} - 12\sqrt{11} = 6\sqrt[3]{11} - 4\sqrt{11}$

33. $-2\sqrt[4]{x^7} + 3\sqrt[4]{16x^7}$
$= -2\sqrt[4]{x^4 \cdot x^3} + 3\sqrt[4]{8x^4 \cdot x^3}$
$= -2x\sqrt[4]{x^3} + 3(2x)\sqrt[4]{x^3}$
$= -2x\sqrt[4]{x^3} + 6x\sqrt[4]{x^3}$
$= 4x\sqrt[4]{x^3}$

35. $\dfrac{4\sqrt{3}}{3} - \dfrac{\sqrt{12}}{3} = \dfrac{4\sqrt{3}}{3} - \dfrac{\sqrt{4 \cdot 3}}{3}$
$= \dfrac{4\sqrt{3} - 2\sqrt{3}}{3} = \dfrac{2\sqrt{3}}{3}$

37. $\dfrac{\sqrt[3]{8x^4}}{7} + \dfrac{3x\sqrt[3]{x}}{7} = \dfrac{\sqrt[3]{8x^3 \cdot x}}{7} + \dfrac{3x\sqrt[3]{x}}{7}$
$= \dfrac{2x\sqrt[3]{x} + 3x\sqrt[3]{x}}{7}$
$= \dfrac{5x\sqrt[3]{x}}{7}$

39. $\sqrt{\dfrac{28}{x^2}} + \sqrt{\dfrac{7}{4x^2}} = \dfrac{\sqrt{4 \cdot 7}}{x} + \dfrac{\sqrt{7}}{2x}$
$= \dfrac{2\sqrt{7}}{x} + \dfrac{\sqrt{7}}{2x}$
$= \dfrac{2(2\sqrt{7}) + \sqrt{7}}{2x}$
$= \dfrac{4\sqrt{7} + \sqrt{7}}{2x} = \dfrac{5\sqrt{7}}{2x}$

41. $\sqrt[3]{\dfrac{16}{27}} - \dfrac{\sqrt[3]{54}}{6} = \dfrac{\sqrt[3]{8 \cdot 2}}{\sqrt[3]{27}} - \dfrac{\sqrt[3]{27 \cdot 2}}{6}$
$= \dfrac{2\sqrt[3]{2}}{3} - \dfrac{3\sqrt[3]{2}}{6}$
$= \dfrac{2(2\sqrt[3]{2}) - 3\sqrt[3]{2}}{6}$
$= \dfrac{4\sqrt[3]{2} - 3\sqrt[3]{2}}{6} = \dfrac{\sqrt[3]{2}}{6}$

43. $-\dfrac{\sqrt[3]{2x^4}}{9} + \sqrt[3]{\dfrac{250x^4}{27}}$
$= -\dfrac{\sqrt[3]{x^3 \cdot 2x}}{9} + \dfrac{\sqrt[3]{125x^3 \cdot 2x}}{\sqrt[3]{27}}$
$= \dfrac{-x\sqrt[3]{2x}}{9} + \dfrac{5x\sqrt[3]{2x}}{3}$
$= \dfrac{-x\sqrt[3]{2x} + 3(5x\sqrt[3]{2x})}{9}$
$= \dfrac{-x\sqrt[3]{2x} + 15x\sqrt[3]{2x}}{9} = \dfrac{14x\sqrt[3]{2x}}{9}$

45. $P = 2\sqrt{12} + \sqrt{12} + 2\sqrt{27} + 3\sqrt{3}$
$= 2\sqrt{4 \cdot 3} + \sqrt{4 \cdot 3} + 2\sqrt{9 \cdot 3} + 3\sqrt{3}$
$= 2(2)\sqrt{3} + 2\sqrt{3} + 2(3)\sqrt{3} + 3\sqrt{3}$
$= 4\sqrt{3} + 2\sqrt{3} + 6\sqrt{3} + 3\sqrt{3}$
$= 15\sqrt{3}$ inches

47. $\sqrt{7}\left(\sqrt{5} + \sqrt{3}\right) = \sqrt{7}\sqrt{5} + \sqrt{7}\sqrt{3}$
$= \sqrt{35} + \sqrt{21}$

49. $\left(\sqrt{5} - \sqrt{2}\right)^2 = \left(\sqrt{5}\right)^2 - 2\sqrt{5}\sqrt{2} + \left(\sqrt{2}\right)^2$
$= 5 - 2\sqrt{10} + 2$
$= 7 - 2\sqrt{10}$

51. $\sqrt{3x}\left(\sqrt{3} - \sqrt{x}\right) = \sqrt{3x}\sqrt{3} - \sqrt{3x}\sqrt{x}$
$= \sqrt{9x} - \sqrt{3x^2}$
$= 3\sqrt{x} - x\sqrt{3}$

53. $\left(2\sqrt{x} - 5\right)\left(3\sqrt{x} + 1\right)$
$= 2\sqrt{x}\left(3\sqrt{x}\right) + 2\sqrt{x} \cdot 1 - 5\left(3\sqrt{x}\right) - 5(1)$
$= 6x + 2\sqrt{x} - 15\sqrt{x} - 5$
$= 6x - 13\sqrt{x} - 5$

55. $\left(\sqrt[3]{a} - 4\right)\left(\sqrt[3]{a} + 5\right)$
$= \sqrt[3]{a}\left(\sqrt[3]{a}\right) + \sqrt[3]{a} \cdot 5 - 4\sqrt[3]{a} - 4(5)$
$= \sqrt[3]{a^2} + 5\sqrt[3]{a} - 4\sqrt[3]{a} - 20$
$= \sqrt[3]{a^2} + \sqrt[3]{a} - 20$

57. $6\left(\sqrt{2} - 2\right) = 6\sqrt{2} - 6(2) = 6\sqrt{2} - 12$

59. $\sqrt{2}\left(\sqrt{2} + x\sqrt{6}\right) = \sqrt{2}\sqrt{2} + \sqrt{2}\left(x\sqrt{6}\right)$
$= 2 + x\sqrt{12}$
$= 2 + x\sqrt{4 \cdot 3}$
$= 2 + 2x\sqrt{3}$

61. $\left(2\sqrt{7} + 3\sqrt{5}\right)\left(\sqrt{7} - 2\sqrt{5}\right)$
$= 2\sqrt{7}\sqrt{7} + 2\sqrt{7}\left(-2\sqrt{5}\right) + 3\sqrt{5}\sqrt{7}$
$\qquad\qquad + 3\sqrt{5}\left(-2\sqrt{5}\right)$
$= 2(7) - 4\sqrt{35} + 3\sqrt{35} - 6(5)$
$= 14 - \sqrt{35} - 30$
$= -16 - \sqrt{35}$

63. $\left(\sqrt{x} - y\right)\left(\sqrt{x} + y\right) = \left(\sqrt{x}\right)^2 - y^2 = x - y^2$

65. $\left(\sqrt{3} + x\right)^2 = \left(\sqrt{3}\right)^2 + 2\sqrt{3} \cdot x + x^2$
$= 3 + 2x\sqrt{3} + x^2$

67. $\left(\sqrt{5x} - 3\sqrt{2}\right)\left(\sqrt{5x} - 3\sqrt{3}\right)$
$= \left(\sqrt{5x}\right)^2 + \sqrt{5x}\left(-3\sqrt{3}\right) - 3\sqrt{2}\left(\sqrt{5x}\right)$
$\qquad\qquad\qquad\qquad - 3\sqrt{2}\left(-3\sqrt{3}\right)$
$= 5x - 3\sqrt{15x} - 3\sqrt{10x} + 9\sqrt{6}$

69. $\left(\sqrt[3]{4} + 2\right)\left(\sqrt[3]{2} - 1\right)$
$= \sqrt[3]{4}\left(\sqrt[3]{2}\right) + \sqrt[3]{4} \cdot (-1) + 2\sqrt[3]{2} + 2(-1)$
$= \sqrt[3]{8} - \sqrt[3]{4} + 2\sqrt[3]{2} - 2$
$= 2 - \sqrt[3]{4} + 2\sqrt[3]{2} - 2$
$= 2\sqrt[3]{2} - \sqrt[3]{4}$

71. $\left(\sqrt[3]{x} + 1\right)\left(\sqrt[3]{x} - 4\sqrt{x} + 7\right)$
$= \left(\sqrt[3]{x}\right)^2 + \sqrt[3]{x}\left(-4\sqrt{x}\right) + \sqrt[3]{x} \cdot 7$
$\qquad\qquad + 1\left(\sqrt[3]{x}\right) + 1\left(-4\sqrt{x}\right) + 1(7)$
$= \sqrt[3]{x^2} - 4x^{1/3}x^{1/2} + 8\sqrt[3]{x} - 4\sqrt{x} + 7$
$= \sqrt[3]{x^2} - 4x^{5/6} + 8\sqrt[3]{x} - 4\sqrt{x} + 7$
$= \sqrt[3]{x^2} - 4\sqrt[6]{x^5} + 8\sqrt[3]{x} - 4\sqrt{x} + 7$

73. $\left(\sqrt{x-1}+5\right)^2$

$= \sqrt{x-1}^2 + 2\sqrt{x-1}\cdot 5 + 5^2$

$= (x-1) + 10\sqrt{x-1} + 25$

$= x + 10\sqrt{x-1} + 24$

75. $\left(\sqrt{2x+5}-1\right)^2$

$= \sqrt{2x+5}^2 - 2\sqrt{2x+5}\cdot 1 + 1^2$

$= (2x+5) - 2\sqrt{2x+5} + 1$

$= 2x - 2\sqrt{2x+5} + 6$

77. $\dfrac{2x-14}{2} = \dfrac{2(x-7)}{2} = x-7$

79. $\dfrac{7x-7y}{x^2-y^2} = \dfrac{7(x-y)}{(x+y)(x-y)} = \dfrac{7}{x+y}$

81. $\dfrac{6a^2b-9ab}{3ab} = \dfrac{3ab(2a-3)}{3ab} = 2a-3$

83. $\dfrac{-4+2\sqrt{3}}{6} = \dfrac{2\left(-2+\sqrt{3}\right)}{6} = \dfrac{-2+\sqrt{3}}{3}$

85. $P = 2l + 2w$

$= 2\left(3\sqrt{20}\right) + 2\left(\sqrt{125}\right)$

$= 6\sqrt{4\cdot 5} + 2\sqrt{25\cdot 5}$

$= 6(2)\sqrt{5} + 2(5)\sqrt{5}$

$= 12\sqrt{5} + 10\sqrt{5}$

$= 22\sqrt{5}$ feet

$A = lw$

$= \left(3\sqrt{20}\right)\left(\sqrt{125}\right)$

$= 3\sqrt{4\cdot 5}\sqrt{25\cdot 5}$

$= 3(2)\sqrt{5}\cdot 5\sqrt{5}$

$= 30\cdot 5$

$= 150$ square feet

87. a. $\sqrt{3} + \sqrt{3} = 2\sqrt{3}$

b. $\sqrt{3}\cdot\sqrt{3} = \sqrt{9} = 3$

c. Answers may vary.

89. Answer may vary.

Section 7.5

Mental Math

1. The conjugate of $\sqrt{2}+x$ is $\sqrt{2}-x$.

2. The conjugate of $\sqrt{3}+y$ is $\sqrt{3}-y$.

3. The conjugate of $5-\sqrt{a}$ is $5+\sqrt{a}$.

4. The conjugate of $6-\sqrt{b}$ is $6+\sqrt{b}$.

5. The conjugate of $7\sqrt{5}+8\sqrt{x}$ is $7\sqrt{5}-8\sqrt{x}$.

6. The conjugate of $9\sqrt{2}-6\sqrt{y}$ is $9\sqrt{2}+6\sqrt{y}$.

Exercise Set 7.5

1. $\dfrac{\sqrt{2}}{\sqrt{7}} = \dfrac{\sqrt{2}\cdot\sqrt{7}}{\sqrt{7}\cdot\sqrt{7}} = \dfrac{\sqrt{14}}{\sqrt{49}} = \dfrac{\sqrt{14}}{7}$

3. $\sqrt{\dfrac{1}{5}} = \dfrac{\sqrt{1}}{\sqrt{5}} = \dfrac{1\cdot\sqrt{5}}{\sqrt{5}\cdot\sqrt{5}} = \dfrac{\sqrt{5}}{5}$

5. $\sqrt[3]{\dfrac{3}{4}} = \dfrac{\sqrt[3]{3}}{\sqrt[3]{4}} = \dfrac{\sqrt[3]{3}\cdot\sqrt[3]{2}}{\sqrt[3]{4}\cdot\sqrt[3]{2}} = \dfrac{\sqrt[3]{6}}{\sqrt[3]{8}} = \dfrac{\sqrt[3]{6}}{2}$

7. $\dfrac{4}{\sqrt[3]{3}} = \dfrac{4\cdot\sqrt[3]{9}}{\sqrt[3]{3}\cdot\sqrt[3]{9}} = \dfrac{4\sqrt[3]{9}}{\sqrt[3]{27}} = \dfrac{4\sqrt[3]{9}}{3}$

9. $\dfrac{3}{\sqrt{8x}} = \dfrac{3\cdot\sqrt{2x}}{\sqrt{8x}\cdot\sqrt{2x}} = \dfrac{3\sqrt{2x}}{\sqrt{16x^2}} = \dfrac{3\sqrt{2x}}{4x}$

11. $\dfrac{3}{\sqrt[3]{4x^2}} = \dfrac{3\cdot\sqrt[3]{2x}}{\sqrt[3]{4x^2}\cdot\sqrt[3]{2x}} = \dfrac{3\sqrt[3]{2x}}{\sqrt[3]{8x^3}} = \dfrac{3\sqrt[3]{2x}}{2x}$

13. $\sqrt{\dfrac{4}{x}} = \dfrac{\sqrt{4}}{\sqrt{x}} = \dfrac{2 \cdot \sqrt{x}}{\sqrt{x} \cdot \sqrt{x}} = \dfrac{2\sqrt{x}}{\sqrt{x^2}} = \dfrac{2\sqrt{x}}{x}$

15. $\dfrac{9}{\sqrt{3a}} = \dfrac{9 \cdot \sqrt{3a}}{\sqrt{3a} \cdot \sqrt{3a}} = \dfrac{9\sqrt{3a}}{3a} = \dfrac{3\sqrt{3a}}{a}$

17. $\dfrac{3}{\sqrt[3]{2}} = \dfrac{3 \cdot \sqrt[3]{4}}{\sqrt[3]{2} \cdot \sqrt[3]{4}} = \dfrac{3\sqrt[3]{4}}{\sqrt[3]{8}} = \dfrac{3\sqrt[3]{4}}{2}$

19. $\dfrac{2\sqrt{3}}{\sqrt{7}} = \dfrac{2\sqrt{3} \cdot \sqrt{7}}{\sqrt{7} \cdot \sqrt{7}} = \dfrac{2\sqrt{21}}{\sqrt{49}} = \dfrac{2\sqrt{21}}{7}$

21. $\sqrt{\dfrac{2x}{5y}} = \dfrac{\sqrt{2x}}{\sqrt{5y}} = \dfrac{\sqrt{2x} \cdot \sqrt{5y}}{\sqrt{5y} \cdot \sqrt{5y}} = \dfrac{\sqrt{10xy}}{5y}$

23. $\sqrt[4]{\dfrac{81}{8}} = \dfrac{\sqrt[4]{81}}{\sqrt[4]{8}} = \dfrac{3 \cdot \sqrt[4]{2}}{\sqrt[4]{8} \cdot \sqrt[4]{2}} = \dfrac{3\sqrt[4]{2}}{\sqrt[4]{16}} = \dfrac{3\sqrt[4]{2}}{2}$

25. $\sqrt[4]{\dfrac{16}{9x^7}} = \dfrac{\sqrt[4]{16}}{\sqrt[4]{9x^7}}$

$= \dfrac{2 \cdot \sqrt[4]{9x}}{\sqrt[4]{9x^7} \cdot \sqrt[4]{9x}} = \dfrac{2\sqrt[4]{9x}}{\sqrt[4]{81x^8}} = \dfrac{2\sqrt[4]{9x}}{3x^2}$

27. $\dfrac{5a}{\sqrt[5]{8a^9b^{11}}} = \dfrac{5a \cdot \sqrt[5]{4ab^4}}{\sqrt[5]{8a^9b^{11}} \cdot \sqrt[5]{4ab^4}}$

$= \dfrac{5a\sqrt[5]{4ab^4}}{\sqrt[5]{32a^{10}b^{15}}} = \dfrac{5a\sqrt[5]{4ab^4}}{2a^2b^3}$

29. $\dfrac{6}{2-\sqrt{7}} = \dfrac{6\left(2+\sqrt{7}\right)}{\left(2-\sqrt{7}\right)\left(2+\sqrt{7}\right)}$

$= \dfrac{6\left(2+\sqrt{7}\right)}{2^2 - \left(\sqrt{7}\right)^2}$

$= \dfrac{6\left(2+\sqrt{7}\right)}{4-7}$

$= \dfrac{6\left(2+\sqrt{7}\right)}{-3} = -2\left(2+\sqrt{7}\right)$

31. $\dfrac{-7}{\sqrt{x}-3} = \dfrac{-7\left(\sqrt{x}+3\right)}{\left(\sqrt{x}-3\right)\left(\sqrt{x}+3\right)}$

$= \dfrac{-7\left(\sqrt{x}+3\right)}{\left(\sqrt{x}\right)^2 - (3)^2}$

$= \dfrac{-7\left(\sqrt{x}+3\right)}{x-9}$

33. $\dfrac{\sqrt{2}-\sqrt{3}}{\sqrt{2}+\sqrt{3}} = \dfrac{\left(\sqrt{2}-\sqrt{3}\right)\left(\sqrt{2}-\sqrt{3}\right)}{\left(\sqrt{2}+\sqrt{3}\right)\left(\sqrt{2}-\sqrt{3}\right)}$

$= \dfrac{\left(\sqrt{2}\right)^2 - 2\sqrt{2}\sqrt{3} + \left(\sqrt{3}\right)^2}{\left(\sqrt{2}\right)^2 - \left(\sqrt{3}\right)^2}$

$= \dfrac{2 - 2\sqrt{6} + 3}{2-3}$

$= \dfrac{5 - 2\sqrt{6}}{-1} = -5 + 2\sqrt{6}$

35. $\dfrac{\sqrt{a}+1}{2\sqrt{a}-\sqrt{b}}$

$= \dfrac{\left(\sqrt{a}+1\right)\left(2\sqrt{a}+\sqrt{b}\right)}{\left(2\sqrt{a}-\sqrt{b}\right)\left(2\sqrt{a}+\sqrt{b}\right)}$

$= \dfrac{\sqrt{a} \cdot 2\sqrt{a} + \sqrt{a}\sqrt{b} + 1 \cdot 2\sqrt{a} + 1 \cdot \sqrt{b}}{\left(2\sqrt{a}\right)^2 - \left(\sqrt{b}\right)^2}$

$= \dfrac{2a + \sqrt{ab} + 2\sqrt{a} + \sqrt{b}}{4a-b}$

37. $\dfrac{8}{1+\sqrt{10}} = \dfrac{8\left(1-\sqrt{10}\right)}{\left(1+\sqrt{10}\right)\left(1-\sqrt{10}\right)}$

$= \dfrac{8\left(1-\sqrt{10}\right)}{1^2 - \left(\sqrt{10}\right)^2}$

$= \dfrac{8\left(1-\sqrt{10}\right)}{1-10} = -\dfrac{8\left(1-\sqrt{10}\right)}{9}$

39. $\dfrac{\sqrt{x}}{\sqrt{x}+\sqrt{y}} = \dfrac{\sqrt{x}\left(\sqrt{x}-\sqrt{y}\right)}{\left(\sqrt{x}+\sqrt{y}\right)\left(\sqrt{x}-\sqrt{y}\right)}$

$\phantom{\dfrac{\sqrt{x}}{\sqrt{x}+\sqrt{y}}} = \dfrac{\sqrt{x}\left(\sqrt{x}-\sqrt{y}\right)}{\left(\sqrt{x}\right)^2 - \left(\sqrt{y}\right)^2}$

$\phantom{\dfrac{\sqrt{x}}{\sqrt{x}+\sqrt{y}}} = \dfrac{\sqrt{x}\left(\sqrt{x}-\sqrt{y}\right)}{x-y}$

$\phantom{\dfrac{\sqrt{x}}{\sqrt{x}+\sqrt{y}}} = \dfrac{x-\sqrt{xy}}{x-y}$

41. $\dfrac{2\sqrt{3}+\sqrt{6}}{4\sqrt{3}-\sqrt{6}} = \dfrac{\left(2\sqrt{3}+\sqrt{6}\right)\left(4\sqrt{3}+\sqrt{6}\right)}{\left(4\sqrt{3}-\sqrt{6}\right)\left(4\sqrt{3}+\sqrt{6}\right)}$

$\phantom{\dfrac{2\sqrt{3}+\sqrt{6}}{4\sqrt{3}-\sqrt{6}}} = \dfrac{8\cdot 3 + 2\sqrt{18} + 4\sqrt{18} + 6}{\left(4\sqrt{3}\right)^2 - \left(\sqrt{6}\right)^2}$

$\phantom{\dfrac{2\sqrt{3}+\sqrt{6}}{4\sqrt{3}-\sqrt{6}}} = \dfrac{30 + 6\sqrt{18}}{16\cdot 3 - 6}$

$\phantom{\dfrac{2\sqrt{3}+\sqrt{6}}{4\sqrt{3}-\sqrt{6}}} = \dfrac{30 + 6(3)\sqrt{2}}{42}$

$\phantom{\dfrac{2\sqrt{3}+\sqrt{6}}{4\sqrt{3}-\sqrt{6}}} = \dfrac{30 + 18\sqrt{2}}{42}$

$\phantom{\dfrac{2\sqrt{3}+\sqrt{6}}{4\sqrt{3}-\sqrt{6}}} = \dfrac{6\left(5 + 3\sqrt{2}\right)}{42}$

$\phantom{\dfrac{2\sqrt{3}+\sqrt{6}}{4\sqrt{3}-\sqrt{6}}} = \dfrac{5 + 3\sqrt{2}}{7}$

43. $\sqrt{\dfrac{5}{3}} = \dfrac{\sqrt{5}}{\sqrt{3}} = \dfrac{\sqrt{5}\cdot\sqrt{5}}{\sqrt{3}\cdot\sqrt{5}} = \dfrac{\sqrt{25}}{\sqrt{15}} = \dfrac{5}{\sqrt{15}}$

45. $\sqrt{\dfrac{18}{5}} = \dfrac{\sqrt{18}}{\sqrt{5}}$

$ = \dfrac{\sqrt{9}\cdot\sqrt{2}}{\sqrt{5}}$

$ = \dfrac{3\sqrt{2}}{\sqrt{5}}$

$ = \dfrac{3\sqrt{2}\cdot\sqrt{2}}{\sqrt{5}\cdot\sqrt{2}}$

$ = \dfrac{3\cdot 2}{\sqrt{10}} = \dfrac{6}{\sqrt{10}}$

47. $\dfrac{\sqrt{4x}}{7} = \dfrac{2\sqrt{x}}{7} = \dfrac{2\sqrt{x}\cdot\sqrt{x}}{7\cdot\sqrt{x}} = \dfrac{2\sqrt{x^2}}{7\sqrt{x}} = \dfrac{2x}{7\sqrt{x}}$

49. $\dfrac{\sqrt[3]{5y^2}}{\sqrt[3]{4x}} = \dfrac{\sqrt[3]{5y^2}\cdot\sqrt[3]{5^2 y}}{\sqrt[3]{4x}\cdot\sqrt[3]{5^2 y}} = \dfrac{\sqrt[3]{5^3 y^3}}{\sqrt[3]{100xy}} = \dfrac{5y}{\sqrt[3]{100xy}}$

51. $\sqrt{\dfrac{2}{5}} = \dfrac{\sqrt{2}}{\sqrt{5}} = \dfrac{\sqrt{2}\cdot\sqrt{2}}{\sqrt{5}\cdot\sqrt{2}} = \dfrac{\sqrt{4}}{\sqrt{10}} = \dfrac{2}{\sqrt{10}}$

53. $\dfrac{\sqrt{2x}}{11} = \dfrac{\sqrt{2x}\cdot\sqrt{2x}}{11\cdot\sqrt{2x}} = \dfrac{\sqrt{4x^2}}{11\sqrt{2x}} = \dfrac{2x}{11\sqrt{2x}}$

55. $\sqrt[3]{\dfrac{7}{8}} = \dfrac{\sqrt[3]{7}}{\sqrt[3]{8}} = \dfrac{\sqrt[3]{7}}{2}$

$ = \dfrac{\sqrt[3]{7}\cdot\sqrt[3]{7^2}}{2\cdot\sqrt[3]{7^2}} = \dfrac{\sqrt[3]{7^3}}{2\sqrt[3]{49}} = \dfrac{7}{2\sqrt[3]{49}}$

57. $\dfrac{\sqrt[3]{3x^5}}{10} = \dfrac{\sqrt[3]{x^3\cdot 3x^2}}{10}$

$ = \dfrac{x\sqrt[3]{3x^2}\cdot\sqrt[3]{3^2 x}}{10\cdot\sqrt[3]{3^2 x}}$

$ = \dfrac{x\sqrt[3]{3^3 x^3}}{10\sqrt[3]{9x}}$

$ = \dfrac{x\cdot 3x}{10\sqrt[3]{9x}}$

$ = \dfrac{3x^2}{10\sqrt[3]{9x}}$

59. $\sqrt{\dfrac{18x^4 y^6}{3z}} = \dfrac{\sqrt{18x^4 y^6}}{\sqrt{3z}}$

$$= \frac{\sqrt{9x^4y^6 \cdot 2}}{\sqrt{3z}}$$

$$= \frac{3x^2y^3\sqrt{2}}{\sqrt{3z}}$$

$$= \frac{3x^2y^3\sqrt{2} \cdot \sqrt{2}}{\sqrt{3z} \cdot \sqrt{2}}$$

$$= \frac{3x^2y^3 \cdot 2}{\sqrt{6z}}$$

$$= \frac{6x^2y^3}{\sqrt{6z}}$$

61. Answers may vary.

63. $\dfrac{2-\sqrt{11}}{6} = \dfrac{\left(2-\sqrt{11}\right)\left(2+\sqrt{11}\right)}{6\left(2+\sqrt{11}\right)}$

$$= \frac{4-11}{12+6\sqrt{11}}$$

$$= \frac{-7}{12+6\sqrt{11}}$$

65. $\dfrac{2-\sqrt{7}}{-5} = \dfrac{\left(2-\sqrt{7}\right)\left(2+\sqrt{7}\right)}{-5\left(2+\sqrt{7}\right)}$

$$= \frac{4-7}{-5\left(2+\sqrt{7}\right)}$$

$$= \frac{-3}{-10-5\sqrt{7}} = \frac{3}{10+5\sqrt{7}}$$

67. $\dfrac{\sqrt{x}+3}{\sqrt{x}} = \dfrac{\left(\sqrt{x}+3\right)\left(\sqrt{x}-3\right)}{\sqrt{x}\left(\sqrt{x}-3\right)}$

$$= \frac{\sqrt{x^2}-9}{\sqrt{x^2}-3\sqrt{x}}$$

$$= \frac{x-9}{x-3\sqrt{x}}$$

69. $\dfrac{\sqrt{2}-1}{\sqrt{2}+1} = \dfrac{\left(\sqrt{2}-1\right)\left(\sqrt{2}+1\right)}{\left(\sqrt{2}+1\right)\left(\sqrt{2}+1\right)}$

$$= \frac{\sqrt{4}-1}{\sqrt{4}+2\sqrt{2}+1}$$

$$= \frac{2-1}{2+2\sqrt{2}+1} = \frac{1}{3+2\sqrt{2}}$$

71. $\dfrac{\sqrt{x}+1}{\sqrt{x}-1} = \dfrac{\left(\sqrt{x}+1\right)\left(\sqrt{x}-1\right)}{\left(\sqrt{x}-1\right)\left(\sqrt{x}-1\right)}$

$$= \frac{\sqrt{x^2}-1}{\sqrt{x^2}-2\sqrt{x}+1}$$

$$= \frac{x-1}{x-2\sqrt{x}+1}$$

73. $2x-7 = 3(x-4)$
$2x-7 = 3x-12$
$-x-7 = -12$
$-x = -5$
$x = 5$
The solution is 5.

75. $(x-6)(2x+1) = 0$
$x-6 = 0$ or $2x+1 = 0$
$x = 6$ or $2x = -1$
$$x = -\frac{1}{2}$$
The solutions are $-\dfrac{1}{2}, 6$.

77. $x^2 - 8x = -12$
$x^2 - 8x + 12 = 0$
$(x-6)(x-2) = 0$
$x-6 = 0$ or $x-2 = 0$
$x = 6$ or $x = 2$
The solutions are 2, 6.

79. $r = \sqrt{\dfrac{A}{4\pi}}$

$\quad = \dfrac{\sqrt{A}}{\sqrt{4\pi}}$

$\quad = \dfrac{\sqrt{A}}{2\sqrt{\pi}}$

$\quad = \dfrac{\sqrt{A} \cdot \sqrt{\pi}}{2\sqrt{\pi} \cdot \sqrt{\pi}} = \dfrac{\sqrt{A\pi}}{2\pi}$

81. Answers may vary.

Integrated Review

1. $\sqrt{81} = 9$

2. $\sqrt[3]{-8} = -2$

3. $\sqrt[4]{\dfrac{1}{16}} = \dfrac{1}{2}$

4. $\sqrt{x^6} = x^3$

5. $\sqrt[3]{y^9} = y^3$

6. $\sqrt{4y^{10}} = 2y^5$

7. $\sqrt[5]{-32y^5} = -2y$

8. $\sqrt[4]{81b^{12}} = 3b^3$

9. $36^{1/2} = \sqrt{36} = 6$

10. $(3y)^{1/4} = \sqrt[4]{3y}$

11. $64^{-2/3} = \dfrac{1}{64^{2/3}} = \dfrac{1}{\left(\sqrt[3]{64}\right)^2} = \dfrac{1}{(4)^2} = \dfrac{1}{16}$

12. $(x+1)^{3/5} = \sqrt[5]{(x+1)^3}$

13. $y^{-1/6} \cdot y^{7/6} = y^{-1/6+7/6} = y^{6/6} = y^1 = y$

14. $\dfrac{(2x^{1/3})^4}{x^{5/6}} = \dfrac{2^4 x^{4/3}}{x^{5/6}}$

$\quad = 16x^{\frac{4}{3}-\frac{5}{6}}$

$\quad = 16x^{\frac{8}{6}-\frac{5}{6}}$

$\quad = 16x^{3/6}$

$\quad = 16x^{1/2}$

15. $\dfrac{x^{1/4}x^{3/4}}{x^{-1/4}} = x^{\frac{1}{4}+\frac{3}{4}-\left(-\frac{1}{4}\right)} = x^{\frac{1+3+1}{4}} = x^{5/4}$

16. $4^{1/3} \cdot 4^{2/5} = 4^{\frac{1}{3}+\frac{2}{5}} = 4^{\frac{5+6}{15}} = 4^{11/15}$

17. $\sqrt[3]{8x^6} = (8x^6)^{1/3}$

$\quad = (2^3 x^6)^{1/3} = 2^{3(1/3)} x^{6(1/3)} = 2x^2$

18. $\sqrt[12]{a^9 b^6} = (a^9 b^6)^{1/12}$

$\quad = a^{9(1/12)} b^{6(1/12)}$

$\quad = a^{3/4} b^{1/2}$

$\quad = a^{3/4} b^{2/4}$

$\quad = (a^3 b^2)^{1/4}$

$\quad = \sqrt[4]{a^3 b^2}$

19. $\sqrt[4]{x} \cdot \sqrt{x} = x^{1/4} \cdot x^{1/2}$

$\quad = x^{\frac{1}{4}+\frac{1}{2}} = x^{\frac{1+2}{4}} = x^{3/4} = \sqrt[4]{x^3}$

20. $\sqrt{5} \cdot \sqrt[3]{2} = 5^{1/2} \cdot 2^{1/3}$

$\quad = 5^{3/6} \cdot 2^{2/6}$

$\quad = (5^3 \cdot 2^2)^{1/6}$

$\quad = \sqrt[6]{125 \cdot 4}$

$\quad = \sqrt[6]{500}$

21. $\sqrt{40} = \sqrt{4 \cdot 10} = \sqrt{4} \cdot \sqrt{10} = 2\sqrt{10}$

22. $\sqrt[4]{16x^7 y^{10}} = \sqrt[4]{16x^4 y^8 \cdot x^3 y^2}$

$\quad = \sqrt[4]{16x^4 y^8} \cdot \sqrt[4]{x^3 y^2}$

$\quad = 2xy^2 \sqrt[4]{x^3 y^2}$

23. $\sqrt[3]{54x^4} = \sqrt[3]{27x^3 \cdot 2x}$

$\quad = \sqrt[3]{27x^3} \cdot \sqrt[3]{2x}$

$\quad = 3x\sqrt[3]{2x}$

24. $\sqrt[5]{-64b^{10}} = \sqrt[5]{-32b^{10} \cdot 2} = -2b^2 \sqrt[5]{2}$

25. $\sqrt{5} \cdot \sqrt{x} = \sqrt{5x}$

26. $\sqrt[3]{8x} \cdot \sqrt[3]{8x^2} = \sqrt[3]{64x^3} = 4x$

27. $\dfrac{\sqrt{98y^6}}{\sqrt{2y}} = \sqrt{\dfrac{98y^6}{2y}}$
$= \sqrt{49y^5}$
$= \sqrt{49y^4 \cdot y}$
$= 7y^2 \sqrt{y}$

28. $\dfrac{\sqrt[4]{48a^9b^3}}{\sqrt[4]{ab^3}} = \sqrt[4]{\dfrac{48a^9b^3}{ab^3}}$
$= \sqrt[4]{48a^8}$
$= \sqrt[4]{16a^8 \cdot 3}$
$= 2a^2 \sqrt[4]{3}$

29. $\sqrt{20} - \sqrt{75} + 5\sqrt{7} = \sqrt{4 \cdot 5} - \sqrt{25 \cdot 3} + 5\sqrt{7}$
$= 2\sqrt{5} - 5\sqrt{3} + 5\sqrt{7}$

30. $\sqrt[3]{54y^4} - y\sqrt[3]{16y} = \sqrt[3]{27y^3 \cdot 2y} - y\sqrt[3]{8 \cdot 2y}$
$= 3y\sqrt[3]{2y} - 2y\sqrt[3]{2y}$
$= y\sqrt[3]{2y}$

31. $\sqrt{3}\left(\sqrt{5} - \sqrt{2}\right) = \sqrt{3}\sqrt{5} - \sqrt{3}\sqrt{2}$
$= \sqrt{15} - \sqrt{6}$

32. $\left(\sqrt{7} + \sqrt{3}\right)^2 = \left(\sqrt{7}\right)^2 + 2\sqrt{7}\sqrt{3} + \left(\sqrt{3}\right)^2$
$= 7 + 2\sqrt{21} + 3$
$= 10 + 2\sqrt{21}$

33. $\left(2x - \sqrt{5}\right)\left(2x + \sqrt{5}\right) = (2x)^2 - \left(\sqrt{5}\right)^2$
$= 4x^2 - 5$

34. $\left(\sqrt{x+1} - 1\right)^2 = \sqrt{x+1}^2 - 2\sqrt{x+1} \cdot 1 + 1^2$
$= x + 1 - 2\sqrt{x+1} + 1$
$= x + 2 - 2\sqrt{x+1}$

35. $\sqrt{\dfrac{7}{3}} = \dfrac{\sqrt{7}}{\sqrt{3}} = \dfrac{\sqrt{7} \cdot \sqrt{3}}{\sqrt{3} \cdot \sqrt{3}} = \dfrac{\sqrt{21}}{\sqrt{9}} = \dfrac{\sqrt{21}}{3}$

36. $\dfrac{5}{\sqrt[3]{2x^2}} = \dfrac{5 \cdot \sqrt[3]{2^2 x}}{\sqrt[3]{2x^2} \cdot \sqrt[3]{2^2 x}} = \dfrac{5\sqrt[3]{4x}}{\sqrt[3]{2^3 x^3}} = \dfrac{5\sqrt[3]{4x}}{2x}$

37. $\dfrac{\sqrt{3} - \sqrt{7}}{2\sqrt{3} + \sqrt{7}}$
$= \dfrac{\left(\sqrt{3} - \sqrt{7}\right)\left(2\sqrt{3} - \sqrt{7}\right)}{\left(2\sqrt{3} + \sqrt{7}\right)\left(2\sqrt{3} - \sqrt{7}\right)}$
$= \dfrac{2\sqrt{9} - \sqrt{3}\sqrt{7} - \sqrt{7} \cdot 2\sqrt{3} + \sqrt{49}}{\left(2\sqrt{3}\right)^2 - \left(\sqrt{7}\right)^2}$
$= \dfrac{2(3) - \sqrt{21} - 2\sqrt{21} + 7}{4 \cdot 3 - 7}$
$= \dfrac{6 - 3\sqrt{21} + 7}{12 - 7} = \dfrac{13 - 3\sqrt{21}}{5}$

38. $\sqrt{\dfrac{7}{3}} = \dfrac{\sqrt{7}}{\sqrt{3}} = \dfrac{\sqrt{7} \cdot \sqrt{7}}{\sqrt{3} \cdot \sqrt{7}} = \dfrac{\sqrt{49}}{\sqrt{21}} = \dfrac{7}{\sqrt{21}}$

39. $\sqrt[3]{\dfrac{9y}{11}} = \dfrac{\sqrt[3]{9y}}{\sqrt[3]{11}}$
$= \dfrac{\sqrt[3]{9y} \cdot \sqrt[3]{3y^2}}{\sqrt[3]{11} \cdot \sqrt[3]{3y^2}} = \dfrac{\sqrt[3]{27y^3}}{\sqrt[3]{33y^2}} = \dfrac{3y}{\sqrt[3]{33y^2}}$

40. $\dfrac{\sqrt{x} - 2}{\sqrt{x}} = \dfrac{\left(\sqrt{x} - 2\right)\left(\sqrt{x} + 2\right)}{\sqrt{x}\left(\sqrt{x} + 2\right)}$
$= \dfrac{\left(\sqrt{x}\right)^2 - 2^2}{\sqrt{x^2} + 2\sqrt{x}} = \dfrac{x - 4}{x + 2\sqrt{x}}$

Section 7.6

Graphing Calculator Explorations

1.

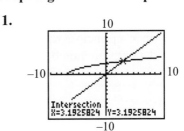

The solution is 3.19.

3.

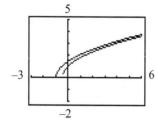

There is no solution.

5.

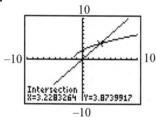

The solution is 3.23.

Exercise Set 7.6

1.
$$\sqrt{2x} = 4$$
$$\left(\sqrt{2x}\right)^2 = 4^2$$
$$2x = 16$$
$$x = 8$$
The solution is 8.

3.
$$\sqrt{x-3} = 2$$
$$\left(\sqrt{x-3}\right)^2 = 2^2$$
$$x - 3 = 4$$
$$x = 7$$
The solution is 7.

5. $\sqrt{2x} = -4$

No solution since a principle square root does not yield a negative number.

7.
$$\sqrt{4x-3} - 5 = 0$$
$$\sqrt{4x-3} = 5$$
$$\left(\sqrt{4x-3}\right)^2 = 5^2$$
$$4x - 3 = 25$$
$$4x = 28$$
$$x = 7$$
The solution is 7.

9.
$$\sqrt{2x-3} - 2 = 1$$
$$\sqrt{2x-3} = 3$$
$$\left(\sqrt{2x-3}\right)^2 = 3^2$$
$$2x - 3 = 9$$
$$2x = 12$$
$$x = 6$$
The solution is 6.

11.
$$\sqrt[3]{6x} = -3$$
$$\left(\sqrt[3]{6x}\right)^3 = (-3)^3$$
$$6x = -27$$
$$x = -\frac{27}{6} = -\frac{9}{2}$$
The solution is $-\frac{9}{2}$.

13.
$$\sqrt[3]{x-2} - 3 = 0$$
$$\sqrt[3]{x-2} = 3$$
$$\left(\sqrt[3]{x-2}\right)^3 = 3^2$$
$$x - 2 = 27$$
$$x = 29$$
The solution is 29.

15.
$$\sqrt{13-x} = x-1$$
$$\left(\sqrt{13-x}\right)^2 = (x-1)^2$$
$$13 - x = x^2 - 2x + 1$$
$$0 = x^2 - x - 12$$
$$0 = (x-4)(x+3)$$
$$x - 4 = 0 \text{ or } x + 3 = 0$$
$$x = 4 \text{ or } \quad x = -3$$
We discard -3 as extraneous.
The solution is 4.

17. $x - \sqrt{4 - 3x} = -8$
$$x + 8 = \sqrt{4 - 3x}$$
$$(x + 8)^2 = \left(\sqrt{4 - 3x}\right)^2$$
$$x^2 + 16x + 64 = 4 - 3x$$
$$x^2 + 19x + 60 = 0$$
$$(x + 4)(x + 15) = 0$$
$$x + 4 = 0 \quad \text{or} \quad x + 15 = 0$$
$$x = -4 \quad \text{or} \quad x = -15$$

We discard –15 as extraneous.
The solution is –4.

19. $\sqrt{y + 5} = 2 - \sqrt{y - 4}$
$$\left(\sqrt{y + 5}\right)^2 = \left(2 - \sqrt{y - 4}\right)^2$$
$$y + 5 = 4 - 4\sqrt{y - 4} + (y - 4)$$
$$y + 5 = y - 4\sqrt{y - 4}$$
$$5 = -4\sqrt{y - 4}$$
$$5^2 = \left(-4\sqrt{y - 4}\right)^2$$
$$25 = 16(y - 4)$$
$$25 = 16y - 64$$
$$89 = 16y$$
$$\frac{89}{16} = y$$

which we discard as extraneous.
There is no solution.

21. $\sqrt{x - 3} + \sqrt{x + 2} = 5$
$$\sqrt{x - 3} = 5 - \sqrt{x + 2}$$
$$\left(\sqrt{x - 3}\right)^2 = \left(5 - \sqrt{x + 2}\right)^2$$
$$x - 3 = 25 - 10\sqrt{x + 2} + (x + 2)$$
$$x - 3 = 27 - 10\sqrt{x + 2} + x$$
$$-30 = -10\sqrt{x + 2}$$
$$3 = \sqrt{x + 2}$$
$$3^2 = \left(\sqrt{x + 2}\right)^2$$
$$9 = x + 2$$
$$7 = x$$

The solution is 7.

23. $\sqrt{3x - 2} = 5$
$$\left(\sqrt{3x - 2}\right)^2 = 5^2$$
$$3x - 2 = 25$$
$$3x = 27$$
$$x = 9$$

The solution is 9.

25. $-\sqrt{2x} + 4 = -6$
$$10 = \sqrt{2x}$$
$$10^2 = \left(\sqrt{2x}\right)^2$$
$$100 = 2x$$
$$50 = x$$

The solution is 50.

27. $\sqrt{3x + 1} + 2 = 0$
$$\sqrt{3x + 1} = -2$$

No solution since a principle square root
does not yield a negative number.

29. $\sqrt[4]{4x + 1} - 2 = 0$
$$\sqrt[4]{4x + 1} = 2$$
$$\left(\sqrt[4]{4x + 1}\right)^4 = 2^4$$
$$4x + 1 = 16$$
$$4x = 15$$
$$x = \frac{15}{4}$$

The solution is $\frac{15}{4}$.

31. $\sqrt{4x - 3} = 7$
$$\left(\sqrt{4x - 3}\right)^2 = 7^2$$
$$4x - 3 = 49$$
$$4x = 52$$
$$x = 13$$

The solution is 13.

33. $\sqrt[3]{6x-3}-3=0$

$\sqrt[3]{6x-3}=3$

$\left(\sqrt[3]{6x-3}\right)^3=3^3$

$6x-3=27$

$6x=30$

$x=5$

The solution is 5.

35. $\sqrt[3]{2x-3}-2=-5$

$\sqrt[3]{2x-3}=-3$

$\left(\sqrt[3]{2x-3}\right)^3=(-3)^3$

$2x-3=-27$

$2x=-24$

$x=-12$

The solution is -12

37. $\sqrt{x+4}=\sqrt{2x-5}$

$\left(\sqrt{x+4}\right)^2=\left(\sqrt{2x-5}\right)^2$

$x+4=2x-5$

$-x=-9$

$x=9$

The solution is 9.

39. $x-\sqrt{1-x}=-5$

$x+5=\sqrt{1-x}$

$(x+5)^2=\left(\sqrt{1-x}\right)^2$

$x^2+10x+25=1-x$

$x^2+11x+24=0$

$(x+8)(x+3)=0$

$x+8=0$ or $x+3=0$

$x=-8$ or $x=-3$

We discard -8 as extraneous.

The solution is -3.

41. $\sqrt[3]{-6x-1}=\sqrt[3]{-2x-5}$

$\left(\sqrt[3]{-6x-1}\right)^3=\left(\sqrt[3]{-2x-5}\right)^3$

$-6x-1=-2x-5$

$-4x=-4$

$x=1$

The solution is 1.

43. $\sqrt{5x-1}-\sqrt{x+2}=3$

$\sqrt{5x-1}=\sqrt{x}+1$

$\left(\sqrt{5x-1}\right)^2=\left(\sqrt{x}+1\right)^2$

$5x-1=x+2\sqrt{x}+1$

$4x-2=2\sqrt{x}$

$2x-1=\sqrt{x}$

$(2x-1)^2=\left(\sqrt{x}\right)^2$

$4x^2-4x+1=x$

$4x^2-5x+1=0$

$(4x-1)(x-1)=0$

$4x-1=0$ or $x-1=0$

$4x=1$ or $x=1$

$x=\dfrac{1}{4}$

We discard $\dfrac{1}{4}$ as extraneous.

The solution is 1.

45. $\sqrt{2x-1}=\sqrt{1-2x}$

$\left(\sqrt{2x-1}\right)^2=\left(\sqrt{1-2x}\right)^2$

$2x-1=1-2x$

$4x=2$

$x=\dfrac{2}{4}=\dfrac{1}{2}$

The solution is $\dfrac{1}{2}$.

47. $\sqrt{3x+4} - 1 = \sqrt{2x+1}$

$\sqrt{3x+4} = \sqrt{2x+1} + 1$

$\left(\sqrt{3x+4}\right)^2 = \left(\sqrt{2x+1}+1\right)^2$

$3x + 4 = (2x+1) + 2\sqrt{2x+1} + 1$

$3x + 4 = 2x + 2 + 2\sqrt{2x+1}$

$x + 2 = 2\sqrt{2x+1}$

$(x+2)^2 = \left(2\sqrt{2x+1}\right)^2$

$x^2 + 4x + 4 = 4(2x+1)$

$x^2 + 4x + 4 = 8x + 4$

$x^2 - 4x = 0$

$x(x-4) = 0$

$x = 0$ or $x - 4 = 0$

$x = 4$

The solutions are 0 and 4.

49. $\sqrt{y+3} - \sqrt{y-3} = 1$

$\sqrt{y+3} = 1 + \sqrt{y-3}$

$\left(\sqrt{y+3}\right)^2 = \left(1 + \sqrt{y-3}\right)^2$

$y + 3 = 1 + 2\sqrt{y-3} + (y-3)$

$y + 3 = -2 + 2\sqrt{y-3} + y$

$5 = 2\sqrt{y-3}$

$(5)^2 = \left(2\sqrt{y-3}\right)^2$

$25 = 4(y-3)$

$25 = 4y - 12$

$37 = 4y$

$\dfrac{37}{4} = y$

The solution is $\dfrac{37}{4}$.

51. Let c = length of the hypotenuse.

$6^2 + 3^2 = c^2$

$36 + 9 = c^2$

$45 = c^2$

$\sqrt{45} = \sqrt{c^2}$

$\sqrt{9 \cdot 5} = c$

$3\sqrt{5} = c$ so $c = 3\sqrt{5}$ feet

53. Let b = length of the unknown leg.

$3^2 + b^2 = 7^2$

$9 + b^2 = 49$

$b^2 = 40$

$\sqrt{b^2} = \sqrt{40}$

$b = \sqrt{4 \cdot 10}$

$b = 2\sqrt{10}$ meters

55. Let b = length of the unknown leg.

$9^2 + b^2 = \left(11\sqrt{5}\right)^2$

$81 + b^2 = 121 \cdot 5$

$81 + b^2 = 605$

$b^2 = 525$

$\sqrt{b^2} = \sqrt{524}$

$b = \sqrt{4 \cdot 131}$

$b = 2\sqrt{131} \approx 22.9$ meters

57. Let c = length of the hypotenuse.

$(7)^2 + 7.2^2 = c^2$

$49 + 51.84 = c^2$

$100.84 = c^2$

$\sqrt{100.84} = \sqrt{c^2}$

$10.04 \approx c$

$c \approx 10.0$ mm

59. Let c = amount of cable needed.
$$15^2 + 8^2 = c^2$$
$$225 + 64 = c^2$$
$$289 = c^2$$
$$\sqrt{289} = \sqrt{c^2}$$
$$17 = c$$
Thus, 17 feet of cable is needed.

61. Let c = length of the ladder.
$$12^2 + 5^2 = c^2$$
$$144 + 25 = c^2$$
$$169 = c^2$$
$$\sqrt{169} = \sqrt{c^2}$$
$$13 = c$$
A 13-foot ladder is needed.

63.
$$r = \sqrt{\frac{A}{4\pi}}$$
$$1080 = \sqrt{\frac{A}{4\pi}}$$
$$(1080)^2 = \left(\sqrt{\frac{A}{4\pi}}\right)^2$$
$$1,166,400 = \frac{A}{4\pi}$$
$$14,657,415 \approx A$$
The surface area is 14,657,415 sq. miles.

65.
$$v = \sqrt{2gh}$$
$$80 = \sqrt{2(32)h}$$
$$(80)^2 = \left(\sqrt{64h}\right)^2$$
$$6400 = 64h$$
$$100 = h$$
The object fell 100 feet.

67.
$$S = 2\sqrt{I} - 9$$
$$11 = 2\sqrt{I} - 9$$
$$20 = 2\sqrt{I}$$
$$10 = \sqrt{I}$$
$$10^2 = \left(\sqrt{I}\right)^2$$
$$100 = I$$
The estimated IQ is 100.

69.
$$P = 2\pi\sqrt{\frac{l}{32}}$$
$$= 2\pi\sqrt{\frac{2}{32}}$$
$$= 2\pi\sqrt{\frac{1}{16}}$$
$$= 2\pi\left(\frac{1}{4}\right)$$
$$= \frac{\pi}{2} \text{ sec} \approx 1.57 \text{ sec}$$

71.
$$P = 2\pi\sqrt{\frac{l}{32}}$$
$$4 = 2\pi\sqrt{\frac{l}{32}}$$
$$\frac{4}{2\pi} = \sqrt{\frac{l}{32}}$$
$$\left(\frac{2}{\pi}\right)^2 = \left(\sqrt{\frac{l}{32}}\right)^2$$
$$\frac{4}{\pi^2} = \frac{l}{32}$$
$$l = 32\left(\frac{4}{\pi^2}\right) \approx 12.97 \text{ feet}$$

73. Answers may vary.

75. $s = \frac{1}{2}(6+10+14) = \frac{1}{2}(30) = 15$

$A = \sqrt{s(s-a)(s-b)(s-c)}$

$\quad = \sqrt{15(15-6)(15-10)(15-14)}$

$\quad = \sqrt{15(9)(5)(1)}$

$\quad = \sqrt{675}$

$\quad = \sqrt{225 \cdot 3}$

$\quad = 15\sqrt{3}$ sq. mi ≈ 25.98 sq. mi.

77. Answers may vary.

79. $\qquad D(h) = 111.7\sqrt{h}$

$\qquad\quad 80 = 111.7\sqrt{h}$

$\qquad \frac{80}{111.7} = \sqrt{h}$

$\qquad \left(\frac{80}{111.7}\right)^2 = \left(\sqrt{h}\right)^2$

$\quad 0.5129483389 = h$

$\qquad\qquad h \approx 0.51$ km

81. Function

83. Function

85. Not a function

87. $\dfrac{\frac{x}{6}}{\frac{2x}{3}+\frac{1}{2}} = \dfrac{\left(\frac{x}{6}\right)6}{\left(\frac{2x}{3}+\frac{1}{2}\right)6} = \dfrac{x}{4x+3}$

89. $\dfrac{\frac{z}{5}+\frac{1}{10}}{\frac{z}{20}-\frac{z}{5}} = \dfrac{\left(\frac{z}{5}+\frac{1}{10}\right)20}{\left(\frac{z}{20}-\frac{z}{5}\right)20}$

$\qquad = \dfrac{4z+2}{z-4z}$

$\qquad = \dfrac{4z+2}{-3z}$

$\qquad = -\dfrac{4z+2}{3z}$

91. $\sqrt{\sqrt{x+3}+\sqrt{x}} = \sqrt{3}$

$\left(\sqrt{\sqrt{x+3}+\sqrt{x}}\right)^2 = \left(\sqrt{3}\right)^2$

$\qquad \sqrt{x+3}+\sqrt{x} = 3$

$\qquad\quad \sqrt{x+3} = 3-\sqrt{x}$

$\qquad \left(\sqrt{x+3}\right)^2 = \left(3-\sqrt{x}\right)^2$

$\qquad x+3 = 9-6\sqrt{x}+x$

$\qquad\qquad -6 = -6\sqrt{x}$

$\qquad \left(-6\right)^2 = \left(-6\sqrt{x}\right)^2$

$\qquad\qquad 36 = 36x$

$\qquad\qquad 1 = x$

93. $C(x) = 80\sqrt[3]{x} + 500$

$\quad 1620 = 80\sqrt[3]{x} + 500$

$\quad 1120 = 80\sqrt[3]{x}$

$\qquad 14 = \sqrt[3]{x}$

$\qquad 14^3 = \left(\sqrt[3]{x}\right)^3$

$\quad 2744 = x$

Thus, 2743 deliveries will keep overhead below $1620.

95. $3\sqrt{x^2-8x} = x^2-8x$

Let $t = x^2 - 8x$. Then

$\qquad 3\sqrt{u} = u$

$\qquad \left(3\sqrt{u}\right)^2 = u^2$

$\qquad\quad 9u = u^2$

$\qquad\quad 0 = u^2 - 9u$

$\qquad\quad 0 = u(u-9)$

$t = 0 \qquad\qquad$ or $\qquad\qquad t = 9$

Replace t with $x^2 - 8x$.

$\quad x^2-8x = 0 \qquad$ or $\qquad x^2-8x = 9$

$\quad x(x-8) = 0 \qquad\qquad x^2-8x-9 = 0$

$\quad x = 0$ or $x = 8 \qquad (x-9)(x+1) = 0$

$\qquad\qquad\qquad\qquad\qquad x = 9$ or $x = -1$

The solutions are -1, 0, 8, and 9.

97. $7 - (x^2 - 3x) = \sqrt{(x^2 - 3x) + 5}$

Let $t = x^2 - 3x$. Then

$$7 - t = \sqrt{t + 5}$$
$$(7 - t)^2 = \left(\sqrt{t + 5}\right)^2$$
$$49 - 14t + t^2 = t + 5$$
$$t^2 - 15t + 44 = 0$$
$$(t - 11)(t - 4) = 0$$
$$t = 11 \text{ or } t = 4$$

Replace t with $x^2 - 3x$.

$x^2 - 3x = 11$	or	$x^2 - 3x = 4$
$x^2 - 3x - 11 = 0$		$x^2 - 3x - 4 = 0$
Can't factor		$(x - 4)(x + 1) = 0$
		$x = 4$ or $x = -1$

The solutions are -1 and 4.

Section 7.7

Mental Math

1. $\sqrt{-81} = 9i$

2. $\sqrt{-49} = 7i$

3. $\sqrt{-7} = i\sqrt{7}$

4. $\sqrt{-3} = i\sqrt{3}$

5. $-\sqrt{16} = -4$

6. $-\sqrt{4} = -2$

7. $\sqrt{-64} = 8i$

8. $\sqrt{-100} = 10i$

Exercise Set 7.7

1. $\sqrt{-24} = \sqrt{-1 \cdot 24}$
$\phantom{\sqrt{-24}} = \sqrt{-1}\sqrt{4 \cdot 6} = i \cdot 2\sqrt{6} = 2i\sqrt{6}$

3. $-\sqrt{-36} = -\sqrt{-1 \cdot 36}$
$\phantom{-\sqrt{-36}} = -\sqrt{-1}\sqrt{36} = -i \cdot 6 = -6i$

5. $8\sqrt{-63} = 8\sqrt{-1 \cdot 63}$
$\phantom{8\sqrt{-63}} = 8\sqrt{-1}\sqrt{9 \cdot 7} = 8i \cdot 3\sqrt{7} = 24i\sqrt{7}$

7. $-\sqrt{54} = -\sqrt{9 \cdot 6} = -3\sqrt{6}$

9. $\sqrt{-2} \cdot \sqrt{-7} = i\sqrt{2} \cdot i\sqrt{7}$
$\phantom{\sqrt{-2} \cdot \sqrt{-7}} = i^2\sqrt{14} = (-1)\sqrt{14} = -\sqrt{14}$

11. $\sqrt{-5} \cdot \sqrt{-10} = i\sqrt{5} \cdot i\sqrt{10}$
$\phantom{\sqrt{-5} \cdot \sqrt{-10}} = i^2\sqrt{50} = (-1)\sqrt{25 \cdot 2}$
$\phantom{\sqrt{-5} \cdot \sqrt{-10}} = -5\sqrt{2}$

13. $\sqrt{16} \cdot \sqrt{-1} = 4i$

15. $\dfrac{\sqrt{-9}}{\sqrt{3}} = \dfrac{i\sqrt{9}}{\sqrt{3}}$

$\phantom{\dfrac{\sqrt{-9}}{\sqrt{3}}} = \dfrac{3i}{\sqrt{3}} = \dfrac{3i \cdot \sqrt{3}}{\sqrt{3} \cdot \sqrt{3}} = \dfrac{3i\sqrt{3}}{3} = i\sqrt{3}$

17. $\dfrac{\sqrt{-80}}{\sqrt{-10}} = \dfrac{i\sqrt{80}}{i\sqrt{10}}$

$\phantom{\dfrac{\sqrt{-80}}{\sqrt{-10}}} = \sqrt{\dfrac{80}{10}} = \sqrt{8} = \sqrt{4 \cdot 2} = 2\sqrt{2}$

19. $(4 - 7i) + (2 + 3i) = (4 + 2) + (-7 + 3)i$
$ = 6 + (-4)i$
$ = 6 - 4i$

21. $(6 + 5i) - (8 - i) = 6 + 5i - 8 + i$
$ = (6 - 8) + (5 + 1)i$
$ = -2 + 6i$

23. $6 - (8 + 4i) = 6 - 8 - 4i$
$ = (6 - 8) - 4i$
$ = -2 - 4i$

25. $6i(2 - 3i) = 12i - 18i^2$
$ = 12i - 18(-1)$
$ = 18 + 12i$

27. $\left(\sqrt{3}+2i\right)\left(\sqrt{3}-2i\right)$
$= \sqrt{3}\cdot\sqrt{3}-\sqrt{3}\cdot 2i+\sqrt{3}\cdot 2i-4i^2$
$= 3-4(-1)$
$= 3+4$
$= 7$

29. $\left(4-2i\right)^2 = (4-2i)(4-2i)$
$= 16-4\cdot 2i-4\cdot 2i+4i^2$
$= 16-8i-8i+4(-1)$
$= 16-16i-4$
$= 12-16i$

31. $\dfrac{4}{i}=\dfrac{4(-i)}{i(-i)}=\dfrac{-4i}{-i^2}=\dfrac{-4i}{-(-1)}=-4i$

33. $\dfrac{7}{4+3i}=\dfrac{7(4-3i)}{(4+3i)(4-3i)}$
$= \dfrac{28-21i}{4^2-9i^2}$
$= \dfrac{28-21i}{16+9}$
$= \dfrac{28-21i}{25}$
$= \dfrac{28}{25}-\dfrac{21}{25}i$

35. $\dfrac{3+5i}{1+i}=\dfrac{(3+5i)(1-i)}{(1+i)(1-i)}$
$= \dfrac{3-3i+5i-5i^2}{1^2-i^2}$
$= \dfrac{3+2i+5}{1+1}$
$= \dfrac{8+2i}{2}=4+i$

37. $\dfrac{5-i}{3-2i}=\dfrac{(5-i)(3+2i)}{(3-2i)(3+2i)}$
$= \dfrac{15+10i-3i-2i^2}{3^2-4i^2}$
$= \dfrac{15+7i+2}{9+4}$
$= \dfrac{17+7i}{13}=\dfrac{17}{13}+\dfrac{7}{13}i$

39. $(7i)(-9i)=-63i^2=-63(-1)=63$

41. $(6-3i)-(4-2i)=6-3i-4+2i=2-i$

43. $(6-2i)(3+i)=18+6i-6i-2i^2$
$= 18+2$
$= 20$

45. $(8-3i)+(2+3i)=8-3i+2+3i=10$

47. $(1-i)+(1+i)=1+i-i-i^2=1+1=2$

49. $\dfrac{16+15i}{-3i}=\dfrac{(16+15i)(3i)}{-3i(3i)}$
$= \dfrac{48i+45i^2}{-9i^2}$
$= \dfrac{-45+48i}{9}$
$= \dfrac{-45}{9}+\dfrac{48}{9}i=-5+\dfrac{16}{3}i$

51. $(9+8i)^2 = 9^2+2(9)(8i)+(8i)^2$
$= 81+144i+64i^2$
$= 81+144i-64$
$= 17+144i$

53. $\dfrac{2}{3+i}=\dfrac{2(3-i)}{(3+i)(3-i)}$
$= \dfrac{6-2i}{3^2-i^2}$
$= \dfrac{6-2i}{9+1}$
$= \dfrac{6-2i}{10}$
$= \dfrac{6}{10}-\dfrac{2}{10}i=\dfrac{3}{5}-\dfrac{1}{5}i$

55. $(5-6i)-4i=5-6i-4i=5-10i$

57. $\dfrac{2-3i}{2+i}=\dfrac{(2-3i)(2-i)}{(2+i)(2-i)}$
$= \dfrac{4-2i-6i+3i^2}{2^2-i^2}$
$= \dfrac{4-8i-3}{4+1}$
$= \dfrac{1-8i}{5}=\dfrac{1}{5}-\dfrac{8}{5}i$

59. $(2+4i)+(6-5i)=2+4i+6-5i=8-i$

61. $i^8=(i^4)^2=1^2=1$

63. $i^{21}=i^{20}\cdot i=(i^4)^5\cdot i=1^5\cdot i=i$

65. $i^{11}=i^8\cdot i^3=(i^4)^2\cdot i^3=1^2\cdot(-i)=-i$

67. $i^{-6}=\dfrac{1}{i^6}=\dfrac{1}{i^4\cdot i^2}=\dfrac{1}{1\cdot(-1)}=-1$

69. $(2i)^6=2^6 i^6=64i^4\cdot i^2=64(1)(-1)=-64$

71. $(-3i)^5=(-3)^5 i^5$
$=-243i^4\cdot i=-243(1)i=-243i$

73. $x+50°+90°=180°$
$x+140°=180°$
$x=40°$

75. $\begin{array}{r|rrrr} 1 & 1 & -6 & 3 & -4 \\ & & 1 & -5 & -2 \\ \hline & 1 & -5 & -2 & -6 \end{array}$

Answer: $x^2-5x-2-\dfrac{6}{x-1}$

77. 5 people

79. $5+9=14$ people

81. $\dfrac{5\text{ people}}{30\text{ people}}=\dfrac{1}{6}\approx 0.1666$

About 16.7% of the people reported an average checking balance of $201 to $300.

83. $i^3+i^4=-i+1=1-i$

85. $i^6+i^8=i^4\cdot i^2+(i^4)^2$
$=1(-1)+1^2=-1+1=0$

87. $2+\sqrt{-9}=2+i\sqrt{9}=2+3i$

89. $\dfrac{6+\sqrt{-18}}{3}=\dfrac{6+i\sqrt{9\cdot 2}}{3}$
$=\dfrac{6+3i\sqrt{2}}{3}$
$=\dfrac{6}{3}+\dfrac{3\sqrt{2}}{3}i$
$=2+i\sqrt{2}$

91. $\dfrac{5-\sqrt{-75}}{10}=\dfrac{5-i\sqrt{25\cdot 3}}{10}$
$=\dfrac{5-5i\sqrt{3}}{10}$
$=\dfrac{5}{10}-\dfrac{5\sqrt{3}}{10}i$
$=\dfrac{1}{2}-\dfrac{\sqrt{3}}{2}i$

93. Answers may vary.

95. $\left(8-\sqrt{-4}\right)-\left(2+\sqrt{-16}\right)=8-2i-2-4i$
$=6-6i$

97. $x^2+2x=-2$
$(-1+i)^2+2(-1+i)=-2$
$(1-2i+i^2)-2+2i=-2$
$1-1-2=-2$
$-2=-2,$ which is true.

Yes, $-1+i$ is a solution.

Chapter 7 Review

1. $\sqrt{81}=9$ because $9^2=81$.

2. $\sqrt[4]{81}=3$ because $3^4=81$.

3. $\sqrt[3]{-8}=-2$ because $(-2)^4=-8$.

4. $\sqrt[4]{16}=2$ because $2^4=16$.

5. $-\sqrt{\dfrac{1}{49}}=-\dfrac{1}{7}$ because $\left(\dfrac{1}{7}\right)^2=\dfrac{1}{49}$.

6. $\sqrt{x^{64}} = x^{32}$ because $(x^{32})^2 = x^{32 \cdot 2} = x^{64}$.

7. $-\sqrt{36} = -6$ because $6^2 = 36$.

8. $\sqrt[3]{64} = 4$ because $4^3 = 64$.

9. $\sqrt[3]{-a^6 b^9} = \sqrt[3]{-1} \sqrt[3]{a^6} \sqrt[3]{b^9}$
$= -1 a^2 b^3$
$= -a^2 b^3$

10. $\sqrt{16 a^4 b^{12}} = \sqrt{16} \sqrt{a^4} \sqrt{b^{12}} = 4a^2 b^6$

11. $\sqrt[5]{32 a^5 b^{10}} = \sqrt[5]{32} \sqrt[5]{a^5} \sqrt[5]{b^{10}} = 2ab^2$

12. $\sqrt[5]{-32 x^{15} y^{20}} = \sqrt[5]{-32} \sqrt[5]{x^{15}} \sqrt[5]{y^{20}} = -2x^3 y^4$

13. $\sqrt{\dfrac{x^{12}}{36 y^2}} = \dfrac{\sqrt{x^{12}}}{\sqrt{36 y^2}} = \dfrac{x^6}{6y}$

14. $\sqrt[3]{\dfrac{27 y^3}{z^{12}}} = \dfrac{\sqrt[3]{27 y^3}}{\sqrt[3]{z^{12}}} = \dfrac{3y}{z^4}$

15. $\sqrt{(-x)^2} = |-x|$

16. $\sqrt[4]{(x^2 - 4)^4} = |x^2 - 4|$

17. $\sqrt[3]{(-27)^3} = -27$

18. $\sqrt[5]{(-5)^5} = -5$

19. $-\sqrt[5]{x^5} = -x$

20. $\sqrt[4]{16(2y + z)^{12}} = \sqrt[4]{16} \sqrt[4]{(2y + z)^{12}}$
$= 2|2y + z|^3$

21. $\sqrt{25(x - y)^{10}} = \sqrt{25} \sqrt{(x - y)^{10}}$
$= 5|(x - y)^5|$

22. $\sqrt[5]{-y^5} = \sqrt[5]{-1} \sqrt[5]{y^5} = -1y = -y$

23. $\sqrt[9]{-x^9} = \sqrt[9]{-1} \sqrt[9]{x^9} = -1x = -x$

24. $f(x) = \sqrt{x} + 3$
Domain: $[0, \infty)$

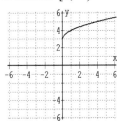

25. $g(x) = \sqrt[3]{x} - 3$
Domain: $(-\infty, \infty)$

x	-5	2	3	4	11
$g(x)$	-2	-1	0	1	2

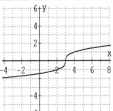

26. $\left(\dfrac{1}{81}\right)^{1/4} = \dfrac{1}{81^{1/4}} = \dfrac{1}{\sqrt[4]{81}} = \dfrac{1}{3}$

27. $\left(-\dfrac{1}{27}\right)^{1/3} = -\dfrac{1}{27^{1/3}} = -\dfrac{1}{\sqrt[3]{27}} = -\dfrac{1}{3}$

28. $(-27)^{-1/3} = \dfrac{1}{(-27)^{1/3}} = \dfrac{1}{\sqrt[3]{-27}} = \dfrac{1}{-3} = -\dfrac{1}{3}$

29. $(-64)^{-1/3} = \dfrac{1}{(-64)^{1/3}} = \dfrac{1}{\sqrt[3]{-64}} = \dfrac{1}{-4} = -\dfrac{1}{4}$

30. $-9^{3/2} = -\left(\sqrt{9}\right)^3 = -3^3 = -27$

31. $64^{-1/3} = \dfrac{1}{64^{1/3}} = \dfrac{1}{\sqrt[3]{64}} = \dfrac{1}{4}$

32. $(-25)^{5/2} = \left(\sqrt{-25}\right)^5$ is not a real number, since there is no real number whose 4th power is negative.

33. $\left(\dfrac{25}{49}\right)^{-3/2} = \dfrac{1}{\left(\dfrac{25}{49}\right)^{3/2}}$

$= \dfrac{1}{\left(\sqrt{\dfrac{25}{49}}\right)^3} = \dfrac{1}{\left(\dfrac{5}{7}\right)^3} = \dfrac{1}{\dfrac{125}{343}} = \dfrac{343}{125}$

34. $\left(\dfrac{8}{27}\right)^{-2/3} = \dfrac{1}{\left(\dfrac{8}{27}\right)^{2/3}}$

$= \dfrac{1}{\left(\sqrt[3]{\dfrac{8}{27}}\right)^2} = \dfrac{1}{\left(\dfrac{2}{3}\right)^2} = \dfrac{1}{\dfrac{8}{9}} = \dfrac{9}{8}$

35. $\left(-\dfrac{1}{36}\right)^{-1/4} = \dfrac{1}{\left(-\dfrac{1}{36}\right)^{1/4}} = \dfrac{1}{\sqrt[4]{-\dfrac{1}{36}}}$ is not a real

number, since there is no real number whose 4th power is negative.

36. $\sqrt[3]{x^2} = x^{2/3}$

37. $\sqrt[5]{5x^2y^3} = (5x^2y^3)^{1/5}$
$= 5^{1/5}(x^2)^{1/5}(y^3)^{1/5}$
$= 5^{1/5}x^{2/5}y^{3/5}$

38. $y^{4/5} = \sqrt[5]{y^4}$

39. $5(xy^2z^5)^{1/3} = 5\sqrt[3]{xy^2z^5}$

40. $(x+2y)^{-1/2} = \dfrac{1}{(x+2y)^{1/2}} = \dfrac{1}{\sqrt{x+2y}}$

41. $a^{1/3}a^{4/3}a^{1/2} = a^{\frac{1}{3}+\frac{4}{3}+\frac{1}{2}} = a^{\frac{2}{6}+\frac{8}{6}+\frac{3}{6}} = a^{13/6}$

42. $\dfrac{b^{1/3}}{b^{4/3}} = b^{1/3-4/3} = b^{-3/3} = b^{-1} = \dfrac{1}{b}$

43. $(a^{1/2}a^{-2})^3 = (a^{1/2-2})^3$
$= (a^{1/2-4/2})^3$
$= (a^{-3/2})^3$
$= a^{-9/2}$
$= \dfrac{1}{a^{9/2}}$

44. $(x^{-3}y^6)^{1/3} = (x^{-3})^{1/3}(y^6)^{1/3} = x^{-1}y^2 = \dfrac{y^2}{x}$

45. $\left(\dfrac{b^{3/4}}{a^{-1/2}}\right)^8 = (a^{1/2}b^{3/4})^8$
$= (a^{1/2})^8(b^{3/4})^8$
$= a^4b^6$

46. $\dfrac{x^{1/4}x^{-1/2}}{x^{2/3}} = x^{1/4+(-1/2)-2/3}$
$= x^{\frac{3}{12}-\frac{6}{12}-\frac{8}{12}}$
$= x^{-11/12}$
$= \dfrac{1}{x^{11/12}}$

47. $\left(\dfrac{49c^{5/3}}{a^{-1/4}b^{5/6}}\right)^{-1} = \dfrac{49^{-1}c^{-5/3}}{a^{1/4}b^{-5/6}}$
$= \dfrac{b^{5/6}}{49a^{1/4}c^{5/3}}$

48. $a^{-1/4}(a^{5/4}-a^{9/4})$
$= a^{-1/4}(a^{5/4})-a^{-1/4}(a^{9/4})$
$= a^{-1/4+5/4}-a^{-1/4+9/4}$
$= a^{4/4}-a^{8/4}$
$= a-a^2$

49. $\sqrt{20} \approx 4.472$

50. $\sqrt[3]{-39} \approx -3.391$

51. $\sqrt[4]{726} \approx 5.191$

52. $56^{1/3} \approx 3.826$

53. $-78^{3/4} \approx -26.246$

54. $105^{-2/3} \approx 0.045$

55. $\sqrt[3]{2} \cdot \sqrt{7} = 2^{1/3} \cdot 7^{1/2}$
$= 2^{2/6} \cdot 7^{3/6}$
$= (2^2 \cdot 7^3)^{1/6}$
$= \sqrt[6]{1372}$

56. $\sqrt[3]{3} \cdot \sqrt[4]{x} = 3^{1/3} \cdot x^{1/4}$
$= 3^{4/12} \cdot x^{3/12}$
$= (3^4 \cdot x^3)^{1/12}$
$= \sqrt[12]{81x^3}$

57. $\sqrt{3} \cdot \sqrt{8} = \sqrt{24} = \sqrt{4 \cdot 6} = 2\sqrt{6}$

58. $\sqrt[3]{7y} \cdot \sqrt[3]{x^2 z} = \sqrt[3]{7y \cdot x^2 z} = \sqrt[3]{7x^2 yz}$

59. $\dfrac{\sqrt{44x^3}}{\sqrt{11x}} = \sqrt{\dfrac{44x^3}{11x}} = \sqrt{4x^2} = 2x$

60. $\dfrac{\sqrt[4]{a^6 b^{13}}}{\sqrt[4]{a^2 b}} = \sqrt[4]{\dfrac{a^6 b^{13}}{a^2 b}} = \sqrt[4]{a^4 b^{12}} = ab^3$

61. $\sqrt{60} = \sqrt{4 \cdot 15} = 2\sqrt{15}$

62. $-\sqrt{75} = -\sqrt{25 \cdot 3} = -5\sqrt{3}$

63. $\sqrt[3]{162} = \sqrt[3]{27 \cdot 6} = 3\sqrt[3]{6}$

64. $\sqrt[3]{-32} = \sqrt[3]{-8 \cdot 4} = -2\sqrt[3]{4}$

65. $\sqrt{36x^7} = \sqrt{36x^6 \cdot x} = 6x^3 \sqrt{x}$

66. $\sqrt[3]{24a^5 b^7} = \sqrt[3]{8a^3 b^6 \cdot 3a^2 b^2} = 2ab^2 \sqrt[3]{3a^2 b^2}$

67. $\sqrt{\dfrac{p^{17}}{121}} = \dfrac{\sqrt{p^{17}}}{\sqrt{121}} = \dfrac{\sqrt{p^{16} \cdot p}}{11} = \dfrac{p^8 \sqrt{p}}{11}$

68. $\sqrt[3]{\dfrac{y^5}{27x^6}} = \dfrac{\sqrt[3]{y^5}}{\sqrt[3]{27x^6}} = \dfrac{\sqrt[3]{y^3 y^2}}{\sqrt[3]{27x^6}} = \dfrac{y\sqrt[3]{y^2}}{3x^2}$

69. $\sqrt[4]{\dfrac{xy^6}{81}} = \dfrac{\sqrt[4]{xy^6}}{\sqrt[4]{81}} = \dfrac{\sqrt[4]{y^4 \cdot xy^2}}{3} = \dfrac{y\sqrt[4]{xy^2}}{3}$

70. $\sqrt{\dfrac{2x^3}{49y^4}} = \dfrac{\sqrt{2x^3}}{\sqrt{49y^4}} = \dfrac{\sqrt{x^2 \cdot 2x}}{7y^2} = \dfrac{x\sqrt{2x}}{7y^2}$

71. $r = \sqrt{\dfrac{A}{\pi}}$

   a. $r = \sqrt{\dfrac{25}{\pi}} = \dfrac{\sqrt{25}}{\sqrt{\pi}} = \dfrac{5}{\sqrt{\pi}}$ meters, or

     $r = \dfrac{5}{\sqrt{\pi}} = \dfrac{5\sqrt{\pi}}{\sqrt{\pi}\sqrt{\pi}} = \dfrac{5\sqrt{\pi}}{\pi}$ meters

   b. $r = \sqrt{\dfrac{104}{\pi}} \approx 5.75$ inches

72. $x\sqrt{75xy} - \sqrt{27x^3 y}$
$= x\sqrt{25 \cdot 3xy} - \sqrt{9x^2 \cdot 3xy}$
$= x \cdot 5\sqrt{3xy} - 3x\sqrt{3xy}$
$= 2x\sqrt{3xy}$

73. $2\sqrt{32x^2 y^3} - xy\sqrt{98y}$
$= 2\sqrt{16x^2 y^2 \cdot 2y} - xy\sqrt{49 \cdot 2y}$
$= 2 \cdot 4xy\sqrt{2y} - xy \cdot 7\sqrt{2y}$
$= 8xy\sqrt{2y} - 7xy\sqrt{2y}$
$= xy\sqrt{2y}$

74. $\sqrt[3]{128} + \sqrt[3]{250} = \sqrt[3]{64 \cdot 2} + \sqrt[3]{125 \cdot 2}$
$= 4\sqrt[3]{2} + 5\sqrt[3]{2}$
$= 9\sqrt[3]{2}$

75. $3\sqrt[4]{32a^5} - a\sqrt[4]{162a}$
$= 3\sqrt[4]{16a^4 \cdot 2a} - a\sqrt[4]{81 \cdot 2a}$
$= 3 \cdot 2a\sqrt[4]{2a} - 3a\sqrt[4]{2a}$
$= 6a\sqrt[4]{2a} - 3a\sqrt[4]{2a}$
$= 3a\sqrt[4]{2a}$

76. $\dfrac{5}{\sqrt{4}} + \dfrac{\sqrt{3}}{3} = \dfrac{5}{2} + \dfrac{\sqrt{3}}{3}$
$= \dfrac{5 \cdot 3 + 2\sqrt{3}}{6} = \dfrac{15 + 2\sqrt{3}}{6}$

77. $\sqrt{\dfrac{8}{x^2}} - \sqrt{\dfrac{50}{16x^2}} = \dfrac{\sqrt{8}}{\sqrt{x^2}} - \dfrac{\sqrt{50}}{\sqrt{16x^2}}$

$\phantom{\sqrt{\dfrac{8}{x^2}}} = \dfrac{\sqrt{4 \cdot 2}}{x} - \dfrac{\sqrt{25 \cdot 2}}{4x}$

$\phantom{\sqrt{\dfrac{8}{x^2}}} = \dfrac{2\sqrt{2} \cdot 4}{x \cdot 4} - \dfrac{5\sqrt{2}}{4x}$

$\phantom{\sqrt{\dfrac{8}{x^2}}} = \dfrac{8\sqrt{2} - 5\sqrt{2}}{4x}$

$\phantom{\sqrt{\dfrac{8}{x^2}}} = \dfrac{3\sqrt{2}}{4x}$

78. $2\sqrt{50} - 3\sqrt{125} + \sqrt{98}$

$= 2\sqrt{25 \cdot 2} - 3\sqrt{25 \cdot 5} + \sqrt{49 \cdot 2}$

$= 2 \cdot 5\sqrt{2} - 3 \cdot 5\sqrt{5} + 7\sqrt{2}$

$= 10\sqrt{2} - 15\sqrt{5} + 7\sqrt{2}$

$= 17\sqrt{2} - 15\sqrt{5}$

79. $2a\sqrt[4]{32b^5} - 3b\sqrt[4]{162a^4b} + \sqrt[4]{2a^4b^5}$

$= 2a\sqrt[4]{16b^4 \cdot 2b} - 3b\sqrt[4]{81a^4 \cdot 2b}$
$ + \sqrt[4]{a^4b^4 \cdot 2b}$

$= 2a \cdot 2b\sqrt[4]{2b} - 3b \cdot 3a\sqrt[4]{2b} + ab\sqrt[4]{2b}$

$= 4ab\sqrt[4]{2b} - 9ab\sqrt[4]{2b} + ab\sqrt[4]{2b}$

$= -4ab\sqrt[4]{2b}$

80. $\sqrt{3}\left(\sqrt{27} - \sqrt{3}\right) = \sqrt{3}\left(\sqrt{9 \cdot 3} - \sqrt{3}\right)$

$\phantom{\sqrt{3}\left(\sqrt{27}\right)} = \sqrt{3}\left(3\sqrt{3} - \sqrt{3}\right)$

$\phantom{\sqrt{3}\left(\sqrt{27}\right)} = \sqrt{3}\left(2\sqrt{3}\right)$

$\phantom{\sqrt{3}\left(\sqrt{27}\right)} = 2\sqrt{9}$

$\phantom{\sqrt{3}\left(\sqrt{27}\right)} = 2(3)$

$\phantom{\sqrt{3}\left(\sqrt{27}\right)} = 6$

81. $\left(\sqrt{x} - 3\right)^2 = \left(\sqrt{x}\right)^2 - 2 \cdot \sqrt{x} \cdot 3 + 3^2$

$\phantom{\left(\sqrt{x} - 3\right)^2} = x - 6\sqrt{x} + 9$

82. $\left(\sqrt{5} - 5\right)\left(2\sqrt{5} + 2\right)$

$= 2\sqrt{25} + 2\sqrt{5} - 10\sqrt{5} - 10$

$= 2(5) - 8\sqrt{5} - 10$

$= 10 - 8\sqrt{5} - 10$

$= -8\sqrt{5}$

83. $\left(2\sqrt{x} - 3\sqrt{y}\right)\left(2\sqrt{x} + 3\sqrt{y}\right)$

$= \left(2\sqrt{x}\right)^2 - \left(3\sqrt{y}\right)^2$

$= 2^2\left(\sqrt{x}\right)^2 - 3^2\left(\sqrt{y}\right)^2$

$= 4x - 9y$

84. $\left(\sqrt{a} - 3\right)\left(\sqrt{a} + 3\right) = \left(\sqrt{a}\right)^2 - (3)^2$

$\phantom{\left(\sqrt{a} - 3\right)\left(\sqrt{a} + 3\right)} = a - 9$

85. $\left(\sqrt[3]{a} + 2\right)^2 = \left(\sqrt[3]{a}\right)^2 + 2 \cdot \sqrt[3]{a} \cdot 2 + 2^2$

$\phantom{\left(\sqrt[3]{a} + 2\right)^2} = \sqrt[3]{a^2} + 4\sqrt[3]{a} + 4$

86. $\left(\sqrt[3]{5x} + 9\right)\left(\sqrt[3]{5x} - 9\right) = \left(\sqrt[3]{5x}\right)^2 - 9^2$

$\phantom{\left(\sqrt[3]{5x} + 9\right)\left(\sqrt[3]{5x} - 9\right)} = \sqrt[3]{(5x)^2} - 81$

$\phantom{\left(\sqrt[3]{5x} + 9\right)\left(\sqrt[3]{5x} - 9\right)} = \sqrt[3]{25x^2} - 81$

87. $\left(\sqrt[3]{a} + 4\right)\left(\sqrt[3]{a^2} - 4\sqrt[3]{a} + 16\right)$

$= \left(\sqrt[3]{a}\right)\left(\sqrt[3]{a^2}\right) - 4 \cdot \left(\sqrt[3]{a}\right)^2 + 16\sqrt[3]{a}$

$ + 4\sqrt[3]{a^2} \quad -16\sqrt[3]{a} + 64$

$= \sqrt[3]{a^3} - 4\sqrt[3]{a^2} + 4\sqrt[3]{a^2} + 64$

$= a + 64$

88. $\dfrac{3}{\sqrt{7}} = \dfrac{3 \cdot \sqrt{7}}{\sqrt{7} \cdot \sqrt{7}} = \dfrac{3\sqrt{7}}{7}$

89.
$$\sqrt{\frac{x}{12}} = \frac{\sqrt{x}}{\sqrt{12}}$$
$$= \frac{\sqrt{x}}{\sqrt{4 \cdot 3}}$$
$$= \frac{\sqrt{x}}{2\sqrt{3}}$$
$$= \frac{\sqrt{x} \cdot \sqrt{3}}{2\sqrt{3} \cdot \sqrt{3}} = \frac{\sqrt{3x}}{2 \cdot 3} = \frac{\sqrt{3x}}{6}$$

90.
$$\frac{5}{\sqrt[3]{4}} = \frac{5 \cdot \sqrt[3]{2}}{\sqrt[3]{4} \cdot \sqrt[3]{2}} = \frac{5\sqrt[3]{2}}{\sqrt[3]{8}} = \frac{5\sqrt[3]{2}}{2}$$

91.
$$\sqrt{\frac{24x^5}{3y^2}} = \sqrt{\frac{8x^5}{y^2}}$$
$$= \frac{\sqrt{8x^5}}{\sqrt{y^2}}$$
$$= \frac{\sqrt{4x^4 \cdot 2x}}{y} = \frac{2x^2\sqrt{2x}}{y}$$

92.
$$\sqrt[3]{\frac{15x^6 y^7}{z^2}} = \frac{\sqrt[3]{15x^6 y^7}}{\sqrt[3]{z^2}}$$
$$= \frac{\sqrt[3]{15x^6 y^7} \cdot \sqrt[3]{z}}{\sqrt[3]{z^2} \cdot \sqrt[3]{z}}$$
$$= \frac{\sqrt[3]{15x^6 y^7 z}}{\sqrt[3]{z^3}}$$
$$= \frac{\sqrt[3]{15x^6 y^6 \cdot yz}}{z} = \frac{x^2 y^2 \sqrt[3]{15yz}}{z}$$

93.
$$\frac{5}{2 - \sqrt{7}} = \frac{5(2 + \sqrt{7})}{(2 - \sqrt{7})(2 + \sqrt{7})}$$
$$= \frac{5(2 + \sqrt{7})}{2^2 - (\sqrt{7})^2}$$
$$= \frac{10 + 5\sqrt{7}}{4 - 7}$$
$$= \frac{10 + 5\sqrt{7}}{-3} = -\frac{10 + 5\sqrt{7}}{3}$$

94.
$$\frac{3}{\sqrt{y} - 2} = \frac{3(\sqrt{y} + 2)}{(\sqrt{y} - 2)(\sqrt{y} + 2)}$$
$$= \frac{3(\sqrt{y} + 2)}{(\sqrt{y})^2 - 2^2} = \frac{3\sqrt{y} + 6}{y - 4}$$

95.
$$\frac{\sqrt{2} - \sqrt{3}}{\sqrt{2} + \sqrt{3}} = \frac{(\sqrt{2} - \sqrt{3})(\sqrt{2} - \sqrt{3})}{(\sqrt{2} + \sqrt{3})(\sqrt{2} - \sqrt{3})}$$
$$= \frac{2 - \sqrt{2}\sqrt{3} - \sqrt{3}\sqrt{2} + 3}{(\sqrt{2})^2 - (\sqrt{3})^2}$$
$$= \frac{5 - 2\sqrt{6}}{-1}$$
$$= -5 + 2\sqrt{6}$$

96.
$$\frac{\sqrt{11}}{3} = \frac{\sqrt{11} \cdot \sqrt{11}}{3 \cdot \sqrt{11}} = \frac{11}{3\sqrt{11}}$$

97.
$$\sqrt{\frac{18}{y}} = \frac{\sqrt{18}}{\sqrt{y}}$$
$$= \frac{3\sqrt{2}}{\sqrt{y}} = \frac{3\sqrt{2} \cdot \sqrt{2}}{\sqrt{y} \cdot \sqrt{2}} = \frac{3 \cdot 2}{\sqrt{2y}} = \frac{6}{\sqrt{2y}}$$

98.
$$\frac{\sqrt[3]{9}}{7} = \frac{\sqrt[3]{9} \cdot \sqrt[3]{3}}{7 \cdot \sqrt[3]{3}} = \frac{\sqrt[3]{27}}{7\sqrt[3]{3}} = \frac{3}{7\sqrt[3]{3}}$$

99.
$$\sqrt{\frac{24x^5}{3y^2}} = \sqrt{\frac{8x^5}{y^2}}$$
$$= \frac{\sqrt{4x^4 \cdot 2x}}{\sqrt{y^2}}$$
$$= \frac{2x^2\sqrt{2x}}{y}$$
$$= \frac{2x^2\sqrt{2x} \cdot \sqrt{2x}}{y \cdot \sqrt{2x}}$$
$$= \frac{2x^2 \cdot 2x}{y\sqrt{2x}} = \frac{4x^3}{y\sqrt{2x}}$$

100. $\sqrt[3]{\dfrac{xy^2}{10z}} = \dfrac{\sqrt[3]{xy^2}}{\sqrt[3]{10z}}$

$= \dfrac{\sqrt[3]{xy^2} \cdot \sqrt[3]{x^2 y}}{\sqrt[3]{10z} \cdot \sqrt[3]{x^2 y}}$

$= \dfrac{\sqrt[3]{x^3 y^3}}{\sqrt[3]{10x^2 yz}}$

$= \dfrac{xy}{\sqrt[3]{10x^2 yz}}$

101. $\dfrac{\sqrt{x}+5}{-3} = \dfrac{\left(\sqrt{x}+5\right)\left(\sqrt{x}-5\right)}{-3\left(\sqrt{x}-5\right)}$

$= \dfrac{\left(\sqrt{x}\right)^2 - 5^2}{-3\sqrt{x}+15}$

$= \dfrac{x-25}{-3\sqrt{x}+15}$

102. $\sqrt{y-7} = 5$

$\left(\sqrt{y-7}\right)^2 = 5^2$

$y-7 = 25$

$y = 32$

The solution is 32.

103. $\sqrt{2x}+10 = 4$

$\sqrt{2x} = -6$

No solution exist since the principle square root of a number is not negative.

104. $\sqrt[3]{2x-6} = 4$

$\left(\sqrt[3]{2x-6}\right)^3 = 4^3$

$2x-6 = 64$

$2x = 70$

$x = 35$

The solution is 35.

105. $\sqrt{x+6} = \sqrt{x+2}$

$\left(\sqrt{x+6}\right)^2 = \left(\sqrt{x+2}\right)^2$

$x+6 = x+2$

$6 = 2,$ which is false.

There is no solution.

106. $2x - 5\sqrt{x} = 3$

$2x - 3 = 5\sqrt{x}$

$(2x-3)^2 = \left(5\sqrt{x}\right)^2$

$4x^2 - 12x + 9 = 25x$

$4x^2 - 37x + 9 = 0$

$(4x-1)(x-9) = 0$

$4x-1 = 0 \ \text{ or } \ x-9 = 0$

$4x = 1 \ \text{ or } \qquad x = 9$

$x = \dfrac{1}{4}$

Discard the solution $\dfrac{1}{4}$ as extraneous.

The solution is 3.

107. $\sqrt{x+9} = 2 + \sqrt{x-7}$

$\left(\sqrt{x+9}\right)^2 = \left(2 + \sqrt{x-7}\right)^2$

$x+9 = 4 + 4\sqrt{x-7} + (x-7)$

$x+9 = x-3 + 4\sqrt{x-7}$

$12 = 4\sqrt{x-7}$

$3 = \sqrt{x-7}$

$3^2 = \left(\sqrt{x-7}\right)^2$

$9 = x-7$

$16 = x$

The solution is 16.

108. Let $c =$ length of the hypotenuse.

$3^2 + 3^2 = c^2$

$18 = c^2$

$\sqrt{18} = \sqrt{c^2}$

$3\sqrt{2} = c$

109. Let c = length of the hypotenuse.

$$7^2 + \left(8\sqrt{3}\right)^2 = c^2$$
$$49 + 64 \cdot 3 = c^2$$
$$241 = c^2$$
$$\sqrt{241} = \sqrt{c^2}$$
$$\sqrt{241} = c$$

110. Let b = width of the lake.

$$a^2 + b^2 = c^2$$
$$40^2 + b^2 = 65^2$$
$$1600 + b^2 = 4225$$
$$b^2 = 2625$$
$$\sqrt{b^2} = \sqrt{2625}$$
$$b = 51.23475$$

The width is about 51.2 feet.

111. Let c = length of the shortest pipe.

$$a^2 + b^2 = c^2$$
$$3^2 + 3^2 = c^2$$
$$18 = c^2$$
$$\sqrt{18} = \sqrt{c^2}$$
$$4.24264 = c$$

The shortest possible pipe is 4.24 feet.

112. $\sqrt{-8} = i\sqrt{4 \cdot 2} = 2i\sqrt{2}$

113. $-\sqrt{-6} = -i\sqrt{6}$

114. $\sqrt{-4} + \sqrt{-16} = 2i + 4i = 6i$

115. $\sqrt{-2} \cdot \sqrt{-5} = i\sqrt{2} \cdot i\sqrt{5}$
$$= i^2\sqrt{10} =$$
$$= -1 \cdot \sqrt{10}$$
$$= -\sqrt{10}$$

116. $(12 - 6i) + (3 + 2i) = (12 + 3) + (-6 + 2)i$
$$= 15 + (-4)i$$
$$= 15 - 4i$$

117. $(-8 - 7i) - (5 - 4i) = -8 - 7i - 5 + 4i$
$$= -13 - 3i$$

118. $\left(\sqrt{3} + \sqrt{2}\right) + \left(3\sqrt{2} - \sqrt{-8}\right)$
$$= \sqrt{3} + \sqrt{2} + 3\sqrt{2} - i\sqrt{4 \cdot 2}$$
$$= \sqrt{3} + 4\sqrt{2} - 2i\sqrt{2}$$

119. $2i(2 - 5i) = 4i - 10i^2$
$$= 4i - 10(-1)$$
$$= 10 + 4i$$

120. $-3i(6 - 4i) = -18i + 12i^2$
$$= -18i + 12(-1)$$
$$= -12 - 18i$$

121. $(3 + 2i)(1 + i) = 3 + 3i + 2i + 2i^2$
$$= 3 + 5i + 2(-1)$$
$$= 1 + 5i$$

122. $(2 - 3i)^2 = 4 - 12i + 9i^2$
$$= 4 - 12i + 9(-1)$$
$$= -5 - 12i$$

123. $\left(\sqrt{6} - 9i\right)\left(\sqrt{6} + 9i\right) = \left(\sqrt{6}\right)^2 - (9i)^2$
$$= 6 - 81i^2$$
$$= 6 + 81$$
$$= 87$$

124. $\dfrac{2 + 3i}{2i} = \dfrac{(2 + 3i) \cdot (-2i)}{2i \cdot (-2i)}$
$$= \dfrac{-4i - 6i^2}{-4i^2}$$
$$= \dfrac{-4i + 6}{4}$$
$$= \dfrac{6}{4} - \dfrac{4}{4}i$$
$$= \dfrac{3}{2} - i$$

125. $\dfrac{1 + i}{-3i} = \dfrac{(1 + i) \cdot (3i)}{-3i \cdot (3i)}$
$$= \dfrac{3i + 3i^2}{-9i^2}$$
$$= \dfrac{3i - 3}{9}$$
$$= \dfrac{-3}{9} - \dfrac{3}{9}i = -\dfrac{1}{3} + \dfrac{1}{3}i$$

Chapter 7 Test

1. $\sqrt{216} = \sqrt{36 \cdot 6} = 6\sqrt{6}$

2. $-\sqrt[4]{x^{64}} = -x^{16}$

3. $\left(\dfrac{1}{125}\right)^{1/3} = \dfrac{1}{125^{1/3}} = \dfrac{1}{\sqrt[3]{125}} = \dfrac{1}{5}$

4. $\left(\dfrac{1}{125}\right)^{-1/3} = \dfrac{1}{\left(\dfrac{1}{125}\right)^{1/3}} = \dfrac{1}{\dfrac{1}{5}} = 5$

5. $\left(\dfrac{8x^3}{27}\right)^{2/3} = \dfrac{(8x^3)^{2/3}}{27^{2/3}}$

$\phantom{\left(\dfrac{8x^3}{27}\right)^{2/3}} = \dfrac{\left(\sqrt[3]{8x^3}\right)^2}{\left(\sqrt[3]{27}\right)^2}$

$\phantom{\left(\dfrac{8x^3}{27}\right)^{2/3}} = \dfrac{(2x)^2}{3^3}$

$\phantom{\left(\dfrac{8x^3}{27}\right)^{2/3}} = \dfrac{4x^2}{9}$

6. $\sqrt[3]{-a^{18}b^9} = \sqrt[3]{-1a^{18}b^9} = (-1)a^6b^3 = -a^6b^3$

7. $\left(\dfrac{64c^{4/3}}{a^{-2/3}b^{5/6}}\right)^{1/2} = \left(\dfrac{64a^{2/3}c^{4/3}}{b^{5/6}}\right)^{1/2}$

$\phantom{\left(\dfrac{64c^{4/3}}{a^{-2/3}b^{5/6}}\right)^{1/2}} = \dfrac{64^{1/2}(a^{2/3})^{1/2}(c^{4/3})^{1/2}}{(b^{5/6})^{1/2}}$

$\phantom{\left(\dfrac{64c^{4/3}}{a^{-2/3}b^{5/6}}\right)^{1/2}} = \dfrac{\sqrt{64}a^{1/3}c^{2/3}}{b^{5/12}}$

$\phantom{\left(\dfrac{64c^{4/3}}{a^{-2/3}b^{5/6}}\right)^{1/2}} = \dfrac{8a^{1/3}c^{2/3}}{b^{5/12}}$

8. $a^{-2/3}(a^{5/4} - a^3) = a^{-2/3}a^{5/4} - a^{-2/3}a^3$

$\phantom{a^{-2/3}(a^{5/4} - a^3)} = a^{-\frac{2}{3}+\frac{5}{4}} - a^{-\frac{2}{3}+3}$

$\phantom{a^{-2/3}(a^{5/4} - a^3)} = a^{-\frac{8}{12}+\frac{15}{12}} - a^{-\frac{2}{3}+\frac{9}{3}}$

$\phantom{a^{-2/3}(a^{5/4} - a^3)} = a^{7/12} - a^{7/3}$

9. $\sqrt[4]{(4xy)^4} = |4xy| = 4|xy|$

10. $\sqrt[3]{(-27)^3} = -27$

11. $\sqrt{\dfrac{9}{y}} = \dfrac{\sqrt{9}}{\sqrt{y}} = \dfrac{3}{\sqrt{y}} = \dfrac{3 \cdot \sqrt{y}}{\sqrt{y} \cdot \sqrt{y}} = \dfrac{3\sqrt{y}}{y}$

12. $\dfrac{4-\sqrt{x}}{4+2\sqrt{x}} = \dfrac{4-\sqrt{x}}{2\left(2+\sqrt{x}\right)}$

$\phantom{\dfrac{4-\sqrt{x}}{4+2\sqrt{x}}} = \dfrac{\left(4-\sqrt{x}\right)\left(2-\sqrt{x}\right)}{2\left(2+\sqrt{x}\right)\left(2-\sqrt{x}\right)}$

$\phantom{\dfrac{4-\sqrt{x}}{4+2\sqrt{x}}} = \dfrac{8 - 4\sqrt{x} - 2\sqrt{x} + x}{2\left[2^2 - \left(\sqrt{x}\right)^2\right]}$

$\phantom{\dfrac{4-\sqrt{x}}{4+2\sqrt{x}}} = \dfrac{8 - 6\sqrt{x} + x}{2(4-x)}$

13. $\dfrac{\sqrt[3]{ab}}{\sqrt[3]{ab^2}} = \sqrt[3]{\dfrac{ab}{ab^2}}$

$\phantom{\dfrac{\sqrt[3]{ab}}{\sqrt[3]{ab^2}}} = \sqrt[3]{\dfrac{1}{b}}$

$\phantom{\dfrac{\sqrt[3]{ab}}{\sqrt[3]{ab^2}}} = \dfrac{1}{\sqrt[3]{b}}$

$\phantom{\dfrac{\sqrt[3]{ab}}{\sqrt[3]{ab^2}}} = \dfrac{1 \cdot \sqrt[3]{b^2}}{\sqrt[3]{b} \cdot \sqrt[3]{b^2}} = \dfrac{\sqrt[3]{b^2}}{b}$

14. $\dfrac{\sqrt{6}+x}{8} = \dfrac{\left(\sqrt{6}+x\right)\left(\sqrt{6}-x\right)}{8\left(\sqrt{6}-x\right)}$

$\phantom{\dfrac{\sqrt{6}+x}{8}} = \dfrac{\left(\sqrt{6}\right)^2 - x^2}{8\left(\sqrt{6}-x\right)}$

$\phantom{\dfrac{\sqrt{6}+x}{8}} = \dfrac{6 - x^2}{8\left(\sqrt{6}-x\right)}$

15.
$$\sqrt{125x^3} - 3\sqrt{20x^3}$$
$$= \sqrt{25x^2 \cdot 5x} - 3\sqrt{4x^2 \cdot 5x}$$
$$= 5x\sqrt{5x} - 3 \cdot 2x\sqrt{5x}$$
$$= 5x\sqrt{5x} - 6x\sqrt{5x}$$
$$= -x\sqrt{5x}$$

16.
$$\sqrt{3}\left(\sqrt{16} - \sqrt{2}\right) = \sqrt{3}\left(4 - \sqrt{2}\right)$$
$$= 4\sqrt{3} - \sqrt{3}\sqrt{2}$$
$$= 4\sqrt{3} - \sqrt{6}$$

17.
$$\left(\sqrt{x} + 1\right)^2 = \left(\sqrt{x}\right)^2 + 2\sqrt{x} + 1^2$$
$$= x + 2\sqrt{x} + 1$$

18.
$$\left(\sqrt{2} - 4\right)\left(\sqrt{3} + 1\right)$$
$$= \sqrt{2}\sqrt{3} + 1 \cdot \sqrt{2} - 4\sqrt{3} - 4$$
$$= \sqrt{6} + \sqrt{2} - 4\sqrt{3} - 4$$

19.
$$\left(\sqrt{5} + 5\right)\left(\sqrt{5} - 5\right) = \left(\sqrt{5}\right)^2 - 5^2$$
$$= 5 - 25$$
$$= -20$$

20. $\sqrt{561} \approx 23.685$

21. $386^{-2/3} \approx 0.019$

22.
$$x = \sqrt{x-2} + 2$$
$$x - 2 = \sqrt{x-2}$$
$$(x-2)^2 = \left(\sqrt{x-2}\right)^2$$
$$x^2 - 4x + 4 = x - 2$$
$$x^2 - 5x + 6 = 0$$
$$(x-2)(x-3) = 0$$
$$x = 2 \text{ or } x = 3$$
The solutions are 2 and 3.

23.
$$\sqrt{x^2 - 7} + 3 = 0$$
$$\sqrt{x^2 - 7} = -3$$
No solution exists since the principle square root of a number is not negative.

24.
$$\sqrt[3]{x+5} = \sqrt[3]{2x-1}$$
$$\left(\sqrt[3]{x+5}\right)^3 = \left(\sqrt[3]{2x-1}\right)^3$$
$$x + 5 = 2x - 1$$
$$-x = -6$$
$$x = 6$$
The solution is 6.

25. $\sqrt{-2} = i\sqrt{2}$

26. $-\sqrt{-8} = -i\sqrt{4 \cdot 2} = 2i\sqrt{2}$

27. $(12 - 6i) - (12 - 3i) = 12 - 6i - 12 + 3i$
$$= -3i$$

28.
$$(6 - 2i)(6 + 2i) = 6^2 - (2i)^2$$
$$= 36 - 4i^2$$
$$= 36 + 4$$
$$= 40$$

29.
$$(4 + 3i)^2 = 4^2 + 2 \cdot 4 \cdot 3i + (3i)^2$$
$$= 16 + 24i + 9i^2$$
$$= 16 + 24i - 9$$
$$= 7 + 24i$$

30.
$$\frac{1+4i}{1-i} = \frac{(1+4i)(1+i)}{(1-i)(1+i)}$$
$$= \frac{1 + i + 4i + 4i^2}{1^2 - i^2}$$
$$= \frac{1 + 5i - 4}{1 - (-1)}$$
$$= \frac{-3 + 5i}{2}$$
$$= -\frac{3}{2} + \frac{5}{2}i$$

31.
$$x^2 + x^2 = 5^2$$
$$2x^2 = 25$$
$$x^2 = \frac{25}{2}$$
$$\sqrt{x^2} = \sqrt{\frac{25}{2}}$$
$$x = \frac{5}{\sqrt{2}} = \frac{5 \cdot \sqrt{2}}{\sqrt{2} \cdot \sqrt{2}} = \frac{5\sqrt{2}}{2}$$

32. $g(x) = \sqrt{x+2}$

Domain: $[-2, \infty)$

x	-2	-1	2	7
$g(x)$	0	1	2	3

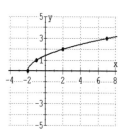

33. $V(r) = \sqrt{2.5r}$

$V(300) = \sqrt{2.5(300)} \approx 27$ mph

34. $V(r) = \sqrt{2.5r}$

$30 = \sqrt{2.5r}$

$30^2 = \left(\sqrt{2.5r}\right)^2$

$900 = 2.5r$

$r = \dfrac{900}{2.5} = 360$ feet

Chapter 7 Cumulative Review

1. a. $3xy - 2xy + 5 - 7 + xy = 2xy - 2$

b. $7x^2 + 3 - 5(x^2 - 4)$
$= 7x^2 + 3 - 5x^2 + 20$
$= 2x^2 + 23$

c. $(2.1x - 5.6) - (-x - 5.3)$
$= 2.1x - 5.6 + x + 5.3$
$= 3.1x - 0.3$

d. $\dfrac{1}{2}(4a - 6b) - \dfrac{1}{3}(9a + 12b - 1) + \dfrac{1}{4}$

$= 2a - 3b - 3a - 4b + \dfrac{1}{3} + \dfrac{1}{4}$

$= 2a - 3a - 3b - 4b + \dfrac{4}{12} + \dfrac{3}{12}$

$= -a - 7b + \dfrac{7}{12}$

2. a. $2(x - 3) + (5x + 3) = 2x - 6 + 5x + 3$
$\qquad\qquad\qquad\qquad = 7x - 3$

b. $4(3x + 2) - 3(5x - 1)$
$= 12x + 8 - 15x + 3$
$= -3x + 11$

c. $7x + 2(x - 7) - 3x$
$= 7x + 2x - 14 - 3x$
$= 6x - 14$

3. $\dfrac{x+5}{2} + \dfrac{1}{2} = 2x - \dfrac{x-3}{8}$

$8\left(\dfrac{x+5}{2} + \dfrac{1}{2}\right) = 8\left(2x - \dfrac{x-3}{8}\right)$

$4(x+5) + 4 = 16x - (x-3)$

$4x + 20 + 4 = 16x - x + 3$

$4x + 24 = 15x + 3$

$-11x = -21$

$x = \dfrac{21}{11}$

4. $\dfrac{a-1}{2} + a = 2 - \dfrac{2a+7}{8}$

$8\left(\dfrac{a-1}{2} + a\right) = 8\left(2 - \dfrac{2a+7}{8}\right)$

$4(a-1) + 8a = 16 - (2a+7)$

$4a - 4 + 8a = 16 - 2a - 7$

$12a - 4 = 9 - 2a$

$14a = 13$

$a = \dfrac{13}{14}$

5. Let x = the sales needed.

$$600 + 0.20x > 1500$$
$$0.20x > 900$$
$$x > \frac{900}{0.20}$$
$$x > 4500$$

They need sales of at least \$4500.

6. Let r = their average speed.

$$t_{going} + t_{returning} = 4.5 \text{ hrs}$$
$$\frac{121.5}{r} + \frac{121.5}{r} = 4.5$$
$$\frac{243}{r} = 4.5$$
$$243 = 4.5r$$
$$r = \frac{243}{4.5} = 54$$

Their average speed was 54 mph.

7. $2|x| + 25 = 23$

$$2|x| = -2$$
$$|x| = -1, \text{ which is impossible.}$$

There is no solution, or the solution is $\varnothing$.

8. $|3x - 2| + 5 = 5$

$$|3x - 2| = 0$$
$$3x - 2 = 0$$
$$3x = 2$$
$$x = \frac{2}{3}$$

9. $\left|\frac{x}{3} - 1\right| - 7 \geq -5$

$$\left|\frac{x}{3} - 1\right| \geq 2$$
$$\frac{x}{3} - 1 \leq -2 \quad \text{or} \quad \frac{x}{3} - 1 \geq 2$$
$$\frac{x}{3} \leq -1 \quad \text{or} \quad \frac{x}{3} \geq 3$$
$$x \leq -3 \quad \text{or} \quad x \geq 9$$
$$(-\infty, -3] \cup [9, \infty)$$

10. $\left|\frac{x}{2} - 1\right| \leq 0$

$$\frac{x}{2} - 1 = 0$$
$$\frac{x}{2} = 1$$
$$x = 2$$

11. $y = |x|$

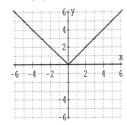

12. $y = |x - 2|$

13. a. Domain: $\{2, 0, 3\}$
    Range: $\{3, 4, -1\}$

b. Domain: $\{-4, -3, -2, -1, 0, 1, 2, 3\}$
    Range: $\{1\}$

c. Domain: $\{$Erie, Miami, Escondido, Waco, Gary$\}$
    Range: $\{109, 359, 117, 104\}$

14. a. Domain: $(-\infty, 0]$, Range: $(-\infty, \infty)$
    not a function

b. Domain: $(-\infty, \infty)$, Range: $(-\infty, \infty)$
    function

c. Domain: $(-\infty, -2] \cup [2, \infty)$
    Range: $(-\infty, \infty)$; not a function

15. $y = -3$

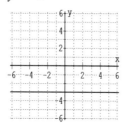

16. $f(x) = -2$
$y = -2$

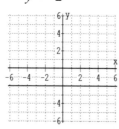

17. $x = -5$ is a vertical line. Thus, the slope is undefined.

18. $y = -3$ is a horizontal line. Thus, the slope is $m = 0$.

19. $\begin{cases} -\dfrac{x}{6} + \dfrac{y}{2} = \dfrac{1}{2} & (1) \\ \dfrac{x}{3} - \dfrac{y}{6} = -\dfrac{3}{4} & (2) \end{cases}$

Multiply E1 by 6 and E2 by 12 to clear fractions.
$\begin{cases} -x + 3y = 3 & (1) \\ 4x - 2y = -9 & (2) \end{cases}$

Solve the new E1 for x.
$-x = -3y + 3$
$x = 3y - 3$

Replace x with $3y - 3$ in the new E2.
$4(3y - 3) - 2y = -9$
$12y - 12 - 2y = -9$
$10y = 3$
$y = \dfrac{3}{10}$

Replace y with $\dfrac{3}{10}$ in the equation $x = 3y - 3$.

$x = 3\left(\dfrac{3}{10}\right) - 3 = \dfrac{9}{10} - \dfrac{30}{10} = -\dfrac{21}{10}$

The solution is $\left(-\dfrac{21}{10}, \dfrac{3}{10}\right)$.

20. $\begin{cases} \dfrac{x}{6} - \dfrac{y}{2} = 1 & (1) \\ \dfrac{x}{3} - \dfrac{y}{4} = 2 & (2) \end{cases}$

Multiply E1 by 6 and E2 by 12 to clear fractions.
$\begin{cases} x - 3y = 6 & (1) \\ 4x - 3y = 24 & (2) \end{cases}$

Solve the new E1 for x.
$x = 3y + 6$

Replace x with $3y + 6$ in the new E2.
$4(3y + 6) - 3y = 24$
$12y + 24 - 3y = 24$
$9y = 0$
$y = 0$

Replace y with 0 in the equation $x = 3y + 6$.
$x = 3(0) + 6 = 6$
The solution is $(6, 0)$.

21. a. $2^2 \cdot 2^5 = 2^{2+5} = 2^7$

b. $x^7 x^3 = x^{7+3} = x^{10}$

c. $y \cdot y^2 \cdot y^4 = y^{1+2+4} = y^7$

22. Let $x =$ number of tee-shirts and $y =$ number of shorts.
$\begin{cases} x + y = 9 & (1) \\ 3.50x + 4.25y = 33.75 & (2) \end{cases}$
Solve E1 for y: $y = 9 - x$.
Replace y with $9 - x$ in E2

$$3.50x + 4.25(9 - x) = 33.75$$
$$3.50x + 38.25 - 4.25x = 33.75$$
$$-0.75x = -4.5$$
$$x = 6$$

Replace x with 6 in the equation
$y = 9 - x$.

$$y = 9 - 6 = 3$$

She bought 6 tee-shirts and 3 shorts.

23. $\dfrac{2000 \times 0.000021}{700} = \dfrac{(2 \times 10^3) \times (2.1 \times 10^{-5})}{7 \times 10^2}$

$$= \dfrac{2 \times 2.1}{7} \times 10^{3 + (-5) - 2}$$
$$= 0.6 \times 10^{-4}$$
$$= 6 \times 10^{-5}$$

24. $\dfrac{0.0000035 \times 4000}{0.28}$

$$= \dfrac{(3.5 \times 10^{-6}) \times (4 \times 10^3)}{2.8 \times 10^{-1}}$$
$$= \dfrac{3.5 \times 4}{2.8} \times 10^{-6 + 3 - (-1)}$$
$$= 5 \times 10^{-2}$$

25. $P(x) = 3x^2 - 2x - 5$

a. $P(1) = 3(1)^2 - 2(1) - 5 = 3 - 2 - 5 = -4$

b. $P(-2) = 3(-2)^2 - 2(-2) - 5$
$$= 3(4) + 4 - 5$$
$$= 12 + 4 - 5$$
$$= 11$$

26. $[(5x^2 - 3x + 6) + (4x^2 + 5x - 3)] - (2x - 5)$
$$= (9x^2 + 2x + 3) - (2x - 5)$$
$$= 9x^2 + 2x + 3 - 2x + 5$$
$$= 9x^2 + 8$$

27. a. $(x + 3)(2x + 5) = 2x^2 + 5x + 6x + 15$
$$= 2x^2 + 11x + 15$$

b. $(2x - 3)(5x^2 - 6x + 7)$
$$= 10x^3 - 12x^2 + 14x$$
$$\underline{\qquad -15x^2 + 18x - 21}$$
$$= 10x^3 - 27x^2 + 32x - 21$$

28. a. $(y - 2)(3y + 4) = 3y^2 + 4y - 6x - 8$
$$= 3y^2 - 6y - 8$$

b. $(3y - 1)(2y^2 + 3y - 1)$
$$= 6y^3 + 9y^2 - 3y$$
$$\underline{\qquad -2y^2 - 3y + 1}$$
$$= 6y^3 + 7y^2 - 6y + 1$$

29. $20x^3 y = 2^2 \cdot 5x^3 y$
$$10x^2 y^2 = 2 \cdot 5x^2 y^2$$
$$35x^3 = 5 \cdot 7x^3$$
$$\text{GCF} = 5x^2$$

30. $x^3 - x^2 + 4x - 4 = (x^3 - x^2) + (4x - 4)$
$$= x^2(x - 1) + 4(x - 1)$$
$$= (x - 1)(x^2 + 4)$$

31. a. $\dfrac{x^3 + 8}{2 + x} = \dfrac{x^3 + 2^3}{x + 2}$
$$= \dfrac{(x + 2)(x^2 - 2x + 4)}{x + 2}$$
$$= x^2 - 2x + 4$$

b. $\dfrac{2y^2 + 2}{y^3 - 5y^2 + y - 5} = \dfrac{2(y^2 + 1)}{y^2(y - 5) + 1(y - 5)}$
$$= \dfrac{2(y^2 + 1)}{(y - 5)(y^2 + 1)}$$
$$= \dfrac{2}{y - 5}$$

32. a. $\dfrac{a^3 - 8}{2 - a} = \dfrac{a^3 - 2^3}{2 - a}$
$$= \dfrac{(a - 2)(a^2 + 2a + 4)}{2 - a}$$
$$= -1(a^2 + 2a + 4)$$
$$= -a^2 - 2a - 4$$

b. $\dfrac{3a^2-3}{a^3+5a^2-a-5} = \dfrac{3(a^2-1)}{a^2(a+5)-1(a+5)}$

$\qquad\qquad = \dfrac{3(a^2-1)}{(a+5)(a^2-1)}$

$\qquad\qquad = \dfrac{3}{a+5}$

33. a. $\dfrac{2}{x^2y}+\dfrac{5}{3x^3y} = \dfrac{2\cdot 3x}{x^2y\cdot 3x}+\dfrac{5}{3x^3y}$

$\qquad\qquad = \dfrac{6x+5}{3x^3y}$

b. $\dfrac{3x}{x+2}+\dfrac{2x}{x-2} = \dfrac{3x(x-2)+2x(x+2)}{(x+2)(x-2)}$

$\qquad\qquad = \dfrac{3x^2-6x+2x^2+4x}{(x+2)(x-2)}$

$\qquad\qquad = \dfrac{5x^2-2x}{(x+2)(x-2)}$

c. $\dfrac{x}{x-1}-\dfrac{4}{1-x} = \dfrac{x}{x-1}+\dfrac{4}{x-1} = \dfrac{x+4}{x-1}$

34. a. $\dfrac{3}{xy^2}-\dfrac{2}{3x^2y} = \dfrac{3\cdot 3x}{xy^2\cdot 3x}-\dfrac{2\cdot y}{3x^2y\cdot y}$

$\qquad\qquad = \dfrac{9x-2y}{3x^2y^2}$

b. $\dfrac{5x}{x+3}-\dfrac{2x}{x-3} = \dfrac{5x(x-3)-2x(x+3)}{(x+3)(x-3)}$

$\qquad\qquad = \dfrac{5x^2-15x-2x^2-6x}{(x+3)(x-3)}$

$\qquad\qquad = \dfrac{3x^2-21x}{(x+3)(x-3)}$

c. $\dfrac{x}{x-2}-\dfrac{5}{2-x} = \dfrac{x}{x-2}+\dfrac{5}{x-2} = \dfrac{x+5}{x-2}$

35. a. $\dfrac{\frac{5x}{x+2}}{\frac{10}{x-2}} = \dfrac{5x}{x+2}\cdot\dfrac{x-2}{10} = \dfrac{x(x-2)}{2(x+2)}$

b. $\dfrac{\frac{x}{y^2}+\frac{1}{y}}{\frac{y}{x^2}+\frac{1}{x}} = \dfrac{\left(\frac{x}{y^2}+\frac{1}{y}\right)x^2y^2}{\left(\frac{y}{x^2}+\frac{1}{x}\right)x^2y^2}$

$\qquad\qquad = \dfrac{x^3+x^2y}{y^3+xy^2}$

$\qquad\qquad = \dfrac{x^2(x+y)}{y^2(y+x)}$

$\qquad\qquad = \dfrac{x^2}{y^2}$

36. a. $\dfrac{\frac{y-2}{16}}{\frac{2y+3}{12}} = \dfrac{y-2}{16}\cdot\dfrac{12}{2y+3} = \dfrac{3(y-2)}{4(2y+3)}$

b. $\dfrac{\frac{x}{16}-\frac{1}{x}}{1-\frac{4}{x}} = \dfrac{\left(\frac{x}{16}-\frac{1}{x}\right)16x}{\left(1-\frac{4}{x}\right)16x}$

$\qquad\qquad = \dfrac{x^2-16}{16x-64}$

$\qquad\qquad = \dfrac{(x+4)(x-4)}{16(x-4)}$

$\qquad\qquad = \dfrac{x+4}{16}$

37. $\dfrac{10x^3-5x^2+20x}{5x} = \dfrac{10x^3}{5x}-\dfrac{5x^2}{5x}+\dfrac{20x}{5x}$

$\qquad\qquad = 2x^2-x+4$

38. $\begin{array}{r} x^2+3 \\ x-2\overline{\smash{)}\,x^3-2x^2+3x-6} \\ \underline{x^3-2x^2} \\ 3x-6 \\ \underline{3x-6} \\ 0 \end{array}$

Answer: x^2+3

39. $\begin{array}{r|rrrr} 3 & 2 & -1 & -13 & 1 \\ & & 6 & 15 & 6 \\ \hline & 2 & 5 & 2 & 7 \end{array}$

Answer: $2x^2+5x+2+\dfrac{7}{x-3}$

40.

$$3 \overline{\smash{)}\begin{array}{rrrr} 4 & -12 & -1 & 12 \\ & 12 & 0 & -3 \end{array}}$$
$$\begin{array}{rrrr} 4 & 0 & -1 & 9 \end{array}$$

Answer: $4y^2 - 1 + \dfrac{9}{y-3}$

41.
$$\frac{x+6}{x-2} = \frac{2(x+2)}{x-2}$$
$$(x-2)\left(\frac{x+6}{x-2}\right) = (x-2)\left[\frac{2(x+2)}{x-2}\right]$$
$$x+6 = 2(x+2)$$
$$x+6 = 2x+4$$
$$-x = -2$$
$$x = 2$$

which we discard as extraneous.
There is no solution.

42.
$$\frac{28}{9-a^2} = \frac{2a}{a-3} + \frac{6}{a+3}$$
$$\frac{28}{-(a^2-9)} = \frac{2a}{a-3} + \frac{6}{a+3}$$
$$\frac{-28}{(a+3)(a-3)} = \frac{2a}{a-3} + \frac{6}{a+3}$$
$$-28 = 2a(a+3) + 6(a-3)$$
$$-28 = 2a^2 + 6a + 6a - 18$$
$$0 = 2a^2 + 12a + 10$$
$$0 = 2(a^2 + 6a + 5)$$
$$0 = 2(a+5)(a+1)$$

$a = -5$ or $a = -1$
The solutions are -5 and -1.

43.
$$\frac{1}{x} + \frac{1}{y} = \frac{1}{z}$$
$$xyz\left(\frac{1}{x} + \frac{1}{y}\right) = xyz\left(\frac{1}{z}\right)$$
$$yz + xz = xy$$
$$yz = xy - xz$$
$$yz = x(y-z)$$
$$x = \frac{yz}{y-z}$$

44.
$$A = \frac{h(a+b)}{2}$$
$$2A = h(a+b)$$
$$2A = ah + bh$$
$$2A - bh = ah$$
$$\frac{2A-bh}{h} = a$$

45. $u = \dfrac{k}{w}$
$$3 = \frac{k}{5}$$
$$k = 3(5) = 15$$
$$u = \frac{15}{w}$$

46.
$$y = kx$$
$$0.51 = k(3)$$
$$k = \frac{0.51}{3} = 0.17$$
$$y = 0.17x$$

47. a. $16^{-3/4} = \dfrac{1}{16^{3/4}} = \dfrac{1}{\left(\sqrt[4]{16}\right)^3} = \dfrac{1}{(2)^3} = \dfrac{1}{8}$

b. $(-27)^{-2/3} = \dfrac{1}{(-27)^{2/3}}$
$$= \frac{1}{\left(\sqrt[3]{-27}\right)^2} = \frac{1}{(-3)^2} = \frac{1}{9}$$

48. a. $81^{-3/4} = \dfrac{1}{81^{3/4}} = \dfrac{1}{\left(\sqrt[4]{81}\right)^3} = \dfrac{1}{(3)^3} = \dfrac{1}{27}$

b. $(-125)^{-2/3} = \dfrac{1}{(-125)^{2/3}}$
$$= \frac{1}{\left(\sqrt[3]{-125}\right)^2} = \frac{1}{(-5)^2} = \frac{1}{25}$$

49. $\dfrac{\sqrt{x}+2}{5} = \dfrac{\left(\sqrt{x}+2\right)\left(\sqrt{x}-2\right)}{5\left(\sqrt{x}-2\right)}$

$\qquad = \dfrac{\left(\sqrt{x}\right)^2 - 2^2}{5\left(\sqrt{x}-2\right)} = \dfrac{x-4}{5\left(\sqrt{x}-2\right)}$

50. a. $\sqrt{36a^3} - \sqrt{144a^3} + \sqrt{4a^3}$

$\qquad = \sqrt{36a^2 \cdot a} - \sqrt{144a^2 \cdot a} + \sqrt{4a^2 \cdot a}$

$\qquad = 6a\sqrt{a} - 12a\sqrt{a} + 2a\sqrt{a}$

$\qquad = -4a\sqrt{a}$

b. $\sqrt[3]{128ab^3} - 3\sqrt[3]{2ab^3} + b\sqrt[3]{16a}$

$\qquad = \sqrt[3]{64b^3 \cdot 2a} - 3\sqrt[3]{b^3 \cdot 2a} + b\sqrt[3]{8 \cdot 2a}$

$\qquad = 4b\sqrt[3]{2a} - 3b\sqrt[3]{2a} + 2b\sqrt[3]{2a}$

$\qquad = 3b\sqrt[3]{2a}$

c. $\dfrac{\sqrt[3]{81}}{10} + \sqrt[3]{\dfrac{192}{125}} = \dfrac{\sqrt[3]{27 \cdot 3}}{10} + \dfrac{\sqrt[3]{192}}{\sqrt[3]{125}}$

$\qquad = \dfrac{3\sqrt[3]{3}}{10} + \dfrac{\sqrt[3]{64 \cdot 3}}{5}$

$\qquad = \dfrac{3\sqrt[3]{3}}{10} + \dfrac{4\sqrt[3]{3}}{5}$

$\qquad = \dfrac{3\sqrt[3]{3}}{10} + \dfrac{4\sqrt[3]{3} \cdot 2}{5 \cdot 2} = \dfrac{11\sqrt[3]{3}}{10}$

Chapter 8

Section 8.1

Graphing Calculator Explorations

1. −1.27, 6.27

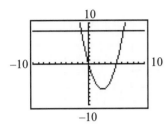

3. −1.10, 0.90

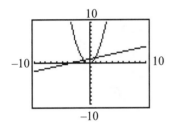

5. No real solutions, or ∅

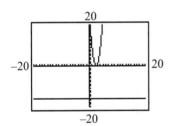

Exercise Set 8.1

1. $x^2 = 16$

$x = \pm\sqrt{16}$

$x = \pm 4$

3. $x^2 - 7 = 0$

$x^2 = 7$

$x = \pm\sqrt{7}$

5. $x^2 = 18$

$x = \pm\sqrt{18}$

$x = \pm\sqrt{9\cdot 2}$

$x = \pm 3\sqrt{2}$

7. $3z^2 - 30 = 0$

$3z^2 = 30$

$z^2 = 10$

$z = \pm\sqrt{10}$

9. $(x+5)^2 = 9$

$x + 5 = \pm\sqrt{9}$

$x + 5 = \pm 3$

$x = -5 \pm 3$

$x = -8$ or $x = -2$

11. $(z-6)^2 = 18$

$z - 6 = \pm\sqrt{18}$

$z - 6 = \pm 3\sqrt{2}$

$z = 6 \pm 3\sqrt{2}$

13. $(2x-3)^2 = 8$

$2x - 3 = \pm\sqrt{8}$

$2x - 3 = \pm 2\sqrt{2}$

$2x = 3 \pm 2\sqrt{2}$

$x = \dfrac{3 \pm 2\sqrt{2}}{2}$

15. $x^2 + 9 = 0$

$x^2 = -9$

$x = \pm\sqrt{-9}$

$x = \pm 3i$

17. $x^2 - 6 = 0$
$$x^2 = 6$$
$$x = \pm\sqrt{6}$$

19. $2z^2 + 16 = 0$
$$2z^2 = -16$$
$$z^2 = -8$$
$$z = \pm\sqrt{-8}$$
$$z = \pm i\sqrt{8}$$
$$z = \pm 2i\sqrt{2}$$

21. $(x-1)^2 = -16$
$$x - 1 = \pm\sqrt{-16}$$
$$x - 1 = \pm 4i$$
$$x = 1 \pm 4i$$

23. $(z+7)^2 = 5$
$$z + 7 = \pm\sqrt{5}$$
$$z = -7 \pm \sqrt{5}$$

25. $(x+3)^2 = -8$
$$x + 3 = \pm\sqrt{-8}$$
$$x + 3 = \pm i\sqrt{8}$$
$$x + 3 = \pm 2i\sqrt{2}$$
$$x = -3 \pm 2i\sqrt{2}$$

27. $x^2 + 16x + \left(\dfrac{16}{2}\right)^2 = x^2 + 16x + 64$
$$= (x+8)^2$$

29. $z^2 - 12z + \left(\dfrac{12}{2}\right)^2 = z^2 - 12z + 36$
$$= (z-6)^2$$

31. $p^2 + 9p + \left(\dfrac{9}{2}\right)^2 = p^2 + 9p + \dfrac{81}{4}$
$$= \left(p + \dfrac{9}{2}\right)^2$$

33. $x^2 + x + \left(\dfrac{1}{2}\right)^2 = x^2 + x + \dfrac{1}{4}$
$$= \left(x + \dfrac{1}{2}\right)^2$$

35. $x^2 + 8x = -15$
$$x^2 + 8x + \left(\dfrac{8}{2}\right)^2 = -15 + 16$$
$$x^2 + 8x + 16 = 1$$
$$(x+4)^2 = 1$$
$$x + 4 = \pm\sqrt{1}$$
$$x = -4 \pm 1$$
$$x = -5 \ \text{ or } \ x = -3$$

37. $x^2 + 6x + 2 = 0$
$$x^2 + 6x = -2$$
$$x^2 + 6x + \left(\dfrac{6}{2}\right)^2 = -2 + 9$$
$$x^2 + 6x + 9 = 7$$
$$(x+3)^2 = 7$$
$$x + 3 = \pm\sqrt{7}$$
$$x = -3 \pm \sqrt{7}$$

39. $x^2 + x - 1 = 0$
$$x^2 + x = 1$$
$$x^2 + x + \left(\dfrac{1}{2}\right)^2 = -2 + \dfrac{1}{4}$$
$$x^2 + x + \dfrac{1}{4} = \dfrac{5}{4}$$
$$\left(x + \dfrac{1}{2}\right)^2 = \dfrac{5}{4}$$
$$x + \dfrac{1}{2} = \pm\sqrt{\dfrac{5}{4}}$$
$$x = -\dfrac{1}{2} \pm \dfrac{\sqrt{5}}{2} = \dfrac{-1 \pm \sqrt{5}}{2}$$

41. $x^2 + 2x - 5 = 0$

$x^2 + 2x = 5$

$x^2 + 2x + \left(\dfrac{2}{2}\right)^2 = 5 + 1$

$x^2 + 2x + 1 = 6$

$(x+1)^2 = 6$

$x + 1 = \pm\sqrt{6}$

$x = -1 \pm \sqrt{6}$

43. $3p^2 - 12p + 2 = 0$

$3p^2 - 12p = -2$

$p^2 - 4p = -\dfrac{2}{3}$

$p^2 - 4p + \left(\dfrac{-4}{2}\right)^2 = -\dfrac{2}{3} + 4$

$(p-2)^2 = \dfrac{10}{3}$

$p - 2 = \pm\sqrt{\dfrac{10}{3}}$

$p - 2 = \pm\dfrac{\sqrt{10}\cdot\sqrt{3}}{\sqrt{3}\cdot\sqrt{3}}$

$p - 2 = \pm\dfrac{\sqrt{30}}{3}$

$p = 2 \pm \dfrac{\sqrt{30}}{3} = \dfrac{6 \pm \sqrt{30}}{3}$

45. $4y^2 - 12y - 2 = 0$

$4y^2 - 12y = 2$

$y^2 - 3y = \dfrac{1}{2}$

$y^2 - 3y + \left(\dfrac{-3}{2}\right)^2 = \dfrac{1}{2} + \dfrac{9}{4}$

$y^2 - 3y + \dfrac{9}{4} = \dfrac{11}{4}$

$\left(y - \dfrac{3}{2}\right)^2 = \dfrac{11}{4}$

$y - \dfrac{3}{2} = \pm\sqrt{\dfrac{11}{4}}$

$y = \dfrac{3}{2} \pm \dfrac{\sqrt{11}}{2} = \dfrac{3 \pm \sqrt{11}}{2}$

47. $2x^2 + 7x = 4$

$x^2 + \dfrac{7}{2}x = 2$

$x^2 + \dfrac{7}{2}x + \left(\dfrac{\frac{7}{2}}{2}\right)^2 = 2 + \dfrac{49}{16}$

$x^2 + \dfrac{7}{2}x + \dfrac{49}{16} = \dfrac{81}{16}$

$\left(x + \dfrac{7}{4}\right)^2 = \dfrac{81}{16}$

$x + \dfrac{7}{4} = \pm\sqrt{\dfrac{81}{16}}$

$x = -\dfrac{7}{4} \pm \dfrac{9}{4} = \dfrac{-7 \pm 9}{4}$

$x = -4, \dfrac{1}{2}$

49. $x^2 - 4x - 5 = 0$

$x^2 - 4x = 5$

$x^2 - 4x + \left(\dfrac{-4}{2}\right)^2 = 5 + 4$

$x^2 - 4x + 4 = 9$

$(x-2)^2 = 9$

$x = 2 \pm 3$

$x = -1, 5$

51. $x^2 + 8x + 1 = 0$

$x^2 + 8x = -1$

$x^2 + 8x + \left(\dfrac{8}{2}\right)^2 = -1 + 16$

$x^2 + 8x + 16 = 15$

$(x+4)^2 = 15$

$x + 4 = \pm\sqrt{15}$

$x = -4 \pm \sqrt{15}$

53.
$$3y^2 + 6y - 4 = 0$$
$$3y^2 + 6y = 4$$
$$y^2 + 2y = \frac{4}{3}$$
$$y^2 + 2y + \left(\frac{2}{2}\right)^2 = \frac{4}{3} + 1$$
$$y^2 + 2y + 1 = \frac{7}{3}$$
$$(y+1)^2 = \frac{7}{3}$$
$$y + 1 = \pm\sqrt{\frac{7}{3}}$$
$$y + 1 = \pm\frac{\sqrt{7} \cdot \sqrt{3}}{\sqrt{3} \cdot \sqrt{3}}$$
$$y + 1 = \pm\frac{\sqrt{21}}{3}$$
$$y = -1 \pm \frac{\sqrt{21}}{3} = \frac{-3 \pm \sqrt{21}}{3}$$

55.
$$2x^2 - 3x - 5 = 0$$
$$2x^2 - 3x = 5$$
$$x^2 - \frac{3}{2}x = \frac{5}{2}$$
$$x^2 - \frac{3}{2}x + \left(\frac{\frac{3}{2}}{2}\right)^2 = \frac{5}{2} + \frac{9}{16}$$
$$x^2 - \frac{3}{2}x + \frac{9}{16} = \frac{49}{16}$$
$$\left(x - \frac{3}{4}\right)^2 = \frac{49}{16}$$
$$x - \frac{3}{4} = \pm\sqrt{\frac{49}{16}}$$
$$x = \frac{3}{4} \pm \frac{7}{4} = \frac{3 \pm 7}{4}$$
$$x = -1, \frac{5}{2}$$

57.
$$y^2 + 2y + 2 = 0$$
$$y^2 + 2y = -2$$
$$y^2 + 2y + \left(\frac{2}{2}\right)^2 = -2 + 1$$
$$y^2 + 2y + 1 = -1$$
$$(y+1)^2 = -1$$
$$y + 1 = \pm\sqrt{-1}$$
$$y = -1 \pm i$$

59.
$$x^2 - 6x + 3 = 0$$
$$x^2 - 6x = -3$$
$$x^2 - 6x + \left(\frac{-6}{2}\right)^2 = -3 + 9$$
$$x^2 - 6x + 9 = 6$$
$$(x-3)^2 = 6$$
$$x - 3 = \pm\sqrt{6}$$
$$x = 3 \pm \sqrt{6}$$

61.
$$2a^2 + 8a = -12$$
$$a^2 + 4a = -6$$
$$a^2 + 4a + \left(\frac{4}{2}\right)^2 = -6 + 4$$
$$a^2 + 4a + 4 = -2$$
$$(a+2)^2 = -2$$
$$a + 2 = \pm\sqrt{-2}$$
$$a + 2 = \pm i\sqrt{2}$$
$$a = -2 \pm i\sqrt{2}$$

63. $5x^2 + 15x - 1 = 0$

$5x^2 + 15x = 1$

$x^2 + 3x = \dfrac{1}{5}$

$x^2 + 3x + \left(\dfrac{3}{2}\right)^2 = \dfrac{1}{5} + \dfrac{9}{4}$

$x^2 + 3x + \dfrac{9}{4} = \dfrac{49}{20}$

$\left(x + \dfrac{3}{2}\right)^2 = \dfrac{49}{20}$

$x + \dfrac{3}{2} = \pm\sqrt{\dfrac{49}{20}}$

$x + \dfrac{3}{2} = \pm\dfrac{7}{\sqrt{20}}$

$x + \dfrac{3}{2} = \pm\dfrac{7}{2\sqrt{5}}$

$x + \dfrac{3}{2} = \pm\dfrac{7 \cdot \sqrt{5}}{2\sqrt{5} \cdot \sqrt{5}}$

$x + \dfrac{3}{2} = \pm\dfrac{7\sqrt{5}}{10}$

$x = -\dfrac{3}{2} \pm \dfrac{7\sqrt{5}}{10} = \dfrac{-15 \pm 7\sqrt{5}}{10}$

65. $2x^2 - x + 6 = 0$

$2x^2 - x = -6$

$x^2 - \dfrac{1}{2}x = -3$

$x^2 - \dfrac{1}{2}x + \left(\dfrac{-\dfrac{1}{2}}{2}\right)^2 = -3 + \dfrac{1}{16}$

$x^2 - \dfrac{1}{2}x + \dfrac{1}{16} = -\dfrac{47}{16}$

$\left(x - \dfrac{1}{4}\right)^2 = -\dfrac{47}{16}$

$x - \dfrac{1}{4} = \pm\sqrt{-\dfrac{47}{16}}$

$x - \dfrac{1}{4} = \pm\dfrac{\sqrt{47}}{4}i$

$x = \dfrac{1}{4} \pm \dfrac{\sqrt{47}}{4}i$

67. $x^2 + 10x + 28 = 0$

$x^2 + 10x = -28$

$x^2 + 10x + \left(\dfrac{10}{2}\right)^2 = -28 + 25$

$(x + 5)^2 = -3$

$x + 5 = \pm\sqrt{-3}$

$x = -5 \pm i\sqrt{3}$

69. $z^2 + 3z - 4 = 0$

$z^2 + 3z = 4$

$z^2 + 3z + \left(\dfrac{3}{2}\right)^2 = 4 + \dfrac{9}{4}$

$z^2 + 3z + \dfrac{9}{4} = \dfrac{25}{4}$

$\left(z + \dfrac{3}{2}\right)^2 = \dfrac{25}{4}$

$z + \dfrac{3}{2} = \pm\sqrt{\dfrac{25}{4}}$

$z = -\dfrac{3}{2} \pm \dfrac{5}{2} = \dfrac{-3 \pm 5}{2}$

$z = -4, 1$

71. $2x^2 - 4x + 3 = 0$

$2x^2 - 4x = -3$

$x^2 - 2x = -\dfrac{3}{2}$

$x^2 - 2x + \left(\dfrac{-2}{2}\right)^2 = -\dfrac{3}{2} + 1$

$x^2 - 2x + 1 = -\dfrac{1}{2}$

$(x - 1)^2 = -\dfrac{1}{2}$

$x - 1 = \pm\sqrt{-\dfrac{1}{2}}$

$x - 1 = \pm\dfrac{1}{\sqrt{2}}i$

$x - 1 = \pm\dfrac{1 \cdot \sqrt{2}}{\sqrt{2} \cdot \sqrt{2}}i$

$x - 1 = \pm\dfrac{\sqrt{2}}{2}i$

$x = 1 \pm \dfrac{\sqrt{2}}{2}i$

73.
$$3x^2 + 3x = 5$$
$$x^2 + x = \frac{5}{3}$$
$$x^2 + x + \left(\frac{1}{2}\right)^2 = \frac{5}{3} + \frac{1}{4}$$
$$x^2 + x + \frac{1}{4} = \frac{23}{12}$$
$$\left(x + \frac{1}{2}\right)^2 = \frac{23}{12}$$
$$x + \frac{1}{2} = \pm\sqrt{\frac{23}{12}}$$
$$x + \frac{1}{2} = \pm\frac{\sqrt{23}}{2\sqrt{3}}$$
$$x + \frac{1}{2} = \pm\frac{\sqrt{23}\cdot\sqrt{3}}{2\sqrt{3}\cdot\sqrt{3}}$$
$$x + \frac{1}{2} = \pm\frac{\sqrt{69}}{6}$$
$$x = -\frac{1}{2} \pm \frac{\sqrt{69}}{6} = \frac{-3 \pm \sqrt{69}}{6}$$

75.
$$A = P(1+r)^t$$
$$4320 = 3000(1+r)^2$$
$$\frac{4320}{3000} = (1+r)^2$$
$$1.44 = (1+r)^2$$
$$\pm\sqrt{1.44} = 1+r$$
$$\pm 1.2 = 1+r$$
$$-1 \pm 1.2 = r$$
$$-2.2 = r \text{ or } 0.2 = r$$
Rate cannot be negative, so the rate is
$r = 0.2 = 20\%.$

77.
$$A = P(1+r)^t$$
$$1000 = 810(1+r)^2$$
$$\frac{1000}{810} = (1+r)^2$$
$$\frac{100}{81} = (1+r)^2$$
$$\pm\sqrt{\frac{100}{81}} = 1+r$$
$$\pm\frac{10}{9} = 1+r$$
$$-1 \pm \frac{10}{9} = r$$
$$-\frac{19}{9} = r \text{ or } \frac{1}{9} = r$$
Rate cannot be negative, so the rate is
$r = \frac{1}{9}$, or $11\frac{1}{9}\%$.

79. Answers may vary.

81. Simple

83. $\dfrac{3}{5} + \sqrt{\dfrac{16}{25}} = \dfrac{3}{5} + \dfrac{4}{5} = \dfrac{7}{5}$

85. $\dfrac{9}{10} - \sqrt{\dfrac{49}{100}} = \dfrac{9}{10} - \dfrac{7}{10} = \dfrac{2}{10} = \dfrac{1}{5}$

87. $\dfrac{10 - 20\sqrt{3}}{2} = \dfrac{10}{2} - \dfrac{20\sqrt{3}}{2} = 5 - 10\sqrt{3}$

89. $\dfrac{12 - 8\sqrt{7}}{16} = \dfrac{12}{16} - \dfrac{8\sqrt{7}}{16}$
$$= \dfrac{3}{4} - \dfrac{\sqrt{7}}{2}$$
$$= \dfrac{3}{4} - \dfrac{2\sqrt{7}}{4} = \dfrac{3 - 2\sqrt{7}}{4}$$

91. $\sqrt{b^2 - 4ac} = \sqrt{(6)^2 - 4(1)(2)}$
$$= \sqrt{36 - 8}$$
$$= \sqrt{28}$$
$$= \sqrt{4\cdot 7}$$
$$= 2\sqrt{7}$$

93. $\sqrt{b^2 - 4ac} = \sqrt{(-3)^2 - 4(1)(-1)}$
$$= \sqrt{9 + 4}$$
$$= \sqrt{13}$$

95. $y^2 + \underline{\quad} + 9$
$$\left(\frac{b}{2}\right)^2 = 9$$
$$\frac{b}{2} = \pm\sqrt{9}$$
$$\frac{b}{2} = \pm 3$$
$$b = \pm 6$$
Answer: $\pm 6y$

97. $x^2 + \underline{\quad} + \frac{1}{4}$
$$\left(\frac{b}{2}\right)^2 = \frac{1}{4}$$
$$\frac{b}{2} = \pm\sqrt{\frac{1}{4}}$$
$$\frac{b}{2} = \pm\frac{1}{2}$$
$$b = \pm 1$$
Answer: $\pm x$

99. $s(t) = 16t^2$
$$1053 = 16t^2$$
$$t^2 = \frac{1053}{16}$$
$$t = \pm\sqrt{\frac{1053}{16}}$$
$$t \approx 8.11 \text{ or } -8.11 \text{ (disregard)}$$
It would take 8.11 seconds.

101. $s(t) = 16t^2$
$$725 = 16t^2$$
$$t^2 = \frac{725}{16}$$
$$t = \pm\sqrt{\frac{725}{16}}$$
$$t \approx 6.73 \text{ or } -6.73 \text{ (disregard)}$$
It would take 6.73 seconds.

103. $A = \pi r^2$
$$36\pi = \pi r^2$$
$$r^2 = \frac{36\pi}{\pi}$$
$$r^2 = 36$$
$$r = \pm\sqrt{36}$$
$$r = 6 \text{ or } -6 \text{ (disregard)}$$
The radius is 6 inches.

105. $a^2 + b^2 = c^2$
$$(4x)^2 + (3x)^2 = 27^2$$
$$16x^2 + 9x^2 = 729$$
$$25x^2 = 729$$
$$x^2 = \frac{729}{25}$$
$$x = \pm\sqrt{\frac{729}{25}} = \pm\frac{27}{5}$$
$$x = 5.4 \text{ or } -5.4 \text{ (disregard)}$$
$3x = 3(5.4) = 16.2$
$4x = 4(5.4) = 21.6$
The sides are 16.2 in. and 21.6 in.

107. $p = -x^2 + 15$
$$7 = -x^2 + 15$$
$$x^2 = 8$$
$$x = \pm\sqrt{8}$$
$$x \approx \pm 2.828$$
Demand cannot be negative. Therefore, the demand is approximately 2.828 thousand (or 2828) units.

Section 8.2

Mental Math

1. $x^2 + 3x + 1$
$a = 1, b = 3, c = 1$

2. $2x^2 - 5x - 7$
$a = 2, b = -5, c = -7$

3. $7x^2 - 4 = 0$
$a = 7, b = 0, c = -4$

4. $x^2 + 9 = 0$
$a = 1, b = 0, c = 9$

5. $6x^2 - x = 0$
$a = 6, b = -1, c = 0$

6. $5x^2 + 3x = 0$
$a = 5, b = 3, c = 0$

Exercise Set 8.2

1. $m^2 + 5m - 6 = 0$
$a = 1, b = 5, c = -6$

$m = \dfrac{-5 \pm \sqrt{(5)^2 - 4(1)(-6)}}{2(1)}$

$= \dfrac{-5 \pm \sqrt{25 + 24}}{2}$

$= \dfrac{-5 \pm \sqrt{49}}{2}$

$= \dfrac{-5 \pm 7}{2} = -6 \text{ or } 1$

The solutions are –6 and 1.

3. $2y = 5y^2 - 3$
$5y^2 - 2y - 3 = 0$
$a = 5, b = -2, c = -3$

$y = \dfrac{2 \pm \sqrt{(-2)^2 - 4(5)(-3)}}{2(5)}$

$= \dfrac{2 \pm \sqrt{4 + 60}}{10}$

$= \dfrac{2 \pm \sqrt{64}}{10}$

$= \dfrac{2 \pm 8}{10} = -\dfrac{3}{5} \text{ or } 1$

The solutions are $-\dfrac{3}{5}$ and 1.

5. $x^2 - 6x + 9 = 0$
$a = 1, b = -6, c = 9$

$x = \dfrac{6 \pm \sqrt{(-6)^2 - 4(1)(9)}}{2(1)}$

$= \dfrac{6 \pm \sqrt{36 - 36}}{2}$

$= \dfrac{6 \pm \sqrt{0}}{2} = \dfrac{6}{2} = 3$

The solution is 3.

7. $x^2 + 7x + 4 = 0$
$a = 1, b = 7, c = 4$

$x = \dfrac{-7 \pm \sqrt{(7)^2 - 4(1)(4)}}{2(1)}$

$= \dfrac{-7 \pm \sqrt{49 - 16}}{2}$

$= \dfrac{-7 \pm \sqrt{33}}{2}$

The solutions are $\dfrac{-7 + \sqrt{33}}{2}$ and $\dfrac{-7 - \sqrt{33}}{2}$.

9. $8m^2 - 2m = 7$
$8m^2 - 2m - 7 = 0$
$a = 8, b = -2, c = -7$

$m = \dfrac{2 \pm \sqrt{(-2)^2 - 4(8)(-7)}}{2(8)}$

$= \dfrac{2 \pm \sqrt{4 + 224}}{16}$

$= \dfrac{2 \pm \sqrt{228}}{16}$

$= \dfrac{2 \pm \sqrt{4 \cdot 57}}{16}$

$= \dfrac{2 \pm 2\sqrt{57}}{16} = \dfrac{1 \pm \sqrt{57}}{8}$

The solutions are $\dfrac{1 + \sqrt{57}}{8}$ and $\dfrac{1 - \sqrt{57}}{8}$.

11. $3m^2 - 7m = 3$

$3m^2 - 7m - 3 = 0$

$a = 3, b = -7, c = -3$

$m = \dfrac{7 \pm \sqrt{(-7)^2 - 4(3)(-3)}}{2(3)}$

$= \dfrac{7 \pm \sqrt{49 + 36}}{6}$

$= \dfrac{7 \pm \sqrt{85}}{6}$

The solutions are $\dfrac{7 + \sqrt{85}}{6}$ and $\dfrac{7 - \sqrt{85}}{6}$.

13. $\dfrac{1}{2}x^2 - x - 1 = 0$

$x^2 - 2x - 2 = 0$

$a = 1, b = -2, c = -2$

$x = \dfrac{2 \pm \sqrt{(-2)^2 - 4(1)(-2)}}{2(1)}$

$= \dfrac{2 \pm \sqrt{4 + 8}}{2}$

$= \dfrac{2 \pm \sqrt{12}}{2}$

$= \dfrac{2 \pm 2\sqrt{3}}{2} = 1 \pm \sqrt{3}$

The solutions are $1 + \sqrt{3}$ and $1 - \sqrt{3}$.

15. $\dfrac{2}{5}y^2 + \dfrac{1}{5}y = \dfrac{3}{5}$

$2y^2 + y - 3 = 0$

$a = 2, b = 1, c = -3$

$y = \dfrac{-1 \pm \sqrt{(1)^2 - 4(2)(-3)}}{2(2)}$

$= \dfrac{-1 \pm \sqrt{1 + 24}}{4}$

$= \dfrac{-1 \pm \sqrt{25}}{4}$

$= \dfrac{-1 \pm 5}{4} = -\dfrac{3}{2}$ or 1

The solutions are $-\dfrac{3}{2}$ and 1.

17. $\dfrac{1}{3}y^2 - y - \dfrac{1}{6} = 0$

$2y^2 - 6y - 1 = 0$

$a = 2, b = -6, c = -1$

$x = \dfrac{6 \pm \sqrt{(-6)^2 - 4(2)(-1)}}{2(2)}$

$= \dfrac{6 \pm \sqrt{36 + 8}}{4}$

$= \dfrac{6 \pm \sqrt{44}}{4}$

$= \dfrac{6 \pm 2\sqrt{11}}{4} = \dfrac{3 \pm \sqrt{11}}{2}$

The solutions are $\dfrac{3 + \sqrt{11}}{2}$ and $\dfrac{3 - \sqrt{11}}{2}$.

19. $m^2 + 5m - 6 = 0$

$(m + 6)(m - 1) = 0$

$m + 6 = 0$ or $m - 1 = 0$

     $m = -6$ or $m = 1$

The results are the same. Answers may vary.

21. $\qquad 6 = -4x^2 + 3x$

$4x^2 - 3x + 6 = 0$

$a = 4, b = -3, c = 6$

$x = \dfrac{3 \pm \sqrt{(-3)^2 - 4(4)(6)}}{2(4)}$

$= \dfrac{3 \pm \sqrt{9 - 96}}{8}$

$= \dfrac{3 \pm \sqrt{-87}}{8}$

$= \dfrac{3 \pm i\sqrt{87}}{8} = \dfrac{3}{8} \pm \dfrac{\sqrt{87}}{8}i$

The solutions are $\dfrac{3 + i\sqrt{87}}{8}$ and $\dfrac{3 - i\sqrt{87}}{8}$.

23. $(x + 5)(x - 1) = 2$

$x^2 + 4x - 5 = 2$

$x^2 + 4x - 7 = 0$

$a = 1, b = 4, c = -7$

$$x = \frac{-4 \pm \sqrt{(4)^2 - 4(1)(-7)}}{2(1)}$$

$$= \frac{-4 \pm \sqrt{16 + 28}}{2}$$

$$= \frac{-4 \pm \sqrt{44}}{2}$$

$$= \frac{-4 \pm 2\sqrt{11}}{2} = -2 \pm \sqrt{11}$$

The solutions are $-2 + \sqrt{11}$ and $-2 - \sqrt{11}$.

25. $10y^2 + 10y + 3 = 0$
$a = 10, b = 10, c = 3$

$$y = \frac{-10 \pm \sqrt{(10)^2 - 4(10)(3)}}{2(10)}$$

$$= \frac{-10 \pm \sqrt{100 - 120}}{20}$$

$$= \frac{-10 \pm \sqrt{-20}}{20}$$

$$= \frac{-10 \pm i\sqrt{4 \cdot 5}}{20}$$

$$= \frac{-10 \pm 2i\sqrt{5}}{20} =$$

$$= \frac{-5 \pm i\sqrt{5}}{10} = -\frac{1}{2} \pm \frac{\sqrt{5}}{10}i$$

The solutions are $\frac{-5 + i\sqrt{5}}{10}$ and $\frac{-5 - i\sqrt{5}}{10}$.

27. $9x - 2x^2 + 5 = 0$
$-2x^2 + 9x + 5 = 0$
$a = -2, b = 9, c = 5$
$b^2 - 4ac = 9^2 - 4(-2)(5)$
$\qquad = 81 + 40$
$\qquad = 121 > 0$
Therefore, there are two real solutions.

29. $4x^2 + 12x = -9$
$4x^2 + 12x + 9 = 0$
$a = 4, b = 12, c = 9$

$b^2 - 4ac = 12^2 - 4(4)(9)$
$\qquad = 144 - 144$
$\qquad = 0$
Therefore, there is one real solution.

31. $3x = -2x^2 + 7$
$2x^2 + 3x - 7 = 0$
$a = 2, b = 3, c = -7$
$b^2 - 4ac = 3^2 - 4(2)(-7)$
$\qquad = 9 + 56$
$\qquad = 65 > 0$
Therefore, there are two real solutions.

33. $6 = 4x - 5x^2$
$5x^2 - 4x + 6 = 0$
$a = 5, b = -4, c = 6$
$b^2 - 4ac = (-4)^2 - 4(5)(6)$
$\qquad = 16 - 120$
$\qquad = -104 < 0$
Therefore, there are two complex but not real solutions.

35. $x^2 + 5x = -2$
$x^2 + 5x + 2 = 0$
$a = 1, b = 5, c = 2$

$$x = \frac{-5 \pm \sqrt{(5)^2 - 4(1)(2)}}{2(1)}$$

$$= \frac{-5 \pm \sqrt{25 - 8}}{2}$$

$$= \frac{-5 \pm \sqrt{17}}{2}$$

The solutions are $\frac{-5 + \sqrt{17}}{2}$ and $\frac{-5 - \sqrt{17}}{2}$.

37. $(m + 2)(2m - 6) = 5(m - 1) - 12$
$2m^2 - 6m + 4m - 12 = 5m - 5 - 12$
$2m^2 - 7m + 5 = 0$
$a = 2, b = -7, c = 5$

$$m = \frac{7 \pm \sqrt{(-7)^2 - 4(2)(5)}}{2(2)}$$

$$= \frac{7 \pm \sqrt{49 - 40}}{4}$$

$$= \frac{7 \pm \sqrt{9}}{4}$$

$$= \frac{7 \pm 3}{4} = 1 \text{ or } \frac{5}{2}$$

The solutions are 1 and $\frac{5}{2}$.

39. $\dfrac{x^2}{3} - x = \dfrac{5}{3}$

$x^2 - 3x = 5$

$x^2 - 3x - 5 = 0$

$a = 1,\, b = -3,\, c = -5$

$$x = \frac{3 \pm \sqrt{(-3)^2 - 4(1)(-5)}}{2(1)}$$

$$= \frac{3 \pm \sqrt{9 + 20}}{2}$$

$$= \frac{3 \pm \sqrt{29}}{2}$$

The solutions are $\dfrac{3 + \sqrt{29}}{2}$ and $\dfrac{3 - \sqrt{29}}{2}$.

41. $x(6x + 2) - 3 = 0$

$6x^2 + 2x - 3 = 0$

$a = 6,\, b = 2,\, c = -3$

$$x = \frac{-2 \pm \sqrt{(2)^2 - 4(6)(-3)}}{2(6)}$$

$$= \frac{-2 \pm \sqrt{4 + 72}}{12}$$

$$= \frac{-2 \pm \sqrt{76}}{12}$$

$$= \frac{-2 \pm \sqrt{4 \cdot 19}}{12}$$

$$= \frac{-2 \pm 2\sqrt{19}}{12} = \frac{-1 \pm \sqrt{19}}{6}$$

The solutions are $\dfrac{-1 + \sqrt{19}}{6}$ and $\dfrac{-1 - \sqrt{19}}{6}$.

43. $x^2 + 6x + 13 = 0$

$a = 1,\, b = 6,\, c = 13$

$$x = \frac{-6 \pm \sqrt{(6)^2 - 4(1)(13)}}{2(1)}$$

$$= \frac{-6 \pm \sqrt{36 - 52}}{2}$$

$$= \frac{-6 \pm \sqrt{-16}}{2}$$

$$= \frac{-6 \pm 4i}{2} = -3 \pm 2i$$

The solutions are $-3 + 2i$ and $-3 - 2i$.

45. $\dfrac{2}{5}y^2 + \dfrac{1}{5}y + \dfrac{3}{5} = 0$

$2y^2 + y + 3 = 0$

$a = 2,\, b = 1,\, c = 3$

$$y = \frac{-1 \pm \sqrt{(1)^2 - 4(2)(3)}}{2(2)}$$

$$= \frac{-1 \pm \sqrt{1 - 24}}{4}$$

$$= \frac{-1 \pm \sqrt{-23}}{4}$$

$$= \frac{-1 \pm i\sqrt{23}}{4} = -\frac{1}{4} \pm \frac{\sqrt{23}}{4}i$$

The solutions are $\dfrac{-1 + i\sqrt{23}}{4}$ and $\dfrac{-1 - i\sqrt{23}}{4}$.

47. $\dfrac{1}{2}y^2 = y - \dfrac{1}{2}$

$y^2 = 2y - 1$

$y^2 - 2y + 1 = 0$

$a = 1,\, b = -2,\, c = 1$

$$y = \frac{2 \pm \sqrt{(-2)^2 - 4(1)(1)}}{2(1)}$$

$$= \frac{2 \pm \sqrt{4 - 4}}{2}$$

$$= \frac{2 \pm \sqrt{0}}{2} = \frac{2}{2} = 1$$

The solution is 1.

49.
$$(n-2)^2 = 15n$$
$$n^2 - 4n + 4 = 15n$$
$$n^2 - 19n + 4 = 0$$
$$a = 1, b = -19, c = 4$$
$$n = \frac{19 \pm \sqrt{(-19)^2 - 4(1)(4)}}{2(1)}$$
$$= \frac{19 \pm \sqrt{361 - 16}}{2}$$
$$= \frac{19 \pm \sqrt{345}}{2}$$

The solutions are $\dfrac{19 + \sqrt{345}}{2}$ and

$\dfrac{19 - \sqrt{345}}{2}$.

51.
$$(x+8)^2 + x^2 = 36^2$$
$$(x^2 + 16x + 64) + x^2 = 1296$$
$$2x^2 + 16x - 1232 = 0$$
$$a = 2, b = 16, c = -1232$$
$$x = \frac{-16 \pm \sqrt{(16)^2 - 4(2)(-1232)}}{2(2)}$$
$$= \frac{-16 \pm \sqrt{10,112}}{4}$$
$$x \approx 21 \text{ or } x \approx -29 \text{ (disregard)}$$
$$x + (x+8) = 21 + 21 + 8 = 50$$
$$50 - 36 = 14$$
They saved about 14 feet of walking distance.

53. Let x = length of leg. Then
$x + 2$ = length of hypotenuse
$$x^2 + x^2 = (x+2)^2$$
$$2x^2 = x^2 + 4x + 4$$
$$x^2 - 4x - 4 = 0$$
$$a = 1, b = -4, c = -4$$

$$x = \frac{4 \pm \sqrt{(-4)^2 - 4(1)(-4)}}{2(1)}$$
$$= \frac{4 \pm \sqrt{32}}{2}$$
$$= \frac{4 \pm 4\sqrt{2}}{2}$$
$$= 2 \pm 2\sqrt{2} \text{ (disregard the negative)}$$
$$= 2 + 2\sqrt{2}$$
The sides measure $2 + 2\sqrt{2}$ cm, $2 + 2\sqrt{2}$ cm, and $4 + 2\sqrt{2}$ cm.

55. Let x = width; then $x + 10$ = length.
Area = length · width
$$400 = (x+10)x$$
$$0 = x^2 + 10x - 400$$
$$a = 1, b = 10, c = -400$$
$$x = \frac{-10 \pm \sqrt{(10)^2 - 4(1)(-400)}}{2(1)}$$
$$= \frac{-10 \pm \sqrt{1700}}{2}$$
$$= \frac{-10 \pm 10\sqrt{17}}{2}$$
$$= -5 \pm 5\sqrt{17}$$
Disregard the negative length. The width is $-5 + 5\sqrt{17}$ ft and the length is $5 + 5\sqrt{17}$ ft.

57. a. Let x = length.
$$x^2 + x^2 = 100^2$$
$$2x^2 - 10,000 = 0$$
$$a = 2, b = 0, c = -10,000$$
$$x = \frac{0 \pm \sqrt{(0)^2 - 4(2)(-10,000)}}{2(2)}$$
$$= \frac{\pm \sqrt{80,000}}{4}$$
$$= \frac{\pm 200\sqrt{2}}{4} = \pm 50\sqrt{2}$$

Disregard the negative length. The side measures $50\sqrt{2}$ meters.

b. Area $= s^2$

$$= \left(50\sqrt{2}\right)^2$$
$$= 2500(2)$$
$$= 5000$$

The area is 5000 square meters.

59. Let $w =$ width; then $w + 1.1 =$ height.

Area = length · width

$$1439.9 = (w + 1.1)w$$
$$0 = w^2 + 1.1w - 1439.9$$
$$a = 1, b = 1.1, c = -1439.9$$

$$w = \frac{-1.1 \pm \sqrt{(1.1)^2 - 4(1)(-1439.9)}}{2(1)}$$

$$= \frac{-1.1 \pm \sqrt{5760.81}}{2}$$

$$= 37.4 \text{ or } -3.608 \text{ (disregard)}$$

Its width is 37.4 ft and its height is 38.5 ft.

61.
$$\frac{x-1}{1} = \frac{1}{x}$$
$$x(x-1) = 1$$
$$x^2 - x - 1 = 0$$
$$a = 1, b = -1, c = -1$$

$$x = \frac{1 \pm \sqrt{(-1)^2 - 4(1)(-1)}}{2(1)}$$

$$= \frac{1 \pm \sqrt{5}}{2} \text{ (disregard the negative)}$$

The value is $\frac{1 + \sqrt{5}}{2}$.

63. $h(t) = -16t^2 + 20t + 1100$
$$0 = -16t^2 - 20t + 1100$$
$$a = -16, b = -20, c = 1100$$

$$t = \frac{-20 \pm \sqrt{(20)^2 - 4(-16)(1100)}}{2(-16)}$$

$$= \frac{-20 \pm \sqrt{70,800}}{-32}$$

$$\approx 8.9 \text{ or } -7.7 \text{ (disregard)}$$

It will take about 8.9 seconds.

65. $h(t) = -16t^2 - 20t + 180$
$$0 = -16t^2 - 20t + 180$$
$$a = -16, b = -20, c = 180$$

$$t = \frac{20 \pm \sqrt{(-20)^2 - 4(-16)(180)}}{2(-16)}$$

$$= \frac{20 \pm \sqrt{11,920}}{-32}$$

$$\approx 2.8 \text{ or } -4.0 \text{ (disregard)}$$

It will take about 2.8 seconds.

67.
$$\sqrt{5x - 2} = 3$$
$$\left(\sqrt{5x - 2}\right)^2 = 3^2$$
$$5x - 2 = 9$$
$$5x = 11$$
$$x = \frac{11}{5}$$

69.
$$\frac{1}{x} + \frac{2}{5} = \frac{7}{x}$$
$$5x\left(\frac{1}{x} + \frac{2}{5}\right) = 5x\left(\frac{7}{x}\right)$$
$$5 + 2x = 35$$
$$2x = 30$$
$$x = 15$$

71. $x^4 + x^2 - 20 = (x^2 + 5)(x^2 - 4)$
$$= (x^2 + 5)(x + 2)(x - 2)$$

73. $z^4 - 13z^2 + 36$
$$= (z^2 - 9)(z^2 - 4)$$
$$= (z + 3)(z - 3)(z + 2)(z - 2)$$

75. $2x^2 - 6x + 3 = 0$
$$a = 2, b = -6, c = 3$$

$$x = \frac{6 \pm \sqrt{(-6)^2 - 4(2)(3)}}{2(2)}$$

$$= \frac{6 \pm \sqrt{12}}{4}$$

$$\approx 0.6 \text{ or } 2.4$$

77. From Sunday to Monday

79. Wednesday

81. $f(x) = 3x^2 - 18x + 56$
$f(4) = 3(4)^2 - 18(4) + 56 = 32$
This answers appears to agree with the graph.

83. $f(x) = 112.5x^2 + 498.7x + 5454$

    a. $x = 2002 - 2000 = 2$
        $f(2) = 112.5(2)^2 + 498.7(2) + 5454$
            $= 6901.4$

    Their net income was \$6901.4 million.

    b. $15,000 = 112.5x^2 + 498.7x + 5454$
          $0 = 112.5x^2 + 498.7x - 9546$
      $a = 112.5, b = 498.7, c = -9546$

$$x = \frac{-498.7 \pm \sqrt{(498.7)^2 - 4(112.5)(-9546)}}{2(112.5)}$$

$$= \frac{-498.7 \pm \sqrt{4,544,401.69}}{225}$$

    ≈ 7.26 or -11.69 (disregard)
    Their income will be \$15,000 million in the year 2007.

85. $\dfrac{-b + \sqrt{b^2 - 4ac}}{2a} + \dfrac{-b - \sqrt{b^2 - 4ac}}{2a}$

$$= \frac{-b + \sqrt{b^2 - 4ac} - b - \sqrt{b^2 - 4ac}}{2a}$$

$$= \frac{-2b}{2a}$$

$$= -\frac{b}{a}$$

87. $3x^2 - \sqrt{12}x + 1 = 0$
$a = 3, b = -\sqrt{12}, c = 1$

$$x = \frac{\sqrt{12} \pm \sqrt{\left(-\sqrt{12}\right)^2 - 4(3)(1)}}{2(3)}$$

$$= \frac{\sqrt{12} \pm \sqrt{12 - 12}}{6}$$

$$= \frac{\sqrt{4 \cdot 3} \pm \sqrt{0}}{6} = \frac{2\sqrt{3}}{6} = \frac{\sqrt{3}}{3}$$

The solution is $\dfrac{\sqrt{3}}{3}$.

89. $x^2 + \sqrt{2}x + 1 = 0$
$a = 1, b = \sqrt{2}, c = 1$

$$x = \frac{-\sqrt{2} \pm \sqrt{\left(\sqrt{2}\right)^2 - 4(1)(1)}}{2(1)}$$

$$= \frac{-\sqrt{2} \pm \sqrt{2 - 4}}{2}$$

$$= \frac{-\sqrt{2} \pm \sqrt{-2}}{2}$$

$$= \frac{-\sqrt{2} \pm i\sqrt{2}}{2} = -\frac{\sqrt{2}}{2} \pm \frac{\sqrt{2}}{2}i$$

The solutions are $\dfrac{-\sqrt{2} + i\sqrt{2}}{2}$ and $\dfrac{-\sqrt{2} - i\sqrt{2}}{2}$.

91. $2x^2 - \sqrt{3}x - 1 = 0$
$a = 2, b = -\sqrt{3}, c = -1$

$$x = \frac{\sqrt{3} \pm \sqrt{\left(-\sqrt{3}\right)^2 - 4(2)(-1)}}{2(2)}$$

$$= \frac{\sqrt{3} \pm \sqrt{3 + 8}}{4}$$

$$= \frac{\sqrt{3} \pm \sqrt{11}}{4}$$

The solutions are $\dfrac{\sqrt{3} + \sqrt{11}}{4}$ and $\dfrac{\sqrt{3} - \sqrt{11}}{4}$.

93. Exercise 63:

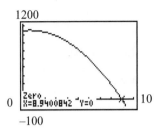

Exercise 65:

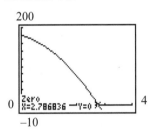

95. $y = 9x - 2x^2 + 5$

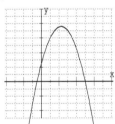

There are two x-intercepts.

Exercise Set 8.3

1.
$$2x = \sqrt{10 + 3x}$$
$$4x^2 = 10 + 3x$$
$$4x^2 - 3x - 10 = 0$$
$$(4x + 5)(x - 2) = 0$$
$$4x + 5 = 0 \quad \text{or } x - 2 = 0$$
$$x = -\frac{5}{4} \quad \text{or} \quad x = 2$$

Discard $-\frac{5}{4}$. The solution is 2.

3.
$$x - 2\sqrt{x} = 8$$
$$x - 8 = 2\sqrt{x}$$
$$(x - 8)^2 = \left(2\sqrt{x}\right)^2$$
$$x^2 - 16x + 64 = 4x$$
$$x^2 - 20x + 64 = 0$$
$$(x - 16)(x - 4) = 0$$
$$x - 16 = 0 \quad \text{or } x - 4 = 0$$
$$x = 16 \quad \text{or} \quad x = 4 \text{ (discard)}$$

The solution is 16.

5.
$$\sqrt{9x} = x + 2$$
$$\left(\sqrt{9x}\right)^2 = (x + 2)^2$$
$$9x = x^2 + 4x + 4$$
$$0 = x^2 - 5x + 4$$
$$0 = (x - 4)(x - 1)$$
$$x - 4 = 0 \quad \text{or } x - 1 = 0$$
$$x = 4 \quad \text{or} \quad x = 1$$

The solutions are 1 and 4.

7. $\dfrac{2}{x} + \dfrac{3}{x-1} = 1$

Multiply each term by $x(x-1)$.
$$2(x - 1) + 3x = x(x - 1)$$
$$2x - 2 + 3x = x^2 - x$$
$$0 = x^2 - 6x + 2$$
$$x = \frac{6 \pm \sqrt{(-6)^2 - 4(1)(2)}}{2(1)}$$
$$= \frac{6 \pm \sqrt{28}}{2}$$
$$= \frac{6 \pm 2\sqrt{7}}{2} = 3 \pm \sqrt{7}$$

The solutions are $3 + \sqrt{7}$ and $3 - \sqrt{7}$.

9. $\dfrac{3}{x} + \dfrac{4}{x+2} = 2$

Multiply each term by $x(x+2)$.

$3(x+2) + 4x = 2x(x+2)$

$3x + 6 + 4x = 2x^2 + 4x$

$0 = 2x^2 - 3x - 6$

$x = \dfrac{3 \pm \sqrt{(-3)^2 - 4(2)(-6)}}{2(2)}$

$= \dfrac{3 \pm \sqrt{57}}{4}$

The solutions are $\dfrac{3+\sqrt{57}}{4}$ and $\dfrac{3-\sqrt{57}}{4}$.

11. $\dfrac{7}{x^2 - 5x + 6} = \dfrac{2x}{x-3} - \dfrac{x}{x-2}$

$\dfrac{7}{(x-3)(x-2)} = \dfrac{2x}{x-3} - \dfrac{x}{x-2}$

Multiply each term by $(x-3)(x-2)$.

$7 = 2x(x-2) - x(x-3)$

$7 = 2x^2 - 4x - x^2 + 3x$

$0 = x^2 - x - 7$

$x = \dfrac{1 \pm \sqrt{(-1)^2 - 4(1)(-7)}}{2(1)}$

$= \dfrac{1 \pm \sqrt{29}}{2}$

The solutions are $\dfrac{1+\sqrt{29}}{2}$ and $\dfrac{1-\sqrt{29}}{2}$.

13. $p^4 - 16 = 0$

$(p^2 - 4)(p^2 + 4) = 0$

$(p+2)(p-2)(p^2 + 4) = 0$

$p + 2 = 0 \quad$ or $p - 2 = 0$ or $p^2 + 4 = 0$

$p = -2$ or $\quad p = 2$ or $\quad p^2 = -4$

$p = \pm\sqrt{-4}$

$p = \pm 2i$

The solutions are -2, 2, $-2i$, and $2i$.

15. $4x^4 + 11x^2 = 3$

$4x^2 + 11x^2 - 3 = 0$

$(4x^2 - 1)(x^2 + 3) = 0$

$(2x+1)(2x-1)(x^2 + 3) = 0$

The solutions are $-\dfrac{1}{2}, \dfrac{1}{2}, -i\sqrt{3},$ and $i\sqrt{3}$.

17. $z^4 - 13z^2 + 36 = 0$

$(z^2 - 9)(z^2 - 4) = 0$

$(z+3)(z-3)(z+2)(z-2) = 0$

$z = -3, z = 3, z = -2, z = 2$

The solutions are -3, 3, -2, and 2.

19. $x^{2/3} - 3x^{1/3} - 10 = 0$

Let $y = x^{1/3}$. Then $y^2 = x^{2/3}$ and

$y^2 - 3y - 10 = 0$

$(y-5)(y+2) = 0$

$y - 5 = 0 \quad$ or $\quad y + 2 = 0$

$y = 5 \quad$ or $\quad\quad y = -2$

$x^{1/3} = 5 \quad$ or $\quad x^{1/3} = -2$

$x = 125$ or $\quad\quad x = -8$

The solutions are -8 and 125.

21. $(5n+1)^2 + 2(5n+1) - 3 = 0$

Let $y = 5n + 1$. Then $y^2 = (5n+1)^2$ and

$y^2 + 2y - 3 = 0$

$(y+3)(y-1) = 0$

$y + 3 = 0 \quad$ or $\quad y - 1 = 0$

$y = -3$ or $\quad\quad y = 1$

$5n + 1 = -3$ or $5n + 1 = 1$

$5n = -4$ or $\quad\quad 5n = 0$

$n = -\dfrac{4}{5} \quad$ or $\quad\quad n = 0$

The solutions are $-\dfrac{4}{5}$ and 0.

23. $2x^{2/3} - 5x^{1/3} = 3$

Let $y = x^{1/3}$. Then $y^2 = x^{2/3}$ and

$$2y^2 - 5y = 3$$
$$2y^2 - 5y - 3 = 0$$
$$(2y+1)(y-3) = 0$$
$$2y+1 = 0 \quad \text{or} \quad y-3 = 0$$
$$y = -\frac{1}{2} \quad \text{or} \quad y = 3$$
$$x^{1/3} = -\frac{1}{2} \quad \text{or} \quad x^{1/3} = 3$$
$$x = -\frac{1}{8} \quad \text{or} \quad x = 27$$

The solutions are $-\frac{1}{8}$ and 27.

25.
$$1 + \frac{2}{3t-2} = \frac{8}{(3t-2)^2}$$
$$(3t-2)^2 + 2(3t-2) = 8$$
$$(3t-2)^2 + 2(3t-2) - 8 = 0$$

Let $y = 3t - 2$. Then $y^2 = (3t-2)^2$ and

$$y^2 + 2y - 8 = 0$$
$$(y+4)(y-2) = 0$$
$$y+4 = 0 \quad \text{or} \quad y-2 = 0$$
$$y = -4 \quad \text{or} \quad y = 2$$
$$3t-2 = -4 \quad \text{or} \quad 3t-2 = 2$$
$$3t = -2 \quad \text{or} \quad 3t = 4$$
$$t = -\frac{2}{3} \quad \text{or} \quad t = \frac{4}{3}$$

The solutions are $-\frac{2}{3}$ and $\frac{4}{3}$.

27. $20x^{2/3} - 6x^{1/3} - 2 = 0$

Let $y = x^{1/3}$. Then $y^2 = x^{2/3}$ and

$$20y^2 - 6y - 2 = 0$$
$$2(10y^2 - 3y - 1) = 0$$
$$2(5y+1)(2y-1) = 0$$

$$5y+1 = 0 \quad \text{or} \quad 2y-1 = 0$$
$$y = -\frac{1}{5} \quad \text{or} \quad y = \frac{1}{2}$$
$$x^{1/3} = -\frac{1}{5} \quad \text{or} \quad x^{1/3} = \frac{1}{2}$$
$$x = -\frac{1}{125} \quad \text{or} \quad x = \frac{1}{8}$$

The solutions are $\frac{1}{8}$ and $-\frac{1}{125}$.

29. $a^4 - 5a^2 + 6 = 0$
$$(a^2 - 3)(a^2 - 2) = 0$$
$$a^2 - 3 = 0 \quad \text{or} \quad a^2 - 2 = 0$$
$$a^2 = 3 \quad \text{or} \quad a^2 = 2$$
$$a = \pm\sqrt{3} \quad \text{or} \quad a = \pm\sqrt{2}$$

The solutions are $-\sqrt{3}, \sqrt{3}, -\sqrt{2}$, and $\sqrt{2}$.

31. $\dfrac{2x}{x-2} + \dfrac{x}{x+3} = -\dfrac{5}{x+3}$

Multiply each term by $(x+3)(x-2)$.

$$2x(x+3) + x(x-2) = -5(x-2)$$
$$2x^2 + 6x + x^2 - 2x = -5x + 10$$
$$3x^2 + 9x - 10 = 0$$
$$x = \frac{-9 \pm \sqrt{(9)^2 - 4(3)(-10)}}{2(3)}$$
$$= \frac{-9 \pm \sqrt{201}}{6}$$

The solutions are $\dfrac{-9 + \sqrt{201}}{6}$ and $\dfrac{-9 - \sqrt{201}}{6}$.

33.
$$(p+2)^2 = 9(p+2) - 20$$
$$(p+2)^2 - 9(p+2) + 20 = 0$$

Let $x = p + 2$. Then $x^2 = (p+2)^2$ and

$$x^2 - 9x + 20 = 0$$
$$(x-5)(x-4) = 0$$
$$x = 5 \quad \text{or} \quad x = 4$$
$$p+2 = 5 \quad \text{or} \quad p+2 = 4$$
$$p = 3 \quad \text{or} \quad p = 2$$

The solutions are 2 and 3.

35.
$$2x = \sqrt{11x+3}$$
$$(2x)^2 = \left(\sqrt{11x+3}\right)^2$$
$$4x^2 = 11x+3$$
$$4x^2 - 11x - 3 = 0$$
$$(4x+1)(x-3) = 0$$
$$x = -\frac{1}{4} \text{ (discard)} \text{ or } x = 3$$
The solution is 3.

37. $x^{2/3} - 8x^{1/3} + 15 = 0$
Let $y = x^{1/3}$. Then $y^2 = x^{2/3}$ and
$$y^2 - 8y + 15 = 0$$
$$(y-5)(y-3) = 0$$
$$y = 5 \quad \text{ or } \quad y = 3$$
$$x^{1/3} = 5 \quad \text{ or } \quad x^{1/3} = 3$$
$$x = 125 \text{ or } \quad x = 27$$
The solutions are 27 and 125.

39.
$$y^3 + 9y - y^2 - 9 = 0$$
$$y(y^2+9) - 1(y^2+9) = 0$$
$$(y^2+9)(y-1) = 0$$

$$y^2 + 9 = 0 \qquad \text{or } y - 1 = 0$$
$$y^2 = -9 \qquad \text{or} \qquad y = 1$$
$$y = \pm\sqrt{-9}$$
$$y = \pm 3i$$
The solutions are 1, $-3i$, and $3i$.

41. $2x^{2/3} + 3x^{1/3} - 2 = 0$
Let $y = x^{1/3}$. Then $y^2 = x^{2/3}$ and
$$2y^2 + 3y - 2 = 0$$
$$(2y-1)(y+2) = 0$$
$$y = \frac{1}{2} \text{ or } \quad y = -2$$
$$x^{1/3} = \frac{1}{2} \text{ or } x^{1/3} = -2$$
$$x = \frac{1}{8} \text{ or } \quad x = -8$$

The solutions are -8 and $\frac{1}{8}$.

43. $x^{-2} - x^{-1} - 6 = 0$
Let $y = x^{-1}$. Then $y^2 = x^{-2}$ and
$$y^2 - y - 6 = 0$$
$$(y-3)(y+2) = 0$$
$$y = 3 \quad \text{ or } \qquad y = -2$$
$$x^{-1} = 3 \quad \text{ or } \quad x^{-1} = -2$$
$$\frac{1}{x} = 3 \quad \text{ or } \qquad \frac{1}{x} = -2$$
$$x = \frac{1}{3} \quad \text{ or } \qquad x = -\frac{1}{2}$$

The solutions are $-\frac{1}{2}$ and $\frac{1}{3}$.

45.
$$x - \sqrt{x} = 2$$
$$x - 2 = \sqrt{x}$$
$$(x-2)^2 = x$$
$$x^2 - 4x + 4 = x$$
$$x^2 - 5x + 4 = 0$$
$$(x-4)(x-1) = 0$$
$$x = 4 \text{ or } x = 1 \text{ (discard)}$$
The solution is 4.

47.
$$\frac{x}{x-1} + \frac{1}{x+1} = \frac{2}{x^2-1}$$
$$\frac{x}{x-1} + \frac{1}{x+1} = \frac{2}{(x+1)(x-1)}$$
$$x(x+1) + (x-1) = 2$$
$$x^2 + x + x - 1 = 2$$
$$x^2 + 2x - 3 = 0$$
$$(x+3)(x-1) = 0$$
$$x = -3 \text{ or } x = 1 \text{ (discard)}$$
The solution is -3.

49.
$$p^4 - p^2 - 20 = 0$$
$$(p^2-5)(p^2+4) = 0$$
$$p^2 - 5 = 0 \qquad \text{or } p^2 + 4 = 0$$
$$p^2 = 5 \qquad \text{or} \qquad p^2 = -4$$
$$p = \pm\sqrt{5} \quad \text{or} \qquad p = \pm 2i$$
The solutions are $-\sqrt{5}, \sqrt{5}, -2i,$ and $2i$.

51.
$$2x^3 = -54$$
$$x^3 = -27$$
$$x^3 + 27 = 0$$
$$(x+3)(x^2 - 3x + 9) = 0$$
$$x + 3 = 0 \quad \text{or} \quad x^2 - 3x + 9 = 0$$
$$x = -3 \quad \text{or}$$
$$x = \frac{3 \pm \sqrt{(-3)^2 - 4(1)(9)}}{2(1)}$$
$$= \frac{3 \pm \sqrt{-27}}{2}$$
$$= \frac{3 \pm 3i\sqrt{3}}{2} = \frac{3}{2} \pm \frac{3\sqrt{3}}{2} i$$

The solutions are -3, $\dfrac{3 + 3i\sqrt{3}}{2}$, and $\dfrac{3 - 3i\sqrt{3}}{2}$.

53.
$$1 = \frac{4}{x-7} + \frac{5}{(x-7)^2}$$
$$(x-7)^2 - 4(x-7) - 5 = 0$$
Let $y = x - 7$. Then $y^2 = (x-7)^2$ and
$$y^2 - 4y - 5 = 0$$
$$(y-5)(y+1) = 0$$
$$y = 5 \quad \text{or} \quad y = -1$$
$$x - 7 = 5 \quad \text{or} \quad x - 7 = -1$$
$$x = 12 \quad \text{or} \quad x = 6$$
The solutions are 6 and 12.

55.
$$27y^4 + 15y^2 = 2$$
$$27y^4 + 15y^2 - 2 = 0$$
$$(9y^2 - 1)(3y^2 + 2) = 0$$
$$(3y+1)(3y-1)(3y^2 + 2) = 0$$
$$y = -\frac{1}{3} \quad \text{or} \quad y = \frac{1}{3} \quad \text{or} \quad y^2 = -\frac{2}{3}$$
$$y = \pm\sqrt{-\frac{2}{3}}$$
$$y = \pm\frac{\sqrt{6}}{3} i$$
The solutions are $-\dfrac{1}{3}, \dfrac{1}{3}, -\dfrac{\sqrt{6}}{3}i$, and $\dfrac{\sqrt{6}}{3}i$.

57. Let x = speed on the first part. Then $x - 1$ = speed on the second part.
$$d = rt \implies t = \frac{d}{r}$$
$$t_{\text{on first part}} + t_{\text{on second part}} = 1\frac{3}{5}$$
$$\frac{3}{x} + \frac{4}{x-1} = \frac{8}{5}$$
$$3 \cdot 5(x-1) + 4 \cdot 5x = 8x(x-1)$$
$$15x - 15 + 20x = 8x^2 - 8x$$
$$0 = 8x^2 - 43x + 15$$
$$0 = (8x - 3)(x - 5)$$
$$8x - 3 = 0 \quad \text{or} \quad x - 5 = 0$$
$$x = \frac{3}{8} \quad \text{or} \quad x = 5$$
$$x - 1 = 4$$

Discard $\dfrac{3}{8}$. Her speeds were 5 mph and 4 mph.

59. Let x = time for hose alone. Then $x - 1$ = time for the inlet pipe alone.
$$\frac{1}{x} + \frac{1}{x-1} = \frac{1}{8}$$
$$8(x-1) + 8x = x(x-1)$$
$$8x - 8 + 8x = x^2 - x$$
$$0 = x^2 - 17x + 8$$
$$x = \frac{17 \pm \sqrt{(-17)^2 - 4(1)(8)}}{2(1)}$$
$$= \frac{17 \pm \sqrt{257}}{2}$$
$$x \approx 0.5 \text{ (discard)} \quad \text{or} \quad x \approx 16.5$$
$$x - 1 \approx 15.5$$

Hose: 16.5 hrs; Inlet pipe: 15.5 hrs

61. Let x = original speed. Then
$x + 11$ = return speed.

$$d = rt \implies t = \frac{d}{r}$$

$$t_{\text{return}} = t_{\text{original}} - 1$$
$$\frac{330}{x+11} = \frac{330}{x} - 1$$
$$330x = 330(x+11) - x(x+11)$$
$$330x = 330x + 3630 - x^2 - 11x$$
$$x^2 + 11x - 3630 = 0$$
$$x = \frac{-11 \pm \sqrt{(11)^2 - 4(1)(-3630)}}{2(1)}$$
$$= \frac{-11 \pm \sqrt{14,641}}{2}$$
$$= \frac{-11 \pm 121}{2} = 55 \text{ or } -66 \text{ (disregard)}$$
$$x + 11 = 55 + 11 = 66$$
Original speed: 55 mph
Return speed: 66 mph

63. Let x = time for son alone. Then
$x - 1$ = time for dad alone.

$$\frac{1}{x} + \frac{1}{x-1} = \frac{1}{4}$$
$$4(x-1) + 4x = x(x-1)$$
$$4x - 4 + 4x = x^2 - x$$
$$0 = x^2 - 9x + 4$$
$$x = \frac{9 \pm \sqrt{(-9)^2 - 4(1)(4)}}{2(1)}$$
$$= \frac{9 \pm \sqrt{65}}{2}$$
$$\approx 0.5 \text{ (discard) or } 8.5$$
It takes his son about 8.5 hours.

65. Let x = the number.
$$x(x-4) = 96$$
$$x^2 - 4x - 96 = 0$$
$$(x-12)(x+8) = 0$$
$$x = 12 \text{ or } x = -8$$
The number is 12 or –8.

67. a. length $= x - 3 - 3 = x - 6$

b. $V = lwh$
$$300 = (x-6)(x-6) \cdot 3$$

c. $300 = 3(x-6)^2$
$$100 = x^2 - 12x + 36$$
$$0 = x^2 - 12x - 64$$
$$0 = (x-16)(x+4)$$
$$x = 16 \text{ or } x = -4 \text{ (discard)}$$
The sheet is 16 cm by 16 cm.
Check: $V = 3(x-6)(x-6)$
$$= 3(16-6)(16-6)$$
$$= 3(10)(10)$$
$$= 300 \text{ cubic cm}$$

69. Let x = length of the side of the square.
Area $= x^2$
$$920 = x^2$$
$$\sqrt{920} = x$$
Adding another radial line to a different corner would yield a right triangle with legs r and hypotenuse x.
$$r^2 + r^2 = x^2$$
$$2r^2 = \left(\sqrt{920}\right)^2$$
$$2r^2 = 920$$
$$r^2 = 460$$
$$r = \pm\sqrt{460} = \pm 21.4476$$
Disregard the negative. The smallest radius would be 22 feet.

71. $\dfrac{5x}{3} + 2 \le 7$
$$\frac{5x}{3} \le 5$$
$$5x \le 15$$
$$x \le 3$$
$$(-\infty, 3]$$

73.
$$\frac{y-1}{15} > -\frac{2}{5}$$
$$15\left(\frac{y-1}{15}\right) > 15\left(-\frac{2}{5}\right)$$
$$y-1 > -6$$
$$y > -5$$
$$(-5, \infty)$$

75. Domain: $\{x \mid x$ is a real number$\}$ or $(-\infty, \infty)$
Range: $\{y \mid y$ is a real number$\}$ or $(-\infty, \infty)$
It is a function.

77. Domain: $\{x \mid x$ is a real number$\}$ or $(-\infty, \infty)$
Range: $\{y \mid y \geq -1\}$ or $[-1, \infty)$
It is a function.

79. Answers may vary.

81. a. Let $x =$ Dominguez's fastest lap speed
and $x + 0.88 =$ Fernandez's fastest lap speed.

Using $t = \frac{d}{r}$, we have

$$t_{\text{Dominguez}} = t_{\text{Fernandez}} + 0.38$$
$$\frac{7920}{x} = \frac{7920}{x + 0.88} + 0.38$$
$$7920(x + 0.88) = 7920x + 0.38x(x + 0.88)$$
$$7920x + 6969.6 = 7920x + 0.38x^2 + 0.3344x$$
$$0 = 0.38x^2 + 0.3344x - 6969.6$$
$$x = \frac{-0.3344 \pm \sqrt{(0.3344)^2 - 4(0.38)(-6969.6)}}{2(0.38)}$$

Using the positive square root, $x \approx 135.0$ feet per second.

b. $x + 0.88 = 135.0 + 0.88$
$= 135.9$ feet per second

c. 5280 ft = 1 mile, and 3600 sec = 1 hr.

Dominguez: $\dfrac{135 \text{ ft}}{\text{sec}} \cdot \dfrac{3600 \text{ sec}}{\text{hr}} \cdot \dfrac{1 \text{ mile}}{5280 \text{ ft}} \approx 92.0$ mph

Fernandez: $\dfrac{135.9 \text{ ft}}{\text{sec}} \cdot \dfrac{3600 \text{ sec}}{\text{hr}} \cdot \dfrac{1 \text{ mile}}{5280 \text{ ft}} \approx 92.7$ mph

Integrated Review

1. $x^2 - 10 = 0$

$\quad x^2 = 10$

$\quad\quad x = \pm\sqrt{10}$

2. $x^2 - 14 = 0$

$\quad x^2 = 14$

$\quad\quad x = \pm\sqrt{14}$

3. $(x-1)^2 = 8$

$\quad x - 1 = \pm\sqrt{8}$

$\quad x - 1 = \pm 2\sqrt{2}$

$\quad\quad x = 1 \pm 2\sqrt{2}$

4. $(x+5)^2 = 12$

$\quad x + 5 = \pm\sqrt{12}$

$\quad x + 5 = \pm 2\sqrt{3}$

$\quad\quad x = -5 \pm 2\sqrt{3}$

5. $x^2 + 2x - 12 = 0$

$\quad x^2 + 2x + \left(\dfrac{2}{2}\right)^2 = 12 + 1$

$\quad x^2 + 2x + 1 = 13$

$\quad (x+1)^2 = 13$

$\quad\quad x + 1 = \pm\sqrt{13}$

$\quad\quad\quad x = -1 \pm \sqrt{13}$

6. $x^2 - 12x + 11 = 0$

$\quad x^2 - 12x + \left(\dfrac{-12}{2}\right)^2 = -11 + 36$

$\quad x^2 - 12x + 36 = 25$

$\quad (x-6)^2 = \pm\sqrt{25}$

$\quad\quad x - 6 = \pm 5$

$\quad\quad\quad x = 6 \pm 5$

$\quad\quad\quad x = 1 \text{ or } x = 11$

7. $3x^2 + 3x = 5$

$\quad x^2 + x = \dfrac{5}{3}$

$\quad x^2 + x + \left(\dfrac{1}{2}\right)^2 = \dfrac{5}{3} + \dfrac{1}{4}$

$\quad x^2 + x + \dfrac{1}{4} = \dfrac{23}{12}$

$\quad \left(x + \dfrac{1}{2}\right)^2 = \dfrac{23}{12}$

$\quad x + \dfrac{1}{2} = \pm\sqrt{\dfrac{23}{12}}$

$\quad x + \dfrac{1}{2} = \pm\dfrac{\sqrt{23}}{2\sqrt{3}}$

$\quad x + \dfrac{1}{2} = \pm\dfrac{\sqrt{23}\cdot\sqrt{3}}{2\sqrt{3}\cdot\sqrt{3}}$

$\quad x + \dfrac{1}{2} = \pm\dfrac{\sqrt{69}}{6}$

$\quad x = -\dfrac{1}{2} \pm \dfrac{\sqrt{69}}{6} = \dfrac{-3 \pm \sqrt{69}}{6}$

8. $16y^2 + 16y = 1$

$\quad y^2 + y = \dfrac{1}{16}$

$\quad y^2 + y + \left(\dfrac{1}{2}\right)^2 = \dfrac{1}{16} + \dfrac{1}{4}$

$\quad y^2 + y + \dfrac{1}{4} = \dfrac{5}{16}$

$\quad \left(y + \dfrac{1}{2}\right)^2 = \dfrac{5}{16}$

$\quad y + \dfrac{1}{2} = \pm\sqrt{\dfrac{5}{16}}$

$\quad y + \dfrac{1}{2} = \pm\dfrac{\sqrt{5}}{4}$

$\quad y = -\dfrac{1}{2} \pm \dfrac{\sqrt{5}}{4} = \dfrac{-2 \pm \sqrt{5}}{4}$

9. $2x^2 - 4x + 1 = 0$

$a = 2, b = -4, c = 1$

$x = \dfrac{4 \pm \sqrt{(-4)^2 - 4(2)(1)}}{2(2)}$

$= \dfrac{4 \pm \sqrt{8}}{4}$

$= \dfrac{4 \pm 2\sqrt{2}}{4} = \dfrac{2 \pm \sqrt{2}}{2}$

10. $\dfrac{1}{2}x^2 + 3x + 2 = 0$

$x^2 + 6x + 4 = 0$

$a = 1, b = 6, c = 4$

$x = \dfrac{-6 \pm \sqrt{(6)^2 - 4(1)(4)}}{2(1)}$

$= \dfrac{-6 \pm \sqrt{20}}{2}$

$= \dfrac{-6 \pm 2\sqrt{5}}{2} = -3 \pm \sqrt{5}$

11. $\quad x^2 + 4x = -7$

$x^2 + 4x + 7 = 0$

$a = 1, b = 4, c = 7$

$x = \dfrac{-4 \pm \sqrt{(4)^2 - 4(1)(7)}}{2(1)}$

$= \dfrac{-4 \pm \sqrt{-12}}{2}$

$= \dfrac{-4 \pm i\sqrt{4 \cdot 3}}{2}$

$= \dfrac{-4 \pm 2i\sqrt{3}}{2} = -2 \pm i\sqrt{3}$

12. $\quad x^2 + x = -3$

$x^2 + x + 3 = 0$

$a = 1, b = 1, c = 3$

$x = \dfrac{-1 \pm \sqrt{(1)^2 - 4(1)(3)}}{2(1)}$

$= \dfrac{-1 \pm \sqrt{-11}}{2}$

$= \dfrac{-1 \pm i\sqrt{11}}{2}$ or $-\dfrac{1}{2} \pm \dfrac{\sqrt{11}}{2}i$

13. $x^2 + 3x + 6 = 0$

$a = 1, b = 3, c = 6$

$x = \dfrac{-3 \pm \sqrt{(3)^2 - 4(1)(6)}}{2(1)}$

$= \dfrac{-3 \pm \sqrt{-15}}{2}$

$= \dfrac{-3 \pm i\sqrt{15}}{2}$ or $-\dfrac{3}{2} \pm \dfrac{\sqrt{15}}{2}i$

14. $2x^2 + 18 = 0$

$2x^2 = -18$

$x^2 = -9$

$x = \pm\sqrt{-9}$

$x = \pm 3i$

15. $\quad x^2 + 17x = 0$

$x(x + 17) = 0$

$x = 0$ or $x + 17 = 0$

$\qquad\qquad\qquad x = -17$

$x = 0, -17$

16. $4x^2 - 2x - 3 = 0$

$a = 4, b = -2, c = -3$

$x = \dfrac{2 \pm \sqrt{(-2)^2 - 4(4)(-3)}}{2(4)}$

$= \dfrac{2 \pm \sqrt{52}}{8}$

$= \dfrac{2 \pm 2\sqrt{13}}{8} = \dfrac{1 \pm \sqrt{13}}{4}$

17. $(x - 2)^2 = 27$

$x - 2 = \pm\sqrt{27}$

$x - 2 = \pm 3\sqrt{3}$

$x = 2 \pm 3\sqrt{3}$

18. $\dfrac{1}{2}x^2 - 2x + \dfrac{1}{2} = 0$

$x^2 - 4x + 1 = 0$

$x^2 - 4x + \left(\dfrac{-4}{2}\right)^2 = -1 + 4$

$x^2 - 4x + 4 = 3$

$(x-2)^2 = 3$

$x - 2 = \pm\sqrt{3}$

$x = 2 \pm \sqrt{3}$

19. $3x^2 + 2x = 8$

$3x^2 + 2x - 8 = 0$

$(3x - 4)(x + 2) = 0$

$3x - 4 = 0$ or $x + 2 = 0$

$x = \dfrac{4}{3}$ or $x = -2$

20. $2x^2 = -5x - 1$

$2x^2 + 5x + 1 = 0$

$a = 2, b = 5, c = 1$

$x = \dfrac{-5 \pm \sqrt{(5)^2 - 4(2)(1)}}{2(2)}$

$= \dfrac{-5 \pm \sqrt{17}}{4}$

21. $x(x - 2) = 5$

$x^2 - 2x = 5$

$x^2 - 2x + \left(\dfrac{-2}{2}\right)^2 = 5 + 1$

$x^2 - 2x + 1 = 6$

$(x - 1)^2 = 6$

$x - 1 = \pm\sqrt{6}$

$x = 1 \pm \sqrt{6}$

22. $x^2 - 31 = 0$

$x^2 = 31$

$x = \pm\sqrt{31}$

23. $5x^2 - 55 = 0$

$5x^2 = 55$

$x^2 = 11$

$x = \pm\sqrt{11}$

24. $5x^2 + 55 = 0$

$5x^2 = -55$

$x^2 = -11$

$x = \pm\sqrt{-11}$

$x = \pm i\sqrt{11}$

25. $x(x + 5) = 66$

$x^2 + 5x = 66$

$x^2 + 5x - 66 = 0$

$(x + 11)(x - 6) = 0$

$x + 11 = 0$ or $x - 6 = 0$

$x = -11$ or $x = 6$

26. $5x^2 + 6x - 2 = 0$

$a = 5, b = 6, c = -2$

$x = \dfrac{-6 \pm \sqrt{(6)^2 - 4(5)(-2)}}{2(5)}$

$= \dfrac{-6 \pm \sqrt{76}}{10}$

$= \dfrac{-6 \pm \sqrt{4 \cdot 19}}{10}$

$= \dfrac{-6 \pm 2\sqrt{19}}{10} = \dfrac{-3 \pm \sqrt{19}}{5}$

27. $2x^2 + 3x = 1$

$2x^2 + 3x - 1 = 0$

$a = 2, b = 3, c = -1$

$x = \dfrac{-3 \pm \sqrt{(3)^2 - 4(2)(-1)}}{2(2)}$

$= \dfrac{-3 \pm \sqrt{17}}{4}$

28. $a^2 + b^2 = c^2$

$$x^2 + x^2 = 20^2$$
$$2x^2 = 400$$
$$x^2 = 200$$
$$x = \pm\sqrt{200}$$
$$= \pm 10\sqrt{2} \approx 14.1421$$

Disregard the negative. A side of the room is $10\sqrt{2}$ feet ≈ 14.1 feet.

29. Let x = time for Jack alone. Then $x - 2$ = time for Lucy alone.

$$\frac{1}{x} + \frac{1}{x-2} = \frac{1}{4}$$
$$4(x-2) + 4x = x(x-2)$$
$$4x - 8 + 4x = x^2 - 2x$$
$$0 = x^2 - 10x + 8$$
$$x = \frac{10 \pm \sqrt{(-10)^2 - 4(1)(8)}}{2(1)}$$
$$= \frac{10 \pm \sqrt{68}}{2} \approx 9.1 \text{ or } 0.9 \text{ (disregard)}$$
$$x - 2 = 9.1 - 2 = 7.1$$

It would take Jack 9.1 hours and Lucy 7.1 hours.

30. Let x = speed on treadmill. Then $x + 1$ = speed running.

$$t_{\text{treadmill}} + t_{\text{running}} = \frac{4}{3}$$
$$\frac{5}{x} + \frac{2}{x+1} = \frac{4}{3}$$
$$5 \cdot 3(x+1) + 2 \cdot 3x = 4x(x+1)$$
$$15x + 15 + 6x = 4x^2 + 4x$$
$$0 = 4x^2 - 17x - 15$$
$$0 = (4x + 3)(x - 5)$$

$x = -\dfrac{4}{3}$ (disregard) or $x = 5$

$x + 1 = 5 + 1 = 6$

Treadmill: 5 mph
Running: 6 mph

Exercise Set 8.4

1. $(x+1)(x+5) > 0$

$x + 1 = 0 \quad$ or $\quad x + 5 = 0$

$x = -1 \quad$ or $\quad x = -5$

Region	Test Point	$(x+1)(x+5) > 0$ Result
A: $(-\infty, -5)$	-6	$(-5)(-11) > 0$ True
B: $(-5, -1)$	-2	$(-1)(4) > 0$ False
C: $(-1, \infty)$	0	$(1)(5) > 0$ True

Solution: $(-\infty, -5) \cup (-1, \infty)$

3. $(x-3)(x+4) \le 0$

$x - 3 = 0 \quad$ or $\quad x + 4 = 0$

$x = 3 \quad$ or $\quad x = -4$

Region	Test Point	$(x-3)(x+4) \le 0$ Result
A: $(-\infty, -4)$	-5	$(-8)(-1) \le 0$ False
B: $(-4, 3)$	0	$(-3)(4) \le 0$ True
C: $(3, \infty)$	4	$(1)(9) \le 0$ False

Solution: $[-4, 3]$

5. $x^2 - 7x + 10 \leq 0$
$(x-5)(x-2) \leq 0$
$x - 5 = 0$ or $x - 2 = 0$
        $x = 5$ or $x = 2$

Region	Test Point	$(x-5)(x-2) \leq 0$ Result
A: $(-\infty, 2)$	0	$(-5)(-2) \leq 0$ False
B: $(2, 5)$	3	$(-2)(1) \leq 0$ True
C: $(5, \infty)$	6	$(1)(4) \leq 0$ False

Solution: $[2, 5]$

7. $3x^2 + 16x < -5$
        $3x^2 + 16x + 5 < 0$
        $(3x+1)(x+5) < 0$
        $3x + 1 = 0$ or $x + 5 = 0$
        $x = -\dfrac{1}{3}$ or $x = -5$

Region	Test Point	$(3x+1)(x+5) < 0$ Result
A: $(-\infty, -5)$	-6	$(-17)(-1) < 0$ False
B: $\left(-5, -\dfrac{1}{3}\right)$	-1	$(-2)(4) < 0$ True
C: $\left(-\dfrac{1}{3}, \infty\right)$	0	$(1)(5) < 0$ False

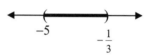

Solution: $\left(-5, -\dfrac{1}{3}\right)$

9. $(x-6)(x-4)(x-2) > 0$
        $x - 6 = 0$ or $x - 4 = 0$ or $x - 2 = 0$
        $x = 6$ or $x = 4$ or $x = 2$

Region	Test Point	$(x-6)(x-4)(x-2) > 0$ Result
A: $(-\infty, 2)$	0	$(-6)(-4)\,(-2) > 0$ False
B: $(2, 4)$	3	$(-3)(-1)\,(1) > 0$ True
C: $(4, 6)$	5	$(-1)(1)(3) > 0$ False
D: $(6, \infty)$	7	$(1)(3)(5) > 0$ True

Solution: $(2, 4) \cup (6, \infty)$

328

11. $x(x-1)(x+4) \le 0$

$x = 0$ or $x - 1 = 0$ or $x + 4 = 0$

$x = 1$ or $x = -4$

Region	Test Point	$x(x-1)(x+4) \le 0$ Result
A: $(-\infty, -4)$	-5	$-5(-6)\,(-1) \le 0$ True
B: $(-4, 0)$	-1	$-1(-2)\,(3) \le 0$ False
C: $(0, 1)$	$\dfrac{1}{2}$	$\dfrac{1}{2}\left(-\dfrac{1}{2}\right)\left(\dfrac{7}{2}\right) \le 0$ True
D: $(1, \infty)$	2	$2(1)(6) \le 0$ False

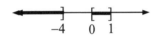

$$-4 \quad 0 \quad 1$$

Solution: $(-\infty, -4] \cup [0, 1]$

13.
$$(x^2 - 9)(x^2 - 4) > 0$$
$$(x+3)(x-3)(x+2)(x-2) > 0$$
$x + 3 = 0$ or $x - 3 = 0$ or $x + 2 = 0$ or $x - 2 = 0$

$x = -3$ or $x = 3$ or $x = -2$ or $x = 2$

Region	Test Point	$(x+3)(x-3)(x+2)(x-2) > 0$ Result
A: $(-\infty, -3)$	-4	$(-1)(-7)\,(-2)(-6) > 0$ True
B: $(-3, -2)$	$-\dfrac{5}{2}$	$\left(\dfrac{1}{2}\right)\left(-\dfrac{11}{2}\right)\left(-\dfrac{1}{2}\right)\left(-\dfrac{9}{2}\right) > 0$ False
C: $(-2, 2)$	0	$(3)(-3)(2)(-2) > 0$ True
D: $(2, 3)$	$\dfrac{5}{2}$	$\left(\dfrac{11}{2}\right)\left(-\dfrac{1}{2}\right)\left(\dfrac{9}{2}\right)\left(\dfrac{1}{2}\right) > 0$ False
E: $(3, \infty)$	4	$(7)(1)(6)(2) > 0$ True

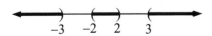

$$-3 \quad -2 \quad 2 \quad 3$$

Solution: $(-\infty, -3) \cup (-2, 2) \cup (3, \infty)$

15. $\dfrac{x+7}{x-2} < 0$

$x + 7 = 0$ or $x - 2 = 0$

$\quad x = -7$ or $x = 2$

Region	Test Point	$\dfrac{x+7}{x-2} < 0$; Result
A: $(-\infty, -7)$	-8	$\dfrac{-1}{-10} < 0$; False
B: $(-7, 2)$	0	$\dfrac{7}{-2} < 0$; True
C: $(2, \infty)$	3	$\dfrac{10}{1} < 0$; False

Solution: $(-7, 2)$

17. $\dfrac{5}{x+1} > 0$

$x + 1 = 0$

$\quad x = -1$

Region	Test Point	$\dfrac{5}{x+1} > 0$; Result
A: $(-\infty, -1)$	-2	$\dfrac{5}{-1} > 0$; False
B: $(-1, \infty)$	0	$\dfrac{5}{1} > 0$; True

Solution: $(-1, \infty)$

19. $\dfrac{x+1}{x-4} \ge 0$

$x + 1 = 0$ or $x - 4 = 0$

$\quad x = -1$ or $x = 4$

Region	Test Point	$\dfrac{x+1}{x-4} \ge 0$; Result
A: $(-\infty, -1)$	-2	$\dfrac{-1}{-6} \ge 0$; True
B: $(-1, 4)$	0	$\dfrac{1}{-4} \ge 0$; False
C: $(4, \infty)$	5	$\dfrac{6}{1} \ge 0$; True

Solution: $(-\infty, -1] \cup (4, \infty)$

21. $\dfrac{3}{x-2} < 4$

The denominator is equal to 0 when

$x - 2 = 0$, or $x = 2$.

$\dfrac{3}{x-2} = 4$

$\quad 3 = 4x - 8$

$\quad 11 = 4x$

$\quad \dfrac{11}{4} = x$

Region	Test Point	$\dfrac{3}{x-2} < 4$; Result
A: $(-\infty, 2)$	0	$\dfrac{3}{-2} < 4$; True
B: $\left(2, \dfrac{11}{4}\right)$	$\dfrac{5}{2}$	$\dfrac{3}{\frac{1}{2}} = 6 < 4$; False
C: $\left(\dfrac{11}{4}, \infty\right)$	4	$\dfrac{3}{2} < 4$; True

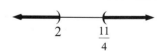

Solution: $(-\infty, 2) \cup \left(\dfrac{11}{4}, \infty\right)$

23. $\dfrac{x^2+6}{5x} \ge 1$

The denominator is equal to 0 when
$5x = 0$, or $x = 0$.

$$\dfrac{x^2+6}{5x} = 1$$
$$x^2 + 6 = 5x$$
$$x^2 - 5x + 6 = 0$$
$$(x-2)(x-3) = 0$$
$$x - 2 = 0 \text{ or } x - 3 = 0$$
$$x = 2 \text{ or } \qquad x = 3$$

Region	Test Point	$\dfrac{x^2+6}{5x} \ge 1$; Result
A: $(-\infty, 0)$	-1	$\dfrac{7}{-5} \ge 1$; F
B: $(0, 2)$	1	$\dfrac{7}{5} \ge 1$; T
C: $(2, 3)$	$\dfrac{5}{2}$	$\dfrac{49/4}{25/2} = \dfrac{49}{50} \ge 1$; F
D: $(3, \infty)$	4	$\dfrac{22}{20} \ge 1$; T

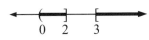

Solution: $(0, 2] \cup [3, \infty)$

25. $(x-8)(x+7) > 0$
$x - 8 = 0 \text{ or } x + 7 = 0$
$\qquad x = 8 \text{ or } \qquad x = -7$

Region	Test Point	$(x-8)(x+7) > 0$ Result
A: $(-\infty, -7)$	-8	$(-16)(-1) > 0$; T
B: $(-7, 8)$	0	$(-8)(7) > 0$; F
C: $(8, \infty)$	9	$(1)(16) > 0$; T

Solution: $(-\infty, -7) \cup (8, \infty)$

27. $(2x-3)(4x+5) \le 0$
$2x - 3 = 0 \text{ or } 4x + 5 = 0$
$\quad x = \dfrac{3}{2} \text{ or } \qquad x = -\dfrac{5}{4}$

Region	Test Point	$(2x-3)(4x+5) \le 0$ Result
A: $\left(-\infty, -\dfrac{5}{4}\right)$	-2	$(-7)(-3) \le 0$ False
B: $\left(-\dfrac{5}{4}, \dfrac{3}{2}\right)$	0	$(-3)(5) \le 0$ True
C: $\left(\dfrac{3}{2}, \infty\right)$	2	$(1)(13) \le 0$ False

Solution: $\left[-\dfrac{5}{4}, \dfrac{3}{2}\right]$

29. $\qquad x^2 > x$
$\quad x^2 - x > 0$
$x(x-1) > 0$
$x = 0 \text{ or } x - 1 = 0$
$\qquad\qquad x = 1$

Region	Test Point	$x(x-1) > 0$; Result
A: $(-\infty, 0)$	-1	$-1(-2) > 0$; True
B: $(0, 1)$	$\dfrac{1}{2}$	$\dfrac{1}{2}\left(-\dfrac{1}{2}\right) > 0$; False
C: $(1, \infty)$	2	$2(1) > 0$; True

Solution: $(-\infty, 0) \cup (1, \infty)$

31. $(2x-8)(x+4)(x-6) \le 0$

$2x-8=0$ or $x+4=0$ or $x-6=0$

   $x=4$ $x=-4$ or $x=6$

Region	Test Point	$(2x-8)(x+4)(x-6) \le 0$ Result
A: $(-\infty, -4)$	-5	$(-18)(-1)\,(-11) \le 0$ True
B: $(-4, 4)$	0	$(-8)(4)\,(-6) \le 0$ False
C: $(4, 6)$	5	$(2)(9)(-1) \le 0$ True
D: $(6, \infty)$	7	$(6)(11)(1) \le 0$ False

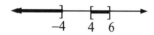

Solution: $(-\infty, -4] \cup [4, 6]$

33. $6x^2 - 5x \ge 6$

      $6x^2 - 5x - 6 \ge 0$

    $(3x+2)(2x-3) \ge 0$

    $3x+2=0$ or $2x-3=0$

      $x = -\dfrac{2}{3}$ or $x = \dfrac{3}{2}$

Region	Test Point	$(3x+2)(2x-3) \ge 0$ Result
A: $\left(-\infty, -\dfrac{2}{3}\right)$	-1	$(-1)(-5) \ge 0$ True
B: $\left(-\dfrac{2}{3}, \dfrac{3}{2}\right)$	0	$(2)(-3) \ge 0$ False
C: $\left(\dfrac{3}{2}, \infty\right)$	2	$(8)(1) \ge 0$ True

Solution: $\left(-\infty, -\dfrac{2}{3}\right] \cup \left[\dfrac{3}{2}, \infty\right)$

35. $4x^3 + 16x^2 - 9x - 36 > 0$

$4x^2(x+4) - 9(x+4) > 0$

$(x+4)(4x^2 - 9) > 0$

$(x+4)(2x+3)(2x-3) > 0$

$x+4 = 0$ or $2x+3 = 0$ or $2x-3 = 0$

$x = -4$ or $x = -\dfrac{3}{2}$ or $x = \dfrac{3}{2}$

Region	Test Point	$(x+4)(2x+3)(2x-3) > 0$ Result
A: $(-\infty, -4)$	-5	$(-1)(-7)(-13) > 0$ False
B: $\left(-4, -\dfrac{3}{2}\right)$	-3	$(1)(-3)(-9) > 0$ True
C: $\left(-\dfrac{3}{2}, \dfrac{3}{2}\right)$	0	$(4)(3)(-3) > 0$ False
D: $\left(\dfrac{3}{2}, \infty\right)$	4	$(8)(11)(5) > 0$ True

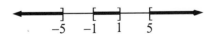

$-4 \quad -\dfrac{3}{2} \qquad \dfrac{3}{2}$

Solution: $\left(-4, -\dfrac{3}{2}\right) \cup \left(\dfrac{3}{2}, \infty\right)$

37. $x^4 - 26x^2 + 25 \geq 0$

$(x^2 - 25)(x^2 - 1) \geq 0$

$(x+5)(x-5)(x+1)(x-1) \geq 0$

$x = -5$ or $x = 5$ or $x = -1$ or $x = 1$

Region	Test Point	$(x+5)(x-5)(x+1)(x-1) \geq 0$ Result
A: $(-\infty, -5)$	-6	$(-1)(-11)(-5)(-7) \geq 0$; True
B: $(-5, -1)$	-2	$(3)(-7)(-1)(-3) \geq 0$; False
C: $(-1, 1)$	0	$(5)(-5)(1)(-1) \geq 0$; True
D: $(1, 5)$	2	$(7)(-3)(3)(1) \geq 0$; False
E: $(5, \infty)$	6	$(11)(1)(6)(2) \geq 0$; True

$-5 \quad -1 \quad 1 \quad 5$

Solution: $(-\infty, -5] \cup [-1, 1] \cup [5, \infty)$

39. $(2x-7)(3x+5) > 0$

$\quad 2x-7 = 0 \quad \text{or} \quad 3x+5 = 0$

$\qquad x = \dfrac{7}{2} \quad \text{or} \qquad x = -\dfrac{5}{3}$

Region	Test Point	$(2x-7)(3x+5) > 0$ Result
A: $\left(-\infty, -\dfrac{5}{3}\right)$	-2	$(-11)(-1) > 0$ True
B: $\left(-\dfrac{5}{3}, \dfrac{7}{2}\right)$	0	$(-7)(5) > 0$ False
C: $\left(\dfrac{7}{2}, \infty\right)$	4	$(1)(17) > 0$ True

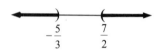

Solution: $\left(-\infty, -\dfrac{5}{3}\right) \cup \left(\dfrac{7}{2}, \infty\right)$

41. $\dfrac{x}{x-10} < 0$

$\quad x = 0 \quad \text{or} \quad x-10 = 0$

$\qquad\qquad\qquad x = 10$

Region	Test Point	$\dfrac{x}{x-10} < 0$ Result
A: $(-\infty, 0)$	-1	$\dfrac{-1}{-11} < 0$ False
B: $(0, 10)$	5	$\dfrac{5}{-5} < 0$ True
C: $(10, \infty)$	11	$\dfrac{11}{1} < 0$ False

Solution: $(0, 10)$

43. $\dfrac{x-5}{x+4} \geq 0$

$\quad x-5 = 0 \quad \text{or} \quad x+4 = 0$

$\qquad x = 5 \quad \text{or} \qquad x = -4$

Region	Test Point	$\dfrac{x-5}{x+4} \geq 0$ Result
A: $(-\infty, -4)$	-5	$\dfrac{-10}{-1} \geq 0$ True
B: $(-4, 5)$	0	$\dfrac{-5}{4} \geq 0$ False
C: $(5, \infty)$	6	$\dfrac{1}{10} \geq 0$ True

Solution: $(-\infty, -4) \cup [5, \infty)$

45. $\dfrac{x(x+6)}{(x-7)(x+1)} \geq 0$

$x = 0$ or $x+6 = 0$ or $x-7 = 0$ or $x+1 = 0$

$\qquad\qquad x = -6$ or $\qquad x = 7 \qquad$ or $x = -1$

Region	Test Point	$\dfrac{x(x+6)}{(x-7)(x+1)} \geq 0$; Result
A: $(-\infty, -6)$	-7	$\dfrac{-7(-1)}{(-14)(-6)} \geq 0$; True
B: $(-6, -1)$	-3	$\dfrac{-3(3)}{(-10)(-2)} \geq 0$; False
C: $(-1, 0)$	$-\dfrac{1}{2}$	$\dfrac{-\dfrac{1}{2}\left(\dfrac{11}{2}\right)}{\left(-\dfrac{13}{2}\right)\left(\dfrac{1}{2}\right)} \geq 0$; True
D: $(0, 7)$	2	$\dfrac{2(8)}{(-5)(3)} \geq 0$; False
E: $(7, \infty)$	8	$\dfrac{8(14)}{(1)(9)} \geq 0$; True

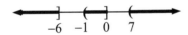

$\qquad\quad -6 \quad -1 \quad 0 \quad 7$

Solution: $(-\infty, -6] \cup (-1, 0] \cup (7, \infty)$

47. $\dfrac{-1}{x-1} > -1$

The denominator is equal to 0 when

$x - 1 = 0$, or $x = 1$.

$\dfrac{-1}{x-1} = -1$

$\quad -1 = -1(x-1)$

$\quad -1 = -x + 1$

$\qquad x = 2$

Region	Test Point	$\dfrac{-1}{x-1} > -1$; Result
A: $(-\infty, 1)$	0	$\dfrac{-1}{-1} > -1$; True
B: $(1, 2)$	$\dfrac{3}{2}$	$\dfrac{-1}{\dfrac{1}{2}} = -2 > -1$; False
C: $(2, \infty)$	3	$\dfrac{-1}{2} > -1$; True

$\qquad\qquad 1 \qquad 2$

Solution: $(-\infty, 1) \cup (2, \infty)$

49. $\dfrac{x}{x+4} \le 2$

The denominator is equal to 0 when
$x+4=0$, or $x=-4$.

$$\frac{x}{x+4}=2$$
$$x=2x+8$$
$$-x=8$$
$$x=-8$$

Region	Test Point	$\dfrac{x}{x+4} \le 2$; Result
A: $(-\infty, -8)$	-9	$\dfrac{-9}{-5} \le 2$; True
B: $(-8, -4)$	-6	$\dfrac{-6}{-2} \le 2$; False
C: $(-4, \infty)$	0	$\dfrac{0}{4} \le 2$; True

Solution: $(-\infty, -8] \cup (-4, \infty)$

51. $\dfrac{z}{z-5} \ge 2z$

The denominator is equal to 0 when
$z-5=0$, or $z=5$.

$$\frac{z}{z-5}=2z$$
$$z=2z(z-5)$$
$$z=2z^2-10z$$
$$0=2z^2-11z$$
$$0=z(2z-11)$$
$$z=0 \text{ or } 2z-11=0$$
$$z=\frac{11}{2}$$

Region	Test Point	$\dfrac{z}{z-5} \ge 2z$; Result
A: $(-\infty, 0)$	-1	$\dfrac{-1}{-6} \ge -2$; True
B: $(0, 5)$	1	$\dfrac{1}{-4} \ge 2$; False
C: $\left(5, \dfrac{11}{2}\right)$	$\dfrac{21}{4}$	$\dfrac{(21/4)}{(1/4)} \ge \dfrac{21}{2}$ $\;21 \ge \dfrac{21}{2}$; True
D: $\left(\dfrac{11}{2}, \infty\right)$	6	$\dfrac{6}{1} \ge 12$; False

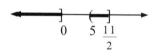

Solution: $(-\infty, 0] \cup \left(5, \dfrac{11}{2}\right]$

53. $\dfrac{(x+1)^2}{5x} > 0$

The denominator is equal to 0 when
$5x=0$, or $x=0$.

$$\frac{(x+1)^2}{5x}=0$$
$$(x+1)^2=0$$
$$x+1=0$$
$$x=-1$$

Region	Test Point	$\dfrac{(x+1)^2}{5x} > 0$; Result
A: $(-\infty, -1)$	-2	$\dfrac{1}{-10} > 0$; False
B: $(-1, 0)$	$-\dfrac{1}{2}$	$\dfrac{(1/4)}{(-5/2)} > 0$; False
C: $(0, \infty)$	1	$\dfrac{4}{5} > 0$; True

Solution: $(0, \infty)$

55. $g(x) = |x| + 2$

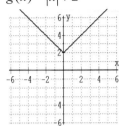

57. $F(x) = |x| - 1$

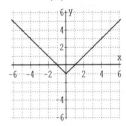

59. $F(x) = x^2 - 3$

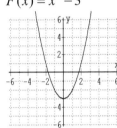

61. $H(x) = x^2 + 1$

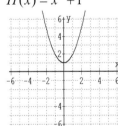

63. Answers may vary.

65. Let $x =$ the number. Then

$\dfrac{1}{x} =$ the reciprocal of the number.

$$x - \frac{1}{x} < 0$$
$$\frac{x^2 - 1}{x} < 0$$
$$\frac{(x+1)(x-1)}{x} < 0$$

$x + 1 = 0$ or $x - 1 = 0$ or $x = 0$
$\qquad x = -1$ or $x = 1$

Region	Test Point	$\dfrac{(x+1)(x-1)}{x} < 0$ Result
A: $(-\infty, -1)$	-2	$\dfrac{(-1)(-3)}{-2} < 0$; True
B: $(-1, 0)$	$-\dfrac{1}{2}$	$\dfrac{\left(\frac{1}{2}\right)\left(-\frac{3}{2}\right)}{\left(-\frac{1}{2}\right)} < 0$; False
C: $(0, 1)$	$\dfrac{1}{2}$	$\dfrac{\left(\frac{3}{2}\right)\left(-\frac{1}{2}\right)}{\left(\frac{1}{2}\right)} < 0$; True
D: $(1, \infty)$	2	$\dfrac{(3)(1)}{2} < 0$; False

The numbers are any number less than -1 or between 0 and 1.

67. $P(x) = -2x^2 + 26x - 44$
$\quad -2x^2 + 26x - 44 > 0$
$\quad -2(x^2 + 13x - 22) > 0$
$\quad\; -2(x - 11)(x - 2) > 0$
$\quad x - 11 = 0$ or $x - 2 = 0$
$\qquad\quad x = 11$ or $x = 2$

Region	Test Point	$-2(x-11)(x-2) > 0$ Result
A: $(0, 2)$	1	$-2(-10)(-3) > 0$ False
B: $(2, 11)$	3	$-2(-8)(1) > 0$ True
C: $(11, \infty)$	12	$-2(1)(10) > 0$ False

The company makes a profit when x is between 2 and 11.

69.

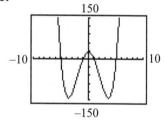

71.

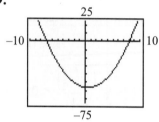

Section 8.5

Graphing Calculator Explorations

1.

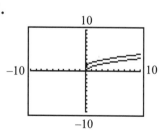

3.

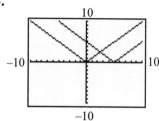

5.

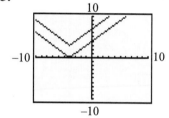

Mental Math

1. $f(x) = x^2$; vertex: (0, 0)

2. $f(x) = -5x^2$; vertex: (0, 0)

3. $g(x) = (x-2)^2$; vertex: (2, 0)

4. $g(x) = (x+5)^2$; vertex: (−5, 0)

5. $f(x) = 2x^2 + 3$; vertex: (0, 3)

6. $h(x) = x^2 - 1$; vertex: (0, −1)

7. $g(x) = (x+1)^2 + 5$; vertex: (−1, 5)

8. $h(x) = (x-10)^2 - 7$; vertex: (10, −7)

Exercise Set 8.5

1. $f(x) = x^2 - 1$

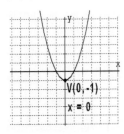

3. $h(x) = x^2 + 5$

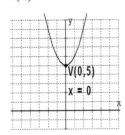

5. $g(x) = x^2 + 7$

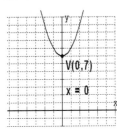

7. $f(x) = (x-5)^2$

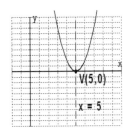

9. $h(x) = (x+2)^2$

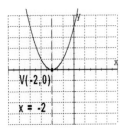

11. $G(x) = (x+3)^2$

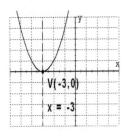

13. $f(x) = (x-2)^2 + 5$

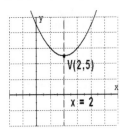

15. $h(x) = (x+1)^2 + 4$

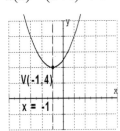

17. $g(x) = (x+2)^2 - 5$

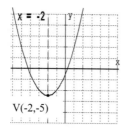

19. $g(x) = -x^2$

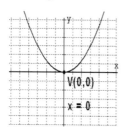

21. $h(x) = \frac{1}{3}x^2$

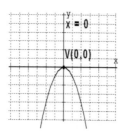

23. $H(x) = 2x^2$

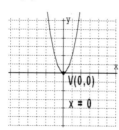

25. $f(x) = 2(x-1)^2 + 3$

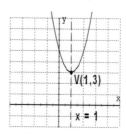

27. $h(x) = -3(x+3)^2 + 1$

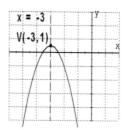

29. $H(x) = \frac{1}{2}(x-6)^2 - 3$

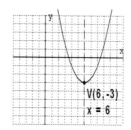

31. $f(x) = -(x-2)^2$

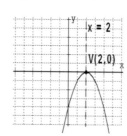

33. $F(x) = -x^2 + 4$

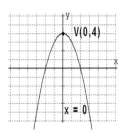

35. $F(x) = 2x^2 - 5$

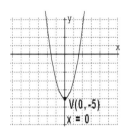

37. $h(x) = (x - 6)^2 + 4$

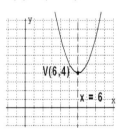

39. $F(x) = \left(x + \dfrac{1}{2}\right)^2 - 2$

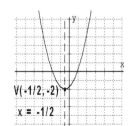

41. $F(x) = \dfrac{3}{2}(x + 7)^2 + 1$

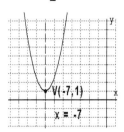

43. $f(x) = \dfrac{1}{4}x^2 - 9$

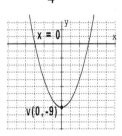

45. $G(x) = 5\left(x + \dfrac{1}{2}\right)^2$

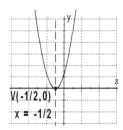

47. $h(x) = -(x - 1)^2 - 1$

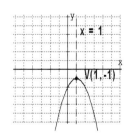

49. $g(x) = \sqrt{3}(x+5)^2 + \dfrac{3}{4}$

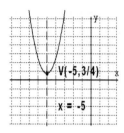

51. $h(x) = 10(x+4)^2 - 6$

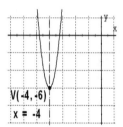

53. $f(x) = -2(x-4)^2 + 5$

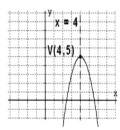

55. $x^2 + 8x$

$$\left[\dfrac{1}{2}(8)\right]^2 = (4)^2 = 16$$

$$x^2 + 8x + 16$$

57. $z^2 - 16z$

$$\left[\dfrac{1}{2}(-16)\right]^2 = (-8)^2 = 64$$

$$z^2 - 16z + 64$$

59. $y^2 + y$

$$\left[\dfrac{1}{2}(1)\right]^2 = \left(\dfrac{1}{2}\right)^2 = \dfrac{1}{4}$$

$$y^2 + y + \dfrac{1}{4}$$

61. $$x^2 + 4x = 12$$

$$x^2 + 4x + \left(\dfrac{4}{2}\right)^2 = 12 + 4$$

$$x^2 + 4x + 4 = 16$$

$$(x+2)^2 = 16$$

$$x + 2 = \pm\sqrt{16}$$

$$x + 2 = \pm 4$$

$$x = -2 \pm 4$$

$$x = -6 \ \text{ or } \ 2$$

63. $$z^2 + 10z - 1 = 0$$

$$z^2 + 10z = 1$$

$$z^2 + 10z + \left(\dfrac{10}{2}\right)^2 = 1 + 25$$

$$z^2 + 10z + 25 = 26$$

$$(z+5)^2 = 26$$

$$z + 5 = \pm\sqrt{26}$$

$$z = -5 \pm \sqrt{26}$$

65. $$z^2 - 8z = 2$$

$$z^2 - 8z + \left(\dfrac{-8}{2}\right)^2 = 2 + 16$$

$$z^2 - 8z + 16 = 18$$

$$(z-4)^2 = 18$$

$$z - 4 = \pm\sqrt{18}$$

$$z - 4 = \pm 3\sqrt{2}$$

$$z = 4 \pm 3\sqrt{2}$$

67. $f(x) = 5(x-2)^2 + 3$

69. $f(x) = 5[x-(-3)]^2 + 6$

$$= 5(x+3)^2 + 6$$

71. $y = f(x) + 1$

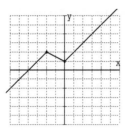

73. $y = f(x-3)$

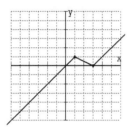

75. $y = f(x+2) + 2$

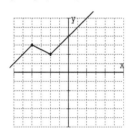

Exercise Set 8.6

1. $f(x) = x^2 + 8x + 7$

$-\dfrac{b}{2a} = \dfrac{-8}{2(1)} = -4$ and

$f(-4) = (-4)^2 + 8(-4) + 7$
$\qquad = 16 - 32 + 7$
$\qquad = -9$

Thus, the vertex is $(-4, -9)$.

3. $f(x) = -x^2 + 10x + 5$

$-\dfrac{b}{2a} = \dfrac{-10}{2(-1)} = 5$ and

$f(5) = -(5)^2 + 10(5) + 5$
$\qquad = -25 + 50 + 5$
$\qquad = 30$

Thus, the vertex is $(5, 30)$.

5. $f(x) = 5x^2 - 10x + 3$

$-\dfrac{b}{2a} = \dfrac{-(-10)}{2(5)} = 1$ and

$f(1) = 5(1)^2 - 10(1) + 3$
$\qquad = 5 - 10 + 3$
$\qquad = -2$

Thus, the vertex is $(1, -2)$.

7. $f(x) = -x^2 + x + 1$

$-\dfrac{b}{2a} = \dfrac{-1}{2(-1)} = \dfrac{1}{2}$ and

$f\left(\dfrac{1}{2}\right) = -\left(\dfrac{1}{2}\right)^2 + \left(\dfrac{1}{2}\right) + 1$
$\qquad\quad = -\dfrac{1}{4} + \dfrac{1}{2} + 1$
$\qquad\quad = \dfrac{5}{4}$

Thus, the vertex is $\left(\dfrac{1}{2}, \dfrac{5}{4}\right)$.

9. $f(x) = x^2 - 4x + 3$

$-\dfrac{b}{2a} = \dfrac{-(-4)}{2(1)} = 2$ and

$f(2) = (2)^2 - 4(2) + 3 = -1$

The vertex is $(2, -1)$, so the graph is D.

11. $f(x) = x^2 - 2x - 3$

$-\dfrac{b}{2a} = \dfrac{-(-2)}{2(1)} = 1$ and

$f(1) = (1)^2 - 2(1) - 3 = -4$

The vertex is $(, -4)$, so the graph is B.

13. $f(x) = x^2 + 4x - 5$

$-\dfrac{b}{2a} = \dfrac{-4}{2(1)} = -2$ and

$f(-2) = (-2)^2 + 4(-2) - 5 = -9$

Thus, the vertex is $(-2, -9)$.

The graph opens upward ($a = 1 > 0$).

$x^2 + 4x - 5 = 0$

$(x+5)(x-1) = 0$

$x + 5 = 0$ or $x - 1 = 0$

$x = -5$ or $x = 1$

x-intercepts: $(-5, 0)$ and $(1, 0)$.

$f(0) = -5$, so the y-intercept is $(0, -5)$.

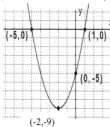

15. $f(x) = -x^2 + 2x - 1$

$-\dfrac{b}{2a} = \dfrac{-2}{2(-1)} = 1$ and

$f(1) = -(1)^2 + 2(1) - 1 = 0$

Thus, the vertex is $(1, 0)$.

The graph opens downward ($a = -1 < 0$).

$-x^2 + 2x - 1 = 0$

$x^2 - 2x + 1 = 0$

$(x-1)^2 = 0$

$x - 1 = 0$

$x = 1$

x-intercept: $(1, 0)$.

$f(0) = -1$, so the y-intercept is $(0, -1)$.

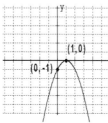

17. $f(x) = x^2 - 4$

$-\dfrac{b}{2a} = \dfrac{-0}{2(1)} = 0$ and

$f(0) = (0)^2 - 4 = -4$

Thus, the vertex is $(0, -4)$.

The graph opens upward ($a = 1 > 0$).

$x^2 - 4 = 0$

$(x+2)(x-2) = 0$

$x + 2 = 0$ or $x - 2 = 0$

$x = -2$ or $x = 2$

x-intercepts: $(-2, 0)$ and $(2, 0)$.

$f(0) = -4$, so the y-intercept is $(0, -4)$.

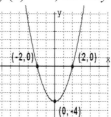

19. $f(x) = 4x^2 + 4x - 3$

$-\dfrac{b}{2a} = \dfrac{-4}{2(4)} = -\dfrac{1}{2}$ and

$f\left(-\dfrac{1}{2}\right) = 4\left(-\dfrac{1}{2}\right)^2 + 4\left(-\dfrac{1}{2}\right) - 3 = -4$

Thus, the vertex is $\left(-\dfrac{1}{2}, -4\right)$.

The graph opens upward ($a = 4 > 0$).

$4x^2 + 4x - 3 = 0$

$(2x+3)(2x-1) = 0$

$2x + 3 = 0$ or $2x - 1 = 0$

$x = -\dfrac{3}{2}$ or $x = \dfrac{1}{2}$

x-intercepts: $\left(-\dfrac{3}{2}, 0\right)$ and $\left(\dfrac{1}{2}, 0\right)$.

$f(0) = -3$, so the y-intercept is $(0, -3)$.

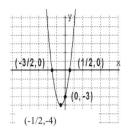

21.
$$f(x) = x^2 + 8x + 15$$
$$y = x^2 + 8x + 15$$
$$y - 15 = x^2 + 8x$$
$$y - 15 + 16 = x^2 + 8x + 16$$
$$y - 1 = (x + 4)^2$$
$$y = (x + 4)^2 + 1$$
$$f(x) = (x + 4)^2 + 1$$

Thus, the vertex is (–4, 1).
The graph opens upward ($a = 1 > 0$).
$$x^2 + 8x + 15 = 0$$
$$(x + 5)(x + 3) = 0$$
$$x + 5 = 0 \quad \text{or} \quad x + 3 = 0$$
$$x = -5 \quad \text{or} \quad x = -3$$

x-intercepts: (–5, 0) and (–3, 0).
$f(0) = 15$, so the y-intercept is (0, 15).

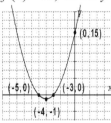

23.
$$f(x) = x^2 - 6x + 5$$
$$y = x^2 - 6x + 5$$
$$y - 5 = x^2 - 6x$$
$$y - 5 + 9 = x^2 - 6x + 9$$
$$y + 4 = (x - 3)^2$$
$$y = (x - 3)^2 - 4$$
$$f(x) = (x - 3)^2 - 4$$

Thus, the vertex is (3, –4).
The graph opens upward ($a = 1 > 0$).

$$x^2 - 6x + 5 = 0$$
$$(x - 5)(x - 1) = 0$$
$$x = -5 \quad \text{or} \quad x = 1$$
x-intercepts: (–5, 0) and (1, 0).
$f(0) = 5$, so the y-intercept is (0, 5).

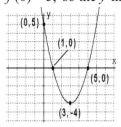

25.
$$f(x) = x^2 - 4x + 5$$
$$y = x^2 - 4x + 5$$
$$y - 5 = x^2 - 4x$$
$$y - 5 + 4 = x^2 - 4x + 4$$
$$y - 1 = (x - 2)^2$$
$$y = (x - 2)^2 + 1$$
$$f(x) = (x - 2)^2 + 1$$

Thus, the vertex is (2, 1).
The graph opens upward ($a = 1 > 0$).
$$x^2 - 4x + 5 = 0$$
$$x = \frac{4 \pm \sqrt{(-4)^2 - 4(1)(5)}}{2(1)} = \frac{4 \pm \sqrt{-4}}{2}$$

which give non-real solutions.
Hence, there are no x-intercepts.
$f(0) = 5$, so the y-intercept is (0, 5).

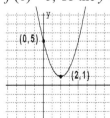

27.
$$f(x) = 2x^2 + 4x + 5$$
$$y = 2x^2 + 4x + 5$$
$$y - 5 = 2(x^2 + 2x)$$
$$y - 5 + 2(1) = 2(x^2 + 2x + 1)$$
$$y - 3 = 2(x + 1)^2$$
$$y = 2(x + 1)^2 + 3$$
$$f(x) = 2(x + 1)^2 + 3$$

Thus, the vertex is $(-1, 3)$.
The graph opens upward ($a = 2 > 0$).
$$2x^2 + 4x + 5 = 0$$
$$x = \frac{-4 \pm \sqrt{(4)^2 - 4(2)(5)}}{2(2)} = \frac{-4 \pm \sqrt{-24}}{4}$$

which give non-real solutions.
Hence, there are no x-intercepts.
$f(0) = 5$, so the y-intercept is $(0, 5)$.

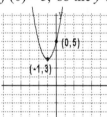

29.
$$f(x) = -2x^2 + 12x$$
$$y = -2(x^2 - 6x)$$
$$y + [-2(9)] = -2(x^2 - 6x + 9)$$
$$y - 18 = -2(x - 3)^2$$
$$y = -2(x - 3)^2 + 18$$
$$f(x) = -2(x - 3)^2 + 18$$

Thus, the vertex is $(3, 18)$.
The graph opens downward ($a = -2 < 0$).
$$-2x^2 + 12x = 0$$
$$-2x(x - 6) = 0$$
$$x = 0 \ \text{ or } \ x - 6 = 0$$
$$x = 6$$

x-intercepts: $(0, 0)$ and $(6, 0)$

$f(0) = 0$, so the y-intercept is $(0, 0)$.

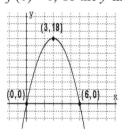

31. $f(x) = x^2 + 1$
$$x = -\frac{b}{2a} = -\frac{0}{2(1)} = 0$$
$$f(0) = (0)^2 + 1 = 1$$

Thus, the vertex is $(0, 1)$.
The graph opens upward ($a = 1 > 0$).
$$x^2 + 1 = 0$$
$$x^2 = -1$$

which give non-real solutions.

Hence, there are no x-intercepts.
$f(0) = 1$, so the y-intercept is $(0, 1)$.

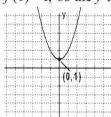

33.
$$f(x) = x^2 - 2x - 15$$
$$y = x^2 - 2x - 15$$
$$y + 15 = x^2 - 2x$$
$$y + 15 + 1 = x^2 - 2x + 1$$
$$y + 16 = (x - 1)^2$$
$$y = (x - 1)^2 - 16$$
$$f(x) = (x - 1)^2 - 16$$

Thus, the vertex is $(1, -16)$.
The graph opens upward ($a = 1 > 0$).
$$x^2 - 2x - 15 = 0$$
$$(x - 5)(x + 3) = 0$$
$$x = 5 \ \text{ or } \ x = -3$$

x-intercepts: $(-3, 0)$ and $(5, 0)$.

$f(0) = -15$ so the y-intercept is $(0, -15)$.

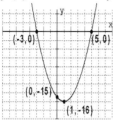

35. $f(x) = -5x^2 + 5x$

$x = -\dfrac{b}{2a} = \dfrac{-5}{2(-5)} = \dfrac{1}{2}$ and

$f\left(\dfrac{1}{2}\right) = -5\left(\dfrac{1}{2}\right)^2 + 5\left(\dfrac{1}{2}\right) = -\dfrac{5}{4} + \dfrac{5}{2} = \dfrac{5}{4}$

Thus, the vertex is $\left(\dfrac{1}{2}, \dfrac{5}{4}\right)$.

The graph opens downward $(a = -5 < 0)$.

$-5x^2 + 5x = 0$

$-5x(x - 1) = 0$

$x = 0$ or $x - 1 = 0$

                  $x = 1$

x-intercepts: $(0, 0)$ and $(1, 0)$

$f(0) = 0$, so the y-intercept is $(0, 0)$.

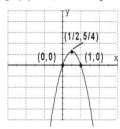

37. $f(x) = -x^2 + 2x - 12$

$x = -\dfrac{b}{2a} = \dfrac{-2}{2(-1)} = 1$ and

$f(1) = -(1)^2 + 2(1) - 12 = -11$

Thus, the vertex is $(1, -11)$.

The graph opens downward $(a = -1 < 0)$.

$-x^2 + 2x - 12 = 0$

$x^2 - 2x + 12 = 0$

$x = \dfrac{2 \pm \sqrt{(-2)^2 - 4(1)(12)}}{2(1)} = \dfrac{2 \pm \sqrt{-44}}{2}$

which yields non-real solutions.

Hence, there are no x-intercepts.

$f(0) = -12$ so the y-intercept is $(0, -12)$.

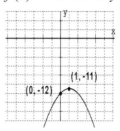

39. $f(x) = 3x^2 - 12 + 15$

$x = -\dfrac{b}{2a} = \dfrac{-(-12)}{2(3)} = \dfrac{12}{6} = 2$ and

$f(2) = 3(2)^2 - 12(2) + 15$

        $= 12 - 24 + 15 = 3$

Thus, the vertex is $(2, 3)$.

The graph opens upward $(a = 3 > 0)$.

$3x^2 - 12x + 15 = 0$

   $x^2 - 4x + 5 = 0$

$x = \dfrac{4 \pm \sqrt{(-4)^2 - 4(1)(5)}}{2(1)} = \dfrac{4 \pm \sqrt{-4}}{2}$

which yields non-real solutions.

Hence, there are no x-intercepts.

$f(0) = 15$, so the y-intercept is $(0, 15)$.

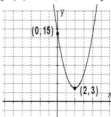

41. $f(x) = x^2 + x - 6$

$x = -\dfrac{b}{2a} = \dfrac{-1}{2(1)} = -\dfrac{1}{2}$ and

$f\left(-\dfrac{1}{2}\right) = \left(-\dfrac{1}{2}\right)^2 + \left(-\dfrac{1}{2}\right) - 6$

$\qquad = \dfrac{1}{4} - \dfrac{1}{2} - 6 = -\dfrac{25}{4}$

Thus, the vertex is $\left(-\dfrac{1}{2}, -\dfrac{25}{4}\right)$.

The graph opens upward ($a = 1 > 0$).

$x^2 + x - 6 = 0$

$(x+3)(x-2) = 0$

$x = 3$ or $x = -2$

x-intercepts: $(-3, 0)$ and $(2, 0)$.

$f(0) = -6$ so the y-intercept is $(0, -6)$.

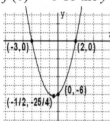

43. $f(x) = -2x^2 - 3x + 35$

$x = -\dfrac{b}{2a} = \dfrac{-(-3)}{2(-2)} = -\dfrac{3}{4}$ and

$f\left(-\dfrac{3}{4}\right) = -2\left(-\dfrac{3}{4}\right)^2 - 3\left(-\dfrac{3}{4}\right) + 35$

$\qquad = -\dfrac{9}{8} + \dfrac{9}{4} + 35 = -\dfrac{289}{8}$

Thus, the vertex is $\left(-\dfrac{3}{4}, \dfrac{289}{8}\right)$.

The graph opens downward ($a = -2 < 0$).

$-2x^2 - 3x + 35 = 0$

$2x^2 + 3x - 35 = 0$

$(2x-7)(x+5) = 0$

$2x - 7 = 0$ or $x + 5 = 0$

$\qquad x = \dfrac{7}{2}$ or $\qquad x = -5$

x-intercepts: $(-5, 0)$ and $\left(\dfrac{7}{2}, 0\right)$.

$f(0) = 35$ so the y-intercept is $(0, 35)$.

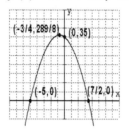

45. $h(t) = -16t^2 + 96t$

$t = -\dfrac{b}{2a} = \dfrac{-96}{2(-16)} = \dfrac{96}{32} = 3$ and

$h(3) = -16(3)^2 + 96(3)$

$\qquad = -144 + 288$

$\qquad = 144$

The maximum height is 144 feet.

47. $h(t) = -16t^2 + 32t$

$t = -\dfrac{b}{2a} = \dfrac{-32}{2(-16)} = \dfrac{32}{32} = 1$ and

$h(1) = -16(1)^2 + 32(1)$

$\qquad = -16 + 32$

$\qquad = 16$

The maximum height is 16 feet.

49. Let x = one number. Then

$60 - x$ = the other number.

$f(x) = x(60 - x)$

$\qquad = 60x - x^2$

$\qquad = -x^2 + 60x$

The maximum will occur at the vertex.

$x = -\dfrac{b}{2a} = \dfrac{-60}{2(-1)} = 30$

$60 - x = 60 - 30 = 30$

The numbers are 30 and 30.

51. Let x = one number. Then

$10 + x$ = the other number.

$f(x) = x(10 + x)$

$\qquad = 10x + x^2$

$\qquad = x^2 + 10x$

The minimum will occur at the vertex.

$$x = -\frac{b}{2a} = \frac{-10}{2(1)} = -5$$

$$10 + x = 10 + (-5) = 5$$

The numbers are −5 and 5.

53. Let x = width. Then
$40 - x$ = the length.
Area = length · width
$$A(x) = (40 - x)x$$
$$= 40x - x^2$$
$$= -x^2 + 40x$$

The maximum will occur at the vertex.
$$x = -\frac{b}{2a} = \frac{-40}{2(-1)} = 20$$
$$40 - x = 40 - 20 = 20$$

The maximum area will occur when the length and width are 20 units each.

55. $f(x) = x^2 + 2$

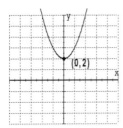

57. $g(x) = x + 2$

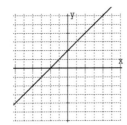

59. $f(x) = (x+5)^2 + 2$

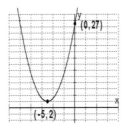

61. $f(x) = 3(x-4)^2 + 1$

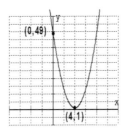

63. $f(x) = -\left(x - 4\right)^2 + \frac{3}{2}$

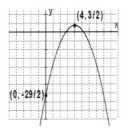

65. $f(x) = x^2 + 10x + 15$
$$x = -\frac{b}{2a} = \frac{-10}{2(1)} = -5 \text{ and}$$
$$f(-5) = (-5)^2 + 10(-5) + 15 = -10$$
Thus, the vertex is (−5, −10).
The graph opens upward ($a = 1 > 0$).
$f(0) = 15$ so the y-intercept is (0, 15).
$$x^2 + 10x + 15 = 0$$
$$x = \frac{-10 \pm \sqrt{(10)^2 - 4(1)(15)}}{2(1)}$$
$$= \frac{-10 \pm \sqrt{40}}{2} \approx -8.2 \text{ or } -1.8$$

The *x*-intercepts are approximately $(-8.2, 0)$ and $(-1.8, 0)$.

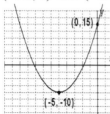

67. $f(x) = 3x^2 - 6x + 7$

$x = -\dfrac{b}{2a} = \dfrac{-(-6)}{2(3)} = 1$ and

$f(1) = 3(1)^2 - 6(1) + 7 = 4$

Thus, the vertex is $(1, 4)$.

The graph opens upward $(a = 1 > 0)$.

$f(0) = 7$ so the *y*-intercept is $(0, 7)$.

$3x^2 - 6x + 7 = 0$

$x = \dfrac{6 \pm \sqrt{(-6)^2 - 4(3)(7)}}{2(1)} = \dfrac{6 \pm \sqrt{-48}}{2}$

which yields non-real solutions.

Hence, there are no *x*-intercepts.

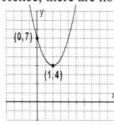

69. $f(x) = 2.3x^2 - 6.1x + 3.2$

minimum ≈ -0.84

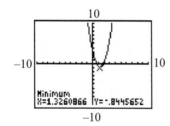

71. $f(x) = -1.9x^2 + 5.6x - 2.7$

maximum ≈ 1.43

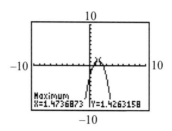

73. $p(x) = -x^2 + 93x + 1128$

a. It will have a maximum; answer may vary (e.g., since $a = -1 < 0$).

b. $x = -\dfrac{b}{2a} = \dfrac{-93}{2(-1)} = 46.5$

$1990 + 46.5 = 2036.5$

In the year 2036.

c. $p(46.5) = -(46.5)^2 + 93(46.5) + 1128$
$= 3290.25$ thousand inmates,

or 32,902,500 inmates

75.

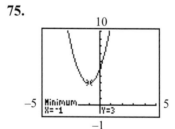

77.

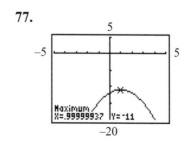

Chapter 8 Review

1. $x^2 - 15x + 14 = 0$
$(x - 14)(x - 1) = 0$
$x - 14 = 0 \quad \text{or} \quad x - 1 = 0$
$\quad x = 14 \quad \text{or} \quad \quad x = 1$
The solutions are 1 and 14.

2. $x^2 - x - 30 = 0$
$(x + 5)(x - 6) = 0$
$x + 5 = 0 \quad \text{or} \quad x - 6 = 0$
$\quad x = -5 \quad \text{or} \quad \quad x = 6$
The solutions are -5 and 6.

3. $\quad\quad 10x^2 = 3x + 4$
$10x^2 - 3x - 4 = 0$
$(5x - 4)(2x + 1) = 0$
$5x - 4 = 0 \quad \text{or} \quad 2x + 1 = 0$
$\quad 5x = 4 \quad \text{or} \quad \quad 2x = -1$
$\quad x = \dfrac{4}{5} \quad \text{or} \quad \quad x = -\dfrac{1}{2}$
The solutions are $-\dfrac{1}{2}$ and $\dfrac{4}{5}$.

4. $\quad\quad 7a^2 = 29a + 30$
$7a^2 - 29a - 30 = 0$
$(7a + 6)(a - 5) = 0$
$7a + 6 = 0 \quad \text{or} \quad a - 5 = 0$
$\quad 7a = -6 \quad \text{or} \quad \quad a = 5$
$\quad a = -\dfrac{6}{7}$
The solutions are $-\dfrac{6}{7}$ and 5.

5. $4m^2 = 196$
$\quad m^2 = 49$
$\quad m = \pm\sqrt{49}$
$\quad m = \pm 7$
The solutions are -7 and 7.

6. $9y^2 = 36$
$\quad y^2 = 4$
$\quad y = \pm\sqrt{4}$
$\quad y = \pm 2$
The solutions are -2 and 2.

7. $(9n + 1)^2 = 9$
$\quad 9n + 1 = \pm\sqrt{9}$
$\quad 9n + 1 = \pm 3$
$\quad 9n = -1 \pm 3$
$\quad n = \dfrac{-1 \pm 3}{9} = \dfrac{2}{9}, -\dfrac{4}{9}$
The solutions are $-\dfrac{4}{9}$ and $\dfrac{2}{9}$.

8. $(5x - 2)^2 = 2$
$\quad 5x - 2 = \pm\sqrt{2}$
$\quad 5x = 2 \pm \sqrt{2}$
$\quad x = \dfrac{2 \pm \sqrt{2}}{5}$
The solutions are $\dfrac{2 + \sqrt{2}}{5}$ and $\dfrac{2 - \sqrt{2}}{5}$.

9. $\quad z^2 + 3z + 1 = 0$
$\quad z^2 + 3z = -1$
$z^2 + 3z + \left(\dfrac{3}{2}\right)^2 = -1 + \dfrac{9}{4}$
$\quad \left(z + \dfrac{3}{2}\right)^2 = \dfrac{5}{4}$
$\quad z + \dfrac{3}{2} = \pm\sqrt{\dfrac{5}{4}}$
$\quad z + \dfrac{3}{2} = \pm\dfrac{\sqrt{5}}{2}$
$\quad z = -\dfrac{3}{2} \pm \dfrac{\sqrt{5}}{2}$
The solutions are $-\dfrac{3}{2} + \dfrac{\sqrt{5}}{2}$ and $-\dfrac{3}{2} - \dfrac{\sqrt{5}}{2}$.

10.
$$x^2 + x + 7 = 0$$
$$x^2 + x = -7$$
$$x^2 + x + \left(\frac{1}{2}\right)^2 = -7 + \frac{1}{4}$$
$$\left(x + \frac{1}{2}\right)^2 = -\frac{27}{4}$$
$$x + \frac{1}{2} = \pm\sqrt{-\frac{27}{4}}$$
$$x + \frac{1}{2} = \pm\frac{\sqrt{9 \cdot 3}}{2}i$$
$$x + \frac{1}{2} = \pm\frac{3\sqrt{3}}{2}i$$
$$x = -\frac{1}{2} \pm \frac{3\sqrt{3}}{2}i$$

The solutions are $-\frac{1}{2} + \frac{3\sqrt{3}}{2}i$ and

$-\frac{1}{2} - \frac{3\sqrt{3}}{2}i$.

11.
$$(2x + 1)^2 = x$$
$$4x^2 + 4x + 1 = x$$
$$4x^2 + 3x = -1$$
$$x^2 + \frac{3}{4}x = -\frac{1}{4}$$
$$x^2 + \frac{3}{4}x + \left(\frac{\frac{3}{4}}{2}\right)^2 = -\frac{1}{4} + \frac{9}{64}$$
$$\left(x + \frac{3}{8}\right)^2 = -\frac{7}{64}$$
$$x + \frac{3}{8} = \pm\sqrt{-\frac{7}{64}}$$
$$x + \frac{3}{8} = \pm\frac{\sqrt{7}}{8}i$$
$$x = -\frac{3}{8} \pm \frac{\sqrt{7}}{8}i$$

The solutions are $-\frac{3}{8} + \frac{\sqrt{7}}{8}i$ and

$-\frac{3}{8} - \frac{\sqrt{7}}{8}i$.

12.
$$(3x - 4)^2 = 10x$$
$$9x^2 - 24x + 16 = 10x$$
$$9x^2 - 34x = -16$$
$$x^2 - \frac{34}{9}x = -\frac{16}{9}$$
$$x^2 - \frac{34}{9}x + \left(\frac{-\frac{34}{9}}{2}\right)^2 = -\frac{16}{9} + \frac{289}{81}$$
$$\left(x - \frac{17}{9}\right)^2 = \frac{145}{81}$$
$$x - \frac{17}{9} = \pm\sqrt{\frac{145}{81}}$$
$$x - \frac{17}{9} = \pm\frac{\sqrt{145}}{9}$$
$$x = \frac{17 \pm \sqrt{145}}{9}$$

The solutions are $\frac{17 + \sqrt{145}}{9}$ and

$\frac{17 - \sqrt{145}}{9}$.

13.
$$A = P(1 + r)^2$$
$$2717 = 2500(1 + r)^2$$
$$\frac{2717}{2500} = (1 + r)^2$$
$$(1 + r)^2 = 1.0868$$
$$1 + r = \pm\sqrt{1.0868}$$
$$1 + r = \pm1.0425$$
$$r = -1 \pm 1.0425$$
$$= 0.0425 \text{ or } -2.0425 \text{ (disregard)}$$
The interest rate is 4.25%.

14. Let x = distance traveled.
$$a^2 + b^2 = c^2$$
$$x^2 + x^2 = (150)^2$$
$$2x^2 = 22,500$$
$$x^2 = 11,250$$
$$x = \pm75\sqrt{2} \approx \pm106.1$$
Disregard the negative.
The ship traveled $75\sqrt{2} \approx 106.1$ miles.

15. Two complex but not real solutions exist.

16. Two real solutions exist.

17. Two real solutions exist.

18. One real solution exist.

19. $x^2 - 16x + 64 = 0$
$a = 1, b = -16, c = 64$
$$x = \frac{16 \pm \sqrt{(-16)^2 - 4(1)(64)}}{2(1)}$$
$$= \frac{16 \pm \sqrt{256 - 256}}{2} = \frac{16 \pm \sqrt{0}}{2} = 8$$
The solution is 8.

20. $x^2 + 5x = 0$
$a = 1, b = 5, c = 0$
$$x = \frac{-5 \pm \sqrt{(5)^2 - 4(1)(0)}}{2(1)}$$
$$= \frac{-5 \pm \sqrt{25}}{2}$$
$$= \frac{-5 \pm 5}{2} = 0 \text{ or } -5$$
The solutions are -5 and 0.

21. $x^2 + 11 = 0$
$a = 1, b = 0, c = 11$
$$x = \frac{0 \pm \sqrt{(0)^2 - 4(1)(11)}}{2(1)}$$
$$= \frac{\pm\sqrt{-44}}{2}$$
$$= \frac{\pm 2i\sqrt{11}}{2} = \pm i\sqrt{11}$$
The solutions are $-i\sqrt{11}$ and $i\sqrt{11}$.

22. $2x^2 + 3x = 5$
$2x^2 + 3x - 5 = 0$
$a = 2, b = 3, c = -5$
$$x = \frac{-3 \pm \sqrt{(3)^2 - 4(2)(-5)}}{2(2)}$$
$$= \frac{-3 \pm \sqrt{49}}{4}$$
$$= \frac{-3 \pm 7}{4} = 1 \text{ or } -\frac{5}{2}$$
The solutions are $-\frac{5}{2}$ and 1.

23. $6x^2 + 7 = 5x$
$6x^2 - 5x + 7 = 0$
$a = 6, b = -5, c = 7$
$$x = \frac{5 \pm \sqrt{(-5)^2 - 4(6)(7)}}{2(6)}$$
$$= \frac{5 \pm \sqrt{25 - 168}}{12}$$
$$= \frac{5 \pm \sqrt{-143}}{12} = \frac{5 \pm i\sqrt{143}}{12} = \frac{5}{12} \pm \frac{\sqrt{143}}{12}i$$
The solutions are $\dfrac{5 + i\sqrt{143}}{12}$ and
$\dfrac{5 - i\sqrt{143}}{12}$.

24. $9a^2 + 4 = 2a$
$9a^2 - 2a + 4 = 0$
$$a = \frac{2 \pm \sqrt{(-2)^2 - 4(9)(4)}}{2(9)}$$
$$= \frac{2 \pm \sqrt{-140}}{18}$$
$$= \frac{2 \pm i\sqrt{4 \cdot 35}}{18}$$
$$= \frac{2 \pm 2i\sqrt{35}}{18} = \frac{1 \pm i\sqrt{35}}{9} = \frac{1}{9} \pm \frac{\sqrt{35}}{9}i$$
The solutions are $\dfrac{1 + i\sqrt{35}}{9}$ and
$\dfrac{1 - i\sqrt{35}}{9}$.

25.
$$(5a-2)^2 - a = 0$$
$$25a^2 - 20a + 4 - a = 0$$
$$25a^2 - 21a + 4 = 0$$

$$a = \frac{21 \pm \sqrt{(-21)^2 - 4(25)(4)}}{2(25)}$$
$$= \frac{21 \pm \sqrt{441 - 400}}{50}$$
$$= \frac{21 \pm \sqrt{41}}{50}$$

The solutions are $\dfrac{21 + \sqrt{41}}{50}$ and $\dfrac{21 - \sqrt{41}}{50}$.

26.
$$(2x-3)^2 = x$$
$$4x^2 - 12x + 9 - x = 0$$
$$4x^2 - 13x + 9 = 0$$
$$a = 4, b = -13, c = 9$$

$$x = \frac{13 \pm \sqrt{(-13)^2 - 4(4)(9)}}{2(4)}$$
$$= \frac{13 \pm \sqrt{169 - 144}}{8}$$
$$= \frac{13 \pm \sqrt{25}}{8}$$
$$= \frac{13 \pm 5}{8} = \frac{9}{4} \text{ or } 1$$

The solutions are 1 and $\dfrac{9}{4}$.

27. $d(t) = -16t^2 + 30t + 6$

a. $d(1) = -16(1)^2 + 30(1) + 6$
$$= -16 + 30 + 6$$
$$= 20 \text{ feet}$$

b.
$$-16t^2 + 30t + 6 = 0$$
$$8t^2 - 15t - 3 = 0$$
$$a = 8, b = -15, c = -3$$

$$t = \frac{15 \pm \sqrt{(-15)^2 - 4(8)(-3)}}{2(8)}$$
$$= \frac{15 \pm \sqrt{225 + 96}}{16}$$
$$= \frac{15 \pm \sqrt{321}}{16}$$

Disregarding the negative, we have
$$t = \frac{15 + \sqrt{321}}{16} \text{ seconds} \approx 2.1 \text{ seconds.}$$

28. Let $x =$ length of the legs. Then
$x + 6 =$ length of the hypotenuse.
$$x^2 + x^2 = (x+6)^2$$
$$2x^2 = x^2 + 12x + 36$$
$$x^2 - 12x - 36 = 0$$
$$a = 1, b = -12, c = -36$$

$$x = \frac{12 \pm \sqrt{(-12)^2 - 4(1)(-36)}}{2(1)}$$
$$= \frac{12 \pm \sqrt{144 + 144}}{2}$$
$$= \frac{12 \pm \sqrt{144 \cdot 2}}{2}$$
$$= \frac{12 \pm 12\sqrt{2}}{2} = 6 \pm 6\sqrt{2}$$

Disregard the negative.
The length of each leg is $\left(6 + 6\sqrt{2}\right)$ cm.

29.
$$x^3 = 27$$
$$x^3 - 27 = 0$$
$$(x-3)(x^2 + 3x + 9) = 0$$
$$x - 3 = 0 \quad \text{or} \quad x^2 + 3x + 9 = 0$$
$$x = 3 \qquad a = 1, b = 3, c = 9$$

$$x = \frac{-3 \pm \sqrt{(3)^2 - 4(1)(9)}}{2(1)}$$

$$= \frac{-3 \pm \sqrt{9 - 36}}{2}$$

$$= \frac{-3 \pm \sqrt{-27}}{2}$$

$$= \frac{-3 \pm 3i\sqrt{3}}{2} \quad \text{or} \quad -\frac{3}{2} \pm \frac{3\sqrt{3}}{2}i$$

The solutions are 3, $-\frac{3}{2} + \frac{3\sqrt{3}}{2}i$, and

$-\frac{3}{2} - \frac{3\sqrt{3}}{2}i$.

30.
$$y^3 = -64$$
$$y^3 + 64 = 0$$
$$(y+4)(y^2 - 4y + 16) = 0$$
$$y + 4 = 0 \quad \text{or} \quad y^2 - 4y + 16 = 0$$
$$y = -4 \qquad a = 1, b = -4, c = 16$$

$$y = \frac{4 \pm \sqrt{(-4)^2 - 4(1)(16)}}{2(1)}$$

$$= \frac{4 \pm \sqrt{16 - 64}}{2}$$

$$= \frac{4 \pm \sqrt{-48}}{2} = \frac{4 \pm 4i\sqrt{3}}{2} = 2 \pm 2i\sqrt{3}$$

The solutions are –4, $2 + 2i\sqrt{3}$, and
$2 - 2i\sqrt{3}$.

31.
$$\frac{5}{x} + \frac{6}{x-2} = 3$$
$$x(x-2)\left(\frac{5}{x} + \frac{6}{x-2}\right) = 3x(x-2)$$
$$5(x-2) + 6x = 3x^2 - 6x$$
$$5x - 10 + 6x = 3x^2 - 6x$$

$$0 = 3x^2 - 17x + 10$$
$$0 = (3x - 2)(x - 5)$$
$$3x - 2 = 0 \quad \text{or} \quad x - 5 = 0$$
$$x = \frac{2}{3} \quad \text{or} \quad x = 5$$

The solutions area $\frac{2}{3}$ and 5.

32.
$$\frac{7}{8} = \frac{8}{x^2}$$
$$7x^2 = 64$$
$$x^2 = \frac{64}{7}$$

$$x = \pm\sqrt{\frac{64}{7}}$$

$$x = \pm\frac{8}{\sqrt{7}} = \pm\frac{8 \cdot \sqrt{7}}{\sqrt{7} \cdot \sqrt{7}} = \pm\frac{8\sqrt{7}}{7}$$

The solutions are $-\frac{8\sqrt{7}}{7}$ and $\frac{8\sqrt{7}}{7}$.

33.
$$x^4 - 21x^2 - 100 = 0$$
$$(x^2 - 25)(x^2 + 4) = 0$$
$$(x+5)(x-5)(x^2 + 4) = 0$$
$$x + 5 = 0 \quad \text{or} \quad x - 5 = 0 \quad \text{or} \quad x^2 + 4 = 0$$
$$x = -5 \quad \text{or} \quad x = 5 \quad \text{or} \quad x^2 = -4$$
$$x = \pm 2i$$

The solutions are –5, 5, –2*i*, and 2*i*.

34.
$$5(x+3)^2 - 19(x+3) = 4$$
$$5(x+3)^2 - 19(x+3) - 4 = 0$$
Let $y = x + 3$. Then $y^2 = (x+3)^2$ and
$$5y^2 - 19y - 4 = 0$$
$$(5y + 1)(y - 4) = 0$$
$$5y + 1 = 0 \quad \text{or} \quad y - 4 = 0$$
$$y = -\frac{1}{5} \quad \text{or} \quad y = 4$$
$$x + 3 = -\frac{1}{5} \quad \text{or} \quad x + 3 = 4$$
$$x = -\frac{16}{5} \quad \text{or} \quad x = 1$$

The solutions are $-\frac{16}{5}$ and 1.

35. $x^{2/3} - 6x^{1/3} + 5 = 0$

Let $y = x^{1/3}$. Then $y^2 = x^{2/3}$ and

$y^2 - 6y + 5 = 0$

$(y-5)(y-1) = 0$

$y - 5 = 0$ or $y - 1 = 0$

$y = 5$ or $y = 1$

$x^{1/3} = 5$ or $x^{1/3} = 1$

$x = 125$ or $x = 1$

The solutions are 1 and 125.

36. $x^{2/3} - 6x^{1/3} = -8$

$x^{2/3} - 6x^{1/3} + 8 = 0$

Let $y = x^{1/3}$. Then $y^2 = x^{2/3}$ and

$y^2 - 6y + 8 = 0$

$(y-4)(y-2) = 0$

$y - 4 = 0$ or $y - 2 = 0$

$y = 4$ or $y = 2$

$x^{1/3} = 4$ or $x^{1/3} = 2$

$x = 64$ or $x = 8$

The solutions are 8 and 64.

37. $a^6 - a^2 = a^4 - 1$

$a^6 - a^4 - a^2 + 1 = 0$

$a^4(a^2 - 1) - 1(a^2 - 1) = 0$

$(a^2 - 1)(a^4 - 1) = 0$

$(a+1)(a-1)(a^2+1)(a^2-1) = 0$

$(a+1)(a-1)(a^2+1)(a+1)(a-1) = 0$

$(a+1)^2(a-1)^2(a^2+1) = 0$

$(a+1)^2 = 0$ or $(a-1)^2 = 0$ or $a^2 + 1 = 0$

$a + 1 = 0$ or $a - 1 = 0$ or $a^2 = -1$

$a = -1$ or $a = 1$ or $a = \pm i$

The solutions are -1, 1, $-i$, and i.

38. $y^{-2} + y^{-1} = 20$

$\dfrac{1}{y^2} + \dfrac{1}{y} = 20$

$1 + y = 20y^2$

$0 = 20y^2 - y - 1$

$0 = (5y+1)(4y-1)$

$5y + 1 = 0$ or $4y - 1 = 0$

$y = -\dfrac{1}{5}$ or $y = \dfrac{1}{4}$

The solutions are $-\dfrac{1}{5}$ and $\dfrac{1}{4}$.

39. Let x = time for Jerome alone. Then $x - 1$ = time for Tim alone.

$\dfrac{1}{x} + \dfrac{1}{x-1} = \dfrac{1}{5}$

$5(x-1) + 5x = x(x-1)$

$5x - 5 + 5x = x^2 - x$

$0 = x^2 - 11x + 5$

$a = 1, b = -11, c = 5$

$x = \dfrac{11 \pm \sqrt{(-11)^2 - 4(1)(5)}}{2(1)}$

$= \dfrac{11 \pm \sqrt{101}}{2}$

≈ 0.475 (disregard) or 10.525

Jerome: 10.5 hours

Tim: 9.5 hours

40. Let x = the number; then

$\dfrac{1}{x}$ = the reciprocal of the number.

$x - \dfrac{1}{x} = -\dfrac{24}{5}$

$5x\left(x - \dfrac{1}{x}\right) = 5x\left(-\dfrac{24}{5}\right)$

$5x^2 - 5 = -24x$

$5x^2 + 24x - 5 = 0$

$(5x-1)(x+5) = 0$

$5x - 1 = 0$ or $x + 5 = 0$

$x = \dfrac{1}{5}$ or $x = -5$

Disregard the positive value as extraneous. The number is -5.

41.
$$2x^2 - 50 \le 0$$
$$2(x^2 - 25) \le 0$$
$$2(x+5)(x-5) \le 0$$
$$x+5 = 0 \quad \text{or} \quad x-5 = 0$$
$$x = -5 \quad \text{or} \quad x = 5$$

Region	Test Point	$2(x+5)(x-5) \le 0$
A: $(-\infty, -5)$	-6	$2(-1)(-11) \le 0$; F
B: $(-5, 5)$	0	$2(5)(-5) \le 0$; T
C: $(5, \infty)$	6	$2(11)(1) \le 0$; F

Solution: $[-5, 5]$

42.
$$\frac{1}{4}x^2 < \frac{1}{16}$$
$$x^2 < \frac{1}{4}$$
$$x^2 - \frac{1}{4} < 0$$
$$\left(x+\frac{1}{2}\right)\left(x-\frac{1}{2}\right) < 0$$
$$x+\frac{1}{2} = 0 \quad \text{or} \quad x-\frac{1}{2} = 0$$
$$x = -\frac{1}{2} \quad \text{or} \quad x = \frac{1}{2}$$

Region	Test Point	$\left(x+\frac{1}{2}\right)\left(x-\frac{1}{2}\right) < 0$
A: $\left(-\infty, -\frac{1}{2}\right)$	-1	$\left(-\frac{1}{2}\right)\left(-\frac{3}{2}\right) < 0$; F
B: $\left(-\frac{1}{2}, \frac{1}{2}\right)$	0	$\left(\frac{1}{2}\right)\left(-\frac{1}{2}\right) < 0$; T
C: $\left(\frac{1}{2}, \infty\right)$	1	$\left(\frac{3}{2}\right)\left(\frac{1}{2}\right) < 0$; F

Solution: $\left(-\frac{1}{2}, \frac{1}{2}\right)$

43.
$$(2x-3)(4x+5) \ge 0$$
$$2x-3 = 0 \quad \text{or} \quad 4x+5 = 0$$
$$x = \frac{3}{2} \quad \text{or} \quad x = -\frac{5}{4}$$

Region	Test Point	$(2x-3)(4x+5) \ge 0$
A: $\left(-\infty, -\frac{5}{4}\right)$	-2	$(-7)(-3) \ge 0$ True
B: $\left(-\frac{5}{4}, \frac{3}{2}\right)$	0	$(-3)(5) \ge 0$ False
C: $\left(\frac{3}{2}, \infty\right)$	3	$(3)(17) \ge 0$ True

Solution: $\left(-\infty, -\frac{5}{4}\right] \cup \left[\frac{3}{2}, \infty\right)$

44.
$$(x^2 - 16)(x^2 - 1) > 0$$
$$(x+4)(x-4)(x+1)(x-1) > 0$$
$$x+4 = 0 \quad \text{or} \quad x-4 = 0 \quad \text{or} \quad x+1 = 0 \quad \text{or} \quad x-1 = 0$$
$$x = -4 \quad \text{or} \quad x = 4 \quad \text{or} \quad x = -1 \quad \text{or} \quad x = 1$$

Region	Test Point	$(x+4)(x-4)(x+1)(x-1) > 0$
A: $(-\infty, -4)$	-5	$(-1)(-9)(-4)(-6) > 0$; True
B: $(-4, -1)$	-2	$(2)(-6)(-1)(-3) > 0$; False
C: $(-1, 1)$	0	$(4)(-4)(1)(-1) > 0$; True
D: $(1, 4)$	2	$(6)(-2)(3)(1) > 0$; False
E: $(4, \infty)$	5	$(9)(1)(6)(4) > 0$; True

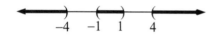

Solution: $(-\infty, -4) \cup (-1, 1) \cup (4, \infty)$

45. $\dfrac{x-5}{x-6} < 0$
$$x - 5 = 0 \quad \text{or} \quad x - 6 = 0$$
$$x = 5 \quad \text{or} \quad x = 6$$

Region	Test Point	$\dfrac{x-5}{x-6} < 0$
A: $(-\infty, 5)$	0	$\dfrac{-5}{-6} < 0$; False
B: $(5, 6)$	$\dfrac{11}{2}$	$\dfrac{\frac{1}{2}}{-\frac{1}{2}} < 0$; True
C: $(6, \infty)$	7	$\dfrac{2}{1} < 0$; False

Solution: $(5, 6)$

46. $\dfrac{x(x+5)}{4x-3} \geq 0$
$$x = 0 \quad \text{or} \quad x+5 = 0 \quad \text{or} \quad 4x-3 = 0$$
$$x = -5 \quad \text{or} \quad x = \frac{3}{4}.$$

Region	Test Point	$\dfrac{x(x+5)}{4x-3} \geq 0$
A: $(-\infty, -5)$	-6	$\dfrac{-6(-1)}{-27} \geq 0$; F
B: $(-5, 0)$	-1	$\dfrac{-1(4)}{-7} \geq 0$; T
C: $\left(0, \dfrac{3}{4}\right)$	$\dfrac{1}{2}$	$\dfrac{\frac{1}{2}\left(\frac{11}{2}\right)}{-1} \geq 0$; F
D: $\left(\dfrac{3}{4}, \infty\right)$	1	$\dfrac{1(6)}{1} \geq 0$; T

Solution: $[-5, 0] \cup \left(\dfrac{3}{4}, \infty\right)$

47. $\dfrac{(4x+3)(x-5)}{x(x+6)} > 0$

$4x+3=0,\ x-5=0,\ x=0,\ \text{or}\ x+6=0$

$x=-\dfrac{3}{4},\ x=5,\ x=0,\ \text{or}\ x=-6$

Region	Test Point	$\dfrac{(4x+3)(x-5)}{x(x+6)} > 0$
A: $(-\infty, -6)$	-7	$\dfrac{(-25)(-12)}{-7(-1)} > 0;\ \text{T}$
B: $\left(-6, -\dfrac{3}{4}\right)$	-3	$\dfrac{(-9)(-8)}{-3(3)} > 0;\ \text{F}$
C: $\left(-\dfrac{3}{4}, 0\right)$	$-\dfrac{1}{2}$	$\dfrac{(1)\left(-\dfrac{11}{2}\right)}{-\dfrac{1}{2}\left(\dfrac{11}{2}\right)} > 0;\ \text{T}$
D: $(0, 5)$	1	$\dfrac{(7)(-4)}{1(7)} > 0;\ \text{F}$
E: $(5, \infty)$	6	$\dfrac{(27)(1)}{6(12)} > 0;\ \text{T}$

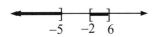

Solution: $(-\infty, -6) \cup \left(-\dfrac{3}{4}, 0\right) \cup (5, \infty)$

48. $(x+5)(x-6)(x+2) \le 0$

$x+5=0\ \text{ or }\ x-6=0\ \text{ or }\ x+2=0$

$x=-5\ \text{ or }\ \ \ x=6\ \text{ or }\ \ \ \ x=-2$

Region	Test Point	$(x+5)(x-6)(x+2) \le 0$
$(-\infty, -5)$	-6	$(-1)(-12)\,(-4) \le 0;\ \text{T}$
$(-5, -2)$	-3	$(2)(-9)\,(-1) \le 0;\ \text{F}$
$(-2, 6)$	0	$(5)(-6)(2) \le 0;\ \text{T}$
$(6, \infty)$	7	$(12)(1)(9) \le 0;\ \text{F}$

Solution: $(-\infty, -5] \cup [-2, 6]$

49. $x^3 + 3x^2 - 25 - 75 > 0$

$x^2(x+3) - 25(x+3) > 0$

$(x+3)(x^2 - 25) > 0$

$(x+3)(x+5)(x-5) > 0$

$x+3=0\ \text{ or }\ x+5=0\ \text{ or }\ x-5=0$

$x=-3\ \text{ or }\ \ \ \ x=-5\ \text{ or }\ \ \ \ x=5$

Region	Test Point	$(x+3)(x+5)(x-5) > 0$
$(-\infty, -5)$	-6	$(-3)(-1)\,(-11) > 0;\ \text{F}$
$(-5, -3)$	-4	$(-1)(1)\,(-9) > 0;\ \text{T}$
$(-3, 5)$	0	$(3)(5)(-5) > 0;\ \text{F}$
$(5, \infty)$	6	$(9)(11)(1) > 0;\ \text{T}$

Solution: $(-5, -3) \cup (5, \infty)$

50. $\dfrac{x^2+4}{3x} \le 1$

The denominator equals 0 when
$3x=0,\ \text{or}\ x=0$.

$\dfrac{x^2+4}{3x} = 1$

$x^2+4 = 3x$

$x^2 - 3x + 4 = 0$

$x = \dfrac{3 \pm \sqrt{(-3)^2 - 4(1)(4)}}{2(1)} = \dfrac{3 \pm \sqrt{-7}}{2(1)}$

which yields non-real solutions.

Region	Test Point	$\dfrac{x^2+4}{3x} \le 1$
A: $(-\infty, 0)$	-1	$\dfrac{5}{-3} \le 1;\ \text{True}$
B: $(0, \infty)$	1	$\dfrac{5}{3} \le 1;\ \text{False}$

Solution: $(-\infty, 0)$

51. $\dfrac{(5x+6)(x-3)}{x(6x-5)} < 0$

$x = -\dfrac{6}{5}$ or $x = 3$ or $x = 0$ or $x = \dfrac{5}{6}$

Region	Test Point	$\dfrac{(5x+6)(x-3)}{x(6x-5)} < 0$
A: $\left(-\infty, -\dfrac{6}{5}\right)$	-2	$\dfrac{(-4)(-5)}{-2(-17)} < 0$; F
B: $\left(-\dfrac{6}{5}, 0\right)$	-1	$\dfrac{(1)(-4)}{-1(-11)} < 0$; T
C: $\left(0, \dfrac{5}{6}\right)$	$\dfrac{1}{2}$	$\dfrac{\left(\dfrac{17}{2}\right)\left(-\dfrac{5}{2}\right)}{\dfrac{1}{2}(-2)} < 0$; F
D: $\left(\dfrac{5}{6}, 3\right)$	2	$\dfrac{(16)(-1)}{2(7)} < 0$; T
E: $(3, \infty)$	4	$\dfrac{(26)(1)}{4(19)} < 0$; F

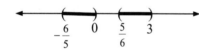

Solution: $\left(-\dfrac{6}{5}, 0\right) \cup \left(\dfrac{5}{6}, 3\right)$

52. $\dfrac{3}{x-2} > 2$

The denominator is equal to 0 when
$x - 2 = 0$, or $x = 2$.

$\dfrac{3}{x-2} = 2$

$3 = 2(x-2)$

$3 = 2x - 4$

$7 = 2x$

$\dfrac{7}{2} = x$

Region	Test Point	$\dfrac{3}{x-2} > 2$
A: $(-\infty, 2)$	0	$\dfrac{3}{-2} > 2$; False
B: $\left(2, \dfrac{7}{2}\right)$	3	$\dfrac{3}{1} > 2$; True
C: $\left(\dfrac{7}{2}, \infty\right)$	5	$\dfrac{3}{3} > 2$; False

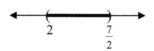

Solution: $\left(2, \dfrac{7}{2}\right)$

53. $f(x) = x^2 - 4$

$x = -\dfrac{b}{2a} = \dfrac{-0}{2(1)} = 0$

$f(0) = (0)^2 - 4 = -4$

Vertex: $(0, -4)$

Axis of symmetry: $x = 0$

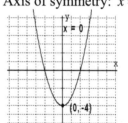

54. $g(x) = x^2 + 7$

$x = -\dfrac{b}{2a} = \dfrac{-0}{2(1)} = 0$

$f(0) = (0)^2 + 7 = 7$

Vertex: $(0, 7)$

Axis of symmetry: $x = 0$

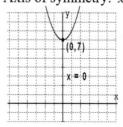

55. $H(x) = 2x^2$

$x = -\dfrac{b}{2a} = \dfrac{-0}{2(2)} = 0$

$f(0) = 2(0)^2 = 0$

Vertex: (0, 0)

Axis of symmetry: $x = 0$

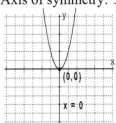

56. $h(x) = -\dfrac{1}{3}x^2$

$x = -\dfrac{b}{2a} = \dfrac{-0}{2(2)} = 0$

$f(0) = -\dfrac{1}{3}(0)^2 = 0$

Vertex: (0, 0)

Axis of symmetry: $x = 0$

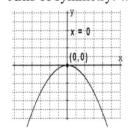

57. $F(x) = (x-1)^2$

Vertex: (1, 0)

Axis of symmetry: $x = 1$

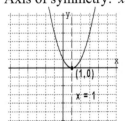

58. $G(x) = (x+5)^2$

Vertex: (−5, 0)

Axis of symmetry: $x = -5$

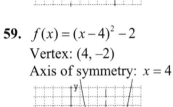

59. $f(x) = (x-4)^2 - 2$

Vertex: (4, −2)

Axis of symmetry: $x = 4$

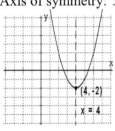

60. $f(x) = -3(x-1)^2 + 1$

Vertex: (1, 1)

Axis of symmetry: $x = 1$

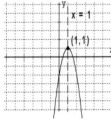

61. $f(x) = x^2 + 10x + 25$

$x = -\dfrac{b}{2a} = \dfrac{-10}{2(1)} = -5$

$f(-5) = (-5)^2 + 10(-5) + 25 = 0$

Vertex: (−5, 0)

$x^2 + 10x + 25 = 0$

$(x+5)^2 = 0$

$x + 5 = 0$

$x = -5$

x-intercept: $(-5, 0)$
$f(0) = 25$ so the y-intercept is $(0, 25)$.

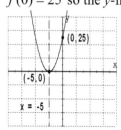

$f(0) = -1$
y-intercept: $(0, -1)$.

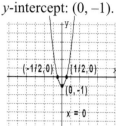

62. $f(x) = -x^2 + 6x - 9$

$x = -\dfrac{b}{2a} = \dfrac{-6}{2(-1)} = 3$

$f(3) = -(3)^2 + 6(3) - 9 = 0$

Vertex: $(3, 0)$

x-intercept: $(3, 0)$

$f(0) = -9$

y-intercept: $(0, -9)$.

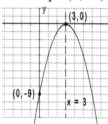

63. $f(x) = 4x^2 - 1$

$x = -\dfrac{b}{2a} = \dfrac{-0}{2(4)} = 0$

$f(0) = 4(0)^2 - 1 = -1$

Vertex: $(0, -1)$

$4x^2 - 1 = 0$

$(2x + 1)(2x - 1) = 0$

$x = -\dfrac{1}{2}$ or $x = \dfrac{1}{2}$

x-intercepts: $\left(-\dfrac{1}{2}, 0\right), \left(\dfrac{1}{2}, 0\right)$

64. $f(x) = -5x^2 + 5$

$x = -\dfrac{b}{2a} = \dfrac{-0}{2(-5)} = 0$

$f(0) = -5(0)^2 + 5 = 5$

Vertex: $(0, 5)$

$-5x^2 + 5 = 0$

$-5x^2 = -5$

$x^2 = 1$

$x = \pm 1$

x-intercepts: $(-1, 0), (1, 0)$

$f(0) = 5$

y-intercept: $(0, 5)$.

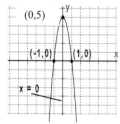

65. $f(x) = -3x^2 - 5x + 4$

$x = -\dfrac{b}{2a} = \dfrac{-(-5)}{2(-3)} = -\dfrac{5}{6}$

$f\left(-\dfrac{5}{6}\right) = -3\left(-\dfrac{5}{6}\right)^2 - 5\left(-\dfrac{5}{6}\right) + 4 = \dfrac{73}{12}$

Vertex: $\left(-\dfrac{5}{6}, \dfrac{73}{12}\right)$

The graph opens downward ($a = -3 < 0$).

$f(0) = 4 \Rightarrow y$-intercept: $(0, 4)$

$$-3x^2 - 5x + 4 = 0$$

$$x = \frac{5 \pm \sqrt{(-5)^2 - 4(-3)(4)}}{2(-3)}$$

$$= \frac{5 \pm \sqrt{73}}{-6} \approx -2.2573 \text{ or } 0.5907$$

x-intercepts: (−2.3, 0), (0.6, 0)

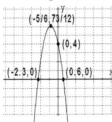

66. $h(t) = -16t^2 + 120t + 300$

a.
$$350 = -16t^2 + 120t + 300$$
$$16t^2 - 120t + 50 = 0$$
$$8t^2 - 60t + 25 = 0$$
$$a = 8, b = -60, c = 25$$
$$t = \frac{60 \pm \sqrt{(-60)^2 - 4(8)(25)}}{2(8)}$$
$$= \frac{60 \pm \sqrt{2800}}{16}$$
$$\approx 0.4 \text{ seconds and } 7.1 \text{ seconds}$$

b. The object will be at 350 feet on the way up and on the way down.

67. Let *x* = one number; then
420 − *x* = the other number.
Let $f(x)$ represent their product.
$$f(x) = x(420 - x)$$
$$= 420x - x^2$$
$$= -x^2 + 420x$$
$$x = -\frac{b}{2a} = \frac{-420}{2(-1)} = 210;$$
$$420 - x = 420 - 210 = 210$$
Therefore, the numbers are both 210.

68. $y = a(x - h)^2 + k$
vertex (−3, 7) gives
$$y = a(x + 3)^2 + 7$$
Passing through the origin gives
$$0 = a(0 + 3)^2 + 7$$
$$-7 = 9a$$
$$-\frac{7}{9} = a$$
Thus, $y = -\frac{7}{9}(x + 3)^2 + 7$.

Chapter 8 Test

1.
$$5x^2 - 2x = 7$$
$$5x^2 - 2x - 7 = 0$$
$$(5x - 7)(x + 1) = 0$$
$$5x - 7 = 0 \text{ or } x + 1 = 0$$
$$x = \frac{7}{5} \text{ or } \quad x = -1$$
The solutions are −1 and $\frac{7}{5}$.

2. $(x + 1)^2 = 10$
$$x + 1 = \pm\sqrt{10}$$
$$x = -1 \pm \sqrt{10}$$
The solutions are $-1 + \sqrt{10}$ and $-1 - \sqrt{10}$.

3. $m^2 - m + 8 = 0$
$$a = 1, b = -1, c = 8$$
$$m = \frac{1 \pm \sqrt{(-1)^2 - 4(1)(8)}}{2(1)}$$
$$= \frac{1 \pm \sqrt{1 - 32}}{2}$$
$$= \frac{1 \pm \sqrt{-31}}{2}$$
$$= \frac{1 \pm i\sqrt{31}}{2} \text{ or } \frac{1}{2} \pm \frac{\sqrt{31}}{2}i$$
The solutions are $\frac{1 + i\sqrt{31}}{2}$ and $\frac{1 - i\sqrt{31}}{2}$.

4. $a^2 - 3a = 5$

$a^2 - 3a - 5 = 0$

$a = 1, b = -3, c = -5$

$a = \dfrac{3 \pm \sqrt{(-3)^2 - 4(1)(-5)}}{2(1)}$

$= \dfrac{3 \pm \sqrt{9 + 20}}{2}$

$= \dfrac{3 \pm \sqrt{29}}{2}$

The solutions are $\dfrac{3 + \sqrt{29}}{2}$ and $\dfrac{3 - \sqrt{29}}{2}$.

5. $\dfrac{4}{x+2} + \dfrac{2x}{x-2} = \dfrac{6}{x^2 - 4}$

$\dfrac{4}{x+2} + \dfrac{2x}{x-2} = \dfrac{6}{(x+2)(x-2)}$

$4(x-2) + 2x(x+2) = 6$

$4x - 8 + 2x^2 + 4x = 6$

$2x^2 + 8x - 14 = 0$

$x^2 + 4x - 7 = 0$

$a = 1, b = 4, c = -7$

$x = \dfrac{-4 \pm \sqrt{(4)^2 - 4(1)(-7)}}{2(1)}$

$= \dfrac{-4 \pm \sqrt{16 + 28}}{2}$

$= \dfrac{-4 \pm \sqrt{44}}{2}$

$= \dfrac{-4 \pm 2\sqrt{11}}{2} = -2 \pm \sqrt{11}$

The solutions are $-2 + \sqrt{11}$ and $-2 - \sqrt{11}$.

6. $x^5 + 3x^4 = x + 3$

$x^5 + 3x^4 - x - 3 = 0$

$x^4(x+3) - 1(x+3) = 0$

$(x+3)(x^4 - 1) = 0$

$(x+3)(x^2 + 1)(x^2 - 1) = 0$

$x + 3 = 0$ or $x^2 + 1 = 0$ or $x^2 - 1 = 0$

$x = -3$ or $x^2 = -1$ or $x^2 = 1$

$x = \pm i$ or $x = \pm 1$

The solutions are $-3, -1, 1, -i,$ and i.

7. $(x+1)^2 - 15(x+1) + 56 = 0$

Let $y = x + 1$. Then $y^2 = (x+1)^2$ and

$y^2 - 15x + 56 = 0$

$(y - 8)(y - 7) = 0$

$y = 8$ or $y = 7$

$x + 1 = 8$ or $x + 1 = 7$

$x = 7$ or $x = 6$

The solutions are 6 and 7.

8. $x^2 - 6x = -2$

$x^2 - 6x + \left(\dfrac{-6}{2}\right)^2 = -2 + 9$

$x^2 - 6x + 9 = 7$

$(x - 3)^2 = 7$

$x - 3 = \pm\sqrt{7}$

$x = 3 \pm \sqrt{7}$

The solutions are $3 + \sqrt{7}$ and $3 - \sqrt{7}$.

9. $2a^2 + 5 = 4a$

$2a^2 - 4a = -5$

$a^2 - 2a = -\dfrac{5}{2}$

$a^2 - 2a + \left(\dfrac{-2}{2}\right)^2 = -\dfrac{5}{2} + 1$

$a^2 - 2a + 1 = -\dfrac{3}{2}$

$(a - 1)^2 = -\dfrac{3}{2}$

$a - 1 = \pm\sqrt{-\dfrac{3}{2}} = \pm\dfrac{\sqrt{3}}{\sqrt{2}}i$

$a - 1 = \pm\dfrac{\sqrt{6}}{2}i$

$a = 1 \pm \dfrac{\sqrt{6}}{2}i$ or $\dfrac{2 \pm i\sqrt{6}}{2}$

The solutions are $1 + \dfrac{\sqrt{6}}{2}i$ and $1 - \dfrac{\sqrt{6}}{2}i$.

10. $2x^2 - 7x > 15$

$2x^2 - 7x - 15 > 0$

$(2x+3)(x-5) > 0$

$2x+3 = 0$ or $x-5 = 0$

$x = -\dfrac{3}{2}$ or $x = 5$

Region	Test Point	$(2x+1)(x-5) > 0$
A: $\left(-\infty, -\dfrac{3}{2}\right)$	-2	$(-3)(-7) > 0$; True
B: $\left(-\dfrac{3}{2}, 5\right)$	0	$(1)(-5) > 0$; False
C: $(5, \infty)$	6	$(13)(1) > 0$; True

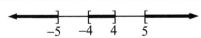

Solution: $\left(-\infty, -\dfrac{3}{2}\right) \cup (5, \infty)$

11. . $(x^2 - 16)(x^2 - 25) \geq 0$

$(x+4)(x-4)(x+5)(x-5) \geq 0$

$x+4 = 0$ or $x-4 = 0$ or $x+5 = 0$ or $x-5 = 0$

$x = -4$ or $x = 4$ or $x = -5$ or $x = 5$

Region	Test Point	$(x+4)(x-4)(x+5)(x-5) \geq 0$
A: $(-\infty, -5)$	-6	$(-2)(-10)\,(-1)(-11) \geq 0$; True
B: $(-5, -4)$	$-\dfrac{9}{2}$	$\left(-\dfrac{1}{2}\right)\left(-\dfrac{17}{2}\right)\left(\dfrac{1}{2}\right)\left(-\dfrac{19}{2}\right) \geq 0$; False
C: $(-4, 4)$	0	$(4)(-4)(5)(-5) \geq 0$; True
D: $(4, 5)$	$\dfrac{9}{2}$	$\left(\dfrac{17}{2}\right)\left(\dfrac{1}{2}\right)\left(\dfrac{19}{2}\right)\left(-\dfrac{1}{2}\right) \geq 0$; False
E: $(5, \infty)$	6	$(10)(2)(11)(1) \geq 0$; True

Solution: $(-\infty, -5] \cup [-4, 4] \cup [5, \infty)$

12. $\dfrac{5}{x+3} < 1$

The denominator is equal to 0 when
$x+3=0$, or $x=-3$.

$\dfrac{5}{x+3} = 1$

$5 = x+3$ so $x=2$

Region	Test Point	$\dfrac{5}{x+3} < 1$
A: $(-\infty, -3)$	-4	$\dfrac{5}{-1} < 1$; True
B: $(-3, 2)$	0	$\dfrac{5}{3} < 1$; False
C: $(2, \infty)$	3	$\dfrac{5}{6} < 1$; True

$$-3 \qquad 2$$

Solution: $(-\infty, -3) \cup (2, \infty)$

13. $\dfrac{7x-14}{x^2-9} \le 0$

$\dfrac{7(x-2)}{(x+3)(x-3)} \le 0$

$x-2=0$ or $x+3=0$ or $x-3=0$

$x=2$ or $x=-3$ or $x=3$

Region	Test Point	$\dfrac{7(x-2)}{(x+3)(x-3)} \le 0$
A: $(-\infty, -3)$	-4	$\dfrac{7(-6)}{(-1)(-7)} \le 0$; T
B: $(-3, 2)$	0	$\dfrac{7(-2)}{(3)(-3)} \le 0$; F
C: $(2, 3)$	$\dfrac{5}{2}$	$\dfrac{7\left(\frac{1}{2}\right)}{\left(\frac{11}{2}\right)\left(-\frac{1}{2}\right)} \le 0$; T
D: $(3, \infty)$	4	$\dfrac{7(2)}{(7)(1)} \le 0$; F

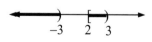
$$-3 \quad 2 \ 3$$

Solution: $(-\infty, -3) \cup [2, 3)$

14. $f(x) = 3x^2$

Vertex: $(0, 0)$

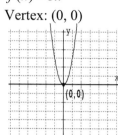
$(0,0)$

15. $G(x) = -2(x-1)^2 + 5$

Vertex: $(1, 5)$

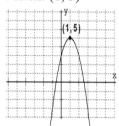
$(1,5)$

16. $h(x) = x^2 - 4x + 4$

$x = -\dfrac{b}{2a} = \dfrac{-(-4)}{2(1)} = 2$

$h(2) = (2)^2 - 4(2) + 4 = 0$

Vertex: $(2, 0)$

$h(0) = 4 \Rightarrow$ y-intercept: $(0, 4)$

x-intercept: $(2, 0)$

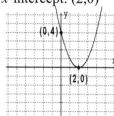
$(0,4)$
$(2,0)$

17. $F(x) = 2x^2 - 8x + 9$

$x = -\dfrac{b}{2a} = \dfrac{-(-8)}{2(2)} = 2$

$F(2) = 2(2)^2 - 8(2) + 9 = 1$

Vertex: (2, 1)

$F(0) = 9 \Rightarrow y\text{-intercept: }(0, 9)$

$2x^2 - 8x + 9 = 0$

$a = 2, b = -8, c = 9$

$x = \dfrac{8 \pm \sqrt{(-8)^2 - 4(2)(9)}}{2(2)}$

$= \dfrac{8 \pm \sqrt{-8}}{4}$

which yields non-real solutons.

Therefore, there are no x-intercepts.

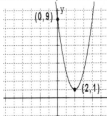

18. Let t = time for Sandy alone. Then $t - 2$ = time for Dave alone.

$\dfrac{1}{t} + \dfrac{1}{t-2} = \dfrac{1}{4}$

$4(t-2) + 4t = t(t-2)$

$4t - 8 + 4t = t^2 - 2t$

$0 = t^2 - 10t + 8$

$a = 1, b = -10, c = 8$

$t = \dfrac{10 \pm \sqrt{(-10)^2 - 4(1)(8)}}{2(1)}$

$= \dfrac{10 \pm \sqrt{68}}{2}$

$= \dfrac{10 \pm 2\sqrt{17}}{2}$

$= 5 \pm \sqrt{17}$

≈ 9.12 or 0.88 (discard)

It takes her about 9.12 hours.

19. $s(t) = -16t^2 + 32t + 256$

a. $t = -\dfrac{b}{2a} = \dfrac{-32}{2(-16)} = 1$

$s(1) = -16(1)^2 + 32(1) + 256 = 272$

Vertex: (1, 272)

The maximum height is 272 feet.

b. $-16t^2 + 32t + 256 = 0$

$t^2 - 2t - 16 = 0$

$a = 1, b = -2, c = -16$

$t = \dfrac{2 \pm \sqrt{(-2)^2 - 4(1)(-16)}}{2(1)}$

$= \dfrac{2 \pm \sqrt{68}}{2}$

$= \dfrac{2 \pm 2\sqrt{17}}{2}$

$= 1 \pm \sqrt{17} \approx -3.12$ and 5.12

Disregard the negative. Hit will hit the water in about 5.12 seconds.

20.
$$a^2 + b^2 = c^2$$
$$x^2 + (x+8)^2 = (20)^2$$
$$x^2 + (x^2 + 16x + 64) = 400$$
$$2x^2 + 16x - 336 = 0$$
$$x^2 + 8x - 168 = 0$$

$a = 1, b = 8, c = -168$

$x = \dfrac{-8 \pm \sqrt{(8)^2 - 4(1)(-168)}}{2(1)}$

$= \dfrac{-8 \pm \sqrt{736}}{2}$

≈ -17.565 or 9.565

Disregard the negative.

$x \approx 9.6$

$x + 8 \approx 9.6 + 8 = 17.6$

$17.6 + 9.6 = 27.2$

$27.2 - 20 = 7.2$

They would save about 7 feet.

Chapter 8 Cumulative Review

1. a. $5 + y \geq 7$

b. $11 \neq z$

c. $20 < 5 - 2x$

2. $|3x - 2| = -5$ which is impossible.
Thus, there is no solution, or $\varnothing$.

3. $m = \dfrac{5-3}{2-0} = \dfrac{2}{2} = 1$
The y-intercept is $(0, 3)$ so $b = 3$.
$y = 1 \cdot x + 3$
$y = x + 3$

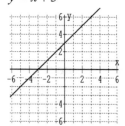

4. $\begin{cases} -6x + \ y = 5 \ (1) \\ \ \ 4x - 2y = 6 \ (2) \end{cases}$
Multiply E1 by 2 and add to E2.
$-12x + 2y = 10$
$\underline{\ \ \ 4x - 2y = 6\ \ \ }$
$-8x \ \ \ \ \ \ = 16$
$x = -2$
Replace x with -2 in E1.
$-6(-2) + y = 5$
$12 + y = 5$
$y = -7$
The solution is $(-2, -7)$.

5. $\begin{cases} \ x - 5y = -12 \ (1) \\ -x + \ y = 4 \ \ \ \ \ \ (2) \end{cases}$
Add E1 and E2.
$-4y = -8$
$y = 2$
Replace y with 2 in E1.

$x - 5(2) = -12$
$x - 10 = -12$
$x = -2$
The solution is $(-2, 2)$.

6. a. $(a^{-2}bc^3)^{-3} = (a^{-2})^{-3}b^{-3}(c^3)^{-3}$
$= a^6 b^{-3} c^{-9}$
$= \dfrac{a^6}{b^3 c^9}$

b. $\left(\dfrac{a^{-4}b^2}{c^3}\right)^{-2} = \dfrac{(a^{-4})^{-2}(b^2)^{-2}}{(c^3)^{-2}}$
$= \dfrac{a^8 b^{-4}}{c^{-6}}$
$= \dfrac{a^8 c^6}{b^4}$

c. $\left(\dfrac{3a^8 b^2}{12a^5 b^5}\right)^{-2} = \left(\dfrac{a^3}{4b^3}\right)^{-2}$
$= \dfrac{(a^3)^{-2}}{4^{-2}(b^3)^{-2}}$
$= \dfrac{4^2 a^{-6}}{b^{-6}}$
$= \dfrac{16 b^6}{a^6}$

7. a. $(2x - 7)(3x - 4) = 6x^2 - 8x - 21x + 28$
$= 6x^2 - 29x + 28$

b. $(3x + y)(5x - 2y)$
$= 15x^2 - 6xy + 5xy - 2y^2$
$= 15x^2 - xy - 2y^2$

8. a. $(4a - 3)(7a - 2) = 28a^2 - 8a - 21a + 6$
$= 28a^2 - 29a + 6$

b. $(2a + b)(3a - 5b)$
$= 6a^2 - 10ab + 3ab - 5b^2$
$= 6a^2 - 7ab - 5b^2$

9. a. $8x^2 + 4 = 4(2x^2 + 1)$

b. $5y - 2z^4$ is a prime polynomial.

c. $6x^2 - 3x^3 = 3x^2(2 - x)$

10. a. $9x^3 + 27x^2 - 15x = 3x(3x^2 + 9x - 5)$

b. $2x(3y - 2) - 5(3y - 2)$
$= (3y - 2)(2x - 5)$

c. $2xy + 6x - y - 3 = 2x(y + 3) - 1(y + 3)$
$= (y + 3)(2x - 1)$

11. $x^2 - 12x + 35 = (x - 5)(x - 7)$

12. $x^2 - 2x - 48 = (x + 6)(x - 8)$

13. $3a^2x - 12ax + 12b^2x = 3x(a^2 - 4a + 4b^2)$
$= 3x(a - 2b)(a - 2b)$
$= 3x(a - 2b)^2$

14. $2ax^2 - 12axy + 18ay^2 = 2a(x^2 - 6xy + 9y^2)$
$= 2a(x - 3y)(x - 3y)$
$= 2a(x - 3y)^2$

15. $3(x^2 + 4) + 5 = -6(x^2 + 2x) + 13$
$3x^2 + 12 + 5 = -6x^2 - 12x + 13$
$9x^2 + 12x + 4 = 0$
$(3x + 2)^2 = 0$
$3x + 2 = 0$
$3x = 2$
$x = -\dfrac{2}{3}$

The solutions is $-\dfrac{2}{3}$.

16. $2(a^2 + 2) - 8 = -2a(a - 2) - 5$
$2a^2 + 4 - 8 = -2a^2 + 4a - 5$
$4a^2 - 4a + 1 = 0$
$(2a - 1)^2 = 0$
$2a - 1 = 0$
$2a = 1$
$a = \dfrac{1}{2}$

The solution is $\dfrac{1}{2}$.

17.
$$x^3 = 4x$$
$$x^3 - 4x = 0$$
$$x(x^2 - 4) = 0$$
$$x(x + 2)(x - 2) = 0$$
$$x = 0 \ \text{ or } \ x + 2 = 0 \quad \text{ or } \ x - 2 = 0$$
$$x = -2 \ \text{ or } \qquad x = 2$$
The solutions are -2, 0, and 2.

18. $f(x) = x^2 + x - 12$
$$x = -\dfrac{b}{2a} = \dfrac{-1}{2(1)} = -\dfrac{1}{2}$$
$$f\left(-\dfrac{1}{2}\right) = \left(-\dfrac{1}{2}\right)^2 + \left(-\dfrac{1}{2}\right) - 12$$
$$= \dfrac{1}{4} - \dfrac{1}{2} - 12$$
$$= -\dfrac{49}{4}$$
Vertex: $\left(-\dfrac{1}{2}, -\dfrac{49}{4}\right)$

To find the *x*-intercepts, set $y = 0$.
$$x^2 + x - 12 = 0$$
$$(x + 4)(x - 3) = 0$$
$$x + 4 = 0 \quad \text{ or } \quad x - 3 = 0$$
$$x = -4 \qquad\qquad x = 3$$
Therefore, *x*-intercepts are $(-4, 0)$ and $(3, 0)$.

To find *y*-intercept, set $x = 0$.
$$y = x^2 + x - 12$$
$$y = 0 + 0 - 12$$
$$y = -12$$
Therefore, the *y*-intercept is $(0, -12)$.

19. $\dfrac{2x^2}{10x^3 - 2x^2} = \dfrac{2x^2}{2x^2(5x - 1)} = \dfrac{1}{5x - 1}$

20. $\dfrac{x^2 - 4x + 4}{2 - x} = \dfrac{(x - 2)^2}{-(x - 2)} = \dfrac{x - 2}{-1} = 2 - x$

21. $\dfrac{2x-1}{2x^2-9x-5}+\dfrac{x+3}{6x^2-x-2}$

$=\dfrac{2x-1}{(2x+1)(x-5)}+\dfrac{x+3}{(2x+1)(3x-2)}$

$=\dfrac{(2x-1)(3x-2)+(x+3)(x-5)}{(2x+1)(x-5)(3x-2)}$

$=\dfrac{(6x^2-4x-3x+2)+(x^2-5x+3x-15)}{(2x+1)(x-5)(3x-2)}$

$=\dfrac{7x^2-9x-13}{(2x+1)(x-5)(3x-2)}$

22. $\dfrac{a+1}{a^2-6a+8}-\dfrac{3}{16-a^2}$

$=\dfrac{a+1}{(a-4)(a-2)}-\dfrac{3}{(4+a)(4-a)}$

$=\dfrac{a+1}{(a-4)(a-2)}+\dfrac{3}{(4+a)(a-4)}$

$=\dfrac{(a+1)(a+4)+3(a-2)}{(a-4)(a-2)(a+4)}$

$=\dfrac{(a^2+4a+a+4)+3a-6}{(a-4)(a-2)(a+4)}$

$=\dfrac{a^2+8a-2}{(a-4)(a-2)(a+4)}$

23. $\dfrac{x^{-1}+2xy^{-1}}{x^{-2}-x^{-2}y^{-1}}=\dfrac{\dfrac{1}{x}+\dfrac{2x}{y}}{\dfrac{1}{x^2}-\dfrac{1}{x^2y}}$

$=\dfrac{\left(\dfrac{1}{x}+\dfrac{2x}{y}\right)x^2y}{\left(\dfrac{1}{x^2}-\dfrac{1}{x^2y}\right)x^2y}$

$=\dfrac{xy+2x^3}{y-1}$

24. $\dfrac{(2a)^{-1}+b^{-1}}{a^{-1}+(2b)^{-1}}=\dfrac{\dfrac{1}{2a}+\dfrac{1}{b}}{\dfrac{1}{a}+\dfrac{1}{2b}}$

$=\dfrac{\left(\dfrac{1}{2a}+\dfrac{1}{b}\right)2ab}{\left(\dfrac{1}{a}+\dfrac{1}{2b}\right)2ab}$

$=\dfrac{b+2a}{2b+a}$

$=\dfrac{2a+b}{a+2b}$

25. $\dfrac{3x^5y^2-15x^3y-x^2y-6x}{x^2y}$

$=\dfrac{3x^5y^2}{x^2y}-\dfrac{15x^3y}{x^2y}-\dfrac{x^2y}{x^2y}-\dfrac{6x}{x^2y}$

$=3x^3y-15x-1-\dfrac{6}{xy}$

26. $x+3\overline{\smash{\big)}\,x^3-3x^2-10x+24}$ quotient x^2-6x+8

$\quad\ \underline{x^3+3x^2}$

$\quad\quad\ -6x^2-10x$

$\quad\quad\ \underline{-6x^2-18x}$

$\quad\quad\qquad\ 8x+24$

$\quad\quad\qquad\ \underline{8x+24}$

$\quad\quad\qquad\qquad\ 0$

Answer: x^2-6x+8

27. $P(x)=2x^3-4x^2+5$

a. $P(2)=2(2)^3-4(2)^2+5$

$\qquad\ =2(8)-4(4)+5$

$\qquad\ =16-16+5$

$\qquad\ =5$

b. $2\,\big|\ 2\quad -4\quad 0\quad\ 5$

$\qquad\qquad\quad\ \underline{4\quad\ \ 0\quad\ 0}$

$\qquad\qquad\ \ 2\quad\ \ 0\quad\ 0\quad\ 5$

Thus, $P(2)=5$.

28. $P(x) = 4x^3 - 2x^2 + 3$

   a. $P(-2) = 4(-2)^3 - 2(-2)^2 + 3$
            $= 4(-8) - 2(4) + 3$
            $= -32 - 8 + 3$
            $= -37$

   b.

$$\begin{array}{r|rrrr} -2 & 4 & -2 & 0 & 3 \\ & & -8 & 20 & -40 \\ \hline & 4 & -10 & 20 & -37 \end{array}$$

     Thus, $P(-2) = -37$.

29. $\dfrac{4x}{5} + \dfrac{3}{2} = \dfrac{3x}{10}$

$$10\left(\frac{4x}{5} + \frac{3}{2}\right) = 10\left(\frac{3x}{10}\right)$$
$$2(4x) + 5(3) = 3x$$
$$8x + 15 = 3x$$
$$5x = -15$$
$$x = -3$$

The solution is –3.

30. $\dfrac{x+3}{x^2 + 5x + 6} = \dfrac{3}{2x+4} - \dfrac{1}{x+3}$

$$\frac{x+3}{(x+3)(x+2)} = \frac{3}{2(x+2)} - \frac{1}{x+3}$$
$$2(x+3) = 3(x+3) - 2(x+2)$$
$$2x + 6 = 3x + 9 - 2x - 4$$
$$2x + 6 = x + 5$$
$$x = -1$$

31. Let x = the number.

$$\frac{9-x}{19+x} = \frac{1}{3}$$
$$3(9-x) = 1(19+x)$$
$$27 - 3x = 19 + x$$
$$-4x = -8$$
$$x = 2$$

The number is 2.

32. Let t = time to roof the house together.

$$\frac{1}{24} + \frac{1}{40} = \frac{1}{t}$$
$$120t\left(\frac{1}{24} + \frac{1}{40}\right) = 120t\left(\frac{1}{t}\right)$$
$$5t + 3t = 120$$
$$8t = 120$$
$$t = \frac{120}{8} = 15$$

It would take them 15 hours to roof the house working together.

33. $y = kx$
    $5 = k(30)$
    $k = \dfrac{5}{30} = \dfrac{1}{6}$ and $y = \dfrac{1}{6}x$

34. $y = \dfrac{k}{x}$
    $8 = \dfrac{k}{24}$
    $k = 8(24) = 192$ and $y = \dfrac{192}{x}$

35. a. $\sqrt{(-3)^2} = |-3| = 3$

   b. $\sqrt{x^2} = |x|$

   c. $\sqrt[4]{(x-2)^4} = |x-2|$

   d. $\sqrt[3]{(-5)^3} = -5$

   e. $\sqrt[5]{(2x-7)^5} = 2x - 7$

   f. $\sqrt{25x^2} = \sqrt{25} \cdot \sqrt{x^2} = 5|x|$

   g. $\sqrt{x^2 + 2x + 1} = \sqrt{(x+1)^2} = |x+1|$

36. a. $\sqrt{(-2)^2} = |-2| = 2$

   b. $\sqrt{y^2} = |y|$

   c. $\sqrt[4]{(a-3)^4} = |a-3|$

d. $\sqrt[3]{(-6)^3} = -6$

e. $\sqrt[5]{(3x-1)^5} = 3x-1$

37. a. $\sqrt[8]{x^4} = x^{4/8} = x^{1/2} = \sqrt{x}$

b. $\sqrt[6]{25} = (25)^{1/6}$
$$= (5^2)^{1/6} = 5^{2/6} = 5^{1/3} = \sqrt[3]{5}$$

c. $\sqrt[4]{r^2 s^6} = (r^2 s^6)^{1/4}$
$$= r^{2/4} s^{6/4}$$
$$= r^{1/2} s^{3/2}$$
$$= (rs^3)^{1/2} = \sqrt{rs^3}$$

38. a. $\sqrt[4]{5^2} = 5^{2/4} = 5^{1/2} = \sqrt{5}$

b. $\sqrt[12]{x^3} = x^{3/12} = x^{1/4} = \sqrt[4]{x}$

c. $\sqrt[6]{x^2 y^4} = (x^2 y^4)^{1/6}$
$$= x^{2/6} y^{4/6}$$
$$= x^{1/3} y^{2/3}$$
$$= (xy^2)^{1/3} = \sqrt[3]{xy^2}$$

39. a. $\sqrt{25x^3} = \sqrt{25x^2 \cdot x} = 5x\sqrt{x}$

b. $\sqrt[3]{54x^6 y^8} = \sqrt[3]{27x^6 y^6 \cdot 2y^2}$
$$= 3x^2 y^2 \sqrt[3]{2y^2}$$

c. $\sqrt[4]{81z^{11}} = \sqrt[4]{81z^8 \cdot z^3} = 3z^2 \sqrt[4]{z^3}$

40. a. $\sqrt{64a^5} = \sqrt{64a^4 \cdot a} = 8a^2 \sqrt{a}$

b. $\sqrt[3]{24a^7 b^9} = \sqrt[3]{8a^6 b^9 \cdot 3a}$
$$= 2a^2 b^3 \sqrt[3]{3a}$$

c. $\sqrt[4]{48x^9} = \sqrt[4]{16x^8 \cdot 3x} = 2x^2 \sqrt[4]{3x}$

41. a. $\dfrac{2}{\sqrt{5}} = \dfrac{2 \cdot \sqrt{5}}{\sqrt{5} \cdot \sqrt{5}} = \dfrac{2\sqrt{5}}{5}$

b. $\dfrac{2\sqrt{16}}{\sqrt{9x}} = \dfrac{2 \cdot 4}{3\sqrt{x}} = \dfrac{8 \cdot \sqrt{x}}{3\sqrt{x} \cdot \sqrt{x}} = \dfrac{8\sqrt{x}}{3x}$

c. $\sqrt[3]{\dfrac{1}{2}} = \dfrac{\sqrt[3]{1}}{\sqrt[3]{2}} = \dfrac{1}{\sqrt[3]{2}} = \dfrac{1 \cdot \sqrt[3]{2^2}}{\sqrt[3]{2} \cdot \sqrt[3]{2^2}} = \dfrac{\sqrt[3]{4}}{2}$

42. a. $\left(\sqrt{3} - 4\right)\left(2\sqrt{3} + 2\right)$
$$= \sqrt{3} \cdot 2\sqrt{3} + 2\sqrt{3} - 4 \cdot 2\sqrt{3} - 4 \cdot 2$$
$$= 2(3) + 2\sqrt{3} - 8\sqrt{3} - 8$$
$$= 6 - 5\sqrt{3} - 8$$
$$= -2 - 6\sqrt{3}$$

b. $\left(\sqrt{5} - x\right)^2 = \left(\sqrt{5}\right)^2 - 2 \cdot \sqrt{5} \cdot x + x^2$
$$= 5 - 2x\sqrt{5} + x^2$$

c. $\left(\sqrt{a} + b\right)\left(\sqrt{a} - b\right) = \left(\sqrt{a}\right)^2 - b^2$
$$= a - b^2$$

43. $\sqrt{2x+5} + \sqrt{2x} = 3$

$$\sqrt{2x+5} = 3 - \sqrt{2x}$$

$$\left(\sqrt{2x+5}\right)^2 = \left(3 - \sqrt{2x}\right)^2$$

$$2x+5 = 9 - 6\sqrt{2x} + 2x$$

$$-4 = -6\sqrt{2x}$$

$$(-4)^2 = \left(-6\sqrt{2x}\right)^2$$

$$16 = 36(2x)$$

$$16 = 72x$$

$$x = \frac{16}{72} = \frac{2}{9}$$

The solution is $\frac{2}{9}$.

44. $\sqrt{x-2} = \sqrt{4x+1} - 3$

$$\left(\sqrt{x-2}\right)^2 = \left(\sqrt{4x+1} - 3\right)^2$$

$$x-2 = (4x+1) - 6\sqrt{4x+1} + 9$$

$$6\sqrt{4x+1} = 3x+12$$

$$2\sqrt{4x+1} = x+4$$

$$\left(2\sqrt{4x+1}\right)^2 = (x+4)^2$$

$$4(4x+1) = x^2 + 8x + 16$$

$$16x+4 = x^2 + 8x + 16$$

$$0 = x^2 - 8x + 12$$

$$0 = (x-6)(x-2)$$

$$x-6 = 0 \quad \text{or} \quad x-2 = 0$$

$$x = 6 \quad \text{or} \qquad x = 2$$

The solutions are 2 and 6.

45. a. $\dfrac{2+i}{1-i} = \dfrac{(2+i)\cdot(1+i)}{(1-i)\cdot(1+i)}$

$$= \frac{2+2i+1i+i^2}{1^2 - i^2}$$

$$= \frac{2+3i-1}{1+1}$$

$$= \frac{1+3i}{2} \quad \text{or} \quad \frac{1}{2} + \frac{3}{2}i$$

b. $\dfrac{7}{3i} = \dfrac{7\cdot(-3i)}{3i\cdot(-3i)} = \dfrac{-21i}{-9i^2} = \dfrac{-21i}{9} = -\dfrac{7}{3}i$

46. a. $3i(5-2i) = 15i - 6i^2$

$$= 15i + 6$$

$$= 6 + 15i$$

b. $(6-5i)^2 = 6^2 - 2(6)(5i) + (5i)^2$

$$= 36 - 60i + 25i^2$$

$$= 36 - 60i - 25$$

$$= 11 - 60i$$

c. $\left(\sqrt{3}+2i\right)\left(\sqrt{3}-2i\right) = \left(\sqrt{3}\right)^2 - (2i)^2$

$$= 3 - 4i^2$$

$$= 3 + 4$$

$$= 7$$

47. $(x+1)^2 = 12$

$$x+1 = \pm\sqrt{12}$$

$$x+1 = \pm 2\sqrt{3}$$

$$x = -1 \pm 2\sqrt{3}$$

The solutions are $-1 + 2\sqrt{3}$ and $-1 - 2\sqrt{3}$.

48. $(y-1)^2 = 24$

$$y-1 = \pm\sqrt{24}$$

$$y-1 = \pm 2\sqrt{6}$$

$$y = 1 \pm 2\sqrt{6}$$

The solutions are $1 + 2\sqrt{6}$ and $1 - 2\sqrt{6}$.

49. $x - \sqrt{x} - 6 = 0$

Let $y = \sqrt{x}$. Then $y^2 = x$ and

$$y^2 - y - 6 = 0$$

$$(y-3)(y+2) = 0$$

$$y-3 = 0 \quad \text{or} \quad y+2 = 0$$

$$y = 3 \quad \text{or} \qquad y = -2$$

$$\sqrt{x} = 3 \quad \text{or} \quad \sqrt{x} = -2 \text{ (can't happen)}$$

$$x = 9$$

The solution is 9.

50.
$$m^2 = 4m + 8$$
$$m^2 - 4m - 8 = 0$$
$$a = 1, b = -4, c = -8$$
$$x = \frac{4 \pm \sqrt{(-4)^2 - 4(1)(-8)}}{2(1)}$$
$$= \frac{4 \pm \sqrt{16 + 32}}{2}$$
$$= \frac{4 \pm \sqrt{48}}{2}$$
$$= \frac{4 \pm 4\sqrt{3}}{2}$$
$$= 2 \pm 2\sqrt{3}$$

The solutions are $2 + 2\sqrt{3}$ and $2 - 2\sqrt{3}$.

Chapter 9

Section 9.1

Mental Math

1. C

2. E

3. F

4. A

5. D

6. B

Exercise Set 9.1

1. **a.** $(f+g)(x) = x-7+2x+1 = 3x-6$

   b. $(f-g)(x) = x-7-(2x+1)$
   $$= x-7-2x-1$$
   $$= -x-8$$

   c. $(f \cdot g)(x) = (x-7)(2x+1)$
   $$= 2x^2 - 13x - 7$$

   d. $\left(\dfrac{f}{g}\right)(x) = \dfrac{x-7}{2x+1}$, where $x \neq \dfrac{1}{2}$

3. **a.** $(f+g)(x) = x^2 + 5x + 1$

   b. $(f-g)(x) = x^2 - 5x + 1$

   c. $(f \cdot g)(x) = (x^2+1)(5x) = 5x^3 + 5x$

   d. $\left(\dfrac{f}{g}\right)(x) = \dfrac{x^2+1}{5x}$, where $x \neq 0$

5. **a.** $(f+g)(x) = \sqrt{x} + x + 5$

   b. $(f-g)(x) = \sqrt{x} - x - 5$

c. $(f \cdot g)(x) = \sqrt{x}(x+5)$
$$= x\sqrt{x} + 5\sqrt{x}$$

d. $\left(\dfrac{f}{g}\right)(x) = \dfrac{\sqrt{x}}{x+5}$, where $x \neq -5$

7. **a.** $(f+g)(x) = -3x + 5x^2$
$$= 5x^2 - 3x$$

   b. $(f-g)(x) = -3x - 5x^2$
$$= -5x^2 - 3x$$

   c. $(f \cdot g)(x) = (-3x)(5x^2) = -15x^3$

   d. $\left(\dfrac{f}{g}\right)(x) = \dfrac{-3x}{5x^2}$
$$= -\dfrac{3}{5x}, \text{ where } x \neq 0.$$

9. $(f \circ g)(2) = f(g(2))$
$$= f(-4)$$
$$= (-4)^2 - 6(-4) + 2$$
$$= 16 + 24 + 2$$
$$= 42$$

11. $(g \circ f)(-1) = g(f(-1))$
$$= g(9)$$
$$= -2(9)$$
$$= -18$$

13. $(g \circ h)(0) = g(h(0))$
$$= g(0)$$
$$= -2(0)$$
$$= 0$$

15. $(f \circ g)(x) = f(g(x))$
$$= f(5x)$$
$$= (5x)^2 + 1$$
$$= 25x^2 + 1$$

$$(g \circ f)(x) = g(f(x))$$
$$= g(x^2 + 1)$$
$$= 5(x^2 + 1)$$
$$= 5x^2 + 5$$

17. $(f \circ g)(x) = f(g(x))$
$$= f(x + 7)$$
$$= 2(x + 7) - 3$$
$$= 2x + 14 - 3$$
$$= 2x + 11$$

$$(g \circ f)(x) = g(f(x))$$
$$= g(2x - 3)$$
$$= (2x - 3) + 7$$
$$= 2x + 4$$

19. $(f \circ g)(x) = f(g(x))$
$$= f(-2x)$$
$$= (-2x)^3 + (-2x) - 2$$
$$= -8x^3 - 2x - 2$$
$$(g \circ f)(x) = g(f(x))$$
$$= g(x^3 + x - 2)$$
$$= -2(x^3 + x - 2)$$
$$= -2x^3 - 2x + 4$$

21. $(f \circ g)(x) = f(g(x))$
$$= f(-5x + 2)$$
$$= \sqrt{-5x + 2} = f(-5x + 2)$$
$$= \sqrt{-5x + 2}$$

$$(g \circ f)(x) = g(f(x))$$
$$= g(\sqrt{x})$$
$$= -5\sqrt{x} + 2$$

23. $H(x) = (g \circ h)(x)$
$$= g(h(x))$$
$$= g(x^2 + 2)$$
$$= \sqrt{x^2 + 2}$$

25. $F(x) = (h \circ f)(x)$
$$= h(f(x))$$
$$= h(3x)$$
$$= (3x)^2 + 2$$
$$= 9x^2 + 2$$

27. $G(x) = (f \circ g)(x)$
$$= f(g(x))$$
$$= f\left(\sqrt{x}\right)$$
$$= 3\sqrt{x}$$

29. Answers may vary. For example,
$g(x) = x + 2$ and $f(x) = x^2$.

31. Answers may vary. For example,
$g(x) = x + 5$ and $f(x) = \sqrt{x} + 2$.

33. Answers may vary. For example,
$g(x) = 2x - 3$ and $f(x) = \dfrac{1}{x}$.

35. $y = x - 2$

37. $y = \dfrac{x}{3}$

39. $y = -\dfrac{x + 7}{2}$

41. $(f + g)(2) = f(2) + g(2)$
$$= 7 + (-1)$$
$$= 6$$

43. $(f \circ g)(2) = f(g(2))$
$$= f(-1)$$
$$= 4$$

45. $(f \cdot g)(7) = f(7) \cdot g(7)$
$$= 1 \cdot 4$$
$$= 4$$

47. $\left(\dfrac{f}{g}\right)(-1) = \dfrac{f(-1)}{g(-1)}$
$$= \dfrac{4}{-4}$$
$$= -1$$

49. Answers may vary

51. $P(x) = R(x) - C(x)$

Exercise Set 9.2

1. $f = \{(-1,-1),(1,1),(0,2),(2,0)\}$

is a one-to-one function.

$f^{-1} = \{(-1,-1),(1,1),(2,0),(0,2)\}$

3. $h = \{(10,10)\}$

is a one-to-one function.

$h^{-1} = \{(10,10)\}$

5. $f = \{(11,12),(4,3),(3,4),(6,6)\}$

is a one-to-one function.

$f^{-1} = \{(12,11),(3,4),(4,3),(6,6)\}$

7. This function is not one-to-one because there are two months with the same output: (May, 6.1) and (September,6.1).

9. This function is one-to-one.

Rank in population (Input)	1	49	12	2	45
State (Output)	California	Vermont	Virginia	Texas	South Dakota

11. $f(x) = x^3 + 2$

    a. $f(1) = 1^3 + 2 = 3$

    b. $f^{-1}(3) = 1$

13. $f(x) = x^3 + 2$

    a. $f(-1) = (-1)^3 + 2 = 1$

    b. $f^{-1}(1) = -1$

15. The graph represents a one-to-one function because it passes the horizontal line test.

17. The graph does not represent a one-to-one function because it does not pass the horizontal line test.

19. The graph represents a one-to-one function because it passes the horizontal line test.

21. The graph does not represent a one-to-one function because it does not pass the horizontal line test.

23. $f(x) = x + 4$

$$y = x + 4$$
$$x = y + 4$$
$$y = x - 4$$
$$f^{-1}(x) = x - 4$$

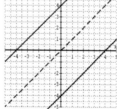

25. $f(x) = 2x - 3$

$$y = 2x - 3$$
$$x = 2y - 3$$
$$2y = x + 3$$
$$y = \frac{x+3}{2}$$
$$f^{-1}(x) = \frac{x+3}{2}$$

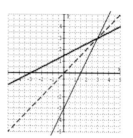

27. $f(x) = \frac{1}{2}x - 1$

$$y = \frac{1}{2}x - 1$$
$$x = \frac{1}{2}y - 1$$
$$\frac{1}{2}y = x + 1$$
$$y = 2x + 2$$
$$f^{-1}(x) = 2x + 2$$

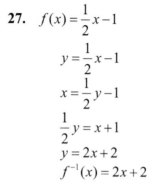

29. $f(x) = x^3$

$$y = x^3$$
$$x = y^3$$
$$y = \sqrt[3]{x}$$
$$f^{-1}(x) = \sqrt[3]{x}$$

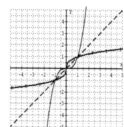

31. $f(x) = 5x + 2$

$\quad y = 5x + 2$

$\quad x = 5y + 2$

$\quad 5y = x - 2$

$\quad y = \dfrac{x - 2}{5}$

$\quad f^{-1}(x) = \dfrac{x - 2}{5}$

33. $f(x) = \dfrac{x - 2}{5}$

$\quad y = \dfrac{x - 2}{5}$

$\quad x = \dfrac{y - 2}{5}$

$\quad 5x = y - 2$

$\quad y = 5x + 2$

$\quad f^{-1}(x) = 5x + 2$

35. $f(x) = \sqrt[3]{x}$

$\quad y = \sqrt[3]{x}$

$\quad x = \sqrt[3]{y}$

$\quad x^3 = y$

$\quad f^{-1}(x) = x^3$

37. $f(x) = \dfrac{5}{3x + 1}$

$\quad y = \dfrac{5}{3x + 1}$

$\quad x = \dfrac{5}{3y + 1}$

$\quad 3y + 1 = \dfrac{5}{x}$

$\quad 3y = \dfrac{5}{x} - 1$

$\quad 3y = \dfrac{5 - x}{x}$

$\quad y = \dfrac{5 - x}{3x}$

$\quad f^{-1}(x) = \dfrac{5 - x}{3x}$

39. $f(x) = (x + 2)^3$

$\quad y = (x + 2)^3$

$\quad x = (y + 2)^3$

$\quad \sqrt[3]{x} = y + 2$

$\quad \sqrt[3]{x} - 2 = y$

$\quad f^{-1}(x) = \sqrt[3]{x} - 2$

41.

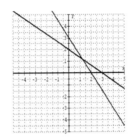

43.

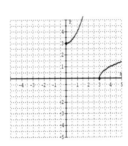

45.

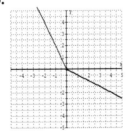

47. $(f \circ f^{-1})(x) = f(f^{-1}(x))$

$\qquad = f\left(\dfrac{x - 1}{2}\right)$

$\qquad = 2\left(\dfrac{x - 1}{2}\right) + 1$

$\qquad = x - 1 + 1$

$\qquad = x$

$$(f^{-1} \circ f)(x) = f^{-1}(f(x))$$
$$= f^{-1}(2x+1)$$
$$= \frac{(2x+1)-1}{2}$$
$$= \frac{2x}{2}$$
$$= x$$

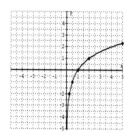

49. $(f \circ f^{-1})(x) = f(f^{-1}(x))$
$$= f\left(\sqrt[3]{x-6}\right)$$
$$= \left(\sqrt[3]{x-6}\right)^3 + 6$$
$$= x - 6 + 6$$
$$= x$$
$$(f^{-1} \circ f)(x) = f^{-1}(f(x))$$
$$= f^{-1}(x^3 + 6)$$
$$= \sqrt[3]{(x^3 + 6) - 6}$$
$$= \sqrt[3]{x^3}$$
$$= x$$

51. 5

53. 8

55. $\dfrac{1}{27}$

57. 9

59. $3^{1/2} \approx 1.73$

61. **a.** $\left(-2, \dfrac{1}{4}\right), \left(-1, \dfrac{1}{2}\right), (0,1), (1,2),$
$(2,5)$

b. $\left(\dfrac{1}{4}, -2\right), \left(\dfrac{1}{2}, -1\right), (1,0), (2,1),$
$(5,2)$

c., d.

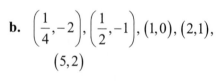

63. Answers may vary.

65. $f(x) = 3x + 1$
$$y = 3x + 1$$
$$x = 3y + 1$$
$$x - 1 = 3y$$
$$y = \frac{x-1}{3}$$
$$f^{-1}(x) = \frac{x-1}{3}$$

67. $f(x) = \sqrt[3]{x+1}$
$$y = \sqrt[3]{x+1}$$
$$x = \sqrt[3]{y+1}$$
$$x^3 = y + 1$$
$$y = x^3 - 1$$
$$f^{-1}(x) = x^3 - 1$$

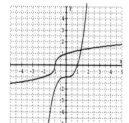

Section 9.3

Graphing Calculator Explorations

1.

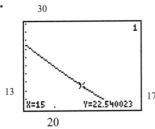

81.98%

3.

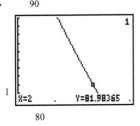

22.54%

Exercise Set 9.3

1. $y = 4^x$

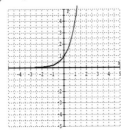

3. $y = 2^x + 1$

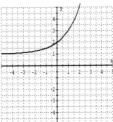

5. $y = \left(\dfrac{1}{4}\right)^x$

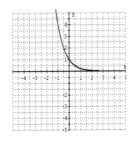

7. $y = \left(\dfrac{1}{2}\right)^x - 2$

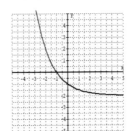

9. $y = -2^x$

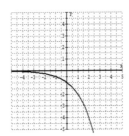

11. $y = -\left(\dfrac{1}{4}\right)^x$

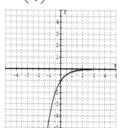

13. $f(x) = 2^{x+1}$

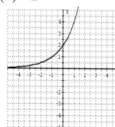

15. $f(x) = 4^{x-2}$

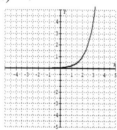

17. C

19. D

21. $3^x = 27$
$3^x = 3^3$
$x = 3$
The solution is 3.

23. $16^x = 8$
$(2^4)^x = 2^3$
$2^{4x} = 2^3$
$4x = 3$
$x = \dfrac{3}{4}$
The solution is $\dfrac{3}{4}$.

25. $32^{2x-3} = 2$
$(2^5)^{2x-3} = 2^1$
$10x - 15 = 1$
$10x = 16$
$x = \dfrac{8}{5}$
The solution is $\dfrac{8}{5}$.

27. $\dfrac{1}{4} = 2^{3x}$
$2^{-2} = 2^{3x}$
$3x = -2$
$x = -\dfrac{2}{3}$
The solution is $-\dfrac{2}{3}$.

29. $5^x = 625$
$5^x = 5^4$
$x = 4$
The solution is 4.

31. $4^x = 8$

$(2^2)^x = 2^3$

$2^{2x} = 2^3$

$2x = 3$

$x = \dfrac{3}{2}$

The solution is $\dfrac{3}{2}$.

33. $27^{x+1} = 9$

$(3^3)^{x+1} = 3^2$

$3^{3x+3} = 3^2$

$3x + 3 = 2$

$3x = -1$

$x = -\dfrac{1}{3}$

The solution is $-\dfrac{1}{3}$.

35. $81^{x-1} = 27^{2x}$

$(3^4)^{x-1} = (3^3)^{2x}$

$3^{4x-4} = 3^{6x}$

$4x - 4 = 6x$

$-4 = 2x$

$x = -2$

The solution is -2.

37. $y = 30(2.7)^{-0.004t}, t = 50$

$= 30(2.7)^{-(0.004)(50)}$

$= 30(2.7)^{-0.2}$

≈ 24.6

Therefore, approximately 24.6 pounds of uranium will remain after 50 days.

39. $y = 260(2.7)^{0.025t}$

$y = 260(2.7)^{0.025(10)}$

$y = 260(2.7)^{0.25}$

$y = 333$

There should be about 333 bison in the park in 10 years.

41. $y = 5(2.7)^{-0.15t}, t = 10$

$= 5(2.7)^{-0.15(10)}$

$= 5(2.7)^{-1.5}$

≈ 1.1 grams

43. **a.** $y = 42.1(1.56)^t$

$= 42.1(1.56)^1$

≈ 65.7

$\$65.7$ billion

b. $y = 42.1(1.56)^t$

$= 42.1(1.56)^9$

≈ 2303.6

$\$2303.6$ billion

45. **a.** $y = 120.882(1.012)^x$

$= 120.882(1.012)^{40}$

≈ 194.8 million people

b. $y = 120.882(1.012)^x$

$= 120.882(1.012)^{90}$

≈ 353.7 million people

47. $A = P\left(1 + \dfrac{r}{n}\right)^{nt}$

$t = 3, P = 6000, r = 0.08,$ and $n = 12$.

$$A = 6000\left(1 + \frac{0.08}{12}\right)^{12(3)}$$
$$= 6000(1.006)^{36}$$
$$\approx 7621.42$$

Erica would owe $7621.42 after 3 years.

49. $A = P\left(1 + \dfrac{r}{n}\right)^{nt}$, where $P = 2000$

$r = 0.06$, $n = 2$, and $t = 12$.

$$A = 2000\left(1 + \frac{0.06}{2}\right)^{2(12)}$$
$$= 2000(1.03)^{24}$$
$$= 4065.59$$

Janina has approximately $4065.59 in her savings account.

51. $y = 34(1.254)^x$

$$= 34(1.254)^{13}$$
$$\approx 645$$

There will be approximately 645 million cellular phone users in 2005.

53. $5x - 2 = 18$

$$5x = 20$$
$$x = 4$$

$$3x - 4 = 3(x + 1)$$

55. $3x - 4 = 3x + 1$

$$-4 = 1$$

Since $-4 \neq 1$, there is no solution.

57.
$$x^2 + 6 = 5x$$
$$x^2 - 5x + 6 = 0$$
$$(x - 3)(x - 2) = 0$$

$$x - 3 = 0 \text{ or } x - 2 = 0$$
$$x = 3 \qquad\quad x = 2$$

59. 3

61. -1

63. answers may vary

65. $y = \left|3^x\right|$

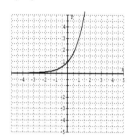

67. $y = 3^{|x|}$

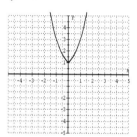

69. The graphs are the same since $\left(\dfrac{1}{2}\right)^{-x} = 2^x$.

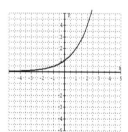

71.

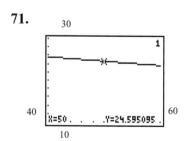

24.60 pounds

73.

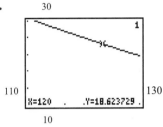

18.62 lbs

75.

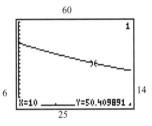

50.41 g

Exercise Set 9.4

1. $\log_6 36 = 2$

$6^2 = 36$

3. $\log_3 \dfrac{1}{27} = -3$

$3^{-3} = \dfrac{1}{27}$

5. $\log_{10} 1000 = 3$

$10^3 = 1000$

7. $\log_e x = 4$

$e^4 = x$

9. $\log_e \dfrac{1}{e^2} = -2$

$e^{-2} = \dfrac{1}{e^2}$

11. $\log_7 \sqrt{7} = \dfrac{1}{2}$

$7^{1/2} = \sqrt{7}$

13. $2^4 = 16$

$\log_2 16 = 4$

15. $10^2 = 100$

$\log_{10} 100 = 2$

17. $e^3 = x$

$\log_e x = 3$

19. $10^{-1} = \dfrac{1}{10}$

$\log_{10} \dfrac{1}{10} = -1$

21. $4^{-2} = \dfrac{1}{16}$

$\log_4 \dfrac{1}{16} = -2$

23. $5^{1/2} = \sqrt{25}$

$\log_5 \sqrt{5} = \dfrac{1}{2}$

25. $\log_2 8 = 3$ since $2^3 = 8$.

27. $\log_3 \dfrac{1}{9} = -2$ since $3^{-2} = \dfrac{1}{9}$.

29. $\log_{25} 5 = \dfrac{1}{2}$ since $25^{1/2} = 5$.

31. $\log_{1/2} 2 = -1$ since $\left(\dfrac{1}{2}\right)^{-1} = 2$.

33. $\log_7 1 = 0$

35. $\log_2 2^4 = 4$

37. $\log_{10} 100 = 2$

39. $3^{\log_3 5} = 5$

41. $\log_3 81 = 4$

43. $\log_4 \left(\dfrac{1}{64} \right) = -3$

45. Answers may vary.

47. $\log_3 9 = x$

$3^x = 9$

$x = 2$

49. $\log_3 x = 4$

$x = 3^4 = 81$

51. $\log_x 49 = 2$

$x^2 = 49$

$x = \pm 7$

We discard the negative base.

$x = 7$

53. $\log_2 \dfrac{1}{8} = x$

$2^x = \dfrac{1}{8}$

$2^x = 2^{-3}$

$x = -3$

55. $\log_3 \left(\dfrac{1}{27} \right) = x$

$\dfrac{1}{27} = 3^x$

$3^{-3} = 3^x$

$-3 = x$

57. $\log_8 x = \dfrac{1}{3}$

$x = 8^{1/3} = 2$

59. $\log_4 16 = x$

$4^x = 4^2$

$x = 2$

61. $\log_{3/4} x = 3$

$\left(\dfrac{3}{4} \right)^3 = x$

$\dfrac{27}{64} = x$

63. $\log_x 100 = 2$ or $x^2 = 100$

$x = \pm 10$ and we discard the negative base.

$x = 10$

65. $\log_5 5^3 = 3$

67. $2^{\log_2 3} = 3$

69. $\log_9 9 = 1$

71. $y = \log_3 x$

$y = 0:$

$\log_3 x = 0$

$x = 3^0 = 1$ is the only x-intercept. No y-intercept exists.

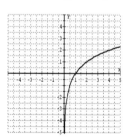

73. $f(x) = \log_{1/4} x$

$0 = \log_{1/4} x$

$x = \left(\dfrac{1}{4} \right)^0 = 1$ is the x-intercept.

No y-intercept exist.

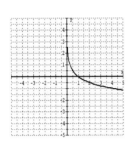

75. $f(x) = \log_5 x$

    $x = 0$:

    $y = \log_5 0$ is undefined so there is no y-intercept.

    $y = 0$:

    $0 = \log_5 x$

    $x = 5^0 = 1$ is the x-intercept.

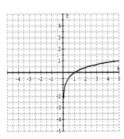

77. $f(x) = \log_{1/16} x$

    $x = 0$:

    $y = \log_{1/6} 0$ is not defined so there is no y-intercept.

    $y = 0$:

    $0 = \log_{1/6} x$

    $x = \left(\dfrac{1}{6}\right)^0 = 1$ is the x-intercept.

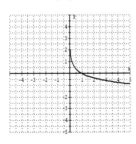

79. 1

81. $\dfrac{x^2 - 8x + 16}{2x - 8} = \dfrac{(x-4)(x-4)}{2(x-4)}$

               $= \dfrac{x-4}{2}$

83. $\dfrac{2}{x} + \dfrac{3}{x^2} = \dfrac{x}{x}\left(\dfrac{2}{x}\right) + \dfrac{3}{x^2}$

            $= \dfrac{2x}{x^2} + \dfrac{3}{x^2}$

            $= \dfrac{2x + 3}{x^2}$

85. $\dfrac{m^2}{m+1} - \dfrac{1}{m+1} = \dfrac{m^2 - 1}{m+1}$

                  $= \dfrac{(m+1)(m-1)}{m+1}$

                  $= m - 1$

87. $\log_7(5x - 2) = 1$

        $5x - 2 = 7$

           $5x = 9$

             $x = \dfrac{9}{5}$

89. $\log_3\left(\log_5 125\right) = \log_3(3) = 1$

91.

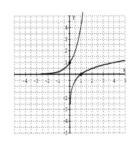

93.

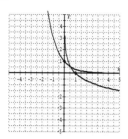

95. $\log_{10}(1-k) = \dfrac{-0.3}{H}$, $H = 8$

$\log_{10}(1-k) = \dfrac{-0.3}{8} = -0.0375$

$1 - k = 10^{-0.0375}$

$1 - 10^{-0.0375} = k$

$k \approx 0.0827$

97. $\log_3 10$ is between 2 and 3 because

$3^2 = 9$ and $3^3 = 27$

Section 9.5

Mental Math

1. A
2. C
3. B
4. C
5. A
6. B

Exercise Set 9.5

1. $\log_5 2 + \log_5 7 = \log_5(2 \cdot 7) = \log_5 14$

3. $\log_4 9 + \log_4 x = \log_4 9x$

5. $\log_{10} 5 + \log_{10} 2 + \log_{10}(x^2 + 2)$

$= \log_{10}\left[5 \cdot 2(x^2 + 2)\right]$

$= \log_{10}(10x^2 + 20)$

7. $\log_5 12 - \log_5 4 = \log_5\left(\dfrac{12}{4}\right) = \log_5 3$

9. $\log_2 x - \log_2 y = \log_2 \dfrac{x}{y}$

11. $\log_4 2 + \log_4 10 - \log_4 5$

$= \log_4 2 \cdot 10 - \log_4 5$

$= \log_4\left(\dfrac{20}{5}\right)$

$= \log_4 4 = 1$

13. $\log_3 x^2 = 2\log_3 x$

15. $\log_4 5^{-1} = -1\log_4 5 = -\log_4 5$

17. $\log_5 \sqrt{y} = \log_5 y^{1/2} = \dfrac{1}{2}\log_5 y$

19. $2\log_2 5 = \log_2 5^2 = \log_2 25$

21. $3\log_5 x + 6\log_5 z = \log_5 x^3 + \log_5 z^6$

$= \log_5 x^3 z^6$

23. $\log_{10} x - \log_{10}(x+1) + \log_{10}(x^2 - 2)$

$= \log_{10}\dfrac{x}{x+1} + \log_{10}(x^2 - 2)$

$= \log_{10}\dfrac{x(x^2 - 2)}{x+1}$

$= \log_{10}\dfrac{x^3 - 2x}{x+1}$

25. $\log_4 5 + \log_4 7 = \log_4(5 \cdot 7) = \log_4 35$

27. $\log_3 8 - \log_3 2 = \log_3\left(\dfrac{8}{2}\right) = \log_3 4$

29. $\log_7 6 + \log_7 3 - \log_7 4$

$= \log_7 (6 \cdot 3) - \log_7 4$

$= \log_7 \left(\dfrac{18}{4} \right)$

$= \log_7 \dfrac{9}{2}$

31. $3 \log_4 2 + \log_4 6 = \log_4 2^3 + \log_4 6$

$\qquad\qquad\qquad = \log_4 8 + \log_4 6$

$\qquad\qquad\qquad = \log_4 (8 \cdot 6)$

$\qquad\qquad\qquad = \log_4 48$

33. $3 \log_2 x + \dfrac{1}{2} \log_2 x - 2 \log_2 (x+1)$

$= \log_2 x^3 + \log_2 x^{1/2} - \log_2 (x+1)^2$

$= \log_2 (x^3 \cdot x^{1/2}) - \log_2 (x+1)^2$

$= \log_2 x^{7/2} - \log_2 (x+1)^2$

$= \log_2 \dfrac{x^{7/2}}{(x+1)^2}$

35. $2 \log_8 x - \dfrac{2}{3} \log_8 x + \log_8 x$

$= \left(2 - \dfrac{2}{3} + 4 \right) \log_8 x$

$= \dfrac{16}{3} \log_8 x$

$= \log_8 x^{16/3}$

37. $\log_2 \dfrac{7 \cdot 11}{3} = \log_2 (7 \cdot 11) - \log_2 3$

$\qquad\qquad\quad = \log_2 7 + \log_2 11 - \log_2 3$

39. $\log_3 \left(\dfrac{4y}{5} \right) = \log_3 4y - \log_3 5$

$\qquad\qquad\quad = \log_3 4 + \log_3 y - \log_3 5$

41. $\log_2 \left(\dfrac{x^3}{y} \right) = \log_2 x^3 - \log_2 y$

$\qquad\qquad\quad = 3 \log_2 x - \log_2 y$

43. $\log_b \sqrt{7x} = \log_b (7x)^{1/2}$

$= \dfrac{1}{2} \log_b (7x)$

$= \dfrac{1}{2} [\log_b 7 + \log_b x]$

$= \dfrac{1}{2} \log_b 7 + \dfrac{1}{2} \log_b x$

45. $\log_7 \left(\dfrac{5x}{4} \right) = \log_7 5x - \log_7 4$

$\qquad\qquad\quad = \log_7 5 + \log_7 x - \log_7 4$

47. $\log_5 x^3 (x+1) = \log_5 x^3 + \log_5 (x+1)$

$\qquad\qquad\qquad = 3 \log_5 x + \log_5 (x+1)$

49. $\log_6 \dfrac{x^2}{x+3} = \log_6 x^2 - \log_6 (x+3)$

$\qquad\qquad\quad = 2 \log_6 x - \log_6 (x+3)$

51. $\log_b \left(\dfrac{5}{3} \right) = \log_b 5 - \log_b 3$

$\qquad\qquad\quad = 0.7 - 0.5 = 0.2$

53. $\log_b 15 = \log_b (5 \cdot 3)$

$\qquad\qquad = \log_b 5 + \log_b 3$

$\qquad\qquad = 0.7 + 0.5 = 1.2$

55. $\log_b \sqrt[3]{5} = \log_b 5^{1/3}$

$= \dfrac{1}{3} \log_b 5$

$= \dfrac{1}{3} (0.7)$

≈ 0.233

57. $\log_b 8 = \log_b 2^3$

$\qquad\quad = 3 \log_b 2$

$\qquad\quad = 3(0.43)$

$\qquad\quad = 1.29$

59. $\log_b \left(\dfrac{3}{9} \right) = \log_b \left(\dfrac{1}{3} \right)$

$$= \log_b 3^{-1}$$
$$= (-1)\log_b 3$$
$$= -(0.68)$$
$$= -0.68$$

61. $\log_b \sqrt{\dfrac{2}{3}} = \log_b \left(\dfrac{2}{3}\right)^{1/2}$

$$= \frac{1}{2}\log_b \frac{2}{3}$$
$$= \frac{1}{2}\left(\log_b 2 - \log_b 3\right)$$
$$= \frac{1}{2}(0.43 - 0.68)$$
$$= \frac{1}{2}(-0.25)$$
$$= -0.125$$

63.

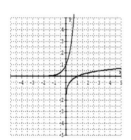

65. $\log_{10} \dfrac{1}{10} = x$

$$10^x = \frac{1}{10}$$
$$10^x = 10^{-1}$$
$$x = -1$$

67. $\log_7 \sqrt{7} = x$

$$7^x = \sqrt{7}$$
$$7^x = 7^{1/2}$$
$$x = \frac{1}{2}$$

69. $\log_3 (x + y) = \log_3 x + \log_3 y$
False, because
$$\log_3 x + \log_3 y = \log_3 (x \cdot y).$$

71. True

73. $(\log_3 6) \cdot (\log_3 4) = \log_3 24$
False, because
$$\log_3 24 = \log_3 (6 \cdot 4).$$

Integrated Review

1. $f(x) + g(x)$
$$x - 6 + x^2 + 1$$
$$x^2 + x - 5$$

2. $f(x) - g(x)$
$$x - 6 - x^2 - 1$$
$$-x^2 + x - 7$$

3. $f(x) \cdot g(x)$
$$(x - 6)\left(x^2 + 1\right)$$
$$x^3 + x - 6x^2 - 6$$
$$x^3 - 6x^2 + x - 6$$

4. $\dfrac{x - 6}{x^2 + 1}$

5. $(f \circ g)(x) = f(g(x))$
$$= f(3x - 1)$$
$$= \sqrt{3x - 1}$$

6. $(g \circ f)(x) = g(f(x))$
$$= g(\sqrt{x})$$
$$= 3\sqrt{x} - 1$$

7. one-to-one;
$$\{(6, -2), (8, 4), (-6, 2), (3, 3)\}$$

8. not one-to-one

9. not one-to-one

10. one-to-one

11. not one-to-one

12. $f(x) = 3x$

$$x = 3y$$

$$\frac{x}{3} = y$$

$$f^{-1}(x) = \frac{x}{3}$$

13. $f(x) = x + 4$

$$x = y + 4$$

$$x - 4 = y$$

$$f^{-1}(x) = x - 4$$

14. $f(x) = 5x - 1$

$$x = 5y - 1$$

$$x + 1 = 5y$$

$$\frac{x+1}{5} = y$$

$$f^{-1}(x) = \frac{x+1}{5}$$

15. $f(x) = 3x + 2$

$$x = 3y + 2$$

$$x - 2 = 3y$$

$$\frac{x-2}{3} = y$$

$$f^{-1}(x) = \frac{x-2}{3}$$

16. $y = \left(\dfrac{1}{2}\right)^x$

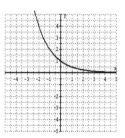

17. $y = 2^x + 1$

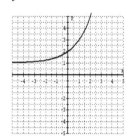

18. $y = \log_3 x$

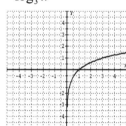

19. $y = \log_{1/3} x$

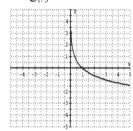

20. $2^x = 8$

$$2^x = 2^3$$

$$x = 3$$

The solution is 3.

21. $9 = 3^{x-5}$
$3^2 = 3^{x-5}$
$2 = x - 5$
$7 = x$
The solution is 7.

22. $4^{x-1} = 8^{x+2}$
$(2^2)^{x-1} = (2^3)^{x+2}$
$2^{2x-2} = 2^{3x+6}$
$2x - 2 = 3x + 6$
$-8 = x$
The solution is –8.

23. $25^x = 125^{x-1}$
$(5^2)^x = (5^3)^{x-1}$
$5^{2x} = 5^{3x-3}$
$2x = 3x - 3$
$3 = x$
The solution is 3.

24. $\log_4 16 = x$
$4^x = 16$
$4^x = 4^2$
$x = 2$
The solution is 2.

25. $\log_{49} 7 = x$
$49^x = 7$
$(7^2)^x = 7$
$7^{2x} = 7$
$2x = 1$
$x = \dfrac{1}{2}$

The solution is $\dfrac{1}{2}$.

26. $\log_2 x = 5$
$2^5 = x$
$32 = x$
The solution is 32.

27. $\log_x 64 = 3$
$x^3 = 64$
$x = \sqrt[3]{64} = 4$

The solution is 4.

28. $\log_x \dfrac{1}{125} = -3$
$x^{-3} = \dfrac{1}{125}$
$\dfrac{1}{x^3} = \dfrac{1}{125}$
$x^3 = 125$
$x = \sqrt[3]{125} = 5$
The solution is 5.

29. $\log_3 x = -2$
$3^{-2} = x$
$x = \dfrac{1}{3^2} = \dfrac{1}{9}$

The solution is $\dfrac{1}{9}$.

30. $\log_2 x^5$

31. $\log_2 5^x$

32. $\log_5 \dfrac{x^3}{y^5}$

33. $\log_5 x^9 y^3$

34. $\log_2 \dfrac{x^2 - 3x}{x^2 + 4}$

35. $\log_3 \dfrac{y\left(y^3 + 11\right)}{y + 2}$

36. $\log_7 9 + 2\log_7 x - \log_7 y$

37. $\log_6 5 + \log_6 y - 2\log_6 z$

Exercise Set 9.6

1. $\log 8 \approx 0.9031$

3. $\log 2.31 \approx 0.3636$

5. $\ln 2 \approx 0.6931$

7. $\ln 0.0716 \approx -2.6367$

9. $\log 12.6 \approx 1.1004$

11. $\ln 5 \approx 1.6094$

13. $\log 41.5 \approx 1.6180$

15. Answers may vary.

17. $\log 100 = \log 10^2 = 2$

19. $\log \dfrac{1}{1000} = \log 10^{-3} = -3$

21. $\ln e^2 = 2$

23. $\ln \sqrt[4]{e} = \ln e^{1/4} = \dfrac{1}{4}$

25. $\log 10^3 = 3$

27. $\ln e^2 = 2$

29. $\log 0.0001 = \log 10^{-4} = -4$

31. $\ln \sqrt{e} = \ln e^{1/2} = \dfrac{1}{2}$

33. $\ln 50$ is larger
Answers may vary.

35. $\log x = 1.3$
$x = 10^{1.3} \approx 19.9526$

37. $\log 2x = 1.1$
$2x = 10^{1.1}$
$x = \dfrac{10^{1.1}}{2} \approx 6.2946$

39. $\ln x = 1.4$
$x = e^{1.4} \approx 4.0552$

41. $\ln(3x - 4) = 2.3$

$3x - 4 = e^{2.3}$
$3x = 4 + e^{2.3}$
$x = \dfrac{4 + e^{2.3}}{3} \approx 4.6581$

43. $\log x = 2.3$
$x = 10^{2.3} \approx 199.5262$

45. $\ln x = -2.3$
$x = e^{-2.3} \approx 0.1003$

47. $\log(2x + 1) = -0.5$
$2x + 1 = 10^{-0.5}$
$2x = 10^{-0.5} - 1$
$x = \dfrac{10^{-0.5} - 1}{2} \approx 0.3419$

49. $\ln 4x = 0.18$
$4x = e^{0.18}$
$x = \dfrac{e^{0.18}}{4} \approx 0.2993$

51. $\log_2 3 = \dfrac{\ln 3}{\ln 2} \approx 1.5850$

53. $\log_{1/2} 5 = \dfrac{\ln 5}{\ln\left(\dfrac{1}{2}\right)} \approx -2.3219$

55. $\log_4 9 = \dfrac{\ln 9}{\ln 4} \approx 1.5850$

57. $\log_3 \left(\dfrac{1}{6}\right) = \log_3 6^{-1}$
$= (-1)\log_3 6$
$= -\dfrac{\ln 6}{\ln 3} \approx -1.6309$

59. $\log_8 6 = \dfrac{\ln 6}{\ln 8} \approx 0.8617$

61. $R = \log\left(\dfrac{a}{T}\right) + B$, $a = 200$, $T = 1.6$

$B = 2.1$

$R = \log\left(\dfrac{200}{1.6}\right) + 2.1 \approx 4.2$

The earthquake measures 4.2 on the Richter scale.

63. $R = \log\left(\dfrac{a}{T}\right) + B$, $a = 400$, $T = 2.6$

$B = 3.1$

$R = \log\left(\dfrac{400}{2.6}\right) + 3.1 \approx 5.3$

The earthquake measures 5.3 on the Richter scale.

65. $A = Pe^{rt}$, $t = 12$, $P = 1400$, $r = 0.08$

$A = 1400e^{(0.08)12} = 1400e^{0.96}$

≈ 3656.38

Dana has \$3656.38 after 12 years.

67. $A = Pe^{rt}$, $t = 4$, $P = 2000$, $r = 0.06$

$A = 2000e^{(0.06)4} = 2000e^{0.24}$

≈ 2542.50

Barbara owes \$2542.50 at the end of 4 years.

69. $6x - 3(2 - 5x) = 6$

$6x - 6 + 15x = 6$

$21x - 6 = 6$

$21x = 12$

$x = \dfrac{12}{21} = \dfrac{4}{7}$

71. $2x + 3y = 6x$

$3y = 4x$

$\dfrac{3y}{4} = x$

$x = \dfrac{3y}{4}$

73. $x^2 + 7x = -6$

$x^2 + 7x + 6 = 0$

$(x + 6)(x + 1) = 0$

$x = -6, -1$

75. $\begin{cases} x + 2y = -4 \\ 3x - y = 9 \end{cases}$

$x + 2y = -4$

$2(3x - y) = 2(9)$

$\overline{}$

$x + 2y = -4$

$6x - 2y = 18$

$\overline{}$

$7x = 14$

$x = 2$

To find y, substitute $x = 2$ in the first equation.

$2 + 2y = -4$

$2y = -6$

$y = -3$

$(2, -3)$

77. $\ln 50$ is larger
Answers may vary.

79. $f(x) = e^x$

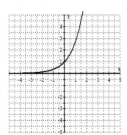

81. $f(x) = e^{-3x}$

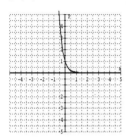

83. $f(x) = e^x + 2$

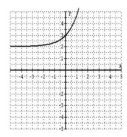

85. $f(x) = e^{x-1}$

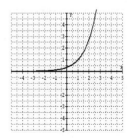

87. $f(x) = 3e^x$

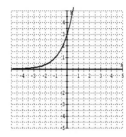

89. $f(x) = \ln x$

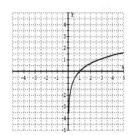

91. $f(x) = -2\log x$

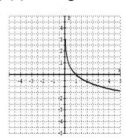

93. $f(x) = \log(x + 2)$

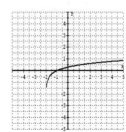

95. $f(x) = \ln x - 3$

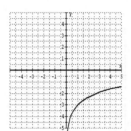

97. $f(x) = e^x$

$f(x) = e^x + 2$

$f(x) = e^x - 3$

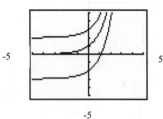

Section 9.7

Graphing Calculator Explorations

1. $Y_1 = 5000\left(1 + \dfrac{0.05}{4}\right)^{4x}, Y_2 = 6000$

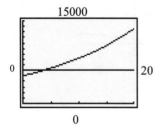

3.67 years, or 3 years and 8 months

3. $Y_1 = 10000\left(1 + \dfrac{0.06}{12}\right)^{12x}, Y_2 = 40000$

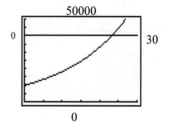

23.16 years, or 23 years and 2 months

Exercise Set 9.7

1. $3^x = 6$

$x = \log_3 6 = \dfrac{\log 6}{\log 3} \approx 1.6309$

3. $3^{2x} = 3.8$

$2x = \log_3 3.8 = \dfrac{\log 3.8}{\log 3}$

$x = \dfrac{\log 3.8}{2\log 3} \approx 0.6076$

5. $2^{x-3} = 5$

$x - 3 = \log_2 5$

$x = 3 + \log_2 5 = 3 + \dfrac{\log 5}{\log 2}$

≈ 5.3219

7. $9^x = 5$

$x = \log_9 5 = \dfrac{\log 5}{\log 9} \approx 0.7325$

9. $4^{x+7} = 3$

$x + 7 = \log_4 3$

$x = -7 + \log_4 3$

$\quad = -7 + \dfrac{\log 3}{\log 4}$

$\quad \approx -6.2075$

11. $7^{3x-4} = 11$

$3x - 4 = \log_7 11$

$3x = 4 + \log_7 11$

$x = \dfrac{1}{3}\left(4 + \dfrac{\log 11}{\log 7}\right) \approx 1.7441$

13. $e^{6x} = 5$

$6x = \ln 5$

$x = \dfrac{\ln 5}{6} \approx 0.2682$

15. $\log_2(x+5) = 4$

$x + 5 = 2^4$

$x + 5 = 16$

$x = 11$

17. $\log_3 x^2 = 4$

$x^2 = 3^4$

$x^2 = 81$

$x = \pm 9$

19. $\log_4 2 + \log_4 x = 0$

$\log_4(2x) = 0$

$2x = 4^0$

$2x = 1$

$x = \dfrac{1}{2}$

21. $\log_2 6 - \log_2 x = 3$

$\log_2\left(\dfrac{6}{x}\right) = 3$

$\dfrac{6}{x} = 2^3$

$\dfrac{6}{x} = 8$

$8x = 6$

$x = \dfrac{3}{4}$

23. $\log_4 x + \log_4(x+6) = 2$

$\log_4 x(x+6) = 2$

$x(x+6) = 4^2$

$x^2 + 6x = 16$

$x^2 + 6x - 16 = 0$

$(x+8)(x-2) = 0$

$x = -8 \text{ or } x = 2$

We discard -8 as extraneous.

25. $\log_5(x+3) - \log_5 x = 2$

$\log_5\left(\dfrac{x+3}{x}\right) = 2$

$\dfrac{x+3}{x} = 5^2$

$\dfrac{x+3}{x} = 25$

$x + 3 = 25x$

$3 = 24x$

$x = \dfrac{1}{8}$

27. $\log_3(x-2) = 2$

$x - 2 = 3^2$

$x - 2 = 9$

$x = 11$

29. $\log_4(x^2 - 3x) = 1$

$x^2 - 3x = 4$

$x^2 - 3x - 4 = 0$

$(x-4)(x+1) = 0$

$x = 4$ or $x = -1$

31. $\ln 5 + \ln x = 0$

$\ln(5x) = 0$

$e^0 = 5x$

$1 = 5x$

$\dfrac{1}{5} = x$

33. $3\log x - \log x^2 = 2$

$3\log x - 2\log x = 2$

$\log x = 2$

$x = 10^2$

$x = 100$

35. $\log_2 x + \log_2(x+5) = 1$

$\log_2 x(x+5) = 1$

$x(x+5) = 2$

$x^2 + 5x - 2 = 0$

$a = 1,\ b = 5,\ c = -2$

$x = \dfrac{-5 \pm \sqrt{5^2 - 4(1)(-2)}}{2(1)}$

$x = \dfrac{-5 \pm \sqrt{33}}{2}$

Discard $\dfrac{-5 - \sqrt{33}}{2}$

37. $\log_4 x - \log_4(2x-3) = 3$

$\log_4\left(\dfrac{x}{2x-3}\right) = 3$

$\dfrac{x}{2x-3} = 64$

$x = 64(2x-3)$

$x = 128x - 192$

$192 = 127x$

$x = \dfrac{192}{127}$

39. $\log_2 x + \log_2(3x+1) = 1$

$\log_2 x(3x+1) = 1$

$x(3x+1) = 2$

$3x^2 + x - 2 = 0$

$(3x-2)(x+1) = 0$

$3x - 2 = 0$ or $x + 1 = 0$

$x = \dfrac{2}{3}$ or $x = -1$

We discard -1 as extraneous.

41. $y = y_0 e^{0.043t}$, $y_0 = 83$, $t = 5$

$y = 83e^{0.043(5)} = 83e^{0.215} \approx 103$

There should be 103 wolves in 5 years.

43. $y = y_0 e^{0.026t}$, $y_0 = 10{,}589{,}571$, $t = 6$

$y = 10{,}589{,}571 e^{0.026(6)}$

$\approx 12{,}377{,}368.06$

There will be approximately 12,380,000 inhabitants in 2005.

$t \approx 1.73$

It would take the investment approximately 1.7 years to earn $200.

45. $y = y_0 e^{-0.005t}$

$y_0 = 144,979, \quad y = 120,000$

$120,000 = 144,979 e^{-0.018t}$

$\dfrac{120,000}{144,979} = e^{-0.018t}$

$t = \dfrac{\ln\left(\dfrac{120,000}{144,979}\right)}{-0.018} \approx 10.5$

It will take approximately 10.5 years to reach 120,000 thousand.

47. $A = P\left(1 + \dfrac{r}{n}\right)^{nt}, P = 600,$

$A = 2(600) = 1200, \ r = 0.07,$

$n = 12$

$1200 = 600\left(1 + \dfrac{0.07}{12}\right)^{12t}$

$2 = (1.00583)^{12t}$

$12t = \log_{1.00583} 2$

$t \approx 9.93$

It would take approximately 9.9 years for the $600 to double.

49. $A = P\left(1 + \dfrac{r}{n}\right)^{nt}, P = 1200,$

$A = P + I = 1200 + 200 = 1400$

$r = 0.009, \ n = 4$

$1400 = 1200\left(1 + \dfrac{0.09}{4}\right)^{4t}$

$\dfrac{7}{6} = (1.0225)^{4t}$

$4t = \log_{1.0225}\left(\dfrac{7}{6}\right)$

51. $A = P\left(1 + \dfrac{r}{n}\right)^{nt}, P = 1000$

$A = 2(1000) = 2000, r = 0.08, n = 2$

$2000 = 1000\left(1 + \dfrac{0.08}{2}\right)^{2t}$

$2 = (1.04)^{2t}$

$2t = \log_{1.04} 2 = \dfrac{\ln 2}{\ln 1.04}$

$t \approx 8.8$

It would take about 8.8 years for $1000 to double.

53. $w = 0.00185 h^{2.67}$, and $h = 35$

$w = 0.00185(35)^{2.67} \approx 24.5$

The expected weight of a boy 35 inches tall is 24.5 pounds.

55. $w = 0.00185 h^{2.67}$, and $w = 85$

$85 = 0.00185 h^{2.67}$

$\dfrac{85}{0.00185} = h^{2.67}$

$h = \left(\dfrac{85}{0.00185}\right)^{1/2.67} \approx 55.7$

The expected height of the boy is approximately 55.7 inches.

57. $P = 14.7e^{-0.21x}$, $x = 1$

$= 14.7e^{-0.21(1)}$

$= 14.7e^{-0.21}$

≈ 11.9

The average atmospheric pressure of Denver is approximately 11.9 pounds per square inch.

59. $P = 14.7e^{-0.21x}$, $P = 7.5$

$7.5 = 14.7e^{-0.21x}$

$\dfrac{7.5}{14.7} = e^{-0.21x}$

$-0.21x = \ln\left(\dfrac{7.5}{14.7}\right)$

$x = -\dfrac{1}{0.21}\ln\left(\dfrac{7.5}{14.7}\right) \approx 3.2$

The elevation of the jet is approximately 3.2 miles.

61. $t = \dfrac{1}{c}\ln\left(\dfrac{A}{A-N}\right)$

$t = \dfrac{1}{0.09}\ln\left(\dfrac{75}{75-50}\right)$

$t = \dfrac{1}{0.09}\ln(3)$

$t \approx 12.21$

It will take 12 weeks.

63. $t = \dfrac{1}{c}\ln\left(\dfrac{A}{A-N}\right)$

$t = \dfrac{1}{0.07}\ln\left(\dfrac{210}{210-150}\right)$

$t = \dfrac{1}{0.07}\ln(3.5)$

$t \approx 17.9$

It will take 18 weeks.

65. $\dfrac{x^2 - y + 2z}{3x} = \dfrac{(-2)^2 - 0 + 2(3)}{3(-2)}$

$= \dfrac{4 - 0 + 6}{-6}$

$= \dfrac{10}{-6} = -\dfrac{5}{3}$

67. $\dfrac{3z - 4x + y}{x + 2z} = \dfrac{3(3) - 4(-2) + 0}{-2 + 2(3)}$

$= \dfrac{9 + 8 + 0}{-2 + 6}$

$= \dfrac{17}{4}$

69. $f(x) = 5x + 2$

$y = 5x + 2$

$x = 5y + 2$

$x - 2 = 5y$

$\dfrac{x - 2}{5} = y$

$f^{-1}(x) = \dfrac{x - 2}{5}$

71. $y = 5,130,632$, $y_0 = 3,665,228$

$$5,130,632 = 3,665,228e^{10k}$$

$$\frac{5,130,632}{3,665,228} = e^{10k}$$

$$\ln\left(\frac{5,130,632}{3,665,228}\right) = 10k$$

$$k \approx 0.0336 \text{ or } 3.4\%$$

73. Answers may vary.

75. $Y_1 = e^{0.3x}$, $Y_2 = 8$

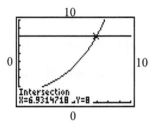

77. $Y_1 = 2\log(-5.6x+1.3)$, $Y_2 = -x-1$

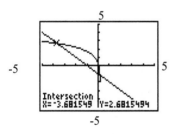

79. $Y_1 = 7^{3x-4}$, $Y_2 = 11$

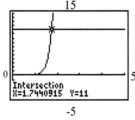

81. $Y_1 = \ln 5 + \ln x$, $Y_2 = 0$

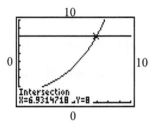

Chapter 9 Review

1. $(f+g)(x) = f(x) + g(x)$
$$= (x-5)+(2x+1)$$
$$= x-5+2x+1$$
$$= 3x-4$$

2. $(f-g)(x) = f(x) - g(x)$
$$= (x-5)+(2x+1)$$
$$= x-5-2x-1$$
$$= -x-6$$

3. $(f \cdot g)(x) = f(x) \cdot g(x)$
$$= (x-5)(2x+1)$$
$$= 2x^2 + x - 10x - 5$$
$$= 2x^2 - 9x - 5$$

4. $\left(\dfrac{g}{f}\right)(x) = \dfrac{g(x)}{f(x)} = \dfrac{2x+1}{x-5}, x \neq 5$

5. $(f \circ g)(x) = f(g(x))$
$$= f(x+1)$$
$$= (x+1)^2 - 2$$
$$= x^2 + 2x - 1$$

6. $(g \circ f)(x) = g(f(x))$
$$= g(x^2 - 2)$$
$$= x^2 - 2 + 1$$
$$= x^2 - 1$$

7. $(h \circ g)(2) = h(g(2))$
$$= h(3)$$
$$= 3^3 - 3^2$$
$$= 18$$

8. $(f \circ f)(x) = f(f(x))$
$$= f(x^2 - 2)$$
$$= (x^2 - 2)^2 - 2$$
$$= x^4 - 4x^2 + 4 - 2$$
$$= x^4 - 4x^2 + 2$$

9. $(f \circ g)(-1) = f(g(-1))$
$$= f(0)$$
$$= 0^2 - 2$$
$$= -2$$

10. $(h \circ h)(2) = h(h(2))$
$$= h(4)$$
$$= 4^3 - 4^2$$
$$= 48$$

11. The function is one-to-one.
$$h^{-1} = \left\{ \begin{array}{l} (14, -9), (8, 6), (12, -11) \\ , (15, 15) \end{array} \right\}$$

12. The function is not one-to-one.

13. The function is one-to-one.

Rank in Auto Thefts (Input)	2	4	1	3
U.S. Region (Output)	West	Midwest	South	Northeast

14. The function is not one-to-one.

15. $f(x) = \sqrt{x+2}$

   a. $f(7) = \sqrt{7+2} = \sqrt{9} = 3$

   b. $f^{-1}(3) = 7$

16. $f(x) = \sqrt{x+2}$

   a. $f(-1) = \sqrt{-1+2} = \sqrt{1} = 1$

   b. $f^{-1}(1) = -1$

17. The graphs does not represent a one-to-one function.

18. The graph is not a one-to-one.

19. The graph is not a one-to-one.

20. The graph is a one-to-one function.

21. $f(x) = x - 9$
$$y = x - 9$$
$$x = y - 9$$
$$y = x + 9$$
$$f^{-1}(x) = x + 9$$

22. $f(x) = x + 8$
$$y = x + 8$$
$$x = y + 8$$
$$y = x - 8$$
$$f^{-1}(x) = x - 8$$

23. $f(x) = 6x - 11$
$$y = 6x + 11$$
$$x = 6y + 11$$
$$6y = x - 11$$
$$y = \frac{x - 11}{6}$$
$$f^{-1}(x) = \frac{x - 11}{6}$$

24. $f(x) = 12x$

$$y = 12x$$

$$x = 12y$$

$$y = \frac{x}{12}$$

$$f^{-1}(x) = \frac{x}{12}$$

25. $f(x) = x^3 - 5$

$$y = x^3 - 5$$

$$x = y^3 - 5$$

$$y^3 = x + 5$$

$$y = \sqrt[3]{x + 5}$$

$$f^{-1}(x) = \sqrt[3]{x + 5}$$

26. $f(x) = \sqrt[3]{x + 2}$

$$y = \sqrt[3]{x + 2}$$

$$x = \sqrt[3]{y + 2}$$

$$x^3 = y + 2$$

$$y = x^3 - 2$$

$$f^{-1}(x) = x^3 - 2$$

27. $g(x) = \dfrac{12x - 7}{6}$

$$y = \frac{12x - 7}{6}$$

$$x = \frac{12y - 7}{6}$$

$$6x = 12y - 7$$

$$y = \frac{6x + 7}{12}$$

$$g^{-1}(x) = \frac{6x + 7}{12}$$

28. $r(x) = \dfrac{13}{2}x - 4$

$$y = \frac{13}{2}x - 4$$

$$x = \frac{13}{2}y - 4$$

$$x + 4 = \frac{13}{2}y$$

$$y = \frac{2(x + 4)}{13}$$

$$r^{-1}(x) = \frac{2(x + 4)}{13}$$

29. $y = g(x) = \sqrt{x}$

$$x = \sqrt{y}$$

$$x^2 = y = g^{-1}(x), \, x \geq 0$$

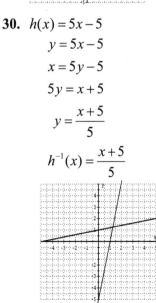

30. $h(x) = 5x - 5$

$$y = 5x - 5$$

$$x = 5y - 5$$

$$5y = x + 5$$

$$y = \frac{x + 5}{5}$$

$$h^{-1}(x) = \frac{x + 5}{5}$$

31. $f(x) = 2x - 3$

$x = 2y - 3$

$y = \dfrac{x + 3}{2}$

$f^{-1}(x) = \dfrac{x + 3}{2}$

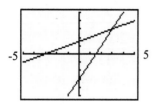

32. $4^x = 64$

$4^x = 4^3$

$x = 3$

33. $3^x = \dfrac{1}{9}$

$3^x = 3^{-2}$

$x = -2$

34. $2^{3x} = \dfrac{1}{16}$

$2^{3x} = 2^{-4}$

$3x = -4$

$x = -\dfrac{4}{3}$

35. $5^{2x} = 125$

$5^{2x} = 5^3$

$2x = 3$

$x = \dfrac{3}{2}$

36. $9^{x+1} = 243$

$(3^2)^{x+1} = 3^5$

$3^{2x+2} = 3^5$

$2x + 2 = 5$

$2x = 3$

$x = \dfrac{3}{2}$

37. $8^{3x-2} = 4$

$(2^3)^{3x-2} = 2^2$

$2^{9x-6} = 2^2$

$9x - 6 = 2$

$9x = 8$

$x = \dfrac{8}{9}$

38. $y = 3^x$

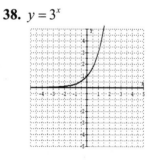

39. $y = \left(\dfrac{1}{3}\right)^x$

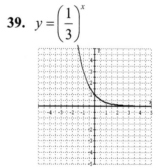

40. $y = 4 \cdot 2^x$

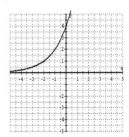

41. $y = 2^x + 4$

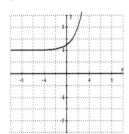

42. $A = P\left(1 + \dfrac{r}{n}\right)^{nt}$

$$A = 1600\left(1 + \dfrac{0.09}{2}\right)^{(2)(7)}$$

$$A = \$2963.11$$

43. $A = P\left(1 + \dfrac{r}{n}\right)^{nt}$

$$A = 800\left(1 + \dfrac{0.07}{4}\right)^{(4)(5)}$$

$$A \approx 1131.82$$

The certificate is worth \$1131.82 at the end of 5 years.

44. $y = 4 \cdot 2^x$

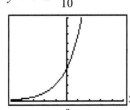

45. $7^2 = 49$

$\log_7 49 = 2$

46. $2^{-4} = \dfrac{1}{16}$

$\log_2 \dfrac{1}{16} = -4$

47. $\log_{1/2} 16 = -4$

$\left(\dfrac{1}{2}\right)^{-4} = 16$

48. $\log_{0.4} 0.064 = 3$

$0.4^3 = 0.064$

49. $\log_4 x = -3$

$x = 4^{-3} = \dfrac{1}{64}$

50. $\log_3 x = 2$

$x = 3^2 = 9$

51. $\log_3 x = 2$

$3^x = 1$

$3^x = 3^0$

$x = 0$

52. $\log_4 64 = x$

$4^x = 64$

$4^x = 4^3$

$x = 3$

53. $\log_x 64 = 2$

$x^2 = 64$

$x = \pm\sqrt{64} = \pm 8$

$x = 8$ since base > 0

54. $\log_x 81 = 4$

$$x^4 = 81$$
$$x = \pm 3$$

We discard the negative base -3..

55. $\log_4 4^5 = x$
　　$x = 5$

56. $\log_7 7^{-2} = x$
　　$x = -2$

57. $5^{\log_5 4} = x$
　　$x = 4$

58. $2^{\log_2 9} = x$
　　$9 = x$

59. $\log_2(3x-1) = 4$
　　$3x - 1 = 2^4 = 16$
　　$3x = 17$
　　$x = \dfrac{17}{3}$

60. $\log_3(2x+5) = 2$
　　$2x + 5 = 3^2$
　　$2x + 5 = 9$
　　$2x = 4$
　　$x = 2$

61. $\log_4(x^2 - 3x) = 1$
　　$x^2 - 3x = 4$
　　$x^2 - 3x - 4 = 0$
　　$(x+1)(x-4) = 0$
　　$x = -1 \text{ or } x = 4$

62. $\log_8(x^2 + 7x) = 1$

$$x^2 + 7x = 8$$
$$x^2 + 7x - 8 = 0$$
$$(x+8)(x-1) = 0$$
$$x = -8 \text{ or } x = 1$$

63. $y = 2^x$ and $y = \log_2 x$

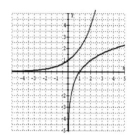

64. $y = \left(\dfrac{1}{2}\right)^x$ and $y = \log_{\frac{1}{2}} x$

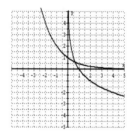

65. $\log_3 8 + \log_3 4 = \log_3 (8)(4) = \log_3 32$

66. $\log_2 6 + \log_2 3 = \log_2 (6 \cdot 3) = \log_2 18$

67. $\log_7 15 - \log_7 20 = \log_7 \dfrac{3}{4}$

68. $\log 18 - \log 12 = \log \dfrac{3}{2}$

69. $\log_{11} 8 + \log_{11} 3 - \log_{11} 6 = \log_{11} \dfrac{(8)(3)}{6}$
　　$= \log_{11} 4$

70. $\log_5 14 + \log_5 3 - \log_5 21$

$= \log_5(14 \cdot 3) - \log_5 21$

$= \log_5\left(\dfrac{42}{21}\right) = \log_5 2$

71. $2\log_5 x - 2\log_5(x+1) + \log_5 x$

$= \log_5 x^2 - \log_5(x+1)^2 + \log_5 x$

$= \log_5 \dfrac{(x^2)(x)}{(x+1)^2}$

$= \log_5 \dfrac{x^3}{(x+1)^2}$

72. $4\log_3 x - \log_3 x + \log_3(x+2)$

$= 3\log_3 x + \log_3(x+2)$

$= \log_3 x^3 + \log_3(x+2)$

$= \log_3\left[x^3(x+2)\right]$

$= \log_3(x^4 + 2x^3)$

73. $\log_3 \dfrac{x^3}{x+2} = \log_3 x^3 - \log_3(x+2)$

$= 3\log_3 x - \log_3(x+2)$

74. $\log_4 \dfrac{x+5}{x^2} = \log_4(x+5) - \log_4 x^2$

$= \log_4(x+5) - 2\log_4 x$

75. $\log_2 \dfrac{3x^2 y}{z}$

$= \log_2 3 + \log_2 x^2 + \log_2 y - \log_2 z$

$= \log_2 3 + 2\log_2 x + \log_2 y - \log_2 z$

76. $\log_7 \dfrac{yz^3}{x} = \log_7(yz^3) - \log_7 x$

$= \log_7 y + \log_7 z^3 - \log_7 x$

$= \log_7 y + 3\log_7 z - \log_7 x$

77. $\log_6 50 = \log_6(5)(5)(2)$

$= \log_6(5) + \log_6(5) + \log_6(2)$

$= 0.83 + 0.83 + 0.36$

$= 2.02$

78. $\log_b \dfrac{4}{5} = \log_b 4 - \log_b 5$

$= \log_b 2^2 - \log_b 5$

$= 2\log_b 2 - \log_b 5$

$= 2(0.36) - 0.83$

$= 0.72 - 0.83$

$= -0.11$

79. $\log 3.6 \approx 0.5563$

80. $\log 0.15 \approx -0.8239$

81. $\ln 1.25 \approx 0.2231$

82. $\ln 4.63 \approx 1.5326$

83. $\log 1000 = 3$

84. $\log \dfrac{1}{10} = \log 10^{-1} = -1$

85. $\ln \dfrac{1}{e} = \ln 1 - \ln e = 0 - 1 = -1$

86. $\ln(e^4) = 4$

87. $\ln(2x) = 2$

$2x = e^2$

$x = \dfrac{e^2}{2}$

88. $\ln(3x) = 1.6$

$3x = e^{1.6}$

$x = \dfrac{e^{1.6}}{3}$

89. $\ln(2x-3) = -1$

$$2x - 3 = e^{-1}$$

$$x = \frac{e^{-1} + 3}{2}$$

90. $\ln(3x + 1) = 2$

$$3x + 1 = e^2$$

$$3x = e^2 - 1$$

$$x = \frac{e^2 - 1}{3}$$

91. $\ln \dfrac{I}{I_0} = -kx$

$$\ln \frac{0.03 I_0}{I_0} = -2.1x$$

$$\ln 0.03 = -2.1x$$

$$\frac{\ln 0.03}{-2.1} = x$$

$$x \approx 1.67 mm$$

92. $\ln \dfrac{I}{I_0} = -kx$

$$\ln \frac{0.02 I_0}{I_0} = -3.2x$$

$$\ln 0.02 = -3.2x$$

$$\frac{\ln 0.02}{-3.2} = x$$

$$x \approx 1.22$$

2% of the original radioactivity will penetrate at a depth of approximately 1.22 millimeters.

93. $\log_5 1.6 = \dfrac{\log 1.6}{\log 5} = 0.2920$

94. $\log_3 4 = \dfrac{\log 4}{\log 3} \approx 1.2619$

95. $A = Pe^{rt}$

$$A = 1450 e^{(0.06)(5)}$$

$$A = \$1957.30$$

96. $A = Pe^{rt}$

$$A = 940 e^{0.11(3)} = 940 e^{0.33} \approx 1307.51$$

97. $3^{2x} = 7$

$$2x \log 3 = \log 7$$

$$x = \frac{\log 7}{2 \log 3} \approx 0.8856$$

98. $6^{3x} = 5$

$$3x = \log_6 5$$

$$x = \frac{1}{3} \log_6 5 = \frac{\log 5}{3 \log 6} \approx 0.2994$$

99. $3^{2x+1} = 6$

$$(2x + 1) \log 3 = \log 6$$

$$2x = \frac{\log 6}{\log 3} - 1$$

$$x = \frac{1}{2} \left(\frac{\log 6}{\log 3} - 1 \right) \approx 0.3155$$

100. $4^{3x+2} = 9$

$$3x + 2 = \log_4 9$$

$$3x = \log_4 9 - 2$$

$$x = \frac{1}{3} \left(\frac{\log 9}{\log 4} - 2 \right) \approx -0.1383$$

101. $5^{3x-5} = 4$

$$(3x - 5) \log 5 = \log 4$$

$$3x = \frac{\log 4}{\log 5} + 5$$

$$x = \frac{1}{3} \left(\frac{\log 4}{\log 5} + 5 \right)$$

$$x \approx 1.9538$$

102. $8^{4x-2} = 3$

$$4x - 2 = \log_8 3$$
$$4x = \log_8 3 + 2$$
$$4x = \frac{\log 3}{\log 8} + 2$$
$$x = \frac{1}{4}\left(\frac{\log 3}{\log 8} + 2\right)$$
$$x \approx 0.6321$$

103. $2 \cdot 5^{x-1} = 1$

$$\log 2 + (x-1)\log 5 = \log 1$$
$$(x-1)\log 5 = -\log 2$$
$$x = -\frac{\log 2}{\log 5} + 1$$
$$x \approx 0.5693$$

104. $3 \cdot 4^{x+5} = 2$

$$4^{x+5} = \frac{2}{3}$$
$$x + 5 = \log_4\left(\frac{2}{3}\right)$$
$$x = \log_4\left(\frac{2}{3}\right) - 5$$
$$x = \frac{\log\left(\frac{2}{3}\right)}{\log 4} - 5$$
$$x \approx -5.2925$$

105. $\log_5 2 + \log_5 x = 2$

$$\log_5 2x = 2$$
$$2x = 5^2 = 25$$
$$x = \frac{25}{2}$$

106. $\log_3 x + \log_3 10 = 2$

$$\log_3(10x) = 2$$
$$10x = 3^2$$
$$10x = 9$$
$$x = \frac{9}{10}$$

107. $\log(5x) - \log(x+1) = 4$

$$\log\frac{5x}{x+1} = 4$$
$$\frac{5x}{x+1} = 10^4 = 10,000$$
$$5x = 10000x + 10000$$
$$x = -1.0005$$

no solution

108. $\ln(3x) - \ln(x-3) = 2$

$$\ln\left(\frac{3x}{x-3}\right) = 2$$
$$\frac{3x}{x-3} = e^2$$
$$3x = e^2 x - 3e^2$$
$$3x - e^2 x = -3e^2$$
$$(3 - e^2)x = -3e^2$$
$$x = \frac{3e^2}{e^2 - 3}$$

109. $\log_2 x + \log_2 2x - 3 = 1$

$$\log_2(x)(2x) = 4$$
$$2x^2 = 16$$
$$x^2 = 8$$
$$x = \pm 2\sqrt{2}$$

Discard $-2\sqrt{2}$

110. $-\log_6(4x+7)+\log_6 x = 1$

$$\log_6 \frac{x}{4x+7} = 1$$

$$\frac{x}{4x+7} = 6$$

$$x = 6(4x+7)$$

$$x = 24x+42$$

$$x = -\frac{42}{23}$$

which is extraneous. No solution.

111. $y = y_0 e^{kt}$

$$y = 155,000 e^{0.06(4)}$$

$$= 197,044 \text{ ducks}$$

112. $y = y_0 e^{kt}$

$$y = 232,073,071 e^{0.015(8)}$$

$$= 232,073,071 e^{0.12}$$

$$= 261,661,656.9$$

The expected population of
Indonesia by the year 2006
is approximately $261,661,657$.

113. $y = y_0 e^{kt}$

$$130,000,000 = 126,975,000 e^{0.001t}$$

$$t = \frac{\ln \dfrac{130,000,000}{126,975,000}}{0.001}$$

$$t \approx 23.54$$

It will take approximately 24 years.

114. $y = y_0 e^{kt}$

$$2(31,902,268) = 31,902,268 e^{0.008t}$$

$$2 = e^{0.008t}$$

$$t = \frac{\ln 2}{0.008}$$

$$t \approx 86.64$$

It will take approximately 87 years.

115. $y = y_0 e^{kt}$

$$2(70,712,345) = 70,712,345 e^{0.016t}$$

$$2 = e^{0.016t}$$

$$t = \frac{\ln 2}{0.016}$$

$$t \approx 43.32$$

It will take approximately 43 years.

116. $A = P\left(1+\dfrac{r}{n}\right)^{nt}$

$$10,000 = 5,000\left(1+\frac{0.08}{4}\right)^{4t}$$

$$2 = (1.02)^{4t}$$

$$\log 2 = 4t \log 1.02$$

$$t = \frac{\log 2}{4\log 1.02} \approx 8.8 \text{ years}$$

117. $A = P\left(1 + \dfrac{r}{n}\right)^{nt}$

$10,000 = 6,000\left(1 + \dfrac{0.06}{12}\right)^{12t}$

$\dfrac{5}{3} = (1.005)^{12t}$

$12t = \log_{1.005}\left(\dfrac{5}{3}\right)$

$t = \dfrac{1}{12}\left(\dfrac{\log\left(\dfrac{5}{3}\right)}{\log(1.005)}\right) \approx 8.5$ years

It was invested for approximately 8.5 years.

118. $x \approx 0.69$

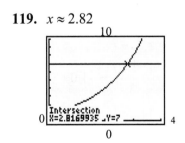

119. $x \approx 2.82$

Chapter 9 Test

1. $(f \circ h)(0) = f(h(0)) = f(5) = 5$

2. $(g \circ f)(x) = g(f(x)) = g(x) = x - 7$

3. $(g \circ h)(x) = g(h(x))$
$= g(x^2 - 6x + 5)$
$= x^2 - 6x + 5 - 7$
$= x^2 - 6x - 2$

4. $f(x) = 7x - 14,\ f^{-1}(x) = \dfrac{x + 14}{7}$

5. The graph is one-to-one.

6. The graph is not one-to-one.

7. $y = 6 - x$ is one-to-one.
$x = 6 - 2y$
$2y = -x + 6$
$y = \dfrac{-x + 6}{2}$
$f^{-1}(x) = \dfrac{-x + 6}{2}$

8. $f = \{(0,0),(2,3),(-1,5)\}$
is one-to-one.
$f^{-1} = \{(0,0),(3,2),(5,-1)\}$

9. The function is not one-to-one.

10. $\log_3 6 + \log_3 4 = \log_3(6 \cdot 4) = \log_3 24$

11. $\log_5 x + 3\log_5 x - \log_5(x+1)$
$= 4\log_5 x - \log_5(x+1)$
$= \log_5 x^4 - \log_5(x+1)$
$= \log_5 \dfrac{x^4}{x+1}$

12. $\log_6 \dfrac{2x}{y^3} = \log_6 2x - \log_6 y^3$

$\qquad\qquad = \log_6 2 + \log_6 x - 3\log_6 y$

13. $\log_b\left(\dfrac{3}{25}\right) = \log_b 3 - \log_b 25$

$\qquad\qquad = \log_b 3 - \log_b 5^2$

$\qquad\qquad = \log_b 3 - 2\log_b 5$

$\qquad\qquad = 0.79 - 2(1.16)$

$\qquad\qquad = -1.53$

14. $\log_7 8 = \dfrac{\ln 8}{\ln 7} \approx 1.0686$

15. $8^{x-1} = \dfrac{1}{64}$

$\qquad 8^{x-1} = 8^{-2}$

$\qquad x - 1 = -2$

$\qquad\quad x = -1$

16. $3^{2x+5} = 4$

$\qquad 2x + 5 = \log_3 4$

$\qquad\quad 2x = \dfrac{\log 4}{\log 3} - 5$

$\qquad\quad x = \dfrac{1}{2}\left(\dfrac{\log 4}{\log 3} - 5\right)$

$\qquad\quad x \approx -1.8691$

17. $\log_3 x = -2$

$\qquad x = 3^{-2}$

$\qquad x = \dfrac{1}{9}$

18. $\ln\sqrt{e} = x$

$\qquad \ln e^{1/2} = x$

$\qquad \dfrac{1}{2} = x$

19. $\log_8(3x - 2) = 2$

$\qquad 3x - 2 = 8^2$

$\qquad 3x - 2 = 64$

$\qquad\quad 3x = 66$

$\qquad\quad x = \dfrac{66}{3} = 22$

20. $\log_5 x + \log_5 3 = 2$

$\qquad \log_5(3x) = 2$

$\qquad\quad 3x = 5^2$

$\qquad\quad 3x = 25$

$\qquad\quad x = \dfrac{25}{3}$

21. $\log_4(x + 1) - \log_4(x - 2) = 3$

$\qquad \log_4 \dfrac{x+1}{x-2} = 3$

$\qquad\quad \dfrac{x+1}{x-2} = 4^3$

$\qquad\quad x + 1 = 64x - 128$

$\qquad\quad x = \dfrac{43}{21}$

22. $\ln(3x + 7) = 1.31$

$\qquad 3x + 7 = e^{1.31}$

$\qquad\quad 3x = e^{1.31} - 7$

$\qquad\quad x = \dfrac{e^{1.31} - 7}{3} \approx -1.0979$

23. $y = \left(\dfrac{1}{2}\right)^x + 1$

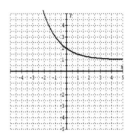

24. $y = 3^x$ and $y = \log_3 x$

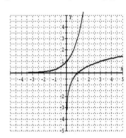

25. $A = \left(1 + \dfrac{r}{n}\right)^{nt}, P = 4000, t = 3, r = 0.09$
and $n = 12$

$$A = 4000\left(1 + \dfrac{0.09}{12}\right)^{12(3)}$$
$$= 4000(1.0075)^{36}$$
$$\approx 5234.58$$

\$5234.58 will be in the account.

26. $A = \left(1 + \dfrac{r}{n}\right)^{nt}, P = 2000, A = 3000$

$r = 0.07, \; n = 2$

$$3000 = 2000\left(1 + \dfrac{0.07}{2}\right)^{2t}$$
$$1.5 = (1.035)^{2t}$$
$$2t = \log_{1.035} 1.5$$
$$t = \dfrac{1}{2}\dfrac{\ln 1.5}{\ln 1.035}$$
$$\approx 5.9$$

27. $y = y_0 e^{kt}$

$$y = 57{,}000 e^{0.026(5)}$$
$$= 57{,}000 e^{0.13}$$
$$\approx 64{,}913$$

There will be approximately 64,913 prairie dogs 5 years from now.

28. $y = y_0 e^{kt}$

$$1000 = 400 e^{0.062(t)}$$
$$2.5 = e^{0.062t}$$
$$0.062t = \ln 2.5$$
$$t = \dfrac{\ln 2.5}{0.062}$$
$$\approx 14.8$$

It will take the naturalists approximately 15 years to reach their goal.

29. $\log(1 + k) = \dfrac{0.3}{D}, D = 56$

$$\log(1 + k) = \dfrac{0.3}{56}$$
$$1 + k = 10^{0.3/56}$$
$$k = -1 + 10^{0.3/56}$$
$$k \approx 0.012$$

The rate of population increase is approximately 1.2%.

Chapter 9 Cumulative Review

1. a. 8

b. $-\dfrac{1}{3}$

c. -9

d. 0

e. $-\dfrac{2}{11}$

f. 42

g. 0

2. $\dfrac{1}{3}(x-2) = \dfrac{1}{4}(x+1)$

$4(x-2) = 3(x+1)$

$4x - 8 = 3x + 3$

$x = 11$

3. $y = x^2$

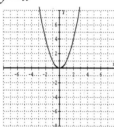

4. $f(x) - 6 = \dfrac{1}{3}(x+2)$

$f(x) - 6 = \dfrac{1}{3}x + \dfrac{2}{3}$

$f(x) = \dfrac{1}{3}x + \dfrac{20}{3} + \dfrac{18}{3}$

$f(x) = \dfrac{1}{3}x + \dfrac{20}{3}$

5. Equation 2 is twice the opposite of equation 1. Therefore, the system is dependent. Thus,
$\{(x, y, z) \mid x - 5y - 2z = 6\}$.

6. $x = 110°, \ y = 70°$

7. a. x^3

   b. 5^6

   c. $5x$

d. $\dfrac{6y^2}{7}$

8. a. $16a^6$

   b. $-\dfrac{8}{27}$

   c. $\dfrac{64a^{15}}{b^9}$

   d. $729x^3$

   e. $\dfrac{a^4 c^8}{b^6}$

9. a. $C(100) = \dfrac{2.6(100) + 10{,}000}{100}$

   $= \$102.60$

   b. $C(1000) = \dfrac{2.6(1000) + 10{,}000}{1000}$

   $= \$12.60$

10. a. $(3x-1)^2 = (3x-1)(3x-1)$

   $= 9x^2 - 6x + 1$

   b. $\left(\dfrac{1}{2}x + 3\right)\left(\dfrac{1}{2}x - 3\right)$

   $= \dfrac{1}{4}x^2 + \dfrac{3}{2}x - \dfrac{3}{2}x - 9$

   $= \dfrac{1}{4}x^2 - 9$

   c. $(2x-5)(6x+7)$

   $= 12x^2 + 14x - 30x - 35$

   $= 12x^2 - 16x - 35$

11. a. $\dfrac{x}{4} + \dfrac{5x}{4} = \dfrac{6x}{4} = \dfrac{3x}{2}$

b. $\dfrac{x^2}{x+7} - \dfrac{49}{x+7} = \dfrac{x^2 - 49}{x+7}$

$\qquad\qquad = \dfrac{(x+7)(x-7)}{(x+7)}$

$\qquad\qquad = x - 7$

c. $\dfrac{x}{3y^2} - \dfrac{x+1}{3y^2} = \dfrac{x-x-1}{3y^2} = -\dfrac{1}{3y^2}$

12. $\dfrac{5}{x-2} + \dfrac{3}{x^2+4x+4} - \dfrac{6}{x+2}$

$\quad = \dfrac{5(x+2)^2 + 3(x-2) - 6(x-2)(x+2)}{(x-2)(x+2)(x+2)}$

$\quad = \dfrac{5x^2 + 20x + 20 + 3x - 6 - 6x^2 + 24}{(x-2)(x+2)(x+2)}$

$\quad = \dfrac{-x^2 + 23x + 38}{(x-2)(x+2)^2}$

13. $x^2 - 1\overline{\smash{\big)}\,3x^4 + 2x^3 + 0x^2 - 8x + 6}$

$\qquad\quad \dfrac{3x^4 \qquad\quad -3x^2}{\quad 2x^3 + 3x^2 - 8x + 6}$

$\qquad\qquad \dfrac{2x^3 \qquad\quad -2x}{\quad 3x^2 - 6x + 6}$

$\qquad\qquad\quad \dfrac{3x^2 \qquad -3}{\quad -6x + 9}$

over the division: $3x^2 + 2x + 3$

$3x^2 + 2x + 3 + \dfrac{-6x+9}{x^2-1}$

14. a. $\dfrac{\dfrac{a}{5}}{\dfrac{a-1}{10}} = \dfrac{a}{5} \cdot \dfrac{10}{a-1} = \dfrac{2a}{a-1}$

b. $\dfrac{\dfrac{3}{2+a} + \dfrac{6}{2-a}}{\dfrac{5}{a+2} - \dfrac{1}{a-2}}$

$\quad = \dfrac{\dfrac{3(2-a)}{(2+a)(2-a)} + \dfrac{6(2+a)}{(2+a)(2-a)}}{\dfrac{5(a-2)}{(a+2)(a-2)} - \dfrac{1(a+2)}{(a+2)(a-2)}}$

$\quad = \dfrac{\dfrac{6-3a}{(2+a)(2-a)} + \dfrac{12+6a}{(2+a)(2-a)}}{\dfrac{5a-10}{(a+2)(a-2)} - \dfrac{a+2}{(a+2)(a-2)}}$

$\quad = \dfrac{\dfrac{18+3a}{(2+a)(2-a)}}{\dfrac{4a-12}{(a+2)(a-2)}}$

$\quad = \dfrac{18+3a}{(2+a)(2-a)} \cdot \dfrac{(a+2)(a-2)}{4a-12}$

$\quad = \dfrac{18+3a}{-1(4a-12)}$

$\quad = \dfrac{-3a-18}{4a-12}$

c. $\left(\dfrac{\dfrac{1}{x} + \dfrac{1}{y}}{xy}\right) = \dfrac{\left(\dfrac{1}{x} + \dfrac{1}{y}\right)xy}{(xy)(xy)} = \dfrac{y+x}{x^2 y^2}$

15. $\dfrac{2x}{2x-1} + \dfrac{1}{x} = \dfrac{1}{2x-1}$

$x(2x-1)\left(\dfrac{2x}{2x-1} + \dfrac{1}{x} = \dfrac{1}{2x-1}\right)$

$2x^2 + 2x - 1 = x$

$2x^2 + x - 1 = 0$

$(2x-1)(x+1) = 0$

$x = \dfrac{1}{2}$ (discard) or $x = -1$

$x = -1$

16. $\dfrac{x^3-8}{x-2} = \dfrac{(x-2)(x^2+2x+4)}{(x-2)}$

$= x^2 + 2x + 4$

17. $\dfrac{72}{30-x} = \dfrac{108}{30+x}$

$72(30+x) = 108(30-x)$

$x = 6$ mph

18. $\underline{2|\;8\quad -12\quad -7}$

$\quad 16\qquad 8$

$\overline{\;8\quad\;\;4\qquad 1}$

$8x + 4 + \dfrac{1}{x-2}$

19. a. 3

b. -3

c. -5

d. not a real number

e. $4x$

20.

$\dfrac{1}{a+5} = \dfrac{1}{3a+6} - \dfrac{a+2}{a^2+7a+10}$

$\dfrac{1}{a+5} = \dfrac{1}{3(a+2)} - \dfrac{a+2}{(a+5)(a+2)}$

$3(a+2)(a+5)\left[\dfrac{1}{a+5}\right] = 3(a+2)(a+5)\left[\dfrac{1}{3(a+2)} - \dfrac{a+2}{(a+5)(a+2)}\right]$

$3(a+2) = 1(a+5) - 3(a+2)$

$3a+6 = a+5-3a-6$

$5a = -7$

$a = -\dfrac{7}{5}$

21. a. $\sqrt[4]{x^3}$

b. $\sqrt[6]{x}$

c. $\sqrt[6]{72}$

22. $\dfrac{1}{2} = 12k$

$k = \dfrac{1}{24}, \; y = \dfrac{1}{24}x$

23. a. $\sqrt{3}\left(5+\sqrt{30}\right)$

$5\sqrt{3} + \sqrt{90}$

$5\sqrt{3} + \sqrt{9\cdot 10}$

$5\sqrt{3} + 3\sqrt{10}$

b. $\left(\sqrt{5}-\sqrt{6}\right)\left(\sqrt{7}+1\right)$

$\sqrt{35} + \sqrt{5} - \sqrt{42} - \sqrt{6}$

c. $\left(7\sqrt{x}+5\right)\left(3\sqrt{x}-\sqrt{5}\right)$

$21x - 7\sqrt{5x} + 15\sqrt{x} - 5\sqrt{5}$

d. $\left(4\sqrt{3}-1\right)^2$

$\left(4\sqrt{3}-1\right)\left(4\sqrt{3}-1\right)$

$16\sqrt{9}-4\sqrt{3}-4\sqrt{3}+1$

$48-8\sqrt{3}+1$

$49-8\sqrt{3}$

e. $\left(\sqrt{2x}-5\right)\left(\sqrt{2x}+5\right)$

$2x+5\sqrt{2x}-5\sqrt{2x}-25$

$2x-25$

f. $\left(\sqrt{x-3}+5\right)^2$

$\left(\sqrt{x-3}+5\right)\left(\sqrt{x-3}+5\right)$

$x-3+5\sqrt{x-3}+5\sqrt{x-3}+25$

$x+22+10\sqrt{x-3}$

24. a. 3

    b. -3

    c. $\dfrac{3}{8}$

    d. x^3

    e. $-5y^2$

25. $\dfrac{\sqrt[4]{x}}{\sqrt[4]{81y^5}}=\dfrac{\sqrt[4]{x}}{\sqrt[4]{81y^5}}\cdot\dfrac{\sqrt[4]{y^3}}{\sqrt[4]{y^3}}=\dfrac{\sqrt[4]{xy^3}}{3y^2}$

26. a. $a^{\frac{1}{4}}\left(a^{\frac{3}{4}}-a^8\right)$

    $a^{\frac{1}{4}+\frac{3}{4}}-a^{\frac{1}{4}+8}$

    $a-a^{\frac{33}{4}}$

b. $\left(x^{\frac{1}{2}}-3\right)\left(x^{\frac{1}{2}}+5\right)$

$x^{\frac{1}{2}+\frac{1}{2}}+5x^{\frac{1}{2}}-3x^{\frac{1}{2}}-15$

$x+2x^{\frac{1}{2}}-15$

27. $\sqrt{4-x}=x-2$

$\left(\sqrt{4-x}\right)^2=\left(x-2\right)^2$

$4-x=x-2$

$x=3$

28. a. $\sqrt{\dfrac{54}{6}}=\sqrt{9}=3$

    b. $\dfrac{\sqrt{108a^2}}{3\sqrt{3}}=\dfrac{1}{3}\sqrt{\dfrac{108a^2}{3}}=2a$

    c.

$3\sqrt[3]{\dfrac{81a^5b^{10}}{3b^4}}=3\sqrt[3]{27a^5b^6}=9ab^2\sqrt[3]{a^2}$

29. $3x^2-9x+8=0$

$3(x^2-3x)=-8$

$3\left(x^2-3x+\dfrac{9}{4}\right)=-8+\dfrac{9}{4}$

$3\left(x-\dfrac{3}{2}\right)^2+\dfrac{5}{4}=0$

$x=\dfrac{9\pm i\sqrt{15}}{6}$

30. a. $\dfrac{\sqrt{20}}{3}+\dfrac{\sqrt{5}}{4}$

$\dfrac{2\sqrt{5}}{3}+\dfrac{\sqrt{5}}{4}$

$\dfrac{4\left(2\sqrt{5}\right)}{12}+\dfrac{3\sqrt{5}}{12}$

$\dfrac{8\sqrt{5}}{12}+\dfrac{3\sqrt{5}}{12}$

$\dfrac{11\sqrt{5}}{12}$

b. $\sqrt[3]{\dfrac{24x}{27}}-\dfrac{\sqrt[3]{3x}}{2}$

$\sqrt[3]{\dfrac{3\cdot 8x}{27}}-\dfrac{\sqrt[3]{3x}}{2}$

$\dfrac{2\sqrt[3]{3x}}{3}-\dfrac{\sqrt[3]{3x}}{2}$

$\dfrac{2\left(2\sqrt[3]{3x}\right)}{6}-\dfrac{3\sqrt[3]{3x}}{6}$

$\dfrac{\sqrt[3]{3x}}{6}$

31. $\dfrac{3x}{x-2}-\dfrac{x+1}{x}=\dfrac{6}{x(x-2)}$

$3x(x)-(x+1)(x-2)=6$

$x=\dfrac{-1\pm\sqrt{33}}{4}$

32. $\sqrt[3]{\dfrac{27}{m^4 n^8}}$

$\dfrac{3}{\sqrt[3]{m^4 n^8}}$

$\dfrac{3}{\sqrt[3]{m^4 n^8}}\cdot\dfrac{\sqrt[3]{m^2 n}}{\sqrt[3]{m^2 n}}$

$\dfrac{3\sqrt[3]{m^2 n}}{\sqrt[3]{m^6 n^9}}$

$\dfrac{3\sqrt[3]{m^2 n}}{m^2 n^3}$

33. $x^2-4x\le 0$

$x(x-4)=0$

$x=0,\ x=4$

See graph

$[0,4]$

34. $4\sqrt{3}$ in.

35. $F(x)=(x-3)^2+1$

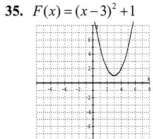

36. a. 1

b. i

c. -1

d. $-i$

37. a. $f(x)+g(x)$

$x-1+2x-3$

$3x-4$

  b. $f(x) - g(x)$

$$x - 1 - 2x + 3$$

$$-x + 2$$

  c. $f(x) \cdot g(x)$

$$(x - 1)(2x - 3)$$

$$2x^2 - 5x + 3$$

  d. $\dfrac{x-1}{2x-3}$, where $x \neq \dfrac{3}{2}$

38. $4x^2 + 8x - 1 = 0$

$$4x^2 + 8x = 1$$

$$4\left(x^2 + 2x + 1\right) = 1 + 4$$

$$4(x+1)^2 = 5$$

$$(x+1)^2 = \frac{5}{4}$$

$$x + 1 = \pm\frac{\sqrt{5}}{2}$$

$$x = -1 \pm \frac{\sqrt{5}}{2}$$

$$x = \frac{-2 \pm \sqrt{5}}{2}$$

39. $f(x) = x + 3$

$$x = y + 3$$

$$y = x - 3$$

$$f^{-1}(x) = x - 3$$

40. $\left(x - \dfrac{1}{2}\right)^2 = \dfrac{1}{2}x$

$$x^2 - x + \frac{1}{4} = \frac{1}{2}x$$

$$x^2 - \frac{3}{2}x + \frac{1}{4} = 0$$

$$a = 1, \; b = -\frac{3}{2}, \; c = \frac{1}{4}$$

$$x = \frac{-b \pm \sqrt{b^2 - 4ac}}{2a}$$

$$x = \frac{3 \pm \sqrt{5}}{4}$$

41. a. 2

   b. -1

   c. $\dfrac{1}{2}$

42. $f(x) = -(x+1)^2 + 1$

    Vertex: $(-1, 1)$

    Axis of symmetry: $x = -1$

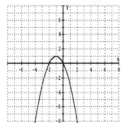

Chapter 10

Section 10.1

Graphing Calculator Explorations

1. $x^2 + y^2 = 55$
$$y^2 = 55 - x^2$$
$$y = \pm\sqrt{55 - x^2}$$

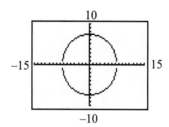

3. $5x^2 + 5y^2 = 50$
$$5y^2 = 50 - 5x^2$$
$$y^2 = 10 - x^2$$
$$y = \pm\sqrt{10 - x^2}$$

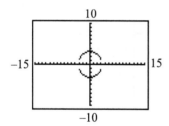

5. $2x^2 + 2y^2 - 34 = 0$
$$2y^2 = 34 - 2x^2$$
$$y^2 = 17 - x^2$$
$$y = \pm\sqrt{17 - x^2}$$

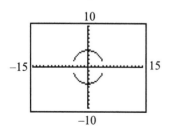

7. $7x^2 + 7y^2 - 89 = 0$
$$7y^2 = 89 - 7x^2$$
$$y^2 = \frac{89 - 7x^2}{7}$$
$$y = \pm\sqrt{\frac{89 - 7x^2}{7}}$$

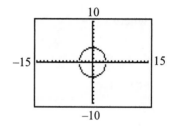

Mental Math

1. $y = x^2 - 7x + 5$; upward

2. $y = -x^2 + 16$; downward

3. $x = -y^2 - y + 2$; to the left

4. $x = 3y^2 + 2y - 5$; to the right

5. $y = -x^2 + 2x + 1$; downward

6. $x = -y^2 + 2y - 6$; to the left

Exercise Set 10.1

1. $x = 3y^2$
$$x = 3(y - 0)^2 + 0$$
Vertex: (0, 0)

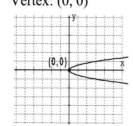

420

3. $x = (y-2)^2 + 3$

Vertex: (3, 2)

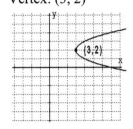

5. $y = 3(x-1)^2 + 5$

Vertex: (1, 5)

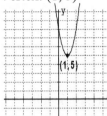

7.
$$x = y^2 + 6y + 8$$
$$x - 8 = y^2 + 6y$$
$$x - 8 + 9 = y^2 + 6y + 9$$
$$x + 1 = (y+3)^2$$
$$x = (y+3)^2 - 1$$

Vertex: (−1, −3)

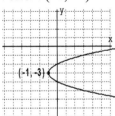

9.
$$y = x^2 + 10x + 20$$
$$y - 20 = x^2 + 10x$$
$$y - 20 + 25 = x^2 + 10x + 25$$
$$y + 5 = (x+5)^2$$
$$y = (x+5)^2 - 5$$

Vertex: (−5, −5)

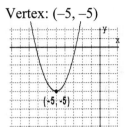

11.
$$x = -2y^2 + 4y + 6$$
$$x - 6 = -2(y^2 - 2y)$$
$$x - 6 + [-2(1)] = -2(y^2 - 2y + 1)$$
$$x - 8 = -2(y-1)^2$$
$$x = -2(y-1)^2 + 8$$

Vertex: (8, 1)

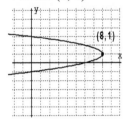

13. (5, 1), (8, 5)
$$d = \sqrt{(8-5)^2 + (5-1)^2}$$
$$= \sqrt{9 + 16}$$
$$= \sqrt{25}$$
$$= 5 \text{ units}$$

15. (−3, 2), (1, −3)
$$d = \sqrt{[1-(-3)]^2 + (-3-2)^2}$$
$$= \sqrt{4^2 + (-5)^2}$$
$$= \sqrt{16 + 25}$$
$$= \sqrt{41} \text{ units}$$

17. (−9, 4), (−8, 1)
$$d = \sqrt{[-8-(-9)]^2 + (1-4)^2}$$
$$= \sqrt{1^2 + (-3)^2}$$
$$= \sqrt{1 + 9}$$
$$= \sqrt{10} \text{ units}$$

19. $\left(0, -\sqrt{2}\right), \left(\sqrt{3}, 0\right)$

$$d = \sqrt{\left(\sqrt{3} - 0\right)^2 + \left[0 - \left(-\sqrt{2}\right)\right]^2}$$
$$= \sqrt{\left(\sqrt{3}\right)^2 + \left(\sqrt{2}\right)^2}$$
$$= \sqrt{3 + 5}$$
$$= \sqrt{5} \text{ units}$$

21. $(1.7, -3.6), (-8.6, 5.7)$

$$d = \sqrt{(-8.6 - 1.7)^2 + [5.7 - (-3.6)]^2}$$
$$= \sqrt{(-10.3)^2 + (9.3)^2}$$
$$= \sqrt{192.58}$$
$$= 13.88 \text{ units}$$

23. $\left(2\sqrt{3}, \sqrt{6}\right), \left(-\sqrt{3}, 4\sqrt{6}\right)$

$$d = \sqrt{\left(-\sqrt{3} - 2\sqrt{3}\right)^2 + \left(4\sqrt{6} - \sqrt{6}\right)^2}$$
$$= \sqrt{\left(-3\sqrt{3}\right)^2 + \left(3\sqrt{6}\right)^2}$$
$$= \sqrt{27 + 54}$$
$$= \sqrt{81}$$
$$= 9 \text{ units}$$

25. $(6, -8), (2, 4)$

$$\left(\frac{6+2}{2}, \frac{-8+4}{2}\right) = \left(\frac{8}{2}, \frac{-4}{2}\right) = (4, -2)$$

The midpoint of the segment is $(4, -2)$.

27. $(-2, -1), (-8, 6)$

$$\left(\frac{-2 + (-8)}{2}, \frac{-1 + 6}{2}\right) = \left(\frac{-10}{2}, \frac{5}{2}\right) = \left(-5, \frac{5}{2}\right)$$

The midpoint of the segment is $\left(-5, \frac{5}{2}\right)$.

29. $(7, 3), (-1, -3)$

$$\left(\frac{7 + (-1)}{2}, \frac{3 + (-3)}{2}\right) = \left(\frac{6}{2}, \frac{0}{2}\right) = (3, 0)$$

The midpoint of the segment is $(3, 0)$.

31. $\left(\frac{1}{2}, \frac{3}{8}\right), \left(-\frac{3}{2}, \frac{5}{8}\right)$

$$\left(\frac{\frac{1}{2} + \left(-\frac{3}{2}\right)}{2}, \frac{\frac{3}{8} + \frac{5}{8}}{2}\right) = \left(\frac{-1}{2}, \frac{1}{2}\right)$$

The midpoint of the segment is $\left(-\frac{1}{2}, \frac{1}{2}\right)$.

33. $\left(\sqrt{2}, 3\sqrt{5}\right), \left(\sqrt{2}, -2\sqrt{5}\right)$

$$\left(\frac{\sqrt{2} + \sqrt{2}}{2}, \frac{3\sqrt{5} + \left(-2\sqrt{5}\right)}{2}\right) = \left(\frac{2\sqrt{2}}{2}, \frac{\sqrt{5}}{2}\right)$$
$$= \left(\sqrt{2}, \frac{\sqrt{5}}{2}\right)$$

The midpoint of the segment is $\left(\sqrt{2}, \frac{\sqrt{5}}{2}\right)$.

35. $(4.6, -3.5), (7.8, -9.8)$

$$\left(\frac{4.6 + 7.8}{2}, \frac{-3.5 + (-9.8)}{2}\right) = \left(\frac{12.4}{2}, \frac{-13.2}{2}\right)$$
$$= (6.2, -6.65)$$

The midpoint of the segment is $(6.2, -6.65)$.

37. $x^2 + y^2 = 9$

$$(x - 0)^2 + (y - 0)^2 = 3^2$$

Center: $(0, 0)$, radius $r = 3$.

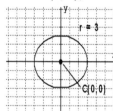

39. $x^2 + (y-2)^2 = 1$
$(x-0)^2 + (y-2)^2 = 1^2$
Center: (0, 2), radius $r = 1$.

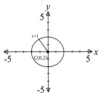

41. $(x-5)^2 + (y+2)^2 = 1$
$(x-5)^2 + (y+2)^2 = 1^2$
Center: (5, –2), radius $r = 1$.

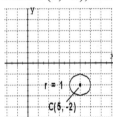

43. $x^2 + y^2 + 6y = 0$
     $x^2 + (y^2 + 6y) = 0$
$x^2 + (y^2 + 6y + 9) = 9$
$(x-0)^2 + (y+3)^2 = 9$
Center: (0, –3), radius $r = 3$.

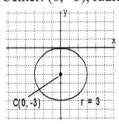

45. $x^2 + y^2 + 2x - 4y = 4$
     $(x^2 + 2x) + (y^2 - 4y) = 4$
$(x^2 + 2x + 1) + (y^2 - 4y + 4) = 4 + 1 + 4$
       $(x+1)^2 + (y-2)^2 = 9$

Center: (–1, 2), radius $r = 3$.

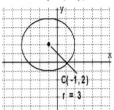

47. $x^2 + y^2 - 4x - 8y - 2 = 0$
     $(x^2 - 4x) + (y^2 - 8y) = 2$
$(x^2 - 4x + 4) + (y^2 - 8y + 16) = 2 + 4 + 16$
       $(x-2)^2 + (y-4)^2 = 22$

Center: (2, 4), radius $r = \sqrt{22}$.

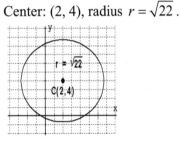

49. Center $(h, k) = (2, 3)$ and radius $r = 6$.
$(x-h)^2 + (y-k)^2 = r^2$
$(x-2)^2 + (y-3)^2 = 6^2$
$(x-2)^2 + (y-3)^2 = 36$

51. Center $(h, k) = (0, 0)$ and radius $r = \sqrt{3}$.
$(x-h)^2 + (y-k)^2 = r^2$
$(x-0)^2 + (y-0)^2 = \left(\sqrt{3}\right)^2$
$x^2 + y^2 = 3$

53. Center $(h, k) = (-5, 4)$ and radius $r = 3\sqrt{5}$.
$(x-h)^2 + (y-k)^2 = r^2$
$[x-(-5)]^2 + (y-4)^2 = \left(3\sqrt{5}\right)^2$
$(x+5)^2 + (y-4)^2 = 45$

55. Answers may vary.

57. $x = y^2 - 3$

$x = (y - 0)^2 - 3$

Vertex: $(-3, 0)$

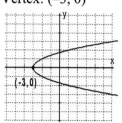

59. $y = (x - 2)^2 - 2$

Vertex: $(2, -2)$

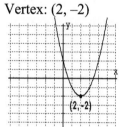

61. $x^2 + y^2 = 1$

Center: $(0, 0)$, radius $r = \sqrt{1} = 1$

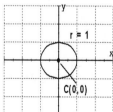

63. $x = (y + 3)^2 - 1$

Vertex: $(-1, -3)$

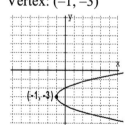

65. $(x - 2)^2 + (y - 2)^2 = 16$

Center: $(2, 2)$, radius $r = \sqrt{16} = 4$

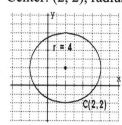

67. $x = -(y - 1)^2$

Vertex: $(0, 1)$

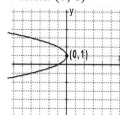

69. $(x - 4)^2 + y^2 = 7$

Center: $(4, 0)$, radius $r = \sqrt{7}$

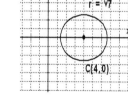

71. $y = 5(x + 5)^2 + 3$

Vertex: $(-5, 3)$

73.
$$\frac{x^2}{8} + \frac{y^2}{8} = 2$$
$$8\left(\frac{x^2}{8} + \frac{y^2}{8}\right) = 8(2)$$
$$x^2 + y^2 = 16$$

Center: (0, 0), radius $r = \sqrt{16} = 4$

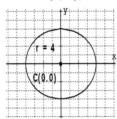

75.
$$y = x^2 + 7x + 6$$
$$y - 6 = x^2 + 7x$$
$$y - 6 + \frac{49}{4} = x^2 + 7x + \frac{49}{4}$$
$$y + \frac{25}{4} = \left(x + \frac{7}{2}\right)^2$$
$$y = \left(x + \frac{7}{2}\right)^2 - \frac{25}{4}$$

Vertex: $\left(-\frac{7}{2}, -\frac{25}{4}\right)$

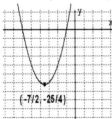

77.
$$x^2 + y^2 + 2x + 12y - 12 = 0$$
$$(x^2 + 2x) + (y^2 + 12y) = 12$$
$$(x^2 + 2x + 1) + (y^2 + 12y + 36) = 12 + 1 + 36$$
$$(x + 1)^2 + (y + 6)^2 = 49$$

Center: (−1, −6), radius $r = \sqrt{49} = 7$

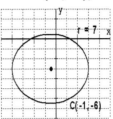

79.
$$x = y^2 + 8y - 4$$
$$x + 4 = y^2 + 8y$$
$$x + 4 + 16 = y^2 + 8y + 16$$
$$x + 20 = (y + 4)^2$$
$$x = (y + 4)^2 - 20$$

Vertex: (−20, −4)

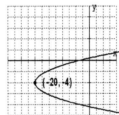

81.
$$x^2 - 10y + y^2 + 4 = 0$$
$$x^2 + (y^2 - 10y) = -4$$
$$x^2 + (y^2 - 10y + 25) = -4 + 25$$
$$x^2 + (y - 5)^2 = 21$$

Center: (0, 5), radius $r = \sqrt{21}$

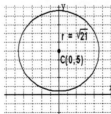

83.
$$x = -3y^2 + 30y$$
$$x = -3(y^2 - 10y)$$
$$x + [-3(25)] = -3(y^2 - 10y + 25)$$
$$x - 75 = -3(y - 5)^2$$
$$x = -3(y - 5)^2 + 75$$
Vertex: (75, 5)

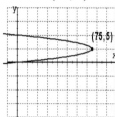

85. $5x^2 + 5y^2 = 25$
$$x^2 + y^2 = 5$$

Center: (0, 0), radius $r = \sqrt{5}$

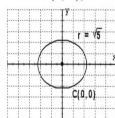

87.
$$y = 5x^2 - 20x + 16$$
$$y - 16 = 5(x^2 - 4x)$$
$$y - 16 + 5(4) = 5(x^2 - 4x + 4)$$
$$y - 4 = 4(x - 2)^2$$
$$y = 4(x - 2)^2 + 4$$
Vertex: (2, −4)

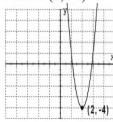

89. $y = -3x + 3$

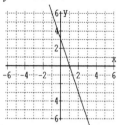

91. $x = -2$

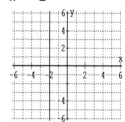

93. $\dfrac{\sqrt{5}}{\sqrt{8}} = \dfrac{\sqrt{5}}{2\sqrt{2}} = \dfrac{\sqrt{5} \cdot \sqrt{2}}{2\sqrt{2} \cdot \sqrt{2}} = \dfrac{\sqrt{10}}{2 \cdot 2} = \dfrac{\sqrt{10}}{4}$

95. $\dfrac{10}{\sqrt{5}} = \dfrac{10 \cdot \sqrt{5}}{\sqrt{5} \cdot \sqrt{5}} = \dfrac{10\sqrt{5}}{5} = 2\sqrt{5}$

97. Height = 264 ft and diameter $d = 250$ ft

  a. radius $= \dfrac{1}{2}d = \dfrac{1}{2}(250) = 125$ ft

  b. 264 − 250 = 14 ft from the ground

  c. 125 + 14 = 139 ft from the ground

  d. center: (0, 139)

  e. $(x - h)^2 + (y - k)^2 = r^2$
$$(x - 0)^2 + (y - 139)^2 = 125^2$$
$$x^2 + (y - 139)^2 = 15,625$$

99. B: (3, 1) and C: (19, 13)
$$d = \sqrt{(13 - 1)^2 + (19 - 3)^2}$$
$$= \sqrt{12^2 + 16^2}$$
$$= \sqrt{144 + 256}$$
$$= \sqrt{400} = 20 \text{ meters}$$

101. $y = a(x-h)^2 + k$

Vertex: (0, 40)

$y = a(x-0)^2 + 40$

$y = ax^2 + 40$

The parabola passes through (50, 0).

$0 = a(50)^2 + 40$

$-40 = 2500a$

$a = \dfrac{-40}{2500} = -\dfrac{2}{125}$

Thus, the equation is $y = -\dfrac{2}{125}x^2 + 40$.

103. $5x^2 + 5y^2 = 25$

$\quad 5y^2 = 25 - 5x^2$

$\quad y^2 = 5 - x^2$

$\quad y = \pm\sqrt{5 - x^2}$

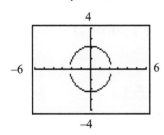

Answers are the same.

105. $y = 5x^2 - 20x + 16$

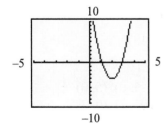

Answers are the same.

Section 10.2

Graphing Calculator Explorations

1. $10x^2 + y^2 = 32$

$\quad y^2 = 32 - 10x^2$

$\quad y = \pm\sqrt{32 - 10x^2}$

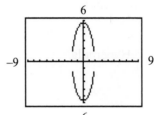

3. $20x^2 + 5y^2 = 100$

$\quad 5y^2 = 100 - 20x^2$

$\quad y^2 = 20 - 4x^2$

$\quad y = \pm\sqrt{20 - 4x^2}$

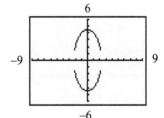

5. $7.3x^2 + 15.5y^2 = 95.2$

$\quad 15.5y^2 = 95.2 - 7.3x^2$

$\quad y^2 = \dfrac{95.2 - 7.3x^2}{15.5}$

$\quad y = \pm\sqrt{\dfrac{95.2 - 7.3x^2}{15.5}}$

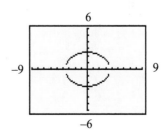

Mental Math

1. $\dfrac{x^2}{16} + \dfrac{y^2}{4} = 1$; Ellipse

2. $\dfrac{x^2}{16} - \dfrac{y^2}{4} = 1$; Hyperbola

3. $x^2 - 5y^2 = 3$; Hyperbola

4. $-x^2 + 5y^2 = 3$
$5y^2 - x^2 = 3$; Hyperbola

5. $-\dfrac{y^2}{25} + \dfrac{x^2}{36} = 1$
$\dfrac{x^2}{36} - \dfrac{y^2}{25} = 1$; Hyperbola

6. $\dfrac{y^2}{25} + \dfrac{x^2}{36} = 1$; Ellipse

Exercise Set 10.2

1. $\dfrac{x^2}{4} + \dfrac{y^2}{25} = 1$
$\dfrac{x^2}{2^2} + \dfrac{y^2}{5^2} = 1$
Center: (0, 0)
x-intercepts: (–2, 0), (2, 0)
y-intercepts: (0, –5), (0, 5)

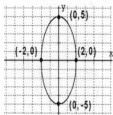

3. $\dfrac{x^2}{16} + \dfrac{y^2}{9} = 1$
$\dfrac{x^2}{4^2} + \dfrac{y^2}{3^2} = 1$

Center: (0, 0)
x-intercepts: (–4, 0), (4, 0)
y-intercepts: (0, –3), (0, 3)

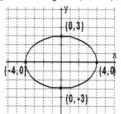

5. $9x^2 + 4y^2 = 36$
$\dfrac{x^2}{4} + \dfrac{y^2}{9} = 1$
$\dfrac{x^2}{2^2} + \dfrac{y^2}{3^2} = 1$
Center: (0, 0)
x-intercepts: (–2, 0), (2, 0)
y-intercepts: (0, –3), (0, 3)

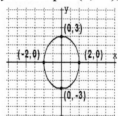

7. $4x^2 + 25y^2 = 100$
$\dfrac{x^2}{25} + \dfrac{y^2}{4} = 1$
$\dfrac{x^2}{5^2} + \dfrac{y^2}{2^2} = 1$
Center: (0, 0)
x-intercepts: (–5, 0), (5, 0)
y-intercepts: (0, –2), (0, 2)

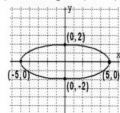

9. $\dfrac{(x+1)^2}{36} + \dfrac{(y-2)^2}{49} = 1$

$\dfrac{(x+1)^2}{6^2} + \dfrac{(y-2)^2}{7^2} = 1$

Center: $(-1, 2)$
Other points:
$(-1-6, 2) = (-7, 2)$
$(-1+6, 2) = (5, 2)$
$(-1, 2-7) = (-1, -5)$
$(-1, 2+7) = (-1, 9)$

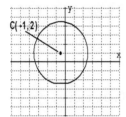

11. $\dfrac{(x-1)^2}{4} + \dfrac{(y-1)^2}{25} = 1$

$\dfrac{(x-1)^2}{2^2} + \dfrac{(y-1)^2}{5^2} = 1$

Center: $(1, 1)$
Other points:
$(1-2, 1) = (-1, 1)$
$(1+2, 1) = (3, 1)$
$(1, 1-5) = (1, -4)$
$(1, 1+5) = (1, 6)$

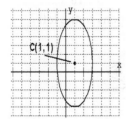

13. $\dfrac{x^2}{4} - \dfrac{y^2}{9} = 1$

$\dfrac{x^2}{2^2} - \dfrac{y^2}{3^2} = 1$

$a = 2, b = 3$

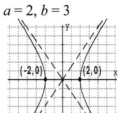

15. $\dfrac{y^2}{25} - \dfrac{x^2}{16} = 1$

$\dfrac{x^2}{5^2} - \dfrac{y^2}{4^2} = 1$

$a = 5, b = 4$

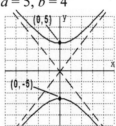

17. $x^2 - 4y^2 = 16$

$\dfrac{x^2}{16} - \dfrac{y^2}{4} = 1$

$\dfrac{x^2}{4^2} - \dfrac{y^2}{2^2} = 1$

$a = 4, b = 2$

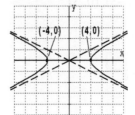

19. $16y^2 - x^2 = 16$

$$\frac{y^2}{1} - \frac{x^2}{16} = 1$$

$$\frac{y^2}{1^2} - \frac{x^2}{4^2} = 1$$

$a = 4$, $b = 1$

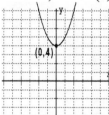

21. Answers may vary.

23. $y = x^2 + 4$

Parabola; vertex (0, 4), opens upward

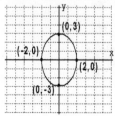

25. $\dfrac{x^2}{4} + \dfrac{y^2}{9} = 1$

$$\frac{x^2}{2^2} + \frac{y^2}{3^2} = 1$$

Ellipse; center: (0, 0)
x-intercepts: (–2, 0), (2, 0)
y-intercepts: (0, –3), (0, 3)

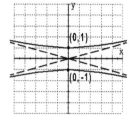

27. $\dfrac{x^2}{16} - \dfrac{y^2}{4} = 1$

$$\frac{x^2}{4^2} - \frac{y^2}{2^2} = 1$$

Hyperbola; center: (0, 0)
$a = 4$, $b = 2$

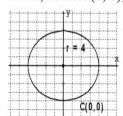

29. $x^2 + y^2 = 16$

Circle; center: (0, 0), radius: $r = \sqrt{16} = 4$

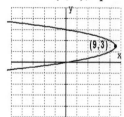

31. $x = -y^2 + 6y$

Parabola: $y = -\dfrac{b}{2a} = \dfrac{-6}{2(-1)} = 3$

$x = -(3)^2 + 6(3) = -9 + 18 = 9$
Vertex: (9, 3), opens to the left

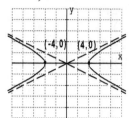

33. $9x^2 + 4y^2 = 36$

$$\frac{x^2}{4} + \frac{y^2}{9} = 1$$

$$\frac{x^2}{2^2} - \frac{y^2}{3^2} = 1$$

Ellipse; center: $(0, 0)$

$a = 2, \ b = 3$

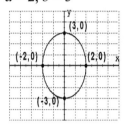

35. $\qquad y^2 = x^2 + 16$

$\qquad y^2 - x^2 = 16$

$$\frac{y^2}{16} - \frac{x^2}{16} = 1$$

$$\frac{y^2}{4^2} - \frac{x^2}{4^2} = 1$$

Hyperbola; center: $(0, 0)$

$a = 4, \ b = 4$

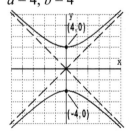

37. $y = -2x^2 + 4x - 3$

Parabola; $x = -\dfrac{b}{2a} = \dfrac{-4}{2(-2)} = 1$

$y = -2(1)^2 + 4(1) - 3 = -1$

Vertex: $(1, -1)$

39. $x < 5$ or $x < 1$

$x < 5$

$(-\infty, 5)$

41. $2x - 1 \ge 7$ and $-3x \le -6$

$\qquad 2x \ge 8$ and $\qquad x \ge 2$

$\qquad x \ge 4$

$x \ge 4$

$[4, \infty)$

43. $2x^3 - 4x^3 = -2x^3$

45. $(-5x^2)(x^2) = -5x^{2+2} = -5x^4$

47. Circles: B, F

Ellipses: C, E, H

Hyperbolas: A, D, G

49. A: $c^2 = 36 + 13 = 49$; $c = \sqrt{49} = 7$

B: $c^2 = 4 - 4 = 0$; $c = \sqrt{0} = 0$

C: $c^2 = |25 - 16| = 9$; $c = \sqrt{9} = 3$

D: $c^2 = 39 + 25 = 64$; $c = \sqrt{64} = 8$

E: $c^2 = |81 - 17| = 64$; $c = \sqrt{64} = 8$

F: $c^2 = |36 - 36| = 0$; $c = \sqrt{0} = 0$

G: $c^2 = 65 + 16 = 81$; $c = \sqrt{81} = 9$

H: $c^2 = |144 - 140| = 4$; $c = \sqrt{4} = 2$

51. A: $e = \dfrac{7}{6}$

B: $e = \dfrac{0}{2} = 0$

C: $e = \dfrac{3}{5}$

D: $e = \dfrac{8}{5}$

E: $e = \dfrac{8}{9}$

F: $e = \dfrac{0}{6} = 0$

G: $e = \dfrac{9}{4}$

H: $e = \dfrac{2}{12} = \dfrac{1}{6}$

53. They are equal to 0.

55. Answers may vary.

57. $a = 130,000,000 \Rightarrow a^2 = (130,000,000)^2$
$$= 1.69 \times 10^{16}$$
$b = 125,000,000 \Rightarrow b^2 = (125,000,000)^2$
$$= 1.5625 \times 10^{16}$$

Thus, the equation is

$$\dfrac{x^2}{1.69 \times 10^{16}} + \dfrac{y^2}{1.5625 \times 10^{16}} = 1.$$

59. $9x^2 + 4y^2 = 36$
$$4y^2 = 36 - 9x^2$$
$$y^2 = \dfrac{36 - 9x^2}{4}$$
$$y = \pm\sqrt{\dfrac{36 - 9x^2}{4}} = \pm\dfrac{\sqrt{36 - 9x^2}}{2}$$

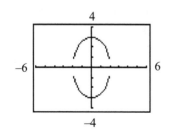

61. $\dfrac{(x-1)^2}{4} - \dfrac{(y+1)^2}{25} = 1$

Center: $(1, -1)$

$a = 2, b = 5$

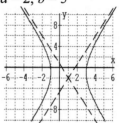

63. $\dfrac{y^2}{16} - \dfrac{(x+3)^2}{9} = 1$

Center: $(-3, 0)$

$a = 3, b = 4$

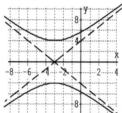

65. $\dfrac{(x+5)^2}{16} - \dfrac{(y+2)^2}{25} = 1$

Center: $(-5, -2)$

$a = 4, b = 5$

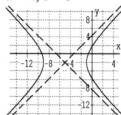

Integrated Review

1. $(x-7)^2 + (y-2)^2 = 4$
Circle; center: (7, 2),
radius: $r = \sqrt{4} = 2$

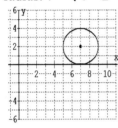

2. $y = x^2 + 4$
Parabola; vertex: (0, 4)

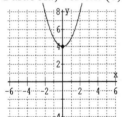

3. $y = x^2 + 12x + 36$

Parabola; $x = -\dfrac{b}{2a} = \dfrac{-12}{2(1)} = -6$

$y = (-6)^2 + 12(-6) + 36 = 0$
Vertex: (−6, 0)

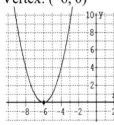

4. $\dfrac{x^2}{4} + \dfrac{y^2}{9} = 1$
Ellipse; center: (0, 0)
$a = 2, b = 3$

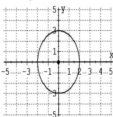

5. $\dfrac{y^2}{9} - \dfrac{x^2}{9} = 1$
Hyperbola; center: (0, 0)
$a = 3, b = 3$

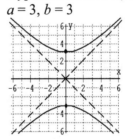

6. $\dfrac{x^2}{16} - \dfrac{y^2}{4} = 1$
Hyperbola; center: (0, 0)
$a = 4, b = 2$

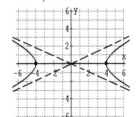

7. $\dfrac{x^2}{16} + \dfrac{y^2}{4} = 1$

Ellipse; center: (0, 0)

$a = 4$, $b = 2$

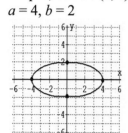

8. $x^2 + y^2 = 16$

Circle; center: (0, 0)

radius: $r = \sqrt{16} = 4$

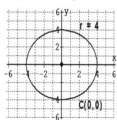

9. $x = y^2 + 4y - 1$

Parabola; $y = -\dfrac{b}{2a} = \dfrac{-4}{2(1)} = -2$

$x = (-2)^2 + 4(-2) - 1 = -5$

Vertex: $(-5, -2)$

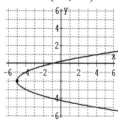

10. $x = -y^2 + 6y$

Parabola; $y = -\dfrac{b}{2a} = \dfrac{-6}{2(-1)} = 3$

$x = -(3)^2 + 6(3) = 9$

Vertex: (9, 3)

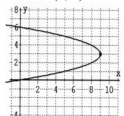

11. $9x^2 - 4y^2 = 36$

$\dfrac{x^2}{4} - \dfrac{y^2}{9} = 1$

Hyperbola; center: (0, 0)

$a = 2$, $b = 3$

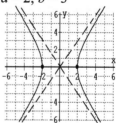

13. $\dfrac{(x-1)^2}{49} + \dfrac{(y+2)^2}{25} = 1$

Ellipse; center: $(1, -2)$,

$a = 7$, $b = 5$

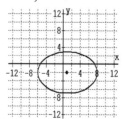

14.
$$y^2 = x^2 + 16$$
$$y^2 - x^2 = 16$$
$$\frac{y^2}{16} - \frac{x^2}{16} = 1$$

Hyperbola; center: (0, 0)

$a = 4$, $b = 4$

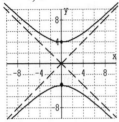

15. $\left(x + \frac{1}{2}\right)^2 + \left(y - \frac{1}{2}\right)^2 = 1$

Circle; center: $\left(-\frac{1}{2}, \frac{1}{2}\right)$,

radius: $r = \sqrt{1} = 1$

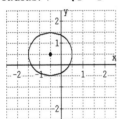

Exercise Set 10.3

1. $\begin{cases} x^2 + y^2 = 25 & (1) \\ 4x + 3y = 0 & (2) \end{cases}$

Solve E2 for y.

$$3y = -4x$$
$$y = -\frac{4x}{3}$$

Substitute into E1.

$$x^2 + \left(-\frac{4x}{3}\right)^2 = 25$$
$$x^2 + \frac{16x^2}{9} = 25$$
$$9\left(x^2 + \frac{16x^2}{9}\right) = 9(25)$$
$$9x^2 + 16x^2 = 225$$
$$25x^2 = 225$$
$$x^2 = 9$$
$$x = \pm\sqrt{9} = \pm 3$$

$x = 3: y = -\dfrac{4(3)}{3} = -4$

$x = -3: y = -\dfrac{4(-3)}{3} = 4$

The solutions are (3, –4) and (–3, 4).

3. $\begin{cases} x^2 + 4y^2 = 10 & (1) \\ y = x & (2) \end{cases}$

Substitute x for y in E1.

$$x^2 + 4x^2 = 10$$
$$5x^2 = 10$$
$$x^2 = 2$$
$$x = \pm\sqrt{2}$$

Replace these values into E2.

$x = \sqrt{2}: y = x = \sqrt{2}$

$x = -\sqrt{2}: y = x = -\sqrt{2}$

The solutions are $\left(\sqrt{2}, \sqrt{2}\right)$ and

$\left(-\sqrt{2}, -\sqrt{2}\right)$.

5. $\begin{cases} y^2 = 4 - x & (1) \\ x - 2y = 4 & (2) \end{cases}$

Solve E2 for x.

$$x = 2y + 4$$

Substitute into E1.

$$y^2 = 4 - (2y + 4)$$
$$y^2 = -2y$$
$$y^2 + 2y = 0$$
$$y(y + 2) = 0$$

$y = 0$ or $y + 2 = 0$

$\qquad y = -2$

Replace these values into the equation

$x = 2y + 4$.

$y = 0 : x = 2(0) + 4 = 4$

$y = -2 : x = 2(-2) + 4 = 0$

The solutions are (4, 0) and (0, –2).

7. $\begin{cases} x^2 + y^2 = 9 & (1) \\ 16x^2 - 4y^2 = 64 & (2) \end{cases}$

Multiply E1 by 4 and add to E2.

$\quad 4x^2 + 4y^2 = 36$

$\quad 16x^2 + 4y^2 = 64$

$\overline{\quad 20x^2 \qquad = 100}$

$\qquad x^2 = 5$

$\qquad x = \pm\sqrt{5}$

Substitute 5 for x^2 into E1.

$5 + y^2 = 9$

$\quad y^2 = 4$

$\quad y = \pm 2$

The solutions are $\left(-\sqrt{5}, -2\right)$, $\left(-\sqrt{5}, 2\right)$,

$\left(\sqrt{5}, -2\right)$, and $\left(\sqrt{5}, 2\right)$.

9. $\begin{cases} x^2 + 2y^2 = 2 & (1) \\ x - y = 2 & (2) \end{cases}$

Solve E2 for x: $x = y + 2$

Substitute into E1.

$(y + 2)^2 + 2y^2 = 2$

$y^2 + 4y + 4 + 2y^2 = 2$

$\quad 3y^2 + 4y + 2 = 0$

$y = \dfrac{-4 \pm \sqrt{(4)^2 - 4(3)(2)}}{2(3)} = \dfrac{-4 \pm \sqrt{-8}}{6}$

which yields no real solutions.

The solution is $\varnothing$.

11. $\begin{cases} y = x^2 - 3 & (1) \\ 4x - y = 6 & (2) \end{cases}$

Substitute $x^2 - 3$ for y in E2.

$4x - (x^2 - 3) = 6$

$4x - x^2 + 3 = 6$

$0 = x^2 - 4x + 3$

$0 = (x - 3)(x - 1)$

$x - 3 = 0$ or $x - 1 = 0$

$x = 3$ or $\quad x = 1$

Substitute these values into E1.

$x = 3 : y = (3)^2 - 3 = 6$

$x = 1 : y = (1)^2 - 3 = -2$

The solutions are (3, 6) and (1, –2).

13. $\begin{cases} y = x^2 & (1) \\ 3x + y = 10 & (2) \end{cases}$

Substitute x^2 for y in E2.

$\quad 3x + x^2 = 10$

$x^2 + 3x - 10 = 0$

$(x + 5)(x - 2) = 0$

$x + 5 = 0$ or $x - 2 = 0$

$x = -5$ or $\quad x = 2$

Substitute these values into E1.

$x = -5 : y = (-5)^2 = 25$

$x = 2 : y = (2)^2 = 4$

The solutions are (–5, 25) and (2, 4).

15. $\begin{cases} y = 2x^2 + 1 & (1) \\ x + y = -1 & (2) \end{cases}$

Substitute $2x^2 + 1$ for y in E2.

$x + 2x^2 + 1 = -1$

$2x^2 + x + 2 = 0$

$x = \dfrac{-1 \pm \sqrt{(1)^2 - 4(2)(2)}}{2(2)} = \dfrac{-1 \pm \sqrt{-15}}{4}$

which yields no real solutions.

The solution is $\varnothing$.

17. $\begin{cases} y = x^2 - 4 & (1) \\ y = x^2 - 4x & (2) \end{cases}$

Substitute $x^2 - 4$ for y in E2.

$x^2 - 4 = x^2 - 4x$

$-4 = -4x$

$1 = x$

Substitute this value into E1.

$y = (1)^2 - 4 = -3$

The solution is $(1, -3)$.

19. $\begin{cases} 2x^2 + 3y^2 = 14 & (1) \\ -x^2 + y^2 = 3 & (2) \end{cases}$

Multiply E2 by 2 and add to E1.

$-2x^2 + 2y^2 = 6$

$\underline{2x^2 + 3y^2 = 14}$

$5y^2 = 20$

$y^2 = 4$

$y = \pm 2$

Substitute 4 for y^2 into E2.

$-x^2 + 4 = 3$

$-x^2 = -1$

$x^2 = 1$

$x = \pm 1$

The solutions are $(-1, -2)$, $(-1, 2)$, $(1, -2)$, and $(1, 2)$.

21. $\begin{cases} x^2 + y^2 = 1 & (1) \\ x^2 + (y+3)^2 = 4 & (2) \end{cases}$

Multiply E1 by -1 and add to E2.

$-x^2 - y^2 = -1$

$\underline{x^2 + (y+3)^2 = 4}$

$(y+3)^3 - y^2 = 3$

$y^2 + 6y + 9 - y^2 = 3$

$6y = -6$

$y = -1$

Replace y with -1 in E1.

$x^2 + (1)^2 = 1$

$x^2 = 0$

$x = 0$

The solution is $(0, -1)$.

23. $\begin{cases} y = x^2 + 2 & (1) \\ y = -x^2 + 4 & (2) \end{cases}$

Add E1 and E2.

$2y = 6$

$y = 3$

Substitute this value into E1.

$3 = x^2 + 2$

$1 = x^2$

$\pm 1 = x$

The solutions are $(-1, 3)$ and $(1, 3)$.

25. $\begin{cases} 3x^2 + y^2 = 9 & (1) \\ 3x^2 - y^2 = 9 & (2) \end{cases}$

Add E1 and E2.

$6x^2 = 18$

$x^2 = 3$

$x = \pm\sqrt{3}$

Substitute 3 for x^2 into E1.

$3(3) + y^2 = 9$

$y^2 = 0$

$y = 0$

The solutions are $\left(-\sqrt{3}, 0\right), \left(\sqrt{3}, 0\right)$.

27. $\begin{cases} x^2 + 3y^2 = 6 & (1) \\ x^2 - 3y^2 = 10 & (2) \end{cases}$

Solve E2 for x^2: $x^2 = 3y^2 + 10$.

Substitute into E1.

$(3y^2 + 10) + 3y^2 = 6$

$6y^2 = -4$

$y^2 = -\dfrac{2}{3}$

which yields no real solutions.

The solution is $\varnothing$.

29. $\begin{cases} x^2 + y^2 = 36 & (1) \\ \quad y = \dfrac{1}{6}x^2 - 6 & (2) \end{cases}$

Solve E1 for x^2: $x^2 = 36 - y^2$.

Substitute into E2.

$$y = \frac{1}{6}(36 - y^2) - 6$$
$$y = 6 - \frac{1}{6}y^2 - 6$$
$$6y = -y^2$$
$$y^2 + 6y = 0$$
$$y(y + 6) = 0$$
$$y = 0 \text{ or } y = -6$$

Replace these values into the equation $x^2 = 36 - y^2$.

$y = 0: x^2 = 36 - (0)^2$
$$x^2 = 36$$
$$x = \pm 6$$

$y = -6: x^2 = 36 - (6)^2$
$$x^2 = 0$$
$$x = 0$$

The solutions are $(-6, 0)$, $(6, 0)$ and $(0, -6)$.

31. $x > -3$

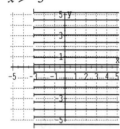

33. $y < 2x - 1$

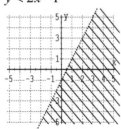

35. $P = x + (2x - 5) + (5x - 20)$
$\quad = (8x - 25)$ inches

37. $P = 2(x^2 + 3x + 1) + 2(x^2)$
$\quad = 2x^2 + 6x + 2 + 2x^2$
$\quad = (4x^2 + 6x + 2)$ meters

39. There are 0, 1, 2, 3, or 4 possible real solutions. Answers may vary.

41. Let x and y represent the numbers.
$\begin{cases} x^2 + y^2 = 130 \\ x^2 - y^2 = 32 \end{cases}$

Add the equations.
$$2x^2 = 162$$
$$x^2 = 81$$
$$x = \pm 9$$

Replace x^2 with 81 in the first equation.
$$81 + y^2 = 130$$
$$y^2 = 49$$
$$y = \pm 7$$

The numbers are -9 and -7, -9 and 7, 9 and -7, and 9 and 7.

43. Let x and y be the length and width.
$\begin{cases} \quad xy = 285 \\ 2x + 2y = 68 \end{cases}$

Solve the first equation for y: $y = \dfrac{285}{x}$.

Substitute into the second equation.
$$2x + 2\left(\frac{285}{x}\right) = 68$$
$$x + \frac{285}{x} = 34$$
$$x^2 + 285 = 34x$$
$$x^2 - 34x + 285 = 0$$
$$(x - 19)(x - 15) = 0$$
$$x = 19 \text{ or } x = 15$$

Using $x = 19$, $y = \dfrac{285}{x} = \dfrac{285}{19} = 15$.

The dimensions are 19 cm by 15 cm.

45. $\begin{cases} p = -0.01x^2 - 0.2x + 9 \\ p = 0.01x^2 - 0.1x + 3 \end{cases}$

Substitute.

$$-0.01x^2 - 0.2x + 9 = 0.01x^2 - 0.1x + 3$$
$$0 = 0.02x^2 + 0.1x - 6$$
$$0 = x^2 + 5x - 300$$
$$0 = (x + 20)(x - 15)$$

$x + 20 = 0$ or $x - 15 = 0$
$\quad\quad x = -20$ or $x = 15$

Disregard the negative.

$$p = -0.01(15)^2 - 0.2(15) + 9$$
$$p = 3.75$$

The equilibrium quantity is 15,000 compact discs, and the corresponding price is $3.75.

47.

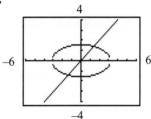

49.

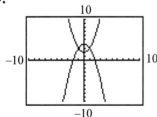

Exercise Set 10.4

1. $y < x^2$

First graph the parabola with dashes.

Test Point	$y < x^2$; Result
(0, 1)	$1 < 0^2$; False

Shade the portion of the graph which does not contain (0, 1).

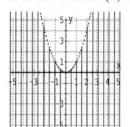

3. $x^2 + y^2 \geq 16$

First graph the circle with a solid curve.

Test Point	$x^2 + y^2 \geq 16$; Result
(0, 0)	$0^2 + 0^2 \geq 16$; False

Shade the portion of the graph which does not contain (0, 0).

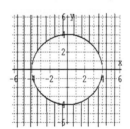

5. $\dfrac{x^2}{4} - y^2 < 1$

First graph the hyperbola with a dashed curve.

Test Points	$\dfrac{x^2}{4} - y^2 < 1$; Result
(−4, 0)	$\dfrac{(-4)^2}{4} - 0^2 < 1$; False
(0, 0)	$\dfrac{(0)^2}{4} - 0^2 < 1$; True
(4, 0)	$\dfrac{(4)^2}{4} - 0^2 < 1$; False

Shade the portion of the graph that contains (0, 0).

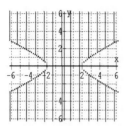

7. $y > (x-1)^2 - 3$

First graph the parabola with a dashed curve.

Test Point	$y > (x-1)^2 - 3$; Result
(0, 0)	$0 > (0-1)^2 - 3$; True

Shade the portion of the graph which contains (0, 0).

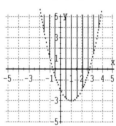

9. $x^2 + y^2 \le 9$

First graph the circle with a solid curve.

Test Point	$x^2 + y^2 \le 9$; Result
(0, 0)	$0^2 + 0^2 \le 9$; True

Shade the portion of the graph which contains (0, 0).

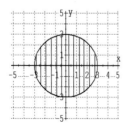

11. $y > -x^2 + 5$

First graph the parabola with a dashed curve.

Test Point	$y > -x^2 + 5$; Result
(0, 0)	$0 > -(0)^2 + 5$; False

Shade the portion of the graph which does not contain (0, 0).

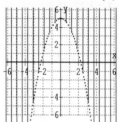

13. $\dfrac{x^2}{4} + \dfrac{y^2}{9} \le 1$

First graph the ellipse with a solid curve.

Test Point	$\dfrac{x^2}{4} + \dfrac{y^2}{9} \le 1$; Result
(0, 0)	$\dfrac{(0)^2}{4} + \dfrac{(0)^2}{9} \le 1$; True

Shade the portion of the graph that contains (0, 0).

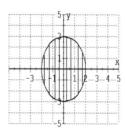

15. $\dfrac{y^2}{4} - x^2 \le 1$

First graph the hyperbola with solid curves.

Test Points	$\dfrac{y^2}{4} - x^2 \le 1$; Result
(0, 4)	$\dfrac{(-4)^2}{4} - 0^2 \le 1$; False
(0, 0)	$\dfrac{(0)^2}{4} - 0^2 \le 1$; True
(0, 4)	$\dfrac{(4)^2}{4} - 0^2 \le 1$; False

Shade the portion of the graph that contains (0, 0).

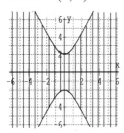

17. $y < (x-2)^2 + 1$

First graph the parabola with a dashed curve.

Test Point	$y < (x-2)^2 + 1$; Result
(0, 0)	$0 < (0-2)^2 + 1$; True

Shade the portion of the graph which contains (0, 0).

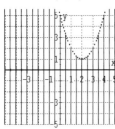

19. $y \le x^2 + x - 2$

First graph the parabola with a solid curve.

Test Point	$y \le x^2 + x - 2$; Result
(0, 0)	$0 \le (0)^2 + (0) - 2$; False

Shade the portion of the graph which contains (0, 0).

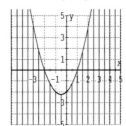

21. $\begin{cases} 2x - y < 2 \\ \quad y \le -x^2 \end{cases}$

First graph $2x - y = 2$ with a dashed line.

Test Point	$2x - y < 2$; Result
(0, 0)	$2(0) - 0 < 2$; True

Shade the portion of the graph which contains (0, 0).

Next, graph the parabola $y = -x^2$ with a solid curve.

Test Point	$y \le -x^2$; Result
(0, 1)	$1 \le -(0)^2$; False

Shade the portion of the graph which does not contain (0, 1).

The solution to the system is the overlapping region.

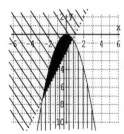

23. $\begin{cases} 4x + 3y \geq 12 \\ x^2 + y^2 < 16 \end{cases}$

First graph $4x + 3y = 12$ with a solid line.

Test Point	$4x + 3y \geq 12$; Result
(0, 0)	$4(0) + 3(0) \geq 12$; False

Shade the portion of the graph which does not contain (0, 0).

Next, graph the circle $x^2 + y^2 = 16$ with a dashed curve.

Test Point	$x^2 + y^2 < 16$; Result
(0, 0)	$0^2 + 0^2 < 16$; True

Shade the portion of the graph which contains (0, 0).

The solution to the system is the overlapping region.

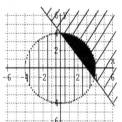

25. $\begin{cases} x^2 + y^2 \leq 9 \\ x^2 + y^2 \geq 1 \end{cases}$

First graph the circle with radius 3 with a solid curve.

Test Point	$x^2 + y^2 \leq 9$; Result
(0, 0)	$0^2 + 0^2 \leq 9$; True

Shade the portion of the graph which contains (0, 0).

Next, graph the circle with 1 with a solid curve.

Test Point	$x^2 + y^2 \geq 1$; Result
(0, 0)	$0^2 + 0^2 \geq 1$; False

Shade the portion of the graph which does not contain (0, 0).

The solution to the system is the overlapping region.

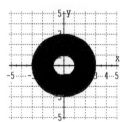

27. $\begin{cases} y > x^2 \\ y \geq 2x + 1 \end{cases}$

First graph the parabola with a dashed curve.

Test Point	$y > x^2$; Result
(0, 1)	$1 > 0^2$; True

Shade the portion of the graph which contains (0, 1).

Next, graph $y = 2x + 1$ with a solid line.

Test Point	$y \geq 2x+1$; Result
(0, 0)	$0 \geq 2(0)+1$; False

Shade the portion of the graph which does not contain (0, 0).

The solution to the system is the overlapping region.

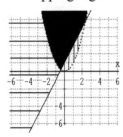

29. $\begin{cases} x > y^2 \\ y > 0 \end{cases}$

First graph the parabola with a dashed curve.

Test Point	$x > y^2$; Result
(1, 0)	$1 > 0^2$; True

Shade the portion of the graph which contains (1, 0).

Next, graph $y = 0$ with a dashed line.

Test Point	$y > x^2$; Result
(0, 1)	$1 \geq 0^2$; True

Shade the portion of the graph which contains (0, 1).

The solution to the system is the overlapping region.

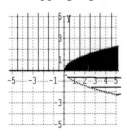

31. $\begin{cases} x^2 + y^2 > 9 \\ y > x^2 \end{cases}$

First graph the circle with a dashed curve.

Test Point	$x^2 + y^2 > 9$; Result
(0, 0)	$0^2 + 0^2 > 9$; False

Shade the portion of the graph which does not contain (0, 0).

Next, graph the parabola with a dashed curve.

Test Point	$y > x^2$; Result
(0, 1)	$1 > 0^2$; True

Shade the portion of the graph which contains (0, 1).

The solution to the system is the overlapping region.

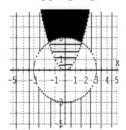

33. $\begin{cases} \dfrac{x^2}{4} + \dfrac{y^2}{9} \geq 1 \\ x^2 + y^2 \geq 4 \end{cases}$

First graph the ellipse with a solid curve.

Test Point	$\dfrac{x^2}{4} + \dfrac{y^2}{9} \geq 1$; Result
(0, 0)	$\dfrac{0^2}{4} + \dfrac{0^2}{9} \geq 1$; False

Shade the portion of the graph which does not contain (0, 0).

Next, graph the circle with a solid curve.

Test Point	$x^2 + y^2 \geq 4$; Result
(0, 0)	$0^2 + 0^2 \geq 4$; False

Shade the portion of the graph which does not contain (0, 0).

The solution to the system is the overlapping region.

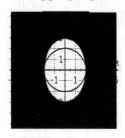

35. $\begin{cases} x^2 - y^2 \geq 1 \\ y \geq 0 \end{cases}$

First graph the hyperbola with solid curves.

Test Point	$x^2 - y^2 \geq 1$; Result
(0, 0)	$0^2 - 0^2 \geq 1$; False

Shade the portion of the graph which does not contain (0, 0).

Next, graph $y = 0$ with a solid line.

Test Point	$y > 0$; Result
(0, 1)	$1 \geq 0$; True

Shade the portion of the graph which contains (0, 1).

The solution to the system is the overlapping region.

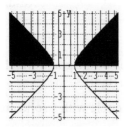

37. $\begin{cases} x + y \geq 1 \\ 2x + 3y < 1 \\ x > -3 \end{cases}$

First graph $x + y = 1$ with a solid line.

Test Point	$x + y \geq 1$; Result
(0, 0)	$0 + 0 \geq 1$; False

Shade the portion of the graph which does not contain (0, 0).

Next, graph $2x + 3y = 1$ with a dashed line.

Test Point	$2x + 3y < 1$; Result
(0, 0)	$2(0) + 3(0) < 1$; True

Shade the portion of the graph which contains (0, 0).

Now graph the line $x = -3$ with a dashed line.

Test Point	$x > -3$; Result
(0, 0)	$0 > -3$; True

Shade the portion of the graph which contains (0, 0).

The solution to the system is the overlapping region.

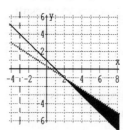

39. $\begin{cases} x^2 - y^2 < 1 \\ \dfrac{x^2}{16} + y^2 \leq 1 \\ x \geq -2 \end{cases}$

First graph the hyperbola with dashed curves.

Test Point	$x^2 - y^2 < 1$; Result
(0, 0)	$0^2 - 0^2 < 1$; True

Shade the portion of the graph which contains (0, 0).

Next, graph the ellipse with a solid curve.

Test Point	$\dfrac{x^2}{16} + y^2 \le 1$; Result
(0, 0)	$\dfrac{0^2}{16} + 0^2 \le 1$; True

Shade the portion of the graph which contains (0, 0).

Now graph the line $x = -2$ with a solid line.

Test Point	$x \ge -2$; Result
(0, 0)	$0 \ge -2$; True

Shade the portion of the graph which contains (0, 0).

The solution to the system is the overlapping region.

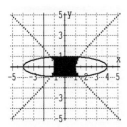

41. This is not a function because a vertical line can cross the graph in more than one place.

43. This is a function because a vertical line can cross the graph no more than one place.

45. $f(x) = 3x^2 - 2$
$f(-1) = 3(-1)^2 - 2 = 3 - 2 = 1$

47. $f(x) = 3x^2 - 2$
$f(a) = 3(a)^2 - 2 = 3a^2 - 2$

49. Answers may vary.

51. $\begin{cases} y \le x^2 \\ y \ge x + 2 \\ x \ge 0 \\ y \ge 0 \end{cases}$

First graph $y = x^2$ with a solid curve.

Test Point	$y \le x^2$; Result
(0, 1)	$1 \le 0^2$; False

Shade the portion of the graph which does not contain (0, 1).

Next, graph $y = x + 2$ with a solid line.

Test Point	$y \ge x + 2$; Result
(0, 0)	$0 \ge 0 + 2$; False

Shade the portion of the graph which does not contain (0, 0).

Next graph the line $x = 0$ with a solid line, and shade to the right.

Now graph the line $y = 0$ with a solid line, and shade above the line.

The solution to the system is the overlapping region.

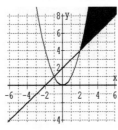

Chapter 10 Review

1. $(-6, 3), (8, 4)$

$$d = \sqrt{(4-3)^2 + [8-(-6)]^2}$$
$$= \sqrt{1^2 + 14^2}$$
$$= \sqrt{1+196}$$
$$= \sqrt{197} \text{ units}$$

2. $(3, 5), (8, 9)$

$$d = \sqrt{(8-3)^2 + (9-5)^2}$$
$$= \sqrt{5^2 + 4^2}$$
$$= \sqrt{25+16}$$
$$= \sqrt{41} \text{ units}$$

3. $(-4, -6), (-1, 5)$

$$d = \sqrt{[-1-(-4)]^2 + [5-(-6)]^2}$$
$$= \sqrt{3^2 + 11^2}$$
$$= \sqrt{9+121}$$
$$= \sqrt{130} \text{ units}$$

4. $(-1, 5), (2, -3)$

$$d = \sqrt{[2-(-1)]^2 + (-3-5)^2}$$
$$= \sqrt{3^2 + (-8)^2}$$
$$= \sqrt{9+64}$$
$$= \sqrt{73} \text{ units}$$

5. $\left(-\sqrt{2}, 0\right), \left(0, -4\sqrt{6}\right)$

$$d = \sqrt{\left[0-\left(-\sqrt{2}\right)\right]^2 + \left(-4\sqrt{6}-0\right)^2}$$
$$= \sqrt{\left(\sqrt{2}\right)^2 + \left(-4\sqrt{6}\right)^2}$$
$$= \sqrt{2+16\cdot6}$$
$$= \sqrt{98}$$
$$= 7\sqrt{2} \text{ units}$$

6. $\left(-\sqrt{5}, -\sqrt{11}\right), \left(-\sqrt{5}, -3\sqrt{11}\right)$

$$d = \sqrt{\left[-\sqrt{5}-\left(-\sqrt{5}\right)\right]^2 + \left[-3\sqrt{11}-\left(-\sqrt{11}\right)\right]^2}$$
$$= \sqrt{0^2 + \left(-2\sqrt{11}\right)^2}$$
$$= \sqrt{4\cdot11}$$
$$= 2\sqrt{11} \text{ units}$$

7. $(7.4, -8.6), (-1.2, 5.6)$

$$d = \sqrt{(-1.2-7.4)^2 + [5.6-(-8.6)]^2}$$
$$= \sqrt{(-8.6)^2 + (14.2)^2}$$
$$= \sqrt{275.6} \approx 16.60 \text{ units}$$

8. $(2.3, 1.8), (10.7, -9.2)$

$$d = \sqrt{(10.7-2.3)^2 + (-9.2-1.8)^2}$$
$$= \sqrt{(8.4)^2 + (-11)^2}$$
$$= \sqrt{191.56} \approx 13.84 \text{ units}$$

9. $(2, 6), (-12, 4)$

$$\left(\frac{2+(-12)}{2}, \frac{6+4}{2}\right) = \left(\frac{-10}{2}, \frac{10}{2}\right) = (-5, 5)$$

The midpoint of is $(-5, 5)$.

10. $(-3, 8), (11, 24)$

$$\left(\frac{-3+11}{2}, \frac{8+24}{2}\right) = \left(\frac{8}{2}, \frac{32}{2}\right) = (4, 16)$$

The midpoint of is $(4, 16)$.

11. $(-6, -5), (-9, 7)$

$$\left(\frac{-6+(-9)}{2}, \frac{-5+7}{2}\right) = \left(\frac{-15}{2}, \frac{2}{2}\right) = -\left(\frac{15}{2}, 1\right)$$

The midpoint of is $\left(-\frac{15}{2}, 1\right)$.

12. $(4, -6), (-15, 2)$

$$\left(\frac{4+(-15)}{2}, \frac{-6+2}{2}\right) = \left(-\frac{11}{2}, -2\right)$$

The midpoint of is $\left(-\frac{11}{2}, -2\right)$.

13. $\left(0, -\frac{3}{8}\right), \left(\frac{1}{10}, 0\right)$

$$\left(\frac{0 + \left(\frac{1}{10}\right)}{2}, \frac{-\frac{3}{8} + 0}{2}\right) = \left(\frac{1}{20}, -\frac{3}{16}\right)$$

The midpoint of is $\left(\frac{1}{20}, -\frac{3}{16}\right)$.

14. $\left(\frac{3}{4}, -\frac{1}{7}\right), \left(-\frac{1}{4}, -\frac{3}{7}\right)$

$$\left(\frac{\frac{3}{4} + \left(-\frac{1}{4}\right)}{2}, \frac{-\frac{1}{7} + \left(-\frac{3}{7}\right)}{2}\right) = \left(\frac{\frac{1}{2}}{2}, -\frac{\frac{4}{7}}{2}\right)$$

$$= \left(\frac{1}{4}, -\frac{2}{7}\right)$$

The midpoint of is $\left(\frac{1}{4}, -\frac{2}{7}\right)$.

15. $\left(\sqrt{3}, -2\sqrt{6}\right), \left(\sqrt{3}, -4\sqrt{6}\right)$

$$\left(\frac{\sqrt{3} + \sqrt{3}}{2}, \frac{-2\sqrt{6} + \left(-4\sqrt{6}\right)}{2}\right)$$

$$= \left(\frac{2\sqrt{3}}{2}, -\frac{6\sqrt{6}}{2}\right) = \left(\sqrt{3}, -3\sqrt{6}\right)$$

The midpoint of is $\left(\sqrt{3}, -3\sqrt{6}\right)$.

16. $\left(-5\sqrt{3}, 2\sqrt{7}\right), \left(-3\sqrt{3}, 10\sqrt{7}\right)$

$$\left(\frac{-5\sqrt{3} + \left(-3\sqrt{3}\right)}{2}, \frac{2\sqrt{7} + 10\sqrt{7}}{2}\right)$$

$$= \left(\frac{-8\sqrt{3}}{2}, \frac{12\sqrt{7}}{2}\right) = \left(-4\sqrt{3}, 6\sqrt{7}\right)$$

The midpoint of is $\left(-4\sqrt{3}, 6\sqrt{7}\right)$.

17. center (–4, 4), radius 3
$$[x - (-4)]^2 + (y - 4)^2 = 3^2$$
$$(x + 4)^2 + (y - 4)^2 = 9$$

18. center (5, 0), radius 5
$$(x - 5)^2 + (y - 0)^2 = 5^2$$
$$(x - 5)^2 + y^2 = 25$$

19. center (–7, –9), radius $\sqrt{11}$
$$[x - (-7)]^2 + [y - (-9)]^2 = \left(\sqrt{11}\right)^2$$
$$(x + 7)^2 + (y + 9)^2 = 11$$

20. center (0, 0), radius $\frac{7}{2}$
$$(x - 0)^2 + (y - 0)^2 = \left(\frac{7}{2}\right)^2$$
$$x^2 + y^2 = \frac{49}{4}$$

21. $x^2 + y^2 = 7$
Circle; center (0, 0), radius $r = \sqrt{7}$

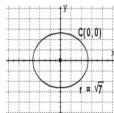

22. $x = 2(y - 5)^2 + 4$
Parabola; vertex: (4, 5)

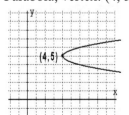

23. $x = -(y+2)^2 + 3$

Parabola; vertex: $(3, -2)$

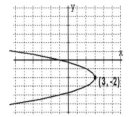

24. $(x-1)^2 + (y-2)^2 = 4$

Circle; center $(1, 2)$, radius $r = \sqrt{4} = 2$

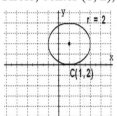

25. $y = -x^2 + 4x + 10$

Parabola; $x = -\dfrac{b}{2a} = \dfrac{-4}{2(-1)} = 2$

$y = -(2)^2 + 4(2) + 10 = 14$

Vertex: $(2, 14)$

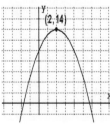

26. $x = -y^2 - 4y + 6$

Parabola; $y = -\dfrac{b}{2a} = \dfrac{-(-4)}{2(-1)} = -2$

$x = -(-2)^2 - 4(-2) + 6 = 10$

Vertex: $(10, -2)$

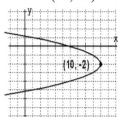

27. $x = \dfrac{1}{2}y^2 + 2y + 1$

Parabola; $y = -\dfrac{b}{2a} = \dfrac{-2}{2\left(\frac{1}{2}\right)} = -2$

$x = \dfrac{1}{2}(-2)^2 + 2(-2) + 1 = -1$

Vertex: $(-1, -2)$

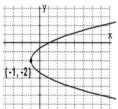

28. $y = -3x^2 + \dfrac{1}{2}x + 4$

Parabola; $x = -\dfrac{b}{2a} = \dfrac{-\frac{1}{2}}{2(-3)} = \dfrac{1}{12}$

$y = -3\left(\dfrac{1}{12}\right)^2 + \dfrac{1}{2}\left(\dfrac{1}{12}\right) + 4 = \dfrac{193}{48}$

Vertex: $\left(\dfrac{1}{12}, \dfrac{193}{48}\right)$

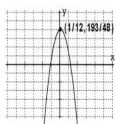

29.
$$x^2 + y^2 + 2x + y = \frac{3}{4}$$
$$(x^2 + 2x) + (y^2 + y) = \frac{3}{4}$$
$$(x^2 + 2x + 1) + \left(y^2 + y + \frac{1}{4}\right) = \frac{3}{4} + 1 + \frac{1}{4}$$
$$(x+1)^2 + \left(y + \frac{1}{2}\right)^2 = 2$$

Circle; center $\left(-1, -\frac{1}{2}\right)$, radius $r = \sqrt{2}$

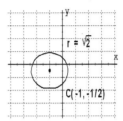

30.
$$x^2 + y^2 + 3y = \frac{7}{4}$$
$$x^2 + \left(y^2 + 3y + \frac{9}{4}\right) = \frac{7}{4} + \frac{9}{4}$$
$$x^2 + \left(y + \frac{3}{2}\right)^2 = 4$$

Circle; center $\left(0, -\frac{3}{2}\right)$, radius $r = 2$

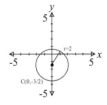

31.
$$4x^2 + 4y^2 + 16x + 8y = 1$$
$$(x^2 + 4x) + (y^2 + 2y) = \frac{1}{4}$$
$$(x^2 + 4x + 4) + (y^2 + 2y + 1) = \frac{1}{4} + 4 + 1$$
$$(x+2)^2 + (y+1)^2 = \frac{21}{4}$$

Circle; center $(-2, -1)$,
radius $r = \sqrt{\frac{21}{4}} = \frac{\sqrt{21}}{2}$

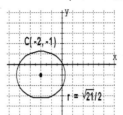

32.
$$3x^2 + 6x + 3y^2 = 9$$
$$x^2 + 2x + y^2 = 3$$
$$(x^2 + 2x + 1) + y^2 = 3 + 1$$
$$(x+1)^2 + y^2 = 4$$

Circle; center $(-1, 0)$, radius $r = \sqrt{4} = 2$

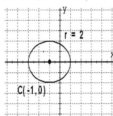

33.
$$y = x^2 + 6x + 9$$
$$= (x+3)^2$$

Parabola; vertex: $(-3, 0)$

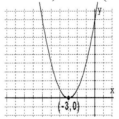

34. $x = y^2 + 6y + 9$
$\qquad = (y + 3)^2$
Parabola; vertex: $(0, -3)$

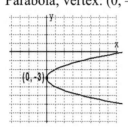

35. Center: $(5.6, -2.4)$, radius $\dfrac{6.2}{2} = 3.1$
$\qquad (x - 5.6)^2 + [y - (-2.4)]^2 = (3.1)^2$
$\qquad (x - 5.6)^2 + (y + 2.4)^2 = 9.61$

36. $x^2 + \dfrac{y^2}{4} = 1$
Center: $(0, 0)$; $a = 1$, $b = 2$

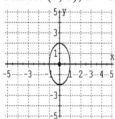

37. $x^2 - \dfrac{y^2}{4} = 1$
Center: $(0, 0)$; $a = 1$, $b = 2$

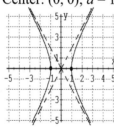

38. $\dfrac{y^2}{4} - \dfrac{x^2}{16} = 1$
Center: $(0, 0)$; $a = 4$, $b = 2$

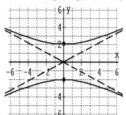

39. $\dfrac{y^2}{4} + \dfrac{x^2}{16} = 1$
Center: $(0, 0)$; $a = 4$, $b = 2$

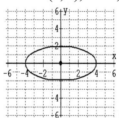

40. $\dfrac{x^2}{5} + \dfrac{y^2}{5} = 1$
$\qquad x^2 + y^2 = 5$
Center: $(0, 0)$; radius $r = \sqrt{5}$

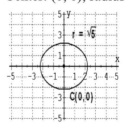

41. $\dfrac{x^2}{5} - \dfrac{y^2}{5} = 1$

Center: (0, 0); $a = \sqrt{5}$, $b = \sqrt{5}$

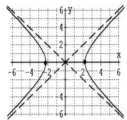

42. $-5x^2 + 25y^2 = 125$

$\dfrac{y^2}{5} - \dfrac{x^2}{25} = 1$

Center: (0, 0); $a = 5$, $b = \sqrt{5}$

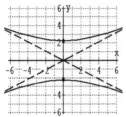

43. $4y^2 + 9x^2 = 36$

$\dfrac{y^2}{9} + \dfrac{x^2}{4} = 1$

Center: (0, 0); $a = 2$, $b = 3$

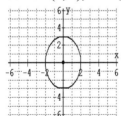

44. $\dfrac{(x-2)^2}{4} + (y-1)^2 = 1$

Center: (2, 1); $a = 2$, $b = 1$

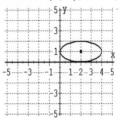

45. $\dfrac{(x+3)^2}{9} + \dfrac{(y-4)^2}{25} = 1$

Center: (–3, 4); $a = 3$, $b = 5$

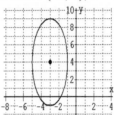

46. $x^2 - y^2 = 1$

Center: (0, 0); $a = 1$, $b = 1$

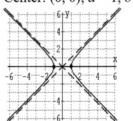

47. $36y^2 - 49x^2 = 1764$

$\dfrac{y^2}{49} - \dfrac{x^2}{36} = 1$

Center: (0, 0); $a = 6$, $b = 7$

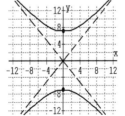

48. $y^2 = x^2 + 9$

$y^2 - x^2 = 9$

$\dfrac{y^2}{9} - \dfrac{x^2}{9} = 1$

Center: (0, 0); $a = 3$, $b = 3$

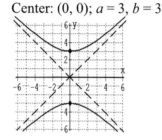

49. $x^2 = 4y^2 - 16$

$16 = 4y^2 - x^2$

$1 = \dfrac{y^2}{4} - \dfrac{x^2}{16}$

Center: (0, 0); $a = 4$, $b = 2$

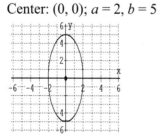

50. $100 - 25x^2 = 4y^2$

$100 = 25x^2 + 4y^2$

$1 = \dfrac{x^2}{4} + \dfrac{y^2}{25}$

Center: (0, 0); $a = 2$, $b = 5$

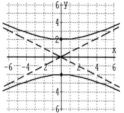

51. $y = x^2 + 4x + 6$

Parabola; $x = -\dfrac{b}{2a} = \dfrac{-4}{2(1)} = -2$

$y = (-2)^2 + 4(-2) + 6 = 2$

Vertex: (−2, 2)

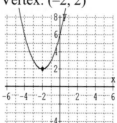

52. $y^2 = x^2 + 6$

$y^2 - x^2 = 6$

$\dfrac{y^2}{6} - \dfrac{x^2}{6} = 1$

Hyperbola; center: (0, 0),

$a = \sqrt{6}$, $b = \sqrt{6}$

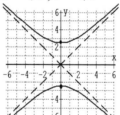

53. $y^2 + x^2 = 4x + 6$

$(x^2 - 4x) + y^2 = 6$

$(x^2 - 4x + 4) + y^2 = 6 + 4$

$(x - 2)^2 + y^2 = 10$

Circle; center: (2, 0), radius $r = \sqrt{10}$

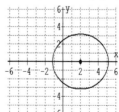

54.
$$y^2 + 2x^2 = 4x + 6$$
$$(2x^2 - 4x) + y^2 = 6$$
$$2(x^2 - 2x + 1) + y^2 = 6 + 2$$
$$2(x-1)^2 + y^2 = 8$$
$$\frac{(x-1)^2}{4} + \frac{y^2}{8} = 1$$
Ellipse; Center: (1, 0);
$a = 2$, $b = \sqrt{8} = 2\sqrt{2}$

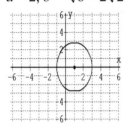

55.
$$x^2 + y^2 - 8y = 0$$
$$x^2 + (y^2 - 8y + 16) = 0 + 16$$
$$x^2 + (y-4)^2 = 16$$

Circle; center: (0, 4), radius $r = \sqrt{16} = 4$

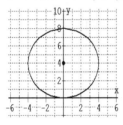

56. $x - 4y = y^2$
$$x = y^2 + 4y$$
$$x + 4 = y^2 + 4y + 4$$
$$x = (y+2)^2 - 4$$
Parabola; vertex: (−4, −2)

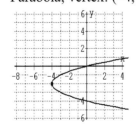

57.
$$x^2 - 4 = y^2$$
$$x^2 - y^2 = 4$$
$$\frac{x^2}{4} - \frac{y^2}{4} = 1$$
Hyperbola; center: (0, 0), $a = 2$, $b = 2$

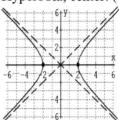

58.
$$x^2 = 4 - y^2$$
$$x^2 + y^2 = 4$$

Circle; center: (0, 0), radius $r = \sqrt{4} = 2$

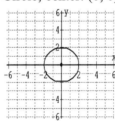

59. $6(x-2)^2 + 9(y+5)^2 = 36$
$$\frac{(x-2)^2}{6} + \frac{(y+5)^2}{4} = 1$$
Ellipse; Center: (2, −5); $a = \sqrt{6}$, $b = 2$

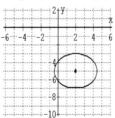

60.
$$36y^2 = 576 + 16x^2$$
$$36y^2 - 16x^2 = 576$$
$$\frac{y^2}{16} - \frac{x^2}{36} = 1$$

Hyperbola: center: $(0, 0)$; $a = 6$, $b = 4$

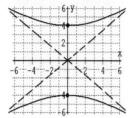

61. $\dfrac{x^2}{16} - \dfrac{y^2}{25} = 1$

Hyperbola; center: $(0, 0)$, $a = 4$, $b = 5$

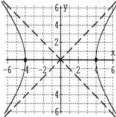

62. $3(x-7)^2 + 3(y+4)^2 = 1$

$(x-7)^2 + (y+4)^2 = \dfrac{1}{3}$

Circle; center: $(7, -4)$;

radius $r = \sqrt{\dfrac{1}{3}} = \dfrac{1}{\sqrt{3}} = \dfrac{\sqrt{3}}{3}$

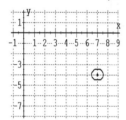

63. $\dfrac{y^2}{4} + \dfrac{x^2}{16} = 1$

$16\left(\dfrac{y^2}{4} + \dfrac{x^2}{16}\right) = 16(1)$

$4y^2 + x^2 = 16$

$4y^2 = 16 - x^2$

$y^2 = \dfrac{16 - x^2}{4}$

$y = \pm\sqrt{\dfrac{16 - x^2}{4}}$

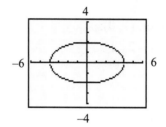

Answers are the same.

64. $\dfrac{x^2}{5} + \dfrac{y^2}{5} = 1$

$x^2 + y^2 = 5$

$y^2 = 5 - x^2$

$y = \pm\sqrt{5 - x^2}$

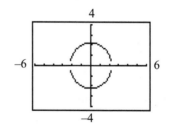

Answers are the same.

65. $y = x^2 + 4x + 6$

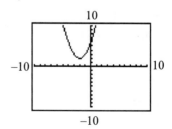

Answers are the same.

454

66. $x^2 = 4 - y^2$

$y^2 = 4 - x^2$

$y = \pm\sqrt{4 - x^2}$

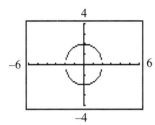

Answers are the same.

67. $\begin{cases} y = 2x - 4 & (1) \\ y^2 = 4x & (2) \end{cases}$

Substitute $2x - 4$ for y into E2.

$(2x - 4)^2 = 4x$

$4x^2 - 16x + 16 = 4x$

$4x^2 - 20x + 16 = 0$

$x^2 - 5x + 4 = 0$

$(x - 4)(x - 1) = 0$

$x = 4$ or $x = 1$

Use these values in E1.

$x = 4 : y = 2(4) - 4 = 4$

$x = 1 : y = 2(1) - 4 = -2$

The solutions are $(4, 4)$ and $(1, -2)$.

68. $\begin{cases} x^2 + y^2 = 2 & (1) \\ x - y = 4 & (2) \end{cases}$

Solve E2 for x: $x = y + 4$.

Substitute into E1.

$(y + 4)^2 + y^2 = 4$

$(y^2 + 8y + 16) + y^2 = 4$

$2y^2 + 8y + 12 = 0$

$y^2 + 4y + 6 = 0$

$y = \dfrac{-4 \pm \sqrt{(4)^2 - 4(1)(6)}}{2(1)} = \dfrac{-4 \pm \sqrt{-8}}{2}$

which yields no real solutions.

The solution is $\varnothing$.

69. $\begin{cases} y = x + 2 & (1) \\ y = x^2 & (2) \end{cases}$

Substitute $x + 2$ for y into E2.

$x + 2 = x^2$

$0 = x^2 - x - 2$

$0 = (x - 2)(x + 1)$

$x = 2$ or $x = -1$

Use these values in E1.

$x = 2 : y = x + 2 = 4$

$x = -1 : y = -1 + 2 = 1$

The solutions are $(2, 4)$ and $(-1, 1)$.

70. $\begin{cases} y = x^2 - 5x + 1 & (1) \\ y = -x + 6 & (2) \end{cases}$

Substitute $-x + 6$ for y into E2.

$-x + 6 = x^2 - 5x + 1$

$0 = x^2 - 4x - 5$

$0 = (x - 5)(x + 1)$

$x = 5$ or $x = -1$

Use these values in E2.

$x = 5 : y = -(5) + 6 = 1$

$x = -1 : y = -(-1) + 6 = 7$

The solutions are $(5, 1)$ and $(-1, 7)$.

71. $\begin{cases} 4x - y^2 = 0 & (1) \\ 2x^2 + y^2 = 16 & (2) \end{cases}$

Solve E1 for y^2: $y^2 = 4x$.

Substitute into E2.

$2x^2 + 4x = 16$

$2x^2 + 4x - 16 = 0$

$x^2 + 2x - 8 = 0$

$(x + 4)(x - 2) = 0$

$x = -4$ or $x = 2$

Use these values in the equation $y^2 = 4x$.

$x = -4 : y^2 = 4(-4)$

$\qquad\quad y^2 = -16$ (no real solutions)

$x = 2 : y^2 = 4(2)$
$$y^2 = 8$$
$$y = \pm\sqrt{8} = \pm 2\sqrt{2}$$

The solutions are $\left(2, -2\sqrt{2}\right)$ and $\left(2, 2\sqrt{2}\right)$.

72. $\begin{cases} x^2 + 4y^2 = 16 & (1) \\ x^2 + y^2 = 4 & (2) \end{cases}$

Multiply E2 by -1 and add to E1.

$-x^2 - y^2 = -4$
$\underline{x^2 + 4y^2 = 16}$
$\qquad 3y^2 = 12$
$\qquad y^2 = 4$
$\qquad y = \pm 2$

Replace y^2 with 4 into E2.

$x^2 + 4 = 4$
$\quad x^2 = 0$
$\quad x = 0$

The solutions are (0, 2) and (0, –2).

73. $\begin{cases} x^2 + y^2 = 10 & (1) \\ 9x^2 + y^2 = 18 & (2) \end{cases}$

Multiply E1 by -1 and add to E2.

$-x^2 - y^2 = -10$
$\underline{9x^2 + y^2 = 18}$
$8x^2 \qquad = 8$
$x^2 = 1$
$x = \pm 1$

Replace x^2 with 1 into E1.

$1 + y^2 = 10$
$\quad y^2 = 9$
$\quad y = \pm 3$

The solutions are (–1, –3), (–1, 3), (1, –3) and (1, 3).

74. $\begin{cases} x^2 + 2y = 9 & (1) \\ 5x - 2y = 5 & (2) \end{cases}$

Add E1 and E2.

$x^2 + 5x = 14$
$x^2 + 5x - 14 = 0$
$(x + 7)(x - 2) = 0$
$x = -7 \ \text{ or } \ x = 2$

Use these values into E1.

$x = -7 : (-7)^2 + 2y = 9$
$\qquad 49 + 2y = 9$
$\qquad 2y = -40$
$\qquad y = -20$
$x = 2 : (2)^2 + 2y = 9$
$\qquad 4 + 2y = 9$
$\qquad 2y = 5$
$\qquad y = \dfrac{5}{2}$

The solutions are (–7, –20) and $\left(2, \dfrac{5}{2}\right)$.

75. $\begin{cases} y = 3x^2 + 5x - 4 & (1) \\ y = 3x^2 - x + 2 & (2) \end{cases}$

Multiply E1 by -1 and add to E2.

$0 = -6x + 6$
$6x = 6$
$x = 1$

Use this value in E1.

$y = 3(1)^2 + 5(1) - 4 = 4$

The solution is (1, 4).

76. $\begin{cases} x^2 - 3y^2 = 1 & (1) \\ 4x^2 + 5y^2 = 21 & (2) \end{cases}$

Multiply E1 by -4 and add to E2.

$-4x^2 + 12y^2 = -4$
$\underline{4x^2 + 5y^2 = 21}$
$\qquad 17y^2 = 17$
$\qquad y^2 = 1$
$\qquad y = \pm 1$

Replace y^2 with 1 into E1.

$x^2 - 3(1) = 1$
$x^2 = 4$
$x = \pm 2$

The solutions are $(-2, -1)$, $(-2, 1)$, $(2, -1)$ and $(2, 1)$.

77. Let x and y be the length and width.

$$\begin{cases} xy = 150 \\ 2x + 2y = 50 \end{cases}$$

Solve the first equation for y: $y = \dfrac{150}{x}$.

Substitute into E2.

$$2x + 2\left(\frac{150}{x}\right) = 50$$
$$x + \frac{150}{x} = 25$$
$$x^2 + 150 = 25x$$
$$x^2 - 25x + 150 = 0$$
$$(x - 15)(x - 10) = 0$$
$$x = 15 \quad \text{or} \quad x = 10$$

Substitute these values into E1.

$$15y = 150 \qquad 10y = 150$$
$$y = 10 \qquad y = 15$$

The room is 15 feet by 10 feet.

78. Four real solutions.

79. $y \le -x^2 + 3$

Graph $y = -x^2 + 3$ with a solid curve.

Test Point	$y \le -x^2 + 3$; Result
$(0, 0)$	$0 \le -(0)^2 + 3$; True

Shade the portion of the graph which contains $(0, 0)$.

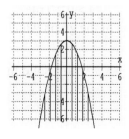

80. $x^2 + y^2 < 9$

First graph the circle with a dashed curve.

Test Point	$x^2 + y^2 < 9$; Result
$(0, 0)$	$0^2 + 0^2 < 9$; True

Shade the portion of the graph which contains $(0, 0)$.

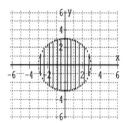

81. $x^2 - y^2 < 1$

First graph the hyperbola with a dashed curve.

Test Points	$x^2 - y^2 < 1$; Result
$(-2, 0)$	$(-2)^2 - 0^2 < 1$; False
$(0, 0)$	$0^2 - 0^2 < 1$; True
$(2, 0)$	$2^2 - 0^2 < 1$; False

Shade the portion of the graph that contains $(0, 0)$.

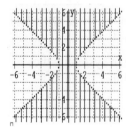

82. $\dfrac{x^2}{4}+\dfrac{y^2}{9}\ge 1$

First graph the ellipse with a solid curve.

Test Point	$\dfrac{x^2}{4}+\dfrac{y^2}{9}\ge 1$; Result
$(0, 0)$	$\dfrac{(0)^2}{4}+\dfrac{(0)^2}{9}\ge 1$; False

Shade the portion of the graph that does not contain $(0, 0)$.

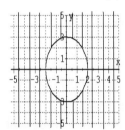

83. $\begin{cases} 2x\le 4 \\ x+y\ge 1 \end{cases}$

First graph $2x=4$, or $x=2$, with a solid line, and shade to the left of the line.

Next, graph $x+y=1$ with a solid line.

Test Point	$x+y\ge 1$; Result
$(0, 0)$	$0+0\ge 1$; False

Shade the portion of the graph which does not contain $(0, 0)$.

The solution to the system is the overlapping region.

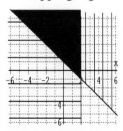

84. $\begin{cases} 3x+4y\le 12 \\ x-2y>6 \end{cases}$

First graph $3x+4y=12$ with a solid line.

Test Point	$3x+4y\le 12$; Result
$(0, 0)$	$3(0)+4(0)\le 12$; True

Shade the portion of the graph which contains $(0, 0)$.

Next, graph $x-2y=6$ with a dashed line.

Test Point	$x-2y>6$; Result
$(0, 0)$	$0-2(0)>6$; False

Shade the portion of the graph which does not contain $(0, 0)$.

The solution to the system is the overlapping region.

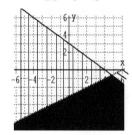

85. $\begin{cases} y>x^2 \\ x+y\ge 3 \end{cases}$

First graph the parabola with a dashed curve.

Test Point	$y>x^2$; Result
$(0, 1)$	$1>0^2$; True

Shade the portion of the graph which contains $(0, 1)$.

Next, graph $x+y=3$ with a solid line.

Test Point	$x + y \geq 3$; Result
$(0, 0)$	$0 + 0 \geq 3$; False

Shade the portion of the graph which does not contain $(0, 0)$.

The solution to the system is the overlapping region.

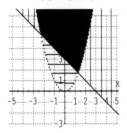

86. $\begin{cases} x^2 + y^2 \leq 16 \\ x^2 + y^2 \geq 4 \end{cases}$

First graph the first circle with a solid curve.

Test Point	$x^2 + y^2 \leq 16$; Result
$(0, 0)$	$0^2 + 0^2 \leq 16$; True

Shade the portion of the graph which contains $(0, 0)$.

Next, graph the second circle with a solid curve.

Test Point	$x^2 + y^2 \geq 4$; Result
$(0, 0)$	$0^2 + 0^2 \geq 4$; False

Shade the portion of the graph which does not contain $(0, 0)$.

The solution to the system is the overlapping region.

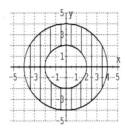

87. $\begin{cases} x^2 + y^2 < 4 \\ x^2 - y^2 \le 1 \end{cases}$

First graph the first circle with a dashed curve.

Test Point	$x^2 + y^2 < 4$; Result
(0, 0)	$0^2 + 0^2 < 4$; True

Shade the portion of the graph which contains (0, 0).

Next, graph the hyperbola with a solid curve.

Test Point	$x^2 - y^2 \le 1$; Result
(0, 0)	$0^2 - 0^2 \le 1$; True

Shade the portion of the graph which contains (0, 0).

The solution to the system is the overlapping region.

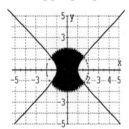

88. $\begin{cases} x^2 + y^2 < 4 \\ y \ge x^2 - 1 \\ x \ge 0 \end{cases}$

First graph the first circle with a dashed curve.

Test Point	$x^2 + y^2 < 4$; Result
(0, 0)	$0^2 + 0^2 < 4$; True

Shade the portion of the graph which contains (0, 0).

Next, graph the parabola with a solid curve.

Test Point	$y^2 \ge x^2 - 1$; Result
(0, 0)	$0 \ge 0^2 - 1$; True

Shade the portion of the graph which contains (0, 0).

Now graph the line $x = 0$ with a solid line, and shade to the right.

The solution to the system is the overlapping region.

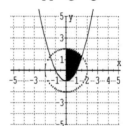

Chapter 10 Test

1. (–6, 3), (–8, –7)

$$d = \sqrt{[-8-(-6)]^2 + (-7-3)^2}$$
$$= \sqrt{(-2)^2 + (-10)^2}$$
$$= \sqrt{4 + 100}$$
$$= \sqrt{104} = 2\sqrt{26} \text{ units}$$

2. $\left(-2\sqrt{5}, \sqrt{10}\right), \left(-\sqrt{5}, 4\sqrt{10}\right)$

$$d = \sqrt{\left[-\sqrt{5}-\left(-2\sqrt{5}\right)\right]^2 + \left(4\sqrt{10} - \sqrt{10}\right)^2}$$
$$= \sqrt{\left(\sqrt{5}\right)^2 + \left(3\sqrt{10}\right)^2}$$
$$= \sqrt{5 + 9 \cdot 10}$$
$$= \sqrt{95} \text{ units}$$

3. (–2, –5), (–6, 12)

$$\left(\frac{-2+(-6)}{2}, \frac{-5+12}{2}\right) = \left(\frac{-8}{2}, \frac{7}{2}\right) = \left(-4, \frac{7}{2}\right)$$

The midpoint is $\left(-4, \frac{7}{2}\right)$.

4. $x^2 + y^2 = 36$

Circle; center: (0, 0), radius $r = \sqrt{36} = 6$

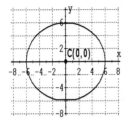

5. $x^2 - y^2 = 36$

$\dfrac{x^2}{36} - \dfrac{y^2}{36} = 1$

Hyperbola; center: (0, 0), $a = 6$, $b = 6$

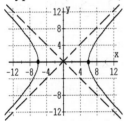

6. $16x^2 + 9y^2 = 144$

$\dfrac{x^2}{9} + \dfrac{y^2}{16} = 1$

Ellipse; center: (0, 0), $a = 3$, $b = 4$

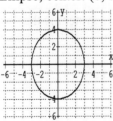

7. $y = x^2 - 8x + 16$

$\quad = (x-4)^2$

Parabola; vertex: (4, 0)

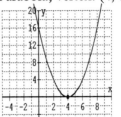

8. $\quad x^2 + y^2 + 6x = 16$

$\quad (x^2 + 6x) + y^2 = 16$

$\quad (x^2 + 6x + 9) + y^2 = 16 + 9$

$\quad\quad (x+3)^2 + y^2 = 25$

Circle; center: (−3, 0),

radius $r = \sqrt{25} = 5$

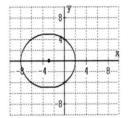

9. $\quad x = y^2 + 8y - 3$

$x + 16 = (y^2 + 8y + 16) - 3$

$\quad x = (y+4)^2 - 19$

Parabola; vertex: (−19, −4)

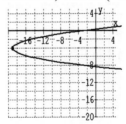

10. $\dfrac{(x-4)^2}{16} + \dfrac{(y-3)^2}{9} = 1$

Ellipse: center: (4, 3), $a = 4$, $b = 3$

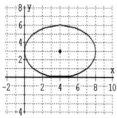

11. $y^2 - x^2 = 1$

Hyperbola: center: (0, 0), $a = 1$, $b = 1$

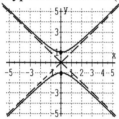

12. $\begin{cases} x^2 + y^2 = 26 \quad (1) \\ x^2 - 2y^2 = 23 \quad (2) \end{cases}$

Solve E1 for x^2: $x^2 = 26 - y^2$.
Substitute into E2.

$(26 - y^2) - 2y^2 = 23$

$-3y^2 = -3$

$y^2 = 1$

$y = \pm 1$

Replace y^2 with 1 in E1.

$x^2 + 1 = 26$

$x^2 = 25$

$x = \pm 5$

The solutions are $(-5, -1)$, $(-5, 1)$, $(5, -1)$, and $(5, 1)$.

13. $\begin{cases} y = x^2 - 5x + 6 \quad (1) \\ y = 2x \qquad\qquad (2) \end{cases}$

Substitute $2x$ for y in E1.

$2x = x^2 - 5x + 6$

$0 = x^2 - 7x + 6$

$0 = (x - 6)(x - 1)$

$x = 6$ or $x = 1$

Use these values in E2.

$x = 6: y = 2(6) = 12$

$x = 1: y = 2(1) = 2$

The solutions are (1, 2) and (6, 12).

14. $\begin{cases} 2x + 5y \geq 10 \\ y \geq x^2 + 1 \end{cases}$

First graph $2x + 5y = 10$ with a solid line.

Test Point	$2x + 5y \geq 10$; Result
(0, 0)	$2(0) + 5(0) \geq 10$; False

Shade the portion of the graph which does not contain (0, 0).

Next, graph $y = x^2 + 1$ with a solid curve.

Test Point	$y \geq x^2 + 1$; Result
(0, 0)	$0 \geq 0^2 + 1$; False

Shade the portion of the graph which does not contain (0, 0).

The solution to the system is the overlapping region.

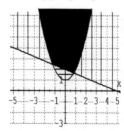

15. $\begin{cases} \dfrac{x^2}{4} + y^2 \leq 1 \\ x + y > 1 \end{cases}$

First graph the ellipse with a solid curve.

Test Point	$\dfrac{x^2}{4} + y^2 \leq 1$; Result
(0, 0)	$\dfrac{0^2}{4} + 0^2 \leq 1$; True

Shade the portion of the graph which contains (0, 0).

Next, graph $x + y = 1$ with a solid line.

Test Point	$x + y > 1$; Result
(0, 0)	$0 + 0 > 1$; False

Shade the portion of the graph which does not contain (0, 0).

The solution to the system is the overlapping region.

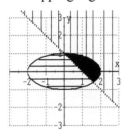

16. $\begin{cases} x^2 + y^2 \geq 4 \\ x^2 + y^2 < 16 \\ \qquad y \geq 0 \end{cases}$

First graph the circle $x^2 + y^2 = 4$ with a solid curve.

Test Point	$x^2 + y^2 \geq 4$; Result
(0, 0)	$0^2 + 0^2 \geq 4$; False

Shade the portion of the graph which does not contain (0, 0).

Next graph the circle $x^2 + y^2 = 16$ with a dashed curve.

Test Point	$x^2 + y^2 < 16$; Result
(0, 0)	$0^2 + 0^2 < 16$; True

Shade the portion of the graph which contains (0, 0).

Now graph the inequality $y = 0$ by shading the region above the *x*-axis.

The solution to the system is the overlapping region.

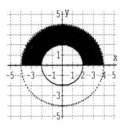

17. $100x^2 + 225y^2 = 22{,}500$

$$\dfrac{x^2}{225} + \dfrac{y^2}{100} = 1$$

$a = \sqrt{225} = 15$
$b = \sqrt{100} = 10$
Width = 15 + 15 = 30 feet
Height = 10 feet

Chapter 10 Cumulative Review

1. $4 \cdot (9y) = (4 \cdot 9)y = 36y$

2. $3x + 4 > 1$ and $2x - 5 \leq 9$
   $\quad 3x > -3$ and $\quad 2x \leq 14$
   $\qquad x > -1$ and $\qquad x \leq 7$
   $-1 < x \leq 7$
   $(-1, 7]$

3. $x = -2y$
   $y = -\dfrac{1}{2}x$

x	$y = -\dfrac{1}{2}x$
-2	1
0	0
4	-2

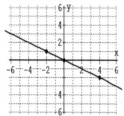

4. $(3, 2)$, $(1, -4)$

$$m = \frac{-4-2}{1-3} = \frac{-6}{-2} = 3$$

5. $\begin{cases} 3x + \dfrac{y}{2} = 2 \ \ (1) \\ 6x + y = 5 \ \ (2) \end{cases}$

Multiply E1 by -2 and add to E2.

$\begin{array}{r} -6x - y = -4 \\ 6x + y = 5 \\ \hline 0 = 1, \end{array}$ which is impossible.

Therefore, the solution is $\varnothing$.

6. Let x = speed of one plane. Then $x + 25$ = speed of the other plane.

$d_{\text{plane 1}} + d_{\text{plane 2}} = 650$ miles

$2x + 2(x + 25) = 650$

$2x + 2x + 50 = 650$

$4x = 600$

$x = 150$

$x + 25 = 150 + 25 = 175$

The planes are traveling at 150 mph and 175 mph.

7. a. $(5x^2)^3 = 5^3(x^2)^3 = 125x^6$

b. $\left(\dfrac{2}{3}\right)^3 = \dfrac{2^3}{3^3} = \dfrac{8}{27}$

c. $\left(\dfrac{3p^4}{q^5}\right)^2 = \dfrac{3^2(p^4)^2}{(q^5)^2} = \dfrac{9p^8}{q^{10}}$

d. $\left(\dfrac{2^{-3}}{y}\right)^{-2} = \left(\dfrac{1}{2^3 y}\right)^{-2}$

$= \dfrac{1^{-2}}{2^{-6}y^{-2}} = \dfrac{2^6 y^2}{1} = 64y^2$

e. $(x^{-5}y^2z^{-1})^7 = x^{-35}y^{14}z^{-7} = \dfrac{y^{14}}{x^{35}z^7}$

8. a. $\dfrac{4^8}{4^3} = 4^{8-3} = 4^5$

b. $\dfrac{y^{11}}{y^5} = y^{11-5} = y^6$

c. $\dfrac{32x^7}{4x^6} = 8x^{7-6} = 8x$

d. $\dfrac{18a^{12}b^6}{12a^8b^6} = \dfrac{3a^{12-8}b^{6-6}}{2} = \dfrac{3a^4b^0}{2} = \dfrac{3a^4}{2}$

9. $\qquad 2x^2 = \dfrac{17}{3}x + 1$

$3(2x^2) = 3\left(\dfrac{17}{3}x + 1\right)$

$6x^2 = 17x + 3$

$6x^2 - 17x - 3 = 0$

$(6x+1)(x-3) = 0$

$6x + 1 = 0 \quad$ or $\quad x - 3 = 0$

$6x = -1 \quad$ or $\qquad x = 3$

$x = -\dfrac{1}{6}$

The solutions are $-\dfrac{1}{6}$ and 3.

10. a. $3y^2 + 14y + 15 = (3y+5)(y+3)$

b. $20a^5 + 54a^4 + 10a^3$

$= 2a^3(10a^2 + 27a + 5)$

$= 2a^3(2a+5)(5a+1)$

c. $(y-3)^2 - 2(y-3) - 8$

Let $u = y - 3$. Then $u^2 = (y-3)^2$ and

$$u^2 - 2u - 8 = (u-4)(u+2)$$
$$= [(y-3)-4][(y-3)+2]$$
$$= (y-7)(y-1)$$

11. $\dfrac{7}{x-1} + \dfrac{10x}{x^2-1} - \dfrac{5}{x+1}$

$= \dfrac{7}{x-1} + \dfrac{10x}{(x+1)(x-1)} - \dfrac{5}{x+1}$

$= \dfrac{7(x+1) + 10x - 5(x-1)}{(x+1)(x-1)}$

$= \dfrac{7x+7+10x-5x+5}{(x+1)(x-1)}$

$= \dfrac{12x+12}{(x+1)(x-1)}$

$= \dfrac{12(x+1)}{(x+1)(x-1)} = \dfrac{12}{x-1}$

12. $\dfrac{2}{3a-15} - \dfrac{a}{25-a^2}$

$= \dfrac{2}{3(a-5)} + \dfrac{a}{a^2-25}$

$= \dfrac{2}{3(a-5)} + \dfrac{a}{(a+5)(a-5)}$

$= \dfrac{2(a+5)+3a}{3(a+5)(a-5)}$

$= \dfrac{2a+10+3a}{3(a+5)(a-5)}$

$= \dfrac{5a+10}{3(a+5)(a-5)}$

13. a. $\dfrac{\dfrac{2x}{27y^2}}{\dfrac{6x^2}{9}} = \dfrac{2x}{27y^2} \cdot \dfrac{9}{6x^2} = \dfrac{1}{3y^2} \cdot \dfrac{1}{3} = \dfrac{1}{9y^2}$

b. $\dfrac{\dfrac{5x}{x+2}}{\dfrac{10}{x-2}} = \dfrac{5x}{x+2} \cdot \dfrac{x-2}{10} = \dfrac{x(x-2)}{2(x+2)}$

c. $\dfrac{\dfrac{x}{y^2} + \dfrac{1}{y}}{\dfrac{y}{x^2} + \dfrac{1}{x}} = \dfrac{\left(\dfrac{x}{y^2} + \dfrac{1}{y}\right)x^2y^2}{\left(\dfrac{y}{x^2} + \dfrac{1}{x}\right)x^2y^2}$

$= \dfrac{x^3 + x^2y}{y^3 + xy^2}$

$= \dfrac{x^2(x+y)}{y^2(y+x)} = \dfrac{x^2}{y^2}$

14. a. $(a^{-1} - b^{-1})^{-1} = \left(\dfrac{1}{a} - \dfrac{1}{b}\right)^{-1}$

$= \left(\dfrac{b-a}{ab}\right)^{-1}$

$= \dfrac{ab}{b-a}$

b. $\dfrac{2 - \dfrac{1}{x}}{4x - \dfrac{1}{x}} = \dfrac{\left(2 - \dfrac{1}{x}\right)x}{\left(4x - \dfrac{1}{x}\right)x}$

$= \dfrac{2x-1}{4x^2-1}$

$= \dfrac{2x-1}{(2x+1)(2x-1)}$

$= \dfrac{1}{2x+1}$

15. $x+2\overline{\smash{\big)}\,2x^2 - x - 10}$

$\underline{2x^2 + 4x}$

$-5x - 10$

$\underline{-5x - 10}$

0

Answer: $2x - 5$

16.
$$\frac{2}{x+3}=\frac{1}{x^2-9}-\frac{1}{x-3}$$
$$\frac{2}{x+3}=\frac{1}{(x+3)(x-3)}-\frac{1}{x-3}$$
$$2(x-3)=1-1(x+3)$$
$$2x-6=1-x-3$$
$$2x-6=-x-2$$
$$3x=4$$
$$x=\frac{4}{3}$$

17.
$$
\begin{array}{r|rrrrrrr}
4 & 4 & -25 & 35 & 0 & 17 & 0 & 0 \\
  &   & 16 & -36 & -4 & -16 & 4 & 16 \\
\hline
  & 4 & -9 & -1 & -4 & 1 & 4 & 16
\end{array}
$$
Thus, $P(4)=16$.

18. $y=\dfrac{k}{x}$

$3=\dfrac{k}{\frac{2}{3}}$

$k=3\left(\dfrac{2}{3}\right)=2$

Thus, the equation is $y=\dfrac{2}{x}$.

19.
$$\frac{2x}{x-3}+\frac{6-2x}{x^2-9}=\frac{x}{x+3}$$
$$\frac{2x}{x-3}+\frac{-2(x-3)}{(x+3)(x-3)}=\frac{x}{x+3}$$
$$\frac{2x}{x-3}-\frac{2}{x+3}=\frac{x}{x+3}$$
$$\frac{2x}{x-3}=\frac{x}{x+3}+\frac{2}{x+3}$$
$$\frac{2x}{x-3}=\frac{x+2}{x+3}$$
$$2x(x+3)=(x+2)(x-3)$$
$$2x^2+6x=x^2-x-6$$
$$x^2+7x+6=0$$
$$(x+6)(x+1)=0$$
$$x+6=0\quad\text{or}\quad x+1=0$$
$$x=-6\quad\text{or}\qquad x=-1$$
The solutions are –6 and –1.

20. a. $\sqrt[5]{-32}=-2$ because $(-2)^5=-32$.

b. $\sqrt[4]{625}=5$ because $5^4=625$.

c. $-\sqrt{36}=-6$ because $6^2=36$.

d. $-\sqrt[3]{-27x^3}=-(-3x)=3x$

e. $\sqrt{144y^2}=12y$

21. Let t = time it will take together.
$$\frac{1}{4}+\frac{1}{5}=\frac{1}{t}$$
$$20t\left(\frac{1}{4}+\frac{1}{5}\right)=20t\left(\frac{1}{t}\right)$$
$$5t+4t=20$$
$$9t=20$$
$$t=\frac{20}{9}=2\frac{2}{9}$$

It will take them $2\frac{2}{9}$ hours. No, they can not finish before the movie starts.

22. a. $\dfrac{\sqrt{32}}{\sqrt{4}}=\sqrt{\dfrac{32}{4}}=\sqrt{8}=\sqrt{4\cdot2}=2\sqrt{2}$

b.
$$\frac{\sqrt[3]{240y^2}}{5\sqrt[3]{3y^{-4}}}=\frac{1}{5}\sqrt[3]{\frac{240y^2}{3y^{-4}}}$$
$$=\frac{1}{5}\sqrt[3]{80y^6}$$
$$=\frac{1}{5}\sqrt[3]{8y^6\cdot10}$$
$$=\frac{2y^3\sqrt[3]{10}}{5}$$

c.
$$\frac{\sqrt[5]{64x^9y^2}}{\sqrt[5]{2x^2y^{-8}}}=\sqrt[5]{\frac{64x^9y^2}{2x^2y^{-8}}}$$
$$=\sqrt[5]{32x^7y^{10}}$$
$$=\sqrt[5]{32x^5y^{10}\cdot x^2}$$
$$=2xy^2\sqrt[5]{x^2}$$

23. a. $\sqrt[3]{1} = 1$

    b. $\sqrt[3]{-64} = -4$

    c. $\sqrt[3]{\dfrac{8}{125}} = \dfrac{\sqrt[3]{8}}{\sqrt[3]{125}} = \dfrac{2}{5}$

    d. $\sqrt[3]{x^6} = x^2$

    e. $\sqrt[3]{-27x^9} = -3x^3$

24. a. $\sqrt{5}\left(2 + \sqrt{15}\right) = 2\sqrt{5} + \sqrt{5} \cdot \sqrt{15}$
$$= 2\sqrt{5} + \sqrt{75}$$
$$= 2\sqrt{5} + 5\sqrt{3}$$

    b. $\left(\sqrt{3} - \sqrt{5}\right)\left(\sqrt{7} - 1\right)$
$$= \sqrt{3} \cdot \sqrt{7} - \sqrt{3} \cdot 1 - \sqrt{5} \cdot \sqrt{7} + \sqrt{5} \cdot 1$$
$$= \sqrt{21} - \sqrt{3} - \sqrt{35} + \sqrt{5}$$

    c. $\left(2\sqrt{5} - 1\right)^2 = \left(2\sqrt{5}\right)^2 - 2 \cdot 2\sqrt{5} \cdot 1 + 1^2$
$$= 4(5) - 4\sqrt{5} + 1$$
$$= 21 - 4\sqrt{5}$$

    d. $\left(3\sqrt{2} + 5\right)\left(3\sqrt{2} - 5\right) = \left(3\sqrt{2}\right)^2 - 5^2$
$$= 9(2) - 25$$
$$= 18 - 25$$
$$= -7$$

25. a. $z^{2/3}\left(z^{1/3} - z^5\right) = z^{2/3 + 1/3} - z^{2/3 + 5}$
$$= z^{3/3} - z^{2/3 + 15/3}$$
$$= z - z^{17/3}$$

    b. $(x^{1/3} - 5)(x^{1/3} + 2)$
$$= x^{1/3} \cdot x^{1/3} + 2x^{1/3} - 5x^{1/3} - 5(2)$$
$$= x^{2/3} - 3x^{1/3} - 10$$

26. $\dfrac{-2}{\sqrt{3} + 3} = \dfrac{-2\left(\sqrt{3} - 3\right)}{\left(\sqrt{3} + 3\right)\left(\sqrt{3} - 3\right)}$
$$= \dfrac{-2\left(\sqrt{3} - 3\right)}{\left(\sqrt{3}\right)^2 - 3^2}$$
$$= \dfrac{-2\left(\sqrt{3} - 3\right)}{3 - 9}$$
$$= \dfrac{-2\left(\sqrt{3} - 3\right)}{-6} = \dfrac{\sqrt{3} - 3}{3}$$

27. a. $\dfrac{\sqrt{20}}{\sqrt{5}} = \sqrt{\dfrac{20}{5}} = \sqrt{4} = 2$

    b. $\dfrac{\sqrt{50x}}{2\sqrt{2}} = \dfrac{1}{2}\sqrt{\dfrac{50x}{2}} = \dfrac{1}{2}\sqrt{25x} = \dfrac{5\sqrt{x}}{2}$

    c. $\dfrac{7\sqrt[3]{48x^4y^8}}{\sqrt[3]{6y^2}} = 7\sqrt[3]{\dfrac{48x^4y^8}{6y^2}}$
$$= 7\sqrt[3]{8x^4y^6}$$
$$= 7\sqrt[3]{8x^3y^6 \cdot x}$$
$$= 7 \cdot 2xy^2\sqrt[3]{x}$$
$$= 14xy^2\sqrt[3]{x}$$

    d. $\dfrac{2\sqrt[4]{32a^8b^6}}{\sqrt[4]{a^{-1}b^2}} = 2\sqrt[4]{\dfrac{32a^8b^6}{a^{-1}b^2}}$
$$= 2\sqrt[4]{32a^8b^4}$$
$$= 2\sqrt[4]{16a^8b^4 \cdot 2}$$
$$= 2 \cdot 2a^2b\sqrt[4]{2}$$
$$= 4a^2b\sqrt[4]{2}$$

28. $\sqrt{2x - 3} = x - 3$
$$2x - 3 = (x - 3)^2$$
$$2x - 3 = x^2 - 6x + 9$$
$$0 = x^2 - 8x + 12$$
$$0 = (x - 6)(x - 2)$$

$x - 6 = 0$ or $x - 2 = 0$

$x = 6$ or $\quad x = 2$

Discard 2 as an extraneous solution.

The solution is 6.

29. a. $\dfrac{\sqrt{45}}{4} - \dfrac{\sqrt{5}}{3} = \dfrac{3\sqrt{5}}{4} - \dfrac{\sqrt{5}}{3}$

$$= \dfrac{9\sqrt{5} - 4\sqrt{5}}{12}$$

$$= \dfrac{5\sqrt{5}}{12}$$

b. $\sqrt[3]{\dfrac{7x}{8}} + 2\sqrt[3]{7x} = \dfrac{\sqrt[3]{7x}}{2} + 2\sqrt[3]{7x}$

$$= \dfrac{\sqrt[3]{7x}}{2} + \dfrac{4\sqrt[3]{7x}}{2}$$

$$= \dfrac{5\sqrt[3]{7x}}{2}$$

30. $\quad 9x^2 - 6x = -4$

$9x^2 - 6x + 4 = 0$

$a = 9, b = -6, c = 4$

$b^2 - 4ac = (-6)^2 - 4(9)(4)$

$\qquad\quad = 36 - 144$

$\qquad\quad = -108$

Two complex but not real solutions

31. $\sqrt{\dfrac{7x}{3y}} = \dfrac{\sqrt{7x}}{\sqrt{3y}} = \dfrac{\sqrt{7x} \cdot \sqrt{3y}}{\sqrt{3y} \cdot \sqrt{3y}} = \dfrac{\sqrt{21xy}}{3y}$

32. $\qquad \dfrac{4}{x-2} - \dfrac{x}{x+2} = \dfrac{16}{x^2-4}$

$\dfrac{4}{x-2} - \dfrac{x}{x+2} = \dfrac{16}{(x+2)(x-2)}$

$4(x+2) - x(x-2) = 16$

$4x + 8 - x^2 + 2x = 16$

$\qquad\qquad 0 = x^2 - 6x + 8$

$\qquad\qquad 0 = (x-4)(x-2)$

$x - 4 = 0$ or $x - 2 = 0$

$\quad x = 4$ or $\quad x = 2$

Discard the solutions 2 as extraneous.

The solution is 4.

33. $\sqrt{2x-3} = 9$

$2x - 3 = 9^2$

$2x - 3 = 81$

$2x = 84$

$x = 42$

The solution is 42.

34. $\qquad x^3 + 2x^2 - 4x \geq 8$

$x^3 + 2x^2 - 4x - 8 \geq 0$

$x^2(x+2) - 4(x+2) \geq 0$

$(x+2)(x^2-4) \geq 0$

$(x+2)(x+2)(x-2) \geq 0$

$(x+2)^2(x-2) \geq 0$

$(x+2)^2 = 0$ or $x - 2 = 0$

$x + 2 = 0$ or $\quad x = 2$

$x = -2$

Region	Test Point	$(x+2)^2(x-2) \geq 0$ Result
A: $(-\infty, -2)$	-3	$(-1)^2(-5) \geq 0$ False
B: $(-2, 2)$	0	$(2)^2(-2) \geq 0$ False
C: $(2, \infty)$	3	$(5)^2(1) \geq 0$ True

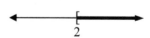

Solution: $[2, \infty)$

35. a. $i^7 = i^4 \cdot i^3 = 1 \cdot (-i) = -i$

b. $i^{20} = (i^4)^5 = 1^5 = 1$

c. $i^{46} = i^{44} \cdot i^2 = (i^4)^{11} \cdot (-1) = 1^{11}(-1) = -1$

d. $i^{-12} = \dfrac{1}{i^{12}} = \dfrac{1}{(i^4)^3} = \dfrac{1}{1^3} = 1$

36. $f(x) = (x+2)^2 - 1$

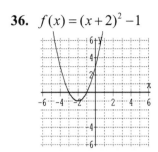

37.
$$p^2 + 2p = 4$$
$$p^2 + 2p + \left(\frac{2}{2}\right)^2 = 4 + 1$$
$$p^2 + 2p + 1 = 5$$
$$(p+1)^2 = 5$$
$$p + 1 = \pm\sqrt{5}$$
$$p = -1 \pm \sqrt{5}$$

The solutions are $-1 + \sqrt{5}$ and $-1 - \sqrt{5}$.

38. $f(x) = -x^2 - 6x + 4$

The maximum will occur at the vertex.
$$x = -\frac{b}{2a} = \frac{-(-6)}{2(-1)} = -3$$
$$f(-3) = -(-3)^2 - 6(-3) + 4 = 13$$
The maximum value is 13.

39.
$$\frac{1}{4}m^2 - m + \frac{1}{2} = 0$$
$$4\left(\frac{1}{4}m^2 - m + \frac{1}{2}\right) = 4(0)$$
$$m^2 - 4m + 2 = 0$$
$$a = 1, b = -4, c = 2$$
$$x = \frac{4 \pm \sqrt{(-4)^2 - 4(1)(2)}}{2(1)}$$
$$= \frac{4 \pm \sqrt{16 - 8}}{2}$$
$$= \frac{4 \pm \sqrt{8}}{2}$$
$$= \frac{4 \pm 2\sqrt{2}}{2} = 2 \pm \sqrt{2}$$

The solutions are $2 + \sqrt{2}$ and $2 - \sqrt{2}$.

40.
$$f(x) = \frac{x+1}{2}$$
$$y = \frac{x+1}{2}$$
$$x = \frac{y+1}{2}$$
$$2x = y + 1$$
$$2x - 1 = y$$
$$f^{-1}(x) = 2x - 1$$

41.
$$p^4 - 3p^2 - 4 = 0$$
$$(p^2 - 4)(p^2 + 1) = 0$$
$$(p+2)(p-2)(p^2 + 1) = 0$$
$$p + 2 = 0 \quad \text{or} \quad p - 2 = 0 \quad \text{or} \quad p^2 + 1 = 0$$
$$p = -2 \quad \text{or} \quad p = 2 \quad \text{or} \quad p^2 = -1$$
$$p = \pm i$$

The solutions are $-2, 2, -i, i$.

42. $f(x) = x^2 - 3x + 2$
$g(x) = -3x + 5$

a. $(f \circ g)(x) = f[g(x)]$
$$= f(-3x + 5)$$
$$= (-3x + 5)^2 - 3(-3x + 5) + 2$$
$$= 9x^2 - 30x + 25 + 9x - 15 + 2$$
$$= 9x^2 - 21x + 12$$

b. $(f \circ g)(-2) = f[g(-2)]$
$$= f[-3(-2) + 5]$$
$$= f(11)$$
$$= (11)^2 - 3(11) + 2$$
$$= 121 - 33 + 2$$
$$= 90$$

c. $(g \circ f)(x) = g[f(x)]$
$$= g(x^2 - 3x + 2)$$
$$= -3(x^2 - 3x + 2) + 5$$
$$= -3x^2 + 9x - 6 + 5$$
$$= -3x^2 + 9x - 1$$

d. $(g \circ f)(5) = g[f(5)]$
$$= g[(5)^2 - 3(5) + 2)]$$
$$= g(12)$$
$$= -3(12) + 5$$
$$= -36 + 5$$
$$= -31$$

43. $\dfrac{x+2}{x-3} \le 0$

$x + 2 = 0$ or $x - 3 = 0$
$\quad x = -2$ or $x = 3$

Region	Test Point	$\dfrac{x+2}{x-3} \le 0$ Result
A: $(-\infty, -2)$	-3	$\dfrac{-1}{-6} \le 0$; False
B: $(-2, 3)$	0	$\dfrac{2}{-3} \le 0$; True
C: $(3, \infty)$	4	$\dfrac{6}{1} \le 0$; False

Solution: $[-2, 3)$

44. $4x^2 + 9y^2 = 36$

$\dfrac{x^2}{9} + \dfrac{y^2}{4} = 1$

Ellipse: center $(0, 0)$, $a = 3$, $b = 2$

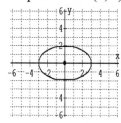

45. $g(x) = \dfrac{1}{2}(x+2)^2 + 5$

Vertex: $(-2, 5)$, axis: $x = -2$

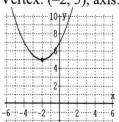

46. a. $64^x = 4$
$$(4^3)^x = 4$$
$$4^{3x} = 4$$
$$3x = 1$$
$$x = \dfrac{1}{3}$$

b. $125^{x-3} = 25$
$$(5^3)^{x-3} = 5^2$$
$$5^{3x-9} = 5^2$$
$$3x - 9 = 2$$
$$3x = 11$$
$$x = \dfrac{11}{3}$$

c. $\dfrac{1}{81} = 3^{2x}$
$$81^{-1} = 3^{2x}$$
$$(3^4)^{-1} = 3^{2x}$$
$$3^{-4} = 3^{2x}$$
$$-4 = 2x$$
$$-\dfrac{4}{2} = x$$
$$-2 = x$$

47. $f(x) = x^2 - 4x - 12$

$x = -\dfrac{b}{2a} = \dfrac{-(-4)}{2(1)} = 2$

$f(2) = (2)^2 - 4(2) - 12 = -16$

Vertex: $(2, -16)$

48. $\begin{cases} x+2y<8 \\ \quad\ y\geq x^2 \end{cases}$

First, graph $x+2y=8$ with a dashed line.

Test Point	$x+2y>8$; Result
(0, 0)	$0+2(0)>8$; False

Shade the portion of the graph which does not contain (0, 0).

Next, graph the parabola $y=x^2$ with a solid curve.

Test Point	$y\geq x^2$; Result
(0, 1)	$1\geq 0^2$; True

Shade the portion of the graph which contains (0, 1).

The solution to the system is the overlapping region.

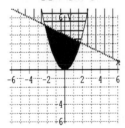

49. $(2, -5), (1, -4)$

$$d=\sqrt{[-4-(-5)]^2+(1-2)^2}$$
$$=\sqrt{1^2+(-1)^2}$$
$$=\sqrt{2}\approx 1.414$$

50. $\begin{cases} x^2+y^2=36 \quad (1) \\ \quad\quad y=x+6 \quad (2) \end{cases}$

Substitute $x+6$ for y in E1.
$$x^2+(x+6)^2=36$$
$$x^2+(x^2+12x+36)=36$$
$$2x^2+12x=0$$
$$2x(x+6)=0$$
$$2x=0 \ \text{ or } \ x+6=0$$
$$x=0 \ \text{ or } \quad\ x=-6$$

Use these values in E2 to find y.
$$x=0: y=0+6=6$$
$$x=-6: y=-6+6=0$$
The solutions are (0, 6) and (–6, 0).

Chapter 11

1. $a_n = n + 4$

$a_1 = 1 + 4 = 5$

$a_2 = 2 + 4 = 6$

$a_3 = 3 + 4 = 7$

$a_4 = 4 + 4 = 8$

$a_5 = 5 + 4 = 9$

Thus, the first five terms of the sequence $a_n = n + 4$ are 5, 6, 7, 8, 9.

3. $a_n = (-1)^n$

$a_1 = (-1)^1 = -1$

$a_2 = (-1)^2 = 1$

$a_3 = (-1)^3 = -1$

$a_4 = (-1)^4 = 1$

$a_5 = (-1)^5 = -1$

Thus, the first five terms of the sequence $a_n = (-1)^n$ are $-1, 1, -1, 1, -1$.

5. $a_n = \dfrac{1}{n+3}$

$a_1 = \dfrac{1}{1+3} = \dfrac{1}{4}$

$a_2 = \dfrac{1}{2+3} = \dfrac{1}{5}$

$a_3 = \dfrac{1}{3+3} = \dfrac{1}{6}$

$a_4 = \dfrac{1}{4+3} = \dfrac{1}{7}$

$a_5 = \dfrac{1}{5+3} = \dfrac{1}{8}$

Thus, the first five terms of the sequence $a_n = \dfrac{1}{n+3}$ are $\dfrac{1}{4}, \dfrac{1}{5}, \dfrac{1}{6}, \dfrac{1}{7}, \dfrac{1}{8}$.

7. $a_n = 2n$

$a_1 = 2(1) = 2$

$a_2 = 2(2) = 4$

$a_3 = 2(3) = 6$

$a_4 = 2(4) = 8$

$a_5 = 2(5) = 10$

or 2, 4, 6, 8, 10.

9. $a_n = -n^2$

$a_1 = -1^2 = -1$

$a_2 = -2^2 = -4$

$a_3 = -3^2 = -9$

$a_4 = -4^2 = -16$

$a_5 = -5^2 = -25$

Thus, the first five terms of the sequence $a_n = -n^2$ are $-1, -4, -9, -16, -25$.

11. $a_n = 2^n$

$a_1 = 2^1 = 2$

$a_2 = 2^2 = 4$

$a_3 = 2^3 = 8$

$a_4 = 2^4 = 16$

$a_5 = 2^5 = 32$

Thus, the first five terms of the sequence $a_n = 2^n$ are 2, 4, 8, 16, 32.

13. $a_n = 2n + 5$

$a_1 = 2(1) + 5 = 2 + 5 = 7$

$a_2 = 2(2) + 5 = 4 + 5 = 9$

$a_3 = 2(3) + 5 = 6 + 5 = 11$

$a_4 = 2(4) + 5 = 8 + 5 = 13$

$a_5 = 2(5) + 5 = 10 + 5 = 15$

Thus, the first five teerms of the sequence $a_n = 2n + 5$ are 7, 9, 11, 13, 15.

15. $a_n = (-1)^n n^2$

$a_1 = (-1)^1 (1)^1 = -1(1) = -1$

$a_2 = (-1)^2 (2)^2 = 1(4) = 4$

$a_3 = (-1)^3 (3)^2 = -1(9) = -9$

$a_4 = (-1)^4 (4)^2 = 1(16) = 16$

$a_5 = (-1)^5 (5)^2 = -1(25) = -25$

Thus, the first five terms of the sequence $a_n = (-1)^n n^2$ are -1, 4, -9, 16, -25.

17. $a_n = 3n^2$

$a_5 = 3(5)^2 = 3(25) = 75$

19. $a_n = 6n - 2$

$a_{20} = 6(20) - 2 = 120 - 2 = 118$

21. $a_n = \dfrac{n+3}{n}$

$a_{15} = \dfrac{15+3}{15} = \dfrac{18}{15} = \dfrac{6}{5}$

23. $a_n = (-3)^n$

$a_6 = (-3)^6 = 729$

25. $a_n = \dfrac{n-2}{n+1}$

$a_6 = \dfrac{6-2}{6+1} = \dfrac{4}{7}$

27. $a_n = \dfrac{(-1)^n}{n}$

$a_8 = \dfrac{(-1)^8}{8} = \dfrac{1}{8}$

29. $a_n = -n^2 + 5$

$a_{10} = -10^2 + 5 = -100 + 5 = -95$

31. $a_n = \dfrac{(-1)^n}{n+6}$

$a_{19} = \dfrac{(-1)^{19}}{19+6} = -\dfrac{1}{25}$

33. 3, 7, 11, 15, or

$4(1) - 1, \ 4(2) - 1, \ 4(3) - 1,$

$4(4) - 1.$ In general, $a_n = 4n - 1$

35. $-2, \ -4, \ -8, \ -16,$ or $-2, \ -2^2,$

$-2^3, -2^4$

In general, $a_n = -2^n$

37. $\dfrac{1}{3}, \dfrac{1}{9}, \dfrac{1}{27}, \dfrac{1}{81},$ or

$\dfrac{1}{3}, \dfrac{1}{3^2}, \dfrac{1}{3^3}, \dfrac{1}{3^4}$

In general, $a_n = \dfrac{1}{3^n}$

39. $a_n = 32n - 16$

$a_2 = 32(2) - 16 = 64 - 16 = 48$ ft

$a_3 = 32(3) - 16 = 96 - 16 = 80$ ft

$a_4 = 32(4) - 16 = 128 - 16 = 112$ ft

41. 0.10, 0.20, 0.40, or

0.10, 0.10(2), $0.10(2)^2$

In general, $a_n = 0.10(2)^{n-1}$

$a_{14} = 0.10(2)^{13} = \$819.20$

43. $a_n = 75(2)^{n-1}$

$a_6 = 75(2)^5 = 75(32) = 2400$ cases

$a_1 = 75(2)^0 = 75(1) = 75$ cases

45. $a_n = \dfrac{1}{2}a_{n-1}$ for $n > 1, a_1 = 800$

In 2000, $n = 1$ and $a_1 = 800$.

In 2001, $n = 2$ and $a_2 = \dfrac{1}{2}(800) = 400$.

In 2002, $n = 3$ and $a_3 = \dfrac{1}{2}(400) = 200$.

In 2003, $n = 4$ and $a_4 = \dfrac{1}{2}(200) = 100$.

In 2004, $n = 5$ and $a_5 = \dfrac{1}{2}(100) = 50$.

The population estimaate for 2004 is 50 sparrows.

Continuing the sequence:

in 2005, $n = 6$ and $a_6 = \dfrac{1}{2}(50) = 25$;

in 2006, $n = 7$ and $a_7 = \dfrac{1}{2}(25) \approx 12$;

in 2007, $n = 8$ and $a_8 = \dfrac{1}{2}(12) = 6$;

in 2008, $n = 9$ and $a_9 = \dfrac{1}{2}(6) = 3$;

in 2009, $n = 10$ and $a_{10} = \dfrac{1}{2}(3) \approx 1$;

in 2010, $n = 11$ and $a_{11} = \dfrac{1}{2}(1) \approx 0$.

The population is estimated to become extinct in 2010.

47. $f(x) = (x-1)^2 + 3$

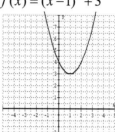

49. $f(x) = 2(x+4)^2 + 2$

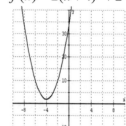

51. $(-4, -1)$ and $(-7, -3)$

$d = \sqrt{[-7 - (-4)]^2 + [-3 - (-1)]^2}$

$d = \sqrt{(-7 + 4)^2 + (-3 + 1)^2}$

$d = \sqrt{(-3)^2 + (-2)^2}$

$d = \sqrt{9 + 4} = \sqrt{13}$ units

53. $(2, -7)$ and $(-3, -3)$

$d = \sqrt{(-3 - 2)^2 + [-3 - (-7)]^2}$

$d = \sqrt{(-5)^2 + (-3 + 7)^2}$

$d = \sqrt{(-5)^2 + (4)^2}$

$d = \sqrt{25 + 16} = \sqrt{41}$ units

55. $a_n = \dfrac{1}{\sqrt{n}}$

$a_1 = \dfrac{1}{\sqrt{1}} = \dfrac{1}{1} = 1$

$a_2 = \dfrac{1}{\sqrt{2}} \approx 0.7071$

$a_3 = \dfrac{1}{\sqrt{3}} \approx 0.5774$

$a_4 = \dfrac{1}{\sqrt{4}} = \dfrac{1}{2} = 0.5$

$a_5 = \dfrac{1}{\sqrt{5}} \approx 0.4472$

Thus, the first five terms of the sequence $a_n = \dfrac{1}{\sqrt{n}}$ are 1, 0.7071, 0.5774, 0.5, 0.4472.

57. $a_n = \left(1 + \dfrac{1}{n}\right)^n$

$a_1 = \left(1 + \dfrac{1}{1}\right)^1 = (2)^1 = 2$

$a_2 = \left(1 + \dfrac{1}{2}\right)^2 = \left(\dfrac{3}{2}\right)^2 = 2.25$

$a_3 = \left(1 + \dfrac{1}{3}\right)^3 = \left(\dfrac{4}{3}\right)^3 \approx 2.3704$

$a_4 = \left(1 + \dfrac{1}{4}\right)^4 = \left(\dfrac{5}{4}\right)^4 \approx 2.4414$

$a_5 = \left(1 + \dfrac{1}{5}\right)^5 = \left(\dfrac{6}{5}\right)^5 \approx 2.4883$

Thus, the first five terms of the sequence $a_n = \left(1 + \dfrac{1}{n}\right)^n$ are 2, 2.25, 2.3704, 2.4414, and 2.4883.

Exercise Set 11.2

1. $a_n = a_1 + (n-1)d$

$a_1 = 4;\ d = 2$

$a_1 = 4$

$a_2 = 4 + (2-1)2 = 6$

$a_3 = 4 + (3-1)2 = 8$

$a_4 = 4 + (4-1)2 = 10$

$a_5 = 4 + (5-1)2 = 12$

The first five terms are 4, 6, 8, 10, 12.

3. $a_n = a_1 + (n-1)d$

$a_1 = 6,\ d = -2$

$a_1 = 6$

$a_2 = 6 + (2-1)(-2) = 4$

$a_3 = 6 + (3-1)(-2) = 2$

$a_4 = 6 + (4-1)(-2) = 0$

$a_5 = 6 + (5-1)(-2) = -2$

The first five terms are $6, 4, 2, 0, -2$.

5. $a_n = a_1 r^{n-1}$

$a_1 = 1,\ r = 3$

$a_1 = 1(3)^{1-1} = 1$

$a_2 = 1(3)^{2-1} = 3$

$a_3 = 1(3)^{3-1} = 9$

$a_4 = 1(3)^{4-1} = 27$

$a_5 = 1(3)^{5-1} = 81$

The first five terms are 1, 3, 9, 27, 81.

7. $a_n = a_1 r^{n-1}$

$a_1 = 48, \ r = \dfrac{1}{2}$

$a_1 = 48\left(\dfrac{1}{2}\right)^{1-1} = 48$

$a_2 = 48\left(\dfrac{1}{2}\right)^{2-1} = 24$

$a_3 = 48\left(\dfrac{1}{2}\right)^{3-1} = 12$

$a_4 = 48\left(\dfrac{1}{2}\right)^{4-1} = 6$

$a_5 = 48\left(\dfrac{1}{2}\right)^{5-1} = 3$

The first five terms are 48, 24, 12, 6, 3.

9. $a_n = a_1 + (n-1)d$

$a_1 = 12, \ d = 3$

$a_n = 12 + (n-1)3$

$a_8 = 12 + 7(3) = 12 + 21 = 33$

11. $a_n = a_1 r^{n-1}$

$a_1 = 7, \ d = 3$

$a_n = a_1 r^{n-1}$

$a_4 = 7(-5)^3 = 7(-125) = -875$

13. $a_n = a_1 + (n-1)d$

$a_1 = -4, \ d = -4$

$a_n = -4 + (n-1)(-4)$

$a_{15} = -4 + 14(-4) = -4 - 56 = -60$

15. 0, 12, 24

$a_1 = 0$ and $d = 12$

$a_n = 0 + (n-1)12$

$a_9 = 8(12) = 96$

17. 20, 18, 16

$a_1 = 20$ and $d = -2$

$a_n = 20 + (n-1)(-2)$

$a_{25} = 20 + 24(-2) = 20 - 48 = -28$

19. 2, -10, 50

$a_1 = 2$ and $r = -5$

$a_n = 2(-5)^{n-1}$

$a_5 = 2(-5)^4 = 2(625) = 1250$

21. $a_4 = 19, \ a_{15} = 52$

$\begin{cases} a_4 = a_1 + (4-1)d \\ a_{15} = a_1 + (15-1)d \end{cases}$ or

$\begin{cases} 19 = a_1 + 3d \\ 52 = a_1 + 14d \end{cases}$

Solving the system gives $d = 3$ and $a_1 = 10$.

$a_n = 10 + (n-1)3$

$= 10 + 3n - 3$

$= 7 + 3n$

and $a_8 = 7 + 3(8)$

$= 7 + 24$

$= 31$

23. $a_2 = -1, a_4 = 5$

$$\begin{cases} a_2 = a_1 + (2-1)d \\ a_4 = a_1 + (4-1)d \end{cases} \text{ or }$$

$$\begin{cases} -1 = a_1 + d \\ 5 = a_1 + 3d \end{cases}$$

Solving the system gives $d = 3$
and $a_1 = -4$.

$a_n = -4(n-1)3$

$\quad = -4 + 3n - 3$

$\quad = -7 + 3n$

and $a_9 = -7 + 3(9)$

$\qquad = -7 + 27$

$\qquad = 20$

25. $a_2 = -\dfrac{4}{3}$ and $a_3 = \dfrac{8}{3}$

Notice that $\dfrac{8}{3} \div \dfrac{-4}{3} = \dfrac{8}{3} \cdot -\dfrac{3}{4} = -2$,

so $r = -2$. Then

$a_2 = a_1(-2)^{2-1}$

$-\dfrac{4}{3} = a_1(-2)$

$\dfrac{2}{3} = a_1.$

The first term is $\dfrac{2}{3}$ and the

common ratio is -2.

27. Answers may vary.

29. $2, 4, 6$ is an arithmetic sequence.

$a_1 = 2$ and $d = 2$

31. $5, 10, 20$ is a geometric sequence.

$a_1 = 5$ and $r = 2$

33. $\dfrac{1}{2}, \dfrac{1}{10}, \dfrac{1}{50}$ is a geometric

sequence.

$a_1 = \dfrac{1}{2}; r = \dfrac{1}{5}$

35. $x, 5x, 25x$ is a geometric

sequence.

$a_1 = x$ and $r = 5$

37. $p, p+4, p+8$ is an arithmetic

sequence.

$a_1 = p$ and $d = 4$

39. $a_1 = 14$ and $d = \dfrac{1}{4}$

$a_n = 14 + (n-1)\dfrac{1}{4}$

$a_{21} = 14 + 20\left(\dfrac{1}{4}\right) = 14 + 5 = 19$

41. $a_1 = 3$ and $r = -\dfrac{2}{3}$

$a_n = 3\left(-\dfrac{2}{3}\right)^{n-1}$

$a_4 = 3\left(-\dfrac{2}{3}\right)^3 = 3\left(-\dfrac{8}{27}\right) = -\dfrac{8}{9}$

43. $\dfrac{3}{2}, 2, \dfrac{5}{2}, \dots$

$a_1 = \dfrac{3}{2}$ and $d = \dfrac{1}{2}$

$a_n = \dfrac{3}{2} + (n-1)\dfrac{1}{2}$

$a_{15} = \dfrac{3}{2} + 14\left(\dfrac{1}{2}\right) = \dfrac{17}{2}$

45. $24, 8, \dfrac{8}{3}, \ldots$

$a_1 = 24$ and $r = \dfrac{1}{3}$

$a_n = 24\left(\dfrac{1}{3}\right)^{n-1}$

$a_6 = 24\left(\dfrac{1}{3}\right)^{5} = 24\left(\dfrac{1}{243}\right) = \dfrac{8}{81}$

47. $a_3 = 2, \; a_{17} = -40$

$\begin{cases} a_3 = a_1 + (3-1)d \\ a_{17} = a_1 + (17-1)d \end{cases}$ or

$\begin{cases} 2 = a_1 + 2d \\ -40 = a_1 + 16d \end{cases}$

Solving the system gives $d = -3$

and $a_1 = 8$.

$a_n = 8 + (n-1)(-3)$

$\quad = 8 - 3n + 3$

$\quad = 11 - 3n$

and $a_8 = 11 - 3(10)$

$\qquad = 11 - 30$

$\qquad = -19$

49. $54, 58, 62$

$a_1 = 54$ and $d = 4$

$a_n = 54 + (n-1)4$

$a_{20} = 54 + 19(4) = 54 + 76 = 130$

The general term of the sequence is

$a_n = 4n + 50$.

There are 130 seats in the twentieth

row.

51. $a_1 = 6$ and $r = 3$

$a_n = 6(3)^{n-1} = 2 \cdot 3 \cdot (3)^{n-1} = 2(3)^n$

The general term of the sequence

is $a_n = 6(3)^{n-1}$ or $a_n = 2(3)^n$.

53. $a_1 = 486$ and $r = \dfrac{1}{3}$

Initial Height $= a_1$

$\qquad = 486\left(\dfrac{1}{3}\right)^{1-1} = 486$

Rebound $1 = a_2$

$\qquad = 486\left(\dfrac{1}{3}\right)^{2-1} = 162$

Rebound $2 = a_3$

$\qquad = 486\left(\dfrac{1}{3}\right)^{3-1} = 54$

Rebound $3 = a_4$

$\qquad = 486\left(\dfrac{1}{3}\right)^{4-1} = 18$

Rebound $4 = a_5$

$\qquad = 486\left(\dfrac{1}{3}\right)^{5-1} = 6$

The first five terms of the

sequence are 486, 162, 54,

18, 6.

It takes 6 bounces.

55. $a_1 = 4000$ and $d = 125$

$a_n = 4000 + (n-1)125$ or

$a_n = 3875 + 125n$

$a_{12} = 4000 + 11(125)$

$a_{12} = 5375$

His salary for his last month of

training is \$5375.

57. $a_1 = 400$ and $r = \dfrac{1}{2}$

12 hrs = 4(3 hrs), so we seek the
fourth term after a_1, namely a_5.

$a_n = a_1 r^{n-1}$

$a_5 = 400\left(\dfrac{1}{2}\right)^4 = \dfrac{400}{16} = 25$

25 grams of the radioactive
material remains after 12 hours.

59. $\dfrac{1}{3(1)} + \dfrac{1}{3(2)} + \dfrac{1}{3(3)}$

$= \dfrac{1}{3} + \dfrac{1}{6} + \dfrac{1}{9}$

$= \dfrac{6}{18} + \dfrac{3}{18} + \dfrac{2}{18}$

$= \dfrac{11}{18}$

61. $3^0 + 3^1 + 3^2 + 3^3$

$= 1 + 3 + 9 + 27$

$= 40$

63. $\dfrac{8-1}{8+1} + \dfrac{8-2}{8+2} + \dfrac{8-3}{8+3}$

$= \dfrac{7}{9} + \dfrac{6}{10} + \dfrac{5}{11}$

$= \dfrac{770}{990} + \dfrac{594}{990} + \dfrac{450}{990}$

$= \dfrac{1814}{990} = \dfrac{907}{495}$

65. $a_1 = \$11,782.40$

$r = 0.5$

$a_2 = (11,782.40)(0.5) = \$5891.20$

$a_3 = (5891.20)(0.5) = \$2945.60$

$a_4 = (2945.60)(0.5) = \$1472.80$

The first four terms of the
sequence are \$11,782.40,
\$5891.20, \$2945.60, \$1472.80.

67. $a_1 = 19.652$ and $d = -0.034$

$a_2 = 19.652 - 0.034 = 19.618$

$a_3 = 19.618 - 0.034 = 19.584$

$a_4 = 19.584 - 0.034 = 19.550.$

69. Answers may vary.

Exercise Set 11.3

1. $\displaystyle\sum_{i=1}^{4}(i-3)$

$= (1-3) + (2-3) + (3-3) + (4-3)$

$= -2 + (-1) + 0 + 1$

$= -2$

3. $\displaystyle\sum_{i=4}^{7}(2i+4)$

$= \left[2(4)+4\right] + \left[2(5)+4\right] + \left[2(6)+4\right]$
$\qquad\qquad + \left[2(7)+4\right]$

$= 12 + 14 + 16 + 18 = 60$

5. $\displaystyle\sum_{i=2}^{4}(i^2 - 3)$

$= (2^2 - 3) + (3^2 - 3) + (4^2 - 3)$

$= 1 + 6 + 13$

$= 20$

7. $\displaystyle\sum_{i=1}^{3}\frac{1}{i+5}=\frac{1}{1+5}+\frac{1}{2+5}+\frac{1}{3+5}$

$\displaystyle=\frac{1}{6}+\frac{1}{7}+\frac{1}{8}=\frac{28}{168}+\frac{24}{168}+\frac{21}{168}$

$\displaystyle=\frac{73}{168}$

9. $\displaystyle\sum_{i=1}^{3}\frac{1}{6i}=\frac{1}{6(1)}+\frac{1}{6(2)}+\frac{1}{6(3)}$

$\displaystyle=\frac{1}{6}+\frac{1}{12}+\frac{1}{18}$

$\displaystyle=\frac{6+3+2}{36}$

$\displaystyle=\frac{11}{36}$

11. $\displaystyle\sum_{i=2}^{6}3i$

$=3(2)+3(3)+3(4)+3(5)+3(6)$

$=6+9+12+15+18$

$=60$

13. $\displaystyle\sum_{i=3}^{5}i(i+2)$

$=3(3+2)+4(4+2)+5(5+2)$

$=15+24+35$

$=74$

15. $\displaystyle\sum_{i=1}^{5}2^{i}=2^{1}+2^{2}+2^{3}+2^{4}+2^{5}$

$=2+4+8+16+32$

$=62$

17. $\displaystyle\sum_{i=1}^{4}\frac{4i}{i+3}=\frac{4(1)}{1+3}+\frac{4(2)}{2+3}+\frac{4(3)}{3+3}+\frac{4(4)}{4+3}$

$\displaystyle=1+\frac{8}{5}+2+\frac{16}{7}=\frac{105}{35}+\frac{56}{35}+\frac{80}{35}$

$\displaystyle=\frac{241}{35}$

19. $1+3+5+7+9$

$\left[(2)-1\right]+\left[2(2)-1\right]+\left[2(3)-1\right]$

$+\left[2(4)-1\right]+\left[2(5)-1\right]$

$\displaystyle=\sum_{i=1}^{5}(2i-1)$

21. $4+12+36+108$

$=4+4(3)+4(3)^{2}+4(3)^{3}$

$\displaystyle=\sum_{i=1}^{4}4(3)^{i-1}$

23. $12+9+6+3+0+(-3)$

$=\left[-3(1)+15\right]+\left[-3(2)+15\right]$

$+\left[-3(3)+15\right]+\left[-3(4)+15\right]$

$+\left[-3(5)+15\right]+\left[-3(6)+15\right]$

$\displaystyle=\sum_{i=1}^{6}(-3i+15)$

25. $\displaystyle 12+4+\frac{4}{3}+\frac{4}{9}=\frac{4}{3^{-1}}+\frac{4}{3^{0}}+\frac{4}{3}+\frac{4}{3^{2}}$

$\displaystyle=\sum_{i=1}^{4}\frac{4}{3^{i-2}}$

27. $1+4+9+16+25+36+49$

$=1^{2}+2^{2}+3^{2}+4^{2}+5^{2}+6^{2}+7^{2}$

$\displaystyle=\sum_{i=1}^{7}i^{2}$

29. $a_{n}=(n+2)(n-5)$

$a_{1}=(1+2)(1-5)=3(-4)=-12$

$a_{2}=(2+2)(2-5)=4(-3)=-12$

$a_{1}+a_{2}=-12+(-12)=-24$

31. $a_{n}=n(n-6)$

$=a_{1}+a_{2}=1(1-6)+2(2-6)$

$=1(-5)+2(-4)=-13$

33. $a_n = (n+3)(n+1)$

$a_1 = (1+3)(1+1) = 4(2) = 8$

$a_2 = (2+3)(2+1) = 5(3) = 15$

$a_3 = (3+3)(3+1) = 6(4) = 24$

$a_4 = (4+3)(4+1) = 7(5) = 35$

$\sum_{i=1}^{4} a_i = 8 + 15 + 24 + 35 = 82$

35. $a_n = -2n$

$\sum_{i=1}^{4} (-2i) =$

$= -2(1) + (-2)(2) + (-2)(3) + (-2)(4)$

$= -2 - 4 - 6 - 8$

$= -20$

37. $a_n = -\dfrac{n}{3}$

$a_1 + a_2 + a_3 = -\dfrac{1}{3} - \dfrac{2}{3} - \dfrac{3}{3} = -2$

39. $1, 2, 3, \ldots, 10$

$a_n = n$

$\sum_{i=1}^{10} i = 1 + 2 + 3 + \ldots + 10$

$= \dfrac{10(11)}{2} = 55$

A total of 55 trees were planted.

41. $a_1 = 6$ and $r = 2$

$a_n = 6 \cdot 2^{n-1}$

$a_5 = 6 \cdot 2^4 = 6 \cdot 16 = 96$

There will be 96 fungus units at the beginning of the 5th day.

43. $a_1 = 50$ and $r = 2$

Since $48 = 4(12)$, we seek the fourth term after a_1, namely a_5.

The general term of the sequence is $a_n = 50(2)^{n-1}$, where n represents the number of 12-hr periods.

$a_5 = 50(2)^4 = 50(16) = 800$

There are 800 bacteria after 48 hours.

45. $a_n = (n+1)(n+2)$

$a_4 = (4+1)(4+2)$

$= 5(6) = 30$ oppossums

$a_1 = (1+1)(1+2) = 2(3) = 6$

$a_2 = (2+1)(2+2) = 3(4) = 12$

$a_3 = (3+1)(3+2) = 4(5) = 20$

$\sum_{i=1}^{4} a_i = 6 + 12 + 20 + 30$

$= 68$ oppossums

47. $a_n = 100(0.5)^n$

$a_4 = 100(0.5)^4 = 6.25$ lbs of decay.

$a_1 = 100(0.5)^1 = 50$

$a_2 = 100(0.5)^2 = 25$

$a_3 = 100(0.5)^3 = 12.5$

$\sum_{i=1}^{4} a_i = 50 + 25 + 12.5 + 6.25$

$= 93.75$ lbs of decay

49. $a_1 = 40$ and $r = \dfrac{4}{5}$

$$a_5 = 40\left(\dfrac{4}{5}\right)^4 = 16.384 \text{ or } 16.4 \text{ in.}$$

$$a_2 = 40\left(\dfrac{4}{5}\right)^1 = 32$$

$$a_3 = 40\left(\dfrac{4}{5}\right)^2 = 25.6$$

$$a_4 = 40\left(\dfrac{4}{5}\right)^3 = 20.48$$

$$\sum_{i=1}^{5} a_i = 40 + 32 + 25.6 + 20.48$$
$$+16.384$$
$$= 134.464 \text{ or } 134.5 \text{ in.}$$

51. $\dfrac{5}{1-\dfrac{1}{2}} = \dfrac{5}{\dfrac{1}{2}} = 5 \cdot \dfrac{2}{1} = 10$

53. $\dfrac{\dfrac{1}{3}}{1-\dfrac{1}{10}} = \dfrac{\dfrac{1}{3}}{\dfrac{9}{10}} = \dfrac{1}{3} \cdot \dfrac{10}{9} = \dfrac{10}{27}$

55. $\dfrac{3(1-2^4)}{1-2} = \dfrac{3(1-16)}{-1}$
$$= \dfrac{3(-15)}{-1} = \dfrac{-45}{-1} = 45$$

57. $\dfrac{10}{2}(3+15) = \dfrac{10}{2}(18) = \dfrac{180}{2} = 90$

59. a. $\sum_{i=1}^{7} i + i^2$

$$= (1+1^2) + (2+2^2) + (3+3^2)$$
$$+ (4+4^2) + (5+5^2) + (6+6^2)$$
$$+ (7+7^2)$$
$$= 2 + 6 + 12 + 20 + 30 + 42 + 56$$

b. $\sum_{i=1}^{7} i + \sum_{i=1}^{7} i^2$

$$= (1+2+3+4+5+6+7) + (1+4$$
$$+9+16+25+36+49)$$

c. They are equal; 168

d. True

Integrated Review

1. $a_n = n - 3$

$$a_1 = 1 - 3 = -2$$
$$a_2 = 2 - 3 = -1$$
$$a_3 = 3 - 3 = 0$$
$$a_4 = 4 - 3 = 1$$
$$a_5 = 5 - 3 = 2$$

Therefore, the first five terms are -2, -1, 0, 1, 2.

2. $a_n = \dfrac{7}{1+n}$

$$a_1 = \dfrac{7}{1+1} = \dfrac{7}{2}$$

$$a_2 = \dfrac{7}{1+2} = \dfrac{7}{3}$$

$$a_3 = \dfrac{7}{1+3} = \dfrac{7}{4}$$

$$a_4 = \dfrac{7}{1+4} = \dfrac{7}{5}$$

$$a_5 = \dfrac{7}{1+5} = \dfrac{7}{6}$$

The first five terms are $\dfrac{7}{2}$, $\dfrac{7}{3}, \dfrac{7}{4}, \dfrac{7}{5}$, and $\dfrac{7}{6}$.

3. $a_n = 3^{n-1}$

$a_1 = 3^{1-1} = 3^0 = 1$

$a_2 = 3^{2-1} = 3^1 = 3$

$a_3 = 3^{3-1} = 3^2 = 9$

$a_4 = 3^{4-1} = 3^3 = 27$

$a_5 = 3^{5-1} = 3^4 = 81$

The first five terms are 1, 3, 9, 27, and 81.

4. $a_n = n^2 - 5$

$a_1 = 1^2 - 5 = 1 - 5 = -4$

$a_2 = 2^2 - 5 = 4 - 5 = -1$

$a_3 = 3^2 - 5 = 9 - 5 = 4$

$a_4 = 4^2 - 5 = 16 - 5 = 11$

$a_5 = 5^2 - 5 = 25 - 5 = 20$

The first five terms are −4, −1, 4, 11, and 20.

5. $(-2)^n; a_6$

$a_6 = (-2)^n = (-2)^6 = 64$

6. $-n^2 + 2; a_4$

$a_4 = -n^2 + 2$

$= -(4)^2 + 2$

$= -16 + 2$

$= -14$

7. $\dfrac{(-1)^n}{n}; a_{40}$

$a_{40} = \dfrac{(-1)^n}{n}$

$= \dfrac{(-1)^{40}}{40}$

$= \dfrac{1}{40}$

8. $\dfrac{(-1)^n}{2n}; a_{41}$

$a_{41} = \dfrac{(-1)^n}{2n}$

$= \dfrac{(-1)^{41}}{2(41)}$

$= \dfrac{-1}{82}$

$= -\dfrac{1}{82}$

9. $a_1 = 7; d = -3$

$a_1 = 7$

$a_2 = 7 - 3 = 4$

$a_3 = 4 - 3 = 1$

$a_4 = 1 - 3 = -2$

$a_5 = -2 - 3 = -5$

The first five terms are $7, 4, 1, -2, -5.$

10. $a_1 = -3; r = 5$

$a_1 = -3$

$a_2 = -3(5) = -15$

$a_3 = -15(5) = -75$

$a_4 = -75(5) = -375$

$a_5 = -375(5) = -1875$

The first five terms are $-3, -15, -75, -375, -1875.$

11. $a_1 = 45; r = \dfrac{1}{3}$

$a_1 = 45$

$a_2 = 45\left(\dfrac{1}{3}\right) = 15$

$a_3 = 15\left(\dfrac{1}{3}\right) = 5$

$a_4 = 5\left(\dfrac{1}{3}\right) = \dfrac{5}{3}$

$a_5 = \dfrac{5}{3}\left(\dfrac{1}{3}\right) = \dfrac{5}{9}$

The first five terms are 45, 15,

5, $\dfrac{5}{3}, \dfrac{5}{9}$.

12. $a_1 = -12; d = 10$

$a_1 = -12$

$a_2 = -12 + 10 = -2$

$a_3 = -2 + 10 = 8$

$a_4 = 8 + 10 = 18$

$a_5 = 18 + 10 = 28$

The first five terms are

$-12, -2, 8, 18, 28$.

13. $a_1 = 20; d = 9$

$a_{10} = a_1 + (n-1)d$

$\qquad = 20 + (10-1)9$

$\qquad = 20 + 81$

$\qquad = 101$

14. $a_1 = 64; r = \dfrac{3}{4}$

$a_6 = a_1 r^{n-1}$

$\quad = 64\left(\dfrac{3}{4}\right)^{6-1}$

$\quad = 64\left(\dfrac{3}{4}\right)^{5}$

$\quad = 64\left(\dfrac{243}{1024}\right)$

$\quad = \dfrac{243}{16}$

15. $a_1 = 6; r = \dfrac{-12}{6} = -2$

$a_n = a_1 r^{n-1}$

$a_7 = 6(-2)^{7-1}$

$a_7 = 6(-2)^{6}$

$a_7 = 6(64)$

$a_7 = 384$

16. $a_1 = -100; d = -85 - (-100) = 15$

$a_n = a_1 + (n-1)d$

$a_{20} = -100 + (20-1)(15)$

$a_{20} = -100 + (19)(15)$

$a_{20} = -100 + 285$

$a_{20} = 185$

17. $a_4 = -5, a_{10} = -35$

$a_n = a_1 + (n-1)d$

$\begin{cases} a_4 = a_1 + (4-1)d \\ a_{10} = a_1 + (10-1)d \end{cases}$

$\begin{cases} -5 = a_1 + 3d \\ -35 = a_1 + 9d \end{cases}$

$\begin{cases} -5 = a_1 + 3d \\ -35 = a_1 + 9d \end{cases}$

Multiply eq. 2 by -1,

then add the equations.

$\begin{cases} -5 = a_1 + 3d \\ (-1) - 35 = -1(a_1 + 9d) \end{cases}$

$\begin{cases} -5 = a_1 + 3d \\ 35 = -a_1 - 9d \end{cases}$

$30 = -6d$

$-5 = d$

To find a_1, let $d = -5$ in

$-5 = a_1 + 3d$

$-5 = a_1 + 3(-5)$

$10 = a_1$

Thus, $a_1 = 10$ and $d = -5$, so

$a_n = 10 + (n-1)(-5)$

$a_n = -5n + 15$

$a_5 = -5(5) + 15$

$a_5 = -10$

18.　　$a_4 = 1; a_7 = \dfrac{1}{125}$

$a_n = a_1 r^{n-1}$

$a_7 = a_4 r^{4-1}$

$\dfrac{1}{125} = 1r^3$

$\dfrac{1}{5} = r$

$a_4(r) = a_5$

$1\left(\dfrac{1}{5}\right) = a_5$

$\dfrac{1}{5} = a_5$

19. $\displaystyle\sum_{i=1}^{4} 5i = 5(1) + 5(2) + 5(3) + 5(4)$

$= 5 + 10 + 15 + 20$

$= 50$

20.

$\displaystyle\sum_{i=1}^{7} (3i + 2)$

$= (3(1) + 2) + (3(2) + 2) + (3(3) + 2) + (3(4) + 2)$

$\quad + (3(5) + 2) + (3(6) + 2) + (3(7) + 2)$

$= 5 + 8 + 11 + 14 + 17 + 20 + 23$

$= 98$

21. $\displaystyle\sum_{i=3}^{7} 2^{i-4}$

$= 2^{3-4} + 2^{4-4} + 2^{5-4} + 2^{6-4} + 2^{7-4}$

$= 2^{-1} + 2^0 + 2^1 + 2^2 + 2^3$

$= \dfrac{1}{2} + 1 + 2 + 4 + 8$

$= 15\dfrac{1}{2} = \dfrac{31}{2}$

22. $\displaystyle\sum_{i=2}^{5} \dfrac{i}{i+1}$

$= \dfrac{2}{2+1} + \dfrac{3}{3+1} + \dfrac{4}{4+1} + \dfrac{5}{5+1}$

$= \dfrac{2}{3} + \dfrac{3}{4} + \dfrac{4}{5} + \dfrac{5}{6}$

$= \dfrac{61}{20}$

23. $S_3 = \displaystyle\sum_{i=1}^{3} n(n-4)$

$= 1(1-4) + 2(2-4) + 3(3-4)$

$= -3 - 4 - 3$

$= -10$

24.

$$S_{10} = \sum_{i=1}^{10} (-1)^n (n+1)$$

$$= (-1)^1 (1+1) + (-1)^2 (2+1) + (-1)^3 (3+1) +$$

$$(-1)^4 (4+1) + (-1)^5 (5+1) + (-1)^6 (6+1) +$$

$$(-1)^7 (7+1) + (-1)^8 (8+1) + (-1)^9 (9+1) +$$

$$(-1)^{10} (10+1)$$

$$= -2 + 3 - 4 + 5 - 6 + 7 - 8 + 9 - 10 + 11$$

$$= 5$$

Exercise Set 11.4

1. $1, 3, 5, 7, ...$

The first term is 1 and the sixth term is 11.

$$S_6 = \frac{6}{2}(1+11) = 3(12) = 36$$

3. $4, 12, 36, ...$

$a_1 = 4, r = 3, n = 5$

$$S_5 = \frac{4(1-3^5)}{1-3} = 484$$

5. $3, 6, 9, ...$

The first term is 3 and the sixth term is 18.

$$S_6 = \frac{6}{2}(3+18)$$

$$= 3(21)$$

$$= 63$$

7. $2, \dfrac{2}{5}, \dfrac{2}{25}, ...$

$$a_1 = 2, r = \frac{1}{5}, n = 4$$

$$S_4 = \frac{2\left[1 - \left(\dfrac{1}{5}\right)^4\right]}{1 - \dfrac{1}{5}} = 2.496$$

9. $1, 2, 3, ..., 10$

The first term is 1 and the tenth term is 10.

$$S_{10} = \frac{10}{2}(1+10)$$

$$= 5(11)$$

$$= 55$$

11. $1, 2, 3, 7$

The first term is 1 and the fourth term is 7.

$$S_4 = \frac{4}{2}(1+7)$$

$$= 2(8)$$

$$= 16$$

13. $12, 6, 3, ...$

$$a_1 = 12, \quad r = \frac{1}{2}$$

$$S_\infty = \frac{12}{1 - \dfrac{1}{2}} = \frac{12}{\dfrac{1}{2}} = 24$$

15. $\dfrac{1}{10}, \dfrac{1}{100}, \dfrac{1}{1000}, ...$

$$a_1 = \frac{1}{10}, \quad r = \frac{1}{10}$$

$$S_\infty = \frac{\dfrac{1}{10}}{1 - \dfrac{1}{10}} = \frac{1}{9}$$

17. $-10, -5, -\dfrac{5}{2}, \ldots$

$$a_1 = -10, \quad r = \dfrac{1}{2}$$

$$S_\infty = \dfrac{-10}{1-\dfrac{1}{2}} = -20$$

19. $2, -\dfrac{1}{4}, \dfrac{1}{32}, \ldots$

$$a_1 = 2, \quad r = -\dfrac{1}{8}$$

$$S_\infty = \dfrac{2}{1-\left(-\dfrac{1}{8}\right)} = \dfrac{16}{9}$$

21. $\dfrac{2}{3}, -\dfrac{1}{3}, \dfrac{1}{6}, \ldots$

$$a_1 = \dfrac{2}{3}, \quad r = -\dfrac{1}{2}$$

$$S_\infty = \dfrac{\dfrac{2}{3}}{1-\left(-\dfrac{1}{2}\right)} = \dfrac{4}{9}$$

23. $-4, 1, 6, \ldots, 41$

The first term is -4 and the tenth term is 41.

$$S_{10} = \dfrac{10}{2}(-4+41)$$
$$= 5(37)$$
$$= 185$$

25. $3, \dfrac{3}{2}, \dfrac{3}{4}, \ldots$

$$a_1 = 3, \ r = \dfrac{1}{2}, \ n = 7$$

$$S_7 = \dfrac{3\left[1-\left(\dfrac{1}{2}\right)^7\right]}{1-\dfrac{1}{2}} = \dfrac{381}{64}$$

27. $-12, 6, -3, \ldots$

$$a_1 = -12, \ r = -\dfrac{1}{2}, \ n = 5$$

$$S_5 = \dfrac{-12\left[1-\left(-\dfrac{1}{2}\right)^5\right]}{1-\left(-\dfrac{1}{2}\right)} = -\dfrac{33}{4}$$

29. $\dfrac{1}{2}, \dfrac{1}{4}, 0, \ldots, -\dfrac{17}{4}$

The first term is $\dfrac{1}{2}$ and the twentieth term is $-\dfrac{17}{4}$.

$$S_{20} = \dfrac{20}{2}\left(\dfrac{1}{2} - \dfrac{17}{4}\right)$$
$$= 10\left(\dfrac{-15}{4}\right)$$
$$= -\dfrac{75}{2}$$

31. $a_1 = 8,\ r = -\dfrac{2}{3},\ n = 3$

$$S_3 = \frac{8\left[1-\left(-\dfrac{2}{3}\right)^3\right]}{1-\left(-\dfrac{2}{3}\right)} = \frac{56}{9}$$

33. The first five terms are 4000, 3950, 3900, 3850, 3800

$a_1 = 4000,\ d = -50,\ n = 12$

$a_{12} = 4000 + 11(-50)$

$\phantom{a_{12}} = 3450$ cars sold in month 12.

$$S_{12} = \frac{12}{2}\left(4000 + 3450\right)$$

$\phantom{S_{12}} = 44,700$ cars sold in the first 12 months.

35. Firm A:

The first term is 22,000 and the tenth term is 31,000.

$$S_{10} = \frac{10}{2}(22000 + 31000)$$

$\phantom{S_{10}} = \$265,000$

Firm B:

The first term is 20,000 and the tenth term is 30,800.

$$S_{10} = \frac{10}{2}(20000 + 30800)$$

$\phantom{S_{10}} = \$254,000$

Thus, Firm A is making the better offer.

37. $a_1 = 30,000,\ r = 1.10,\ n = 4$

$a_4 = 30000(1.10)^{4-1}$

$a_4 = \$39,930$ made during her fourth year of business

$$S_4 = \frac{30000(1 - 1.10^4)}{1 - 1.10}$$

$ = \$139,230$ made during the first four years of business.

39. $a_1 = 30,\ r = 0.9,\ n = 5$

$a_5 = 30(0.9)^{5-1} = 19.63$

Approximately 20 minutes to assemble the first computer.

$$S_5 = \frac{30(1 - 0.9^5)}{1 - 0.9} = 122.853$$

Approximately 123 minutes to assemble the first 5 computers.

41. $a_1 = 20,\ r = \dfrac{4}{5}$

$$S_\infty = \frac{20}{1 - \dfrac{4}{5}} = 100$$

We double the number (to account for the flight up as well as down) and subtract 20 (since the first bounce was preceded by only a downward flight). Thus, the ball 2(100) - 20 = 180 feet

43. Player A:

The first term is 1 and the

ninth term is 9.

$$S_9 = \frac{9}{2}(1+9)$$
$$= 45 \text{ points}$$

Player B:

The first term is 10 and the

sixth term is 15.

$$S_6 = \frac{6}{2}(10+15)$$
$$= 75 \text{ points}$$

45. The first term is 200 and the

twentieth is 105.

$$S_{20} = \frac{20}{2}(200+105)$$
$$= 3050$$

Thus, $3050 rent is paid for 20 days

during the holiday rush.

47. $a_1 = 0.01$, $r = 2$, $n = 30$

$$S_3 = \frac{0.01\left[1-2^{30}\right]}{1-2} = 10,737,418.23$$

He would pay $10,737,418.23

in room and board for the 30 days.

49. 720

51. 3

53. $x^2 + 10x + 25$

55. $8x^3 - 12x^2 + 6x - 1$

57. $0.8\overline{88} = 0.8 + 0.08 + 0.008 + \cdots$

$$= \frac{8}{10} + \frac{8}{100} + \frac{8}{1000} + \cdots$$

This is a geometric series with

$$a_1 = \frac{8}{10}, \ r = \frac{1}{10}$$

$$S_\infty = \frac{\dfrac{8}{10}}{1-\dfrac{1}{10}} = \frac{8}{9}$$

59. Answers may vary.

Exercise Set 11.5

1. $(m+n)^3$

$$= m^3 + 3m^2n + 3mn^2 + n^3$$

3. $(c+d)^5$

$$= c^5 + 5c^4d + 10c^3d^2 + 10c^2d^3 + 5cd^4$$
$$+ d^5$$

5. $(y-x)^5 = \left[y+(-x)\right]^5$

$$= y^5 - 5y^4x + 10y^3x^2 - 10y^2x^3 + 5yx^4$$
$$- x^5$$

7. Answers may vary.

9. $\dfrac{8!}{7!} = \dfrac{8 \cdot 7!}{7!} = 8$

11. $\dfrac{7!}{5!} = \dfrac{7 \cdot 6 \cdot 5!}{5!} = 7 \cdot 6 = 42$

13. $\dfrac{10!}{7!2!} = \dfrac{10 \cdot 9 \cdot 8 \cdot 7!}{7!2!} = \dfrac{10 \cdot 9 \cdot 8}{2 \cdot 1} = 360$

15. $\dfrac{8!}{6!0!} = \dfrac{8 \cdot 7 \cdot 6!}{6! \cdot 1} = 56$

17. $(a+b)^7$

$$= a^7 + 7a^6b + \frac{7\cdot6}{2!}a^5b^2 + \frac{7\cdot6\cdot5}{3!}a^4b^3 + \frac{7\cdot6\cdot5\cdot4}{4!}a^3b^4 + \frac{7\cdot6\cdot5\cdot4\cdot3}{5!}a^2b^5$$

$$+ \frac{7\cdot6\cdot5\cdot4\cdot3\cdot2}{6!}ab^6 + b^7$$

$$= a^7 + 7a^6b + 21a^5b^2 + 35a^4b^3 + 35a^3b^4 + 21a^2b^5 + 7ab^6 + b^7$$

19. $(a+2b)^5$

$$= a^5 + 5a^4(2b) + \frac{5\cdot4}{2!}a^3(2b)^2 + \frac{5\cdot4\cdot3}{3!}a^2(2b)^3 + \frac{5\cdot4\cdot3\cdot2}{4!}a(2b)^4$$

$$+ (2b)^5$$

$$= a^5 + 10a^4b + 40a^3b^2 + 80a^2b^3 + 80ab^4 + 32b^5$$

21. $(q+r)^9$

$$= q^9 + \frac{9}{1!}q^8r + \frac{9\cdot8}{2!}q^7r^2 + \frac{9\cdot8\cdot7}{3!}q^6r^3 + \frac{9\cdot8\cdot7\cdot6}{4!}q^5r^4 + \frac{9\cdot8\cdot7\cdot6\cdot5}{5!}q^4r^5$$

$$+ \frac{9\cdot8\cdot7\cdot6\cdot5\cdot4}{6!}q^3r^6 + \frac{9\cdot8\cdot7\cdot6\cdot5\cdot4\cdot3}{7!}q^2r^7 + \frac{9\cdot8\cdot7\cdot6\cdot5\cdot4\cdot3\cdot2}{8!}qr^8 + r^9$$

$$= q^9 + 9q^8r + 36q^7r^2 + 84q^6r^3 + 126q^5r^4 + 126q^4r^5 + 84q^3r^6 + 36q^2r^7 + 9qr^8 + r^9$$

23. $(4a+b)^5$

$$= (4a)^5 + \frac{5}{1!}(4a)^4 b + \frac{5\cdot4}{2!}(4a)^3 b^2 + \frac{5\cdot4\cdot3}{3!}(4a)^2 b^3 + \frac{5\cdot4\cdot3\cdot2}{4!}(4a)b^4 + b^5$$

$$= 1024a^5 + 1280a^4b + 640a^3b^2 + 160a^2b^3 + 20ab^4 + b^5$$

25. $(5a-2b)^4$

$$= (5a)^4 + \frac{4}{1!}(5a)^3(-2b) + \frac{4\cdot3}{2!}(5a)^2(-2b)^2 + \frac{4\cdot3\cdot2}{3!}(5a)(-2b)^3 + (-2b)^4$$

$$= 625a^4 - 1000a^3b + 600a^2b^2 - 160ab^3 + 16b^4$$

27. $(2a+3b)^3$

$$= (2a)^3 + \frac{3}{1!}(2a)^2(3b) + \frac{3\cdot2}{2!}(2a)(3b)^2 + (3b)^3$$

$$= 8a^3 + 36a^2b + 54ab^2 + 27b^3$$

29. $(x+2)^5$

$$= x^5 + \frac{5}{1!}x^4(2) + \frac{5\cdot 4}{2!}x^3(2)^2 + \frac{5\cdot 4\cdot 3}{3!}x^2(2)^3 + \frac{5\cdot 4\cdot 3\cdot 2}{4!}x(2)^4 + (2)^5$$

$$= x^5 + 10x^4 + 40x^3 + 80x^2 + 80x + 32$$

31. 5th term of $(c-d)^5$ corresponds to
$r = 4$:

$$\frac{5!}{4!(5-4)!}c^{5-4}(-d)^4 = 5cd^4$$

33. 8th term of $(2c+d)^7$ corresponds to
$r = 7$:

$$\frac{7!}{7!(7-7)!}(2c)^{7-7}(d)^7 = d^{7\cdot}$$

35. 4th term of $(2r-s)^5$ corresponds to
$r = 3$:

$$\frac{5!}{3!(5-3)!}(2r)^{5-3}(-s)^3 = -40r^2s^3$$

37. 3rd term of $(x+y)^4$ corresponds to
$r = 2$:

$$\frac{4!}{2!(4-2)!}(x)^{4-2}(y)^2 = 6x^2y^2$$

39. 2nd term of $(a+3b)^{10}$ corresponds to $r=1$:

$$\frac{10!}{1!(10-1)!}(a)^{10-1}(3b)^1 = 30a^9b$$

41. $f(x) = |x|$

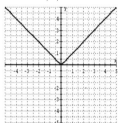

Not one-to-one

43. $H(x) = 2x+3$

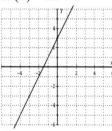

One-to-one

45. $f(x) = x^2 + 3$

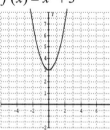

Not one-to-one

47.

$$\left(\sqrt{x}+\sqrt{3}\right)^5$$

$$= \left(\sqrt{x}\right)^5 + \frac{5}{1!}\left(\sqrt{x}\right)^4\left(\sqrt{3}\right) + \frac{5\cdot4}{2!}\left(\sqrt{x}\right)^3\left(\sqrt{3}\right)^2 +$$

$$\frac{5\cdot4\cdot3}{3!}\left(\sqrt{x}\right)^2\left(\sqrt{3}\right)^3 + \frac{5\cdot4\cdot3\cdot2}{4!}\left(\sqrt{x}\right)\left(\sqrt{3}\right)^4 + \left(\sqrt{3}\right)^5$$

$$= x^2\sqrt{x} + 5\sqrt{3}x^2 + 30x\sqrt{x} + 30\sqrt{3}x + 45\sqrt{x} + 9\sqrt{3}$$

49. $\binom{9}{5} = \dfrac{9!}{5!(9-5)!}$

$$= \frac{9!}{5!4!}$$

$$= \frac{9\cdot8\cdot7\cdot6\cdot5\cdot4\cdot3\cdot2\cdot1}{(5\cdot4\cdot3\cdot2\cdot1)\cdot(4\cdot3\cdot2\cdot1)} = 126$$

51. $\binom{8}{2} = \dfrac{8!}{2!(8-2)!}$

$$= \frac{8!}{2!6!}$$

$$= \frac{8\cdot7\cdot6\cdot5\cdot4\cdot3\cdot2\cdot1}{(2\cdot1)\cdot(6\cdot5\cdot4\cdot3\cdot2\cdot1)} = 28$$

53. answers may vary

Chapter 11 Review

1. $a_n = -3n^2$

$$a_1 = -3(1)^2 = -3$$
$$a_2 = -3(2)^2 = -12$$
$$a_3 = -3(3)^2 = -27$$
$$a_4 = -3(4)^2 = -48$$
$$a_5 = -3(5)^2 = -75$$

2. $a_n = n^2 + 2n$

$a_1 = 1^2 + 2(1) = 3$

$a_2 = 2^2 + 2(2) = 8$

$a_3 = 3^2 + 2(3) = 15$

$a_4 = 4^2 + 2(4) = 24$

$a_5 = 5^2 + 2(5) = 35$

3. $a_n = \dfrac{(-1)^n}{100}$

$a_{100} = \dfrac{(-1)^{100}}{100} = \dfrac{1}{100}$

4. $a_n = \dfrac{2n}{(-1)^2}$

$a_{50} = \dfrac{2(50)}{(-1)^2} = 100$

5. $\dfrac{1}{6\cdot1}, \dfrac{1}{6\cdot2}, \dfrac{1}{6\cdot3}, \cdots$

In general, $a_n = \dfrac{1}{6n}$

6. $-1, 4, -9, 16, \ldots$

$a_n = (-1)^n n^2$

7. $a_n = 32n - 16$

$a_5 = 32(5) - 16 = 144$ ft

$a_6 = 32(6) - 16 = 176$ ft

$a_7 = 32(7) - 16 = 208$ ft

8. $a_n = 100(2)^{n-1}$

$10,000 = 100(2)^{n-1}$

$100 = 2^{n-1}$

$\log 100 = (n-1)\log 2$

$n = \dfrac{\log 100}{\log 2} + 1 \approx 7.6$

Eighth day culture will be at least 10,000. Originally, $a_1 = 100(2)^{-1} = 100\left(\dfrac{1}{2}\right) = 50$.

9. $a_1 = 450$

$a_2 = 3(450) = 1350$

$a_3 = 3(1350) = 4050$

$a_4 = 3(4050) = 12,150$

$a_5 = 3(12,150) = 36,450$

In 2007, the number of infected people should be 36,450.

10. $a_n = 50 + (n-1)8$

$a_1 = 50$

$a_2 = 50 + 8 = 58$

$a_3 = 50 + 2(8) = 66$

$a_4 = 50 + 4(8) = 74$

$a_5 = 50 + 4(8) = 82$

$a_6 = 50 + 5(8) = 90$

$a_7 = 50 + 6(8) = 98$

$a_8 = 50 + 7(8) = 106$

$a_9 = 50 + 8(8) = 114$

$a_{10} = 50 + 9(8) = 122$

There are 122 seats in the tenth row.

11. $a_1 = -2$, $r = \dfrac{2}{3}$

$a_1 = -2$

$a_2 = -2\left(\dfrac{2}{3}\right) = -\dfrac{4}{3}$

$a_3 = \left(-\dfrac{4}{3}\right)\left(\dfrac{2}{3}\right) = -\dfrac{8}{9}$

$a_4 = \left(-\dfrac{8}{9}\right)\left(\dfrac{2}{3}\right) = -\dfrac{16}{27}$

$a_5 = \left(-\dfrac{16}{27}\right)\left(\dfrac{2}{3}\right) = -\dfrac{32}{81}$

The first 5 terms of the sequence are

$-2, -\dfrac{4}{3}, -\dfrac{8}{9}, -\dfrac{16}{27}, -\dfrac{32}{81}$.

12. $a_n = 12 + (n-1)(-1.5)$

$a_1 = 12$

$a_2 = 12 + (1)(-1.5) = 10.5$

$a_3 = 12 + 2(-1.5) = 9$

$a_4 = 12 + 3(-1.5) = 7.5$

$a_5 = 12 + 4(-1.5) = 6$

13. $a_1 = -5$, $d = 4$, $n = 30$

$a_{30} = 5 + (30-1)4 = 111$

14. $a_n = 2 + (n-1)\dfrac{3}{4}$

$a_{11} = 2 + 10\left(\dfrac{3}{4}\right) = \dfrac{19}{2}$

15. 12, 7, 2,...

$a_1 = 12$, $d = -5$, $n = 20$

$a_{20} = 12 + (20-1)(-5) = -83$

16. $a_n = a_1 r^{n-1}$

$a_6 = 4\left(\dfrac{3}{2}\right)^{6-1} = \dfrac{243}{8}$

17. $a_4 = 18$, $a_{20} = 98$

Use the relationship:

$a_4 + 16d = a_{20}$

$18 + 16d = 98$

$d = 5$

Now use the relationship:

$a_4 = a_1 + 3d$

$18 = a_1 + 3(5)$

$a_1 = 3$

18. $-48 = a_3 = a_1 r^{3-1}$

$192 = a_4 = a_1 r^{4-1}$

$-48 = a_1 r^2$

$192 = a_1 r^3$

$192 = a_1 r^2 \cdot r$

$192 = -48r$

$-4 = r$

$-48 = a_1 r^2$

$-48 = a_1(-4)^2$

$-3 = a_1$

$r = -4$, $a_1 = -3$.

19. $\dfrac{3}{10}, \dfrac{3}{10^2}, \dfrac{3}{10^3}, ...$

In general, $a_n = \dfrac{3}{10^n}$

20. 50, 58, 66, ...

$a_n = 50 + (n-1)8$

or $a_n = 42 = 8n$

21. $\dfrac{8}{3}, 4, 6, \ldots$

Geometric, $a_1 = \dfrac{8}{3}$,

$r = \dfrac{4}{\frac{8}{3}} = 4 \cdot \dfrac{3}{8} = \dfrac{12}{8} = \dfrac{3}{2}$

22. arithmetic; $a_1 = -10.5$,

$d = -6.1 - (10.5) = 4.4$

23. $7x, -14x, 28x$

Geometric: $a_1 = 7x, \ r = -2$

24. neither

25. $a_1 = 8, \ r = 0.75$

8, 6, 4.5, 3.4, 2.5, 1.9

Yes, a ball that rebounds to a height
of 2.5 feet after the fifth bounce is
good, since $2.5 \geq 1.9$.

26. $a_1 = 25 \quad d = -4$

$a_n = a_1 + (n-1)d$

$a_n = 25 + (n-1)(-4)$

$a_n = 25 + 6(-4) = 1$

Continuing the progression as far
as possible leaves 1 can in the top
row.

27. $a_1 = 1, \ r = 2$

$a_n = 2^{n-1}$

$a_{10} = 2^9 = \$512$

$a_{30} = 2^{29} = \$536,870,912$

28. $a_n = a_1 r^{n-1}$

$a_5 = 30(0.7)^4 = 7.203$ in.

29. $a_1 = 900, \ d = 150$

$a_n = 900 + (n-1)150$

$a_6 = 900 + (6-1)150$

$\quad = \$1650/\text{month}$

30. $\dfrac{1}{512}, \dfrac{1}{256}, \dfrac{1}{128}, \ldots$

first fold: $a_1 = \dfrac{1}{256}, \ r = 2$

$a_{15} = \dfrac{1}{256}(2)^{15-1} = 64$ inches

31. $\displaystyle\sum_{i=1}^{5}(2i-1)$

$= \left[2(1)-1\right] + \left[2(2)-1\right] + \left[2(3)-1\right]$

$\quad + \left[2(4)-1\right] + \left[2(5)-1\right]$

$= 1 + 3 + 5 + 7 + 9$

$= 25$

32. $\displaystyle\sum_{i=1}^{5}i(i+2)$

$= 1(1+2) + 2(2+2) + 3(3+2)$

$\quad + 4(4+2) + 5(5+2)$

$= 3 + 8 + 15 + 24 + 35$

$= 85$

33. $\displaystyle\sum_{i=2}^{4}\dfrac{(-1)^i}{2i} = \dfrac{(-1)^2}{2(2)} + \dfrac{(-1)^3}{2(3)} + \dfrac{(-1)^4}{2(4)}$

$= \dfrac{1}{4} - \dfrac{1}{6} + \dfrac{1}{8} = \dfrac{5}{24}$

34. $\displaystyle\sum_{i=3}^{5}5(-1)^{i-1}$

$= 5(-1)^{3-1} + 5(-1)^{4-1} + 5(-1)^{5-1}$

$= 5(1) + 5(-1) + 5(1) = 5 - 5 + 5 = 5$

35. $a_n = (n-3)(n+2)$

$S_4 = (1-3)(1+2)+(2-3)(2+2)+$

$\quad (3-3)(3+2)+(4-3)(4+2)$

$\quad = -6-4+0+6$

$\quad = -4$

36. $a_n = n^2$

$S_6 = (1)^2 + (2)^2 + (3)^2 + (4)^2 + (5)^2$

$\quad + (6)^2$

$\quad = 91$

37. $a_n = -8 + (n-1)3$

$a_1 = -8 + (1-1)3 = -8$

$a_2 = -8 + (2-1)3 = -5$

$a_3 = -8 + (3-1)3 = -2$

$a_4 = -8 + (4-1)3 = 1$

$a_5 = -8 + (5-1)3 = 4$

So $S_5 = -10$

38. $a_n = 5(4)^{n-1}$

$S_3 = 5(4)^0 + 5(4)^1 + 5(4)^2 = 105$

39. $1+3+9+27+81+243$

$\quad = 3^0 + 3^1 + 3^2 + 3^3 + 3^4 + 3^5$

$\quad = \sum_{i=1}^{6} 3^{i-1}$

40. $6+2+(-2)+(-6)+(-10)+(-14)$

$\quad +(-18)$

$a_1 = 4 \quad d = -4$

$a_n = 6 + (n-1)(-4)$

$\sum_{i=1}^{7} 6 + (i-1)(-4)$

41. $\dfrac{1}{4} + \dfrac{1}{16} + \dfrac{1}{64} + \dfrac{1}{256} = \sum_{i=1}^{4} \dfrac{1}{4^i}$

42. $1 + \left(-\dfrac{3}{2}\right) + \dfrac{9}{4} = \sum_{i=1}^{3} \left(-\dfrac{3}{2}\right)^{i-1}$

43. $a_1 = 20, \ r = 2$

$a_n = 20(2)^n$ represents the number of yeast, where n represents the number of 8-hr periods. Since $48 = 6(8)$ here, $n = 6$.

$a_6 = 20(2)^6 = 1280$ yeast

44. $a_n = n^2 + 2n - 1$

$a_4 = (4)^2 + 2(4) - 1 = 23$ cranes

$\sum_{i=1}^{4} i^2 + 2i - 1$

$\quad = (1+2-1)+(4+4-1)+$

$\quad\quad +(9+6-1)+(16+8-1)$

$\quad = 46$ cranes

45. For Job A: $a_1 = 39{,}500, \ d = 2200;$

$a_5 = 39{,}500 + (5-1)2200 = \$48{,}330$

For Job B: $a_1 = 41{,}000, \ d = 1400$

$a_5 = 41{,}000 + (5-1)1400 = \$46{,}600$

For the fifth year, Job A has a higher salary.

46. $a_n = 200(0.5)^n$

$a_3 = 200(0.5)^3 = 25$ kg

$\sum_{i=1}^{3} 200(0.5)^i$

$\quad = 200(0.5) + 200(0.5)^2 + 200(0.5)^3$

$\quad = 175$ kg

47. 15, 19, 23, ...

$a_1 = 15, \ d = 4$

$S_6 = \dfrac{6}{2}\left[2(15) + (6-1)4\right] = 150$

48. $5, -10, 20, \ldots$

$a_1 = 5, \; r = -2$

$$S_n = \frac{a_1(1-r^n)}{1-r}$$

$$S_9 = \frac{5(1-(-2)^9)}{1-(-2)} = 855$$

49. $a_1 = 1, \; d = 2, \; n = 30$

$$S_{30} = \frac{30}{2}\big[2(1) + (30-1)2\big] = 900$$

50. $7, 14, 21, 28, \ldots$

$a_n = 7 + (n-1)7$

$a_{20} = 7 + (20-1)7 = 140$

$$S_{20} = \frac{20}{2}(7 + 140) = 1470$$

51. $8, 5, 2, \ldots$

$a_1 = 8, \; d = -3, \; n = 20$

$$S_n = \frac{n}{2}\big[2a_1 + (n-1)d\big]$$

$$S_{20} = \frac{20}{2}\big[2(8) + (20-1)(-3)\big]$$

$$= -410$$

52. $\dfrac{3}{4}, \dfrac{9}{4}, \dfrac{27}{4}, \ldots$

$a_1 = \dfrac{3}{4}, \; r = 3$

$$S_8 = \frac{\frac{3}{4}(1-3^8)}{1-3} = 2460$$

53. $a_1 = 6, \; r = 5$

$$S_4 = \frac{6(1-5^4)}{1-5} = 936$$

54. $a_1 = -3, \; d = -6$

$a_n = -3 + (n-1)(-6)$

$a_{100} = -3 + (100-1)(-6) = -597$

$$S_{100} = \frac{100}{2}(-3 + (-597)) = -30,000$$

55. $5, \dfrac{5}{2}, \dfrac{5}{4}, \ldots$

$a_1 = 5, \; r = \dfrac{1}{2}$

$$S_\infty = \frac{5}{1-\frac{1}{2}} = 10$$

56. $18, -2, \dfrac{2}{9}, \ldots$

$a_1 = 18, \; r = -\dfrac{1}{9}$

$$S_\infty = \frac{18}{1+\frac{1}{9}} = \frac{81}{5}$$

57. $-20, -4, -\dfrac{4}{5}, \ldots$

$a_1 = -20, \; r = \dfrac{1}{5}$

$$S_\infty = \frac{-20}{1-\frac{1}{5}} = -25$$

58. $0.2, 0.02, 0.002, \ldots$

$a_1 = 0.2, \; r = \dfrac{1}{10}$

$$S_\infty = \frac{0.2}{1-\frac{1}{10}} = \frac{2}{9}$$

59. $a_1 = 20,000, r = 1.15, n = 4$

$a_4 = 20,000(1.15)^{4-1} = 30,418$

Earned in his fourth year.

$S_4 = \dfrac{20,000(1-1.15^4)}{1-1.15} = \$99,868$

earned in his first four years.

60. $a_n = 40(0.8)^{n-1}$

$a_4 = 40(0.8)^{4-1} = 20.48$ min

$S_4 = \dfrac{40(1-0.8^4)}{1-0.8} = 118$ min

61. $a_1 = 100, d = -7, n = 7$

$a_7 = 100 + (7-1)(-7)$

$= \$58$ rent paid for the seventh day.

$S_7 = \dfrac{7}{2}\left[2(100) + (7-1)(-7)\right]$

$= \$553$ rent paid for the first seven days.

62. $a_1 = 15, r = 0.8$

$S_\infty = \dfrac{15}{1-0.8} = 75$ feet downward

$a_1 = 12, r = 0.8$

$S_\infty = \dfrac{12}{1-0.8} = 60$ feet upward

The total is 135 feet.

66. 27, 30, 33, ...

$a_n = 27 + (n-1)(3)$

$a_{20} = 27 + (20-1)(3) = 84$

$S_{20} = \dfrac{20}{2}(27+84) = 1110$

1110 seats

63. 1800, 600, 200, ...

$a_1 = 1800, r = \dfrac{1}{3}, n = 6$

$S_6 = 1800\dfrac{\left(1-\left(\dfrac{1}{3}\right)^6\right)}{1-\dfrac{1}{3}}$

≈ 2696 mosquitoes killed during the first six days after the spraying.

64. 1800, 600, 200, ...

For which n is $a_n > 1$?

$a_n = 1800\left(\dfrac{1}{3}\right)^{n-1} > 1$

$(n-1)\log\left(\dfrac{1}{3}\right) > \log\dfrac{1}{1800}$

$n < 7.8$

No longer effective on the 8th day.

About 2700 mosquitoes were killed.

65. $0.5\overline{55} = 0.5 + 0.05 + 0.005 + \cdots$

$a_1 = 0.5, r = 0.1$

$S_\infty = \dfrac{0.5}{1-0.1} = \dfrac{5}{9}$

67. $(x+z)^5 = x^5 + 5x^4z + 10x^3z^2 + 10x^2z^3 + 5xz^4 + z^5$

68. $(y-r)^6 = y^6 - 6y^5r + 15y^4r^2 - 20y^3r^3 + 15y^2r^4 - 6yr^5 + r^6$

69. $(2x+y)^4 = 16x^4 + 32x^3y + 24x^2y^2 + 8xy^3 + y^4$

70. $(3y-z)^4 = 81y^4 - 108y^3z + 54y^2z^2 - 12yz^3 + z^4$

71. $(b+c)^8$

$$= b^8 + \frac{8}{1!}b^7c + \frac{8\cdot 7}{2!}b^6c^2 + \frac{8\cdot 7\cdot 6}{3!}b^5c^3 + \frac{8\cdot 7\cdot 6\cdot 5}{4!}b^4c^4 + \frac{8\cdot 7\cdot 6\cdot 5\cdot 4}{5!}b^3c^5$$

$$+ \frac{8\cdot 7\cdot 6\cdot 5\cdot 4\cdot 3}{6!}b^2c^6 + \frac{8\cdot 7\cdot 6\cdot 5\cdot 4\cdot 3\cdot 2}{7!}bc^7 + c^8$$

$$= b^8 + 8b^7c + 28b^6c^2 + 56b^5c^3 + 70b^4c^4 + 56b^3c^5 + 28b^2c^6 + 8bc^7 + c^8$$

72. $(x-w)^7$

$$= x^7 + \frac{7}{1!}x^6(-w) + \frac{7\cdot 6}{2!}x^5(-w)^2 + \frac{7\cdot 6\cdot 5}{3!}x^4(-w)^3 + \frac{7\cdot 6\cdot 5\cdot 4}{4!}x^3(-w)^4$$

$$+ \frac{7\cdot 6\cdot 5\cdot 4\cdot 3}{5!}x^2(-w)^5 + \frac{7\cdot 6\cdot 5\cdot 4\cdot 3\cdot 2}{6!}x(-w)^6 + (-w)^7$$

$$= x^7 - 7x^6w + 21x^5w^2 - 35x^4w^3 + 35x^3w^4 - 21x^2w^5 + 7xw^6 - w^7$$

73. $(4m-n)^4 = (4m+(-n))^4$

$$= (4m)^4 + \frac{4}{1!}(4m)^3(-n) + \frac{4\cdot 3}{2!}(4m)^2(-n)^2 + \frac{4\cdot 3\cdot 2}{3!}(4m)(-n)^3 + (-n)^4$$

$$= 256m^4 - 256m^3n + 96m^2n^2 - 16mn^3 + n^4$$

74. $(p-2r)^5$

$$= p^5 + \frac{5}{1!}p^4(-2r) + \frac{5\cdot 4}{2!}p^3(-2r)^2 + \frac{5\cdot 4\cdot 3}{3!}p^2(-2r)^3 + \frac{5\cdot 4\cdot 3\cdot 2}{4!}p(-2r)^4$$

$$+ (-2r)^5$$

$$= p^5 - 10p^4r + 40p^3r^2 - 80p^2r^3 + 80pr^4 - 32r^5$$

75. The 4th term corresponds to $r = 3$.

$$\frac{7!}{3!(7-3)!}a^{7-3}b^3 = 35a^4b^3$$

76. The 11th term is $\dfrac{10!}{10!0!}y^{10-10}(2z)^{10} = 1024z^{10}$

Chapter 11 Test

1. $a_n = \dfrac{(-1)^n}{n+4}$

$a_1 = \dfrac{(-1)^1}{1+4} = -\dfrac{1}{5}$

$a_2 = \dfrac{(-1)^2}{2+4} = \dfrac{1}{6}$

$a_3 = \dfrac{(-1)^3}{3+4} = -\dfrac{1}{7}$

$a_4 = \dfrac{(-1)^4}{4+4} = \dfrac{1}{8}$

$a_5 = \dfrac{(-1)^5}{5+4} = -\dfrac{1}{9}$

2. $a_n = 10 + 3(n-1)$

$a_{80} = 10 + 3(80-1) = 247$

3. $\dfrac{2}{5}, \dfrac{2}{25}, \dfrac{2}{125}, \ldots$

In general, $a_n = \dfrac{2}{5^n}$

4. $(-1)^1 9 \cdot 1, \ (-1)^2 9 \cdot 2, \ldots, a_n = (-1)^n 9n$

5. $a_n = 5(2)^{n-1}, S_5 = \dfrac{5(1-2^5)}{1-2} = 155$

6. $a_n = 18 + (n-1)(-2)$

$a_1 = 18, \ d = -2$

$S_{30} = \dfrac{30}{2}\left[2(18) + (30-1)(-2)\right] = -330$

7. $a_1 = 24, \ r = \dfrac{1}{6}$

$S_\infty = \dfrac{24}{1-\dfrac{1}{6}} = \dfrac{144}{5}$

8. $\dfrac{3}{2}, -\dfrac{3}{4}, \dfrac{3}{8}, \ldots$

$a_1 = \dfrac{3}{2}, \ r = -\dfrac{1}{2}$

$S_\infty = \dfrac{\dfrac{3}{2}}{1-\left(-\dfrac{1}{2}\right)} = 1$

9. $\displaystyle\sum_{i=1}^{4} i(i-2)$

$= 1(1-2) + 2(2-2) + 3(3-2) +$

$\quad 4(4-2)$

$= 10$

10. $\displaystyle\sum_{i=2}^{4} 5(2)^i (-1)^{i-1}$

$= 5(2)^2(-1)^{2-1} + 5(2)^3(-1)^{3-1} + 5(2)^4(-1)^{4-1} = -60$

11. $(a-b)^6 = a^6 - 6a^5 b + 15a^4 b^2 - 20a^3 b^3 + 15a^2 b^4 - 6ab^5 + b^6$

12. $(2x+y)^5$

$$= (2x)^5 + \frac{5}{1!}(2x)^4 y + \frac{5 \cdot 4}{2!}(2x)^3 y^2 + \frac{5 \cdot 4 \cdot 3}{3!}(2x)^2 y^3 + \frac{5 \cdot 4 \cdot 3 \cdot 2}{4!}(2x)y^4 + y^5$$

$$= 32x^5 + 80x^4 y + 80x^3 y^2 + 40x^2 y^3 + 10xy^4 + y^5$$

13. $a_n = 250 + 75(n-1)$

$a_{10} = 250 + 75(10-1) = 925$

There were 925 people in the town at the beginning of the tenth year.

$a_1 = 250 + 75(1-1) = 250$

There were 250 people in the town at the beginning of the first year.

14. $1, 3, 5, \cdots$

$a_1 = 1,\ d = 2,\ n = 8$

$a_8 = 1 + (8-1)2 = 15$

We want $1+3+5+\ldots+15$

$$S_8 = \frac{8}{2}[1+15] = 64$$

There were 64 shrubs planted in the 8 rows.

15. $a_1 = 80,\ r = \frac{3}{4},\ n = 4$

$$a_4 = 80\left(\frac{3}{4}\right)^{4-1} = 33.75$$

The arc length is 33.75 cm on the 4th swing.

$$S_4 = \frac{80\left(1-\left(\frac{3}{4}\right)^4\right)}{1-\frac{3}{4}} = 218.75$$

The total of the arc lengths is 218.75 cm for the first 4 swings.

16. $a_1 = 80,\ r = \frac{3}{4}$

$$S_\infty = \frac{80}{1-\frac{3}{4}} = 320$$

The total of the arc lengths is 320 cm before the pendulum comes to rest.

17. 16, 48, 80,...

$$a_{10} = 16 + (10 - 1)32 = 304$$

He falls 304 feet during the 10th second.

$$S_{10} = \frac{10}{2}[16 + 304] = 1600$$

He falls 1600 feet during the first 10 seconds.

18. $0.42\overline{42} = 0.42 + 0.0042 + 0.000042$

$$S_\infty = \frac{0.42}{1 - 0.01} = \frac{14}{33}$$

Thus, $0.42\overline{42} = \frac{14}{33}$

Chapter 11 Cumulative Review

1. a. $\dfrac{20}{-4} = -5$

b. $\dfrac{-9}{-3} = 3$

c. $-\dfrac{3}{8} \div 3 = -\dfrac{1}{8}$

d. $\dfrac{-40}{10} = -4$

e. $\dfrac{-1}{10} \div \dfrac{-2}{3} = \dfrac{1}{4}$

f. Undefined

2. a. $3a - (4a + 3) =$
$$= 3a - 4a - 3$$
$$= -a - 3$$

b. $(5x - 3) + (2x + 6) =$
$$= 7x + 3$$

c. $4(2x - 5) - 3(5x + 1) =$
$$= 8x - 20 - 15x - 3$$
$$= -7x - 23$$

3. Let $x =$ the original price, then
$$x - 0.08x = 2162$$
$$0.92x = 2162$$
$$x = \$2350$$
The original price is $\$2,350$.

4. Let $x =$ the price before taxes, then
$$x + 0.06x = 344.50$$
$$1.06x = 344.50$$
$$x = 325$$
The price before taxes was $\$325$.

5. a. $\begin{cases} 3x + 4y = -7 \\ x - 2y = -9 \end{cases}$

$$D = \begin{vmatrix} 3 & 4 \\ 1 & -2 \end{vmatrix} = 3(-2) - 4(1) = -10$$

$$D_x = \begin{vmatrix} -7 & 4 \\ -9 & -2 \end{vmatrix} = 50$$

$$D_y = \begin{vmatrix} 3 & -7 \\ 1 & -9 \end{vmatrix} = -20$$

$$x = \frac{D_x}{D} = \frac{50}{-10} = -5$$

$$y = \frac{D_y}{D} = \frac{-20}{-10} = 2$$

The solution is $(-5, 2)$.

b. $\begin{cases} 5x + y = 5 \\ -7x - 2y = -7 \end{cases}$

$D = \begin{vmatrix} 5 & 1 \\ -7 & -2 \end{vmatrix} = -3$

$D_x = \begin{vmatrix} 5 & 1 \\ -7 & -2 \end{vmatrix} = -3$

$D_y = \begin{vmatrix} 5 & 5 \\ -7 & -7 \end{vmatrix} = 0$

$x = \dfrac{D_x}{D} = \dfrac{50}{-10} = 1$

$y = \dfrac{D_y}{D} = \dfrac{-20}{-10} = 0$

The solution is (1, 0).

6. If the line is to be parallel,
then the slope of the new has to
the same slope as the given line.

Therefore $m = \dfrac{3}{2}$

and using the point-slope with $(3, -2)$;

$(y - (-2)) = \dfrac{3}{2}(x - 3)$

$y + 2 = \dfrac{3}{2}(x - 3)$

$y = \dfrac{3}{2}x - \dfrac{13}{2}$

$f(x) + 2 = \dfrac{3}{2}(x - 3)$

$f(x) = \dfrac{3}{2}x - \dfrac{13}{2}$

7. a. $\left(3x^6\right)\left(5x\right) = 15x^7$

  b. $\left(-2x^3p^2\right)\left(4xp^{10}\right) = -8x^4p^{12}$

8. $y^3 + 5y^2 - y - 5 = 0$

$(y^3 + 5y^2) + (-y - 5) = 0$

$y^2(y + 5) - 1(y + 5) = 0$

$(y^2 - 1)(y + 5) = 0$

$y = -5, -1, 1$

9. $-2 \big|\ \begin{array}{rrrrr} 1 & -2 & -11 & 5 & 34 \\ & -2 & 8 & 6 & -22 \end{array}$

$\quad\ \ \overline{\begin{array}{rrrrr} 1 & -4 & -3 & 11 & 12 \end{array}}$

Answer: $x^3 - 4x^2 - 3x + 11 + \dfrac{12}{x+2}$

10. $\dfrac{5}{3a-6} - \dfrac{a}{a-2} + \dfrac{3+2a}{5a-10}$

$= \dfrac{5}{3a-6}\left(\dfrac{5}{5}\right) - \dfrac{a}{a-2}\left(\dfrac{15}{15}\right) + \dfrac{3+2a}{5a-10}\left(\dfrac{3}{3}\right)$

$= \dfrac{25 - 15a + 9 + 6a}{15(a-2)}$

$= \dfrac{34 - 9a}{15(a-2)}$

11. a. $\sqrt{50} = \sqrt{2}\sqrt{25} = 5\sqrt{2}$

  b. $\sqrt[3]{24} = \sqrt[3]{8}\sqrt[3]{3} = 2\sqrt[3]{3}$

  c. $\sqrt{26}$

  d. $\sqrt[4]{32}$

$\qquad = \sqrt[4]{16}\sqrt[4]{2}$

$\qquad = 2\sqrt[4]{2}$

12. $\sqrt{3x+6} - \sqrt{7x-6} = 0$

$\left(\sqrt{3x+6}\right)^2 = \left(\sqrt{7x-6}\right)^2$

$3x + 6 = 7x - 6$

$x = 3$

13. $\quad 2420 = 2000(1+r)^2$

$\sqrt{1.21} = \sqrt{(1+r)^2}$

$r = 10\%$

14. a. $\sqrt[3]{\dfrac{4}{3x}} = \sqrt[3]{\dfrac{4}{3x}}\left(\dfrac{\sqrt[3]{9x^2}}{\sqrt[3]{9x^2}}\right) = \dfrac{\sqrt[3]{36x^2}}{3x}$

b. $\dfrac{\sqrt{2}+1}{\sqrt{2}-1} =$

$= \dfrac{\sqrt{2}+1}{\sqrt{2}-1}\left(\dfrac{\sqrt{2}+1}{\sqrt{2}+1}\right)$

$= \dfrac{2+2\sqrt{2}+1}{2-1}$

$= 3+2\sqrt{2}$

15. $(x-3)^2 - 3(x-3) - 4 = 0$

$x^2 - 6x + 9 - 3x + 9 - 4 = 0$

$x^2 - 9x + 14 = 0$

$x = 2, 7$

16. $\dfrac{10}{(2x+4)^2} - \dfrac{1}{2x+4} = 3$

$\left((2x+4)^2\right)\left(\dfrac{10}{(2x+4)^2} - \dfrac{1}{2x+4} = 3\right)$

$\dfrac{10}{(2x+4)^2} - \dfrac{1}{2x+4} = 3$

$\left((2x+4)^2\right)\left(\dfrac{10}{(2x+4)^2} - \dfrac{1}{2x+4} = 3\right)$

$10 - (2x+4) = 3(2x+4)^2$

$10 - 2x - 4 = 3(4x^2 + 16x + 16)$

$-2x + 6 = 12x^2 + 48x + 48$

$12x^2 + 50x + 42 = 0$

$6x^2 + 25x + 21 = 0$

$(6x+7)(x+3) = 0$

$x = \dfrac{-7}{6}, -3$

17. $\dfrac{5}{x+1} < -2$

Set denominator $= 0$.

$x + 1 = 0$

$x = -1$

Next solve:

$\dfrac{5}{x+1} = -2$

$5 = -2(x+1)$

$5 = -2x - 2$

$-2x = 7$

$x = -\dfrac{7}{2}$

$x = -\dfrac{7}{2}, -1$

$\left(\dfrac{-7}{2}, -1\right)$

18. Axis of symmetry: $x = -2$
vertex: $(-2, -6)$

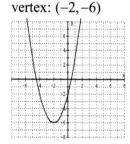

19. $-16t^2 + 20t = 0$

$-16(t - \dfrac{5}{8})^2 + \dfrac{25}{4} = 0$

$t = \dfrac{25}{4}$ feet, $\dfrac{5}{2}$ sec

20. $f(x) = x^2 + 3x - 18$

$f(x) = \left(x - \dfrac{3}{2}\right)^2 - \dfrac{81}{4}$

$\left(-\dfrac{3}{2}, \dfrac{-81}{4}\right)$

21. a. $(f \circ g)(2) = (2+3)^2 = 25$
$(g \circ f)(2) = 2^2 + 3 = 7$

b. $(f \circ g)(x) = (x+3)^2 = x^2 + 6x + 9$
$(g \circ f)(x) = x^2 + 3$

22. $f(x) = -2x + 3$
$x = -2y + 3$
$x - 3 = -2y$
$f^{-1}(x) = -\dfrac{x-3}{2}$

23. $f^{-1} = \{(1,0), (7,-2), (-6,3), (4,4)\}$

24. a. $(f \circ g)(2) = (2+1)^2 - 2 = 7$
$(g \circ f)(2) = 2^2 + 1 - 2 = 3$

b. $(f \circ g)(x) = (x+1)^2 - 2$
$= x^2 + 2x - 1$
$(g \circ f)(x) = x^2 - 1$

25. a. $2^x = 16$
$2^x = 2^4$
$x = 4$

b. $9^x = 27$
$\left(3^2\right)^x = 3^3$
$2x = 3$
$x = \dfrac{3}{2}$

c. $4^{x+3} = 8^x$
$\left(2^2\right)^{x+3} = \left(2^3\right)^x$
$2^{2x+6} = 2^{3x}$
$2x + 6 = 3x$
$x = 6$

26. a. $\log_2 32 = x$
$2^x = 32$
$x = 5$

b. $\log_4 \dfrac{1}{64} = \log_4 4^{-3} = -3$

c. $\log_{\frac{1}{2}} x = 5$
$\left(\dfrac{1}{2}\right)^5 = x$
$x = \dfrac{1}{32}$

27. a. $\log_3 3^2 = 2$

b. $\log_7 7^{-1} = -1$

c. $5^{\log_5 3} = 3$

d. $2^{\log_2 6} = 6$

28. a. $4^x = 64$
$\left(2^2\right)^x = 2^6$
$2x = 6$
$x = 3$

b. $8^x = 32$
$\left(2^3\right)^x = 2^5$
$3x = 5$
$x = \dfrac{5}{3}$

c. $9^{x+4} = 243^x$

$3^{2(x+4)} = 3^{5x}$

$2x + 8 = 5x$

$x = \dfrac{8}{3}$

29. a. $\log_{11} 30$

   b. $\log_3 6$

   c. $\log_2 (x^2 + 2x)$

30. a. $\log_{10} 10^5 = 5$

   b. $\log_{10} 10^{-3} = -3$

   c. $\ln e^{\frac{1}{5}} = \dfrac{1}{5}$

   d. $\ln e^4 = 4$

31. $A = Pe^{rt}$

   $A = 1600e^{0.9(5)}$

   $A = \$2509.30$

32 a. $\log_6 5 + \log_6 4 = \log_6 20$

   b. $\log_8 12 - \log_8 4 = \log_8 3$

   c. $2\log_2 x + 3\log_2 x - 2\log_2 (x-1)$

   $= \log_2 \dfrac{x^5}{(-x-1)^2}$

33. $3^x = 7$

   $x \log 3 = \log 7$

   $x = \dfrac{\log 7}{\log 3} \approx 1.7712$

34. $10000 = 5000\left(1 + \dfrac{0.02}{4}\right)^{4t}$

   $\ln 2 = 4t \ln\left(1 + \dfrac{0.02}{4}\right)$

   $t \approx 34.7$ years

35. $\log_4 (x-2) = 2$

   $4^2 = x - 2$

   $x - 2 = 16$

   $x = 18$

36. $\log_4 \dfrac{10}{x} = 2$

   $4^2 = \dfrac{10}{x}$

   $16 = \dfrac{10}{x}$

   $16x = 10$

   $x = \dfrac{5}{8}$

37. $\dfrac{x^2}{16} - \dfrac{y^2}{25} = 1$

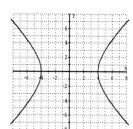

38. $d = \sqrt{(-2-8)^2 + (4-5)^2} = \sqrt{101}$

39. $\begin{cases} y = \sqrt{x} \\ x^2 + y^2 = 6 \end{cases}$

$(x)^2 + \left(\sqrt{x}\right)^2 = 6$

$x^2 + x - 6 = 0$

$(x+3)(x-2) = 0$

$x = -3 \text{(discard) or } x = 2$

$\left(2, \sqrt{2}\right)$

40. $\begin{cases} x^2 + y^2 = 36 \\ x - y = 6 \Rightarrow x = y + 6 \end{cases}$

$(y+6)^2 + y^2 = 36$

$2y^2 + 12y = 0$

$2y(y+6) = 0$

$y = 0 \qquad \text{or} \qquad y = -6$

$x = 0 + 6 = 6 \qquad x = -6 + 6 = 0$

$(0, -6); (6, 0)$

41. $\dfrac{x^2}{9} + \dfrac{y^2}{16} \le 1$

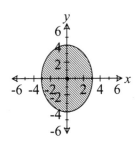

42.

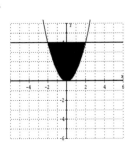

43. 0, 3, 8, 15, 24

44. $a_n = \dfrac{n}{n+4} = \dfrac{8}{8+4} = \dfrac{8}{12} = \dfrac{2}{3}$

45. $a_{11} = 2 + (11-1)(7) = 72$

46. $S_6 = \dfrac{2(1-5^6)}{-4} = \dfrac{2(-15624)}{-4} = 6250$

47. a. $\displaystyle\sum_{i=0}^{6} \dfrac{i-2}{2} = -1 - \dfrac{1}{2} + 0 + \cdots + \dfrac{3}{2} = \dfrac{7}{2}$

b. $\displaystyle\sum_{i=3}^{5} 2^i = 8 + 16 + 32 = 56$

48. a. $\displaystyle\sum_{i=0}^{4} i^2 + i = 0 + 2 + 6 + \cdots + 20 = 40$

b. $\displaystyle\sum_{i=0}^{3} 2^i = 1 + 2 + 4 + 8 = 15$

49. $S_n = \dfrac{n}{2}(a_1 + a_n) = \dfrac{30}{2}(1+30) = 465$

50. $(x-y)^6$ where, $a = x$, $b = -y$,
$n = 6$, and $r = 2$

$\dfrac{6!}{2!(6-2)!} x^{6-2} y^2 = 15x^4 y^2$

The third term in the expansion of $(x-y)^6$ is $15x^4 y^2$.

Appendices

Appendix A Exercise Set

Section 2.1

1. $3x - 4 = 3(2x - 1) + 7$
$3x - 4 = 6x - 3 + 7$
$3x - 4 = 6x + 4$
$-3x = 8$
$x = -\dfrac{8}{3}$

The solution is $-\dfrac{8}{3}$.

3. $5 + 2x = 5(x + 1)$
$5 + 2x = 5x + 5$
$-3x = 0$
$x = 0$
The solution is 0.

Section 2.4

1. $\dfrac{x + 3}{2} > 1$
$x + 3 > 2$
$x > -1$
The solution is $(-1, \infty)$.

3. $\dfrac{x - 2}{2} - \dfrac{x - 4}{3} = \dfrac{5}{6}$
$6\left(\dfrac{x - 2}{2} - \dfrac{x - 4}{3}\right) = 6\left(\dfrac{5}{6}\right)$
$3(x - 2) - 2(x - 4) = 5$
$3x - 6 - 2x + 8 = 5$
$x + 2 = 5$
$x = 3$
The solution is 3.

Section 2.5

1. $x - 2 \le 1$ and $3x - 1 \ge -4$
$x \le 3$ and $3x \ge -3$
$x \ge -1$
$-1 \le x \le 3$
The solution is $[-1, 3]$.

3. $-2x + 2.5 = -7.7$
$-2x = -10.2$
$x = \dfrac{-10.2}{-2} = 5.1$
The solution is 5.1.

5. $x \le -3$ or $x \le -5$
$x \le -3$
The solution is $(-\infty, -3]$.

Section 2.6

1. $|2 + 3x| = 7$
$2 + 3x = 7$ or $2 + 3x = -7$
$3x = 5$ or $3x = -9$
$x = \dfrac{5}{3}$ or $x = -3$

The solutions are -3 and $\dfrac{5}{3}$.

3. $\dfrac{5t}{2} - \dfrac{3t}{4} = 7$
$4\left(\dfrac{5t}{2} - \dfrac{3t}{4}\right) = 4(7)$
$2(5t) - 3t = 28$
$10t - 3t = 28$
$7t = 28$
$t = 4$
The solution is 4.

5. $5(x-3)+x+2 \geq 3(x+2)+2x$
$5x-15+x+2 \geq 3x+6+2x$
$6x-13 \geq 5x+6$
$x \geq 19$
The solution is $[19, \infty)$.

Section 2.7

1. $|x-11| \geq 7$
$x-11 \leq -7$ or $x-11 \geq 7$
$x \leq 4$ or $x \geq 18$
The solution is $(-\infty, 4] \cup [18, \infty)$.

3. $-5 < x-(2x+3) < 0$
$-5 < x-2x-3 < 0$
$-5 < -x-3 < 0$
$-2 < -x < 3$
$2 > x > -3$
$-3 < x < 2$
The solution is $(-3, 2)$.

5. $\dfrac{4x}{5}-1 = \dfrac{x}{2}+2$
$10\left(\dfrac{4x}{5}-1\right) = 10\left(\dfrac{x}{2}+2\right)$
$2(4x)-10 = 5x+20$
$8x-10 = 5x+20$
$3x = 30$
$x = 10$
The solution is 10.

Section 5.8

1. $2x^2-17x = 9$
$2x^2-17x-9 = 0$
$(2x+1)(x-9) = 0$

$2x+1=0$ or $x-9=0$
$2x = -1$ or $x = 9$
$x = -\dfrac{1}{2}$
The solutions are $-\dfrac{1}{2}$ and 9.

3. $|4x+7| = |-35|$
$|4x+7| = 35$
$4x+7 = 35$ or $4x+7 = -35$
$4x = 28$ or $4x = -42$
$x = 7$ or $x = \dfrac{-42}{4} = -\dfrac{21}{2}$
The solutions are $-\dfrac{21}{2}$ and 9.

5. $3(2x-1) < 9$ and $-4x > -12$
$6x-3 < 9$ and $x < 3$
$6x < 12$
$x < 2$
$x < 2$
The solution is $(-\infty, 2)$.

Section 6.6

1. $\dfrac{x}{10}-\dfrac{1}{2} = \dfrac{7}{5x}$
$10x\left(\dfrac{x}{10}-\dfrac{1}{2}\right) = 10x\left(\dfrac{7}{5x}\right)$
$x^2-5x = 2(7)$
$x^2-5x-14 = 0$
$(x+2)(x-7) = 0$
$x+2=0$ or $x-7=0$
$x = -2$ or $x = 7$
The solutions are -2 and 7.

3. $x+2 \leq 0$ or $5x \leq 0$
$x \leq -2$ or $x \leq 0$
$x \leq 0$
The solution is $(-\infty, 0]$.

5. $-8 + |2x - 4| \le -2$
$$|2x - 4| \le 6$$
$$-6 \le 2x - 4 \le 6$$
$$-2 \le 2x \le 10$$
$$-1 \le x \le 5$$
The solution is $[-1, 5]$.

Section 7.6

1. $x(3x + 14) = 5$
$$3x^2 + 14x = 5$$
$$3x^2 + 14x - 5 = 0$$
$$(3x - 1)(x + 5) = 0$$
$$3x - 1 = 0 \ \text{ or } \ x + 5 = 0$$
$$3x = 1 \ \text{ or } \ \ \ \ \ x = -5$$
$$x = \frac{1}{3}$$

The solutions are -5 and $\frac{1}{3}$.

3. $|5x - 4| = |4x + 1|$
$$5x - 4 = 4x + 1 \ \text{ or } \ 5x - 4 = -(4x + 1)$$
$$x = 5 \ \ \ \ \ \ \ \ \text{ or } \ 5x - 4 = -4x - 1$$
$$9x = 3$$
$$x = \frac{1}{3}$$

The solutions are $\frac{1}{3}$ and 5.

5. $-2(x - 4) + 3x \le -3(x + 2) - 2$
$$-2x + 8 + 3x \le -3x - 6 - 2$$
$$x + 8 \le -3x - 8$$
$$4x \le -16$$
$$x \le -4$$
The solution is $(-\infty, -4]$.

Section 8.2

1. $(x - 2)^2 = 17$
$$x - 2 = \pm\sqrt{17}$$
$$x = 2 \pm \sqrt{17}$$

3. $x^2 - 5x + 6 = 0$
$$(x - 2)(x - 3) = 0$$
$$x - 2 = 0 \ \text{ or } \ x - 3 = 0$$
$$x = 2 \ \text{ or } \ \ \ \ \ x = 3$$
The solutions are 2 and 3.

5. $\sqrt{2x + 30} = x + 3$
$$2x + 30 = (x + 3)^2$$
$$2x + 30 = x^2 + 6x + 9$$
$$0 = x^2 + 4x - 21$$
$$0 = (x + 7)(x - 3)$$
$$x + 7 = 0 \ \ \text{ or } \ x - 3 = 0$$
$$x = -7 \ \text{ or } \ \ \ \ \ x = 3$$
Discard -7 as extraneous.
The solution is 3.

7. $\dfrac{3x^2 - 7}{3x^2 - 8x - 3} = \dfrac{1}{x - 3} + \dfrac{2}{3x + 1}$
$$\frac{3x^2 - 7}{(3x + 1)(x - 3)} = \frac{1}{x - 3} + \frac{2}{3x + 1}$$
$$3x^2 - 7 = 1(3x + 1) + 2(x - 3)$$
$$3x^2 - 7 = 3x + 1 + 2x - 6$$
$$3x^2 - 7 = 5x - 5$$
$$3x^2 - 5x - 2 = 0$$
$$(3x + 1)(x - 2) = 0$$
$$3x + 1 = 0 \ \ \ \text{ or } \ x - 2 = 0$$
$$3x = -1 \ \text{ or } \ \ \ \ \ x = 2$$
$$x = -\frac{1}{3} \ \text{ (discard as extraneous)}$$
The solution is 2.

Section 8.4

1. $x^2 - 3x - 10 = 0$
$(x-5)(x+2) = 0$
$x-5 = 0$ or $x+2 = 0$
$\quad x = 5$ or $\quad x = -2$
The solutions are –2 and 5.

3. $\dfrac{x+4}{x-10} = 0$
$x+4 = 0$
$\quad x = -4$
The solution is –4.

5. $\sqrt{x-7} - 12 = -8$
$\quad \sqrt{x-7} = 4$
$\quad x-7 = 4^2$
$\quad x-7 = 16$
$\quad\quad x = 23$
The solution is 23.

7. $\left|\dfrac{3x+5}{2}\right| = -9$ is impossible.
There is no solution, or $\varnothing$.

9. $-4(x-3) + 2x < 6x + 4$
$-4x + 12 + 2x < 6x + 4$
$\quad -2x + 12 < 6x + 4$
$\quad\quad\quad -8x < -8$
$\quad\quad\quad\quad x > 1$
The solution is $(1, \infty)$.

Section 10.7

1. $4^x = 8^{x-1}$
$(2^2)^x = (2^3)^{x-1}$
$2^{2x} = 2^{3x-3}$
$2x = 3x - 3$
$-x = -3$
$\quad x = 3$
The solution is 3.

3. $x(x-9) > 0$
$x = 0$ or $x - 9 = 0$
$\quad\quad\quad\quad x = 9$

Region	Test Point	$x(x-9) > 0$ Result
A: $(-\infty, 0)$	-1	$(-1)(-10) > 0$ True
B: $(0, 9)$	2	$(2)(-7) > 0$ False
C: $(9, \infty)$	10	$(10)(1) > 0$ True

The solution is $(-\infty, 0) \cup (9, \infty)$.

5. $\log_4(x^2 - 3x) = 1$
$x^2 - 3x = 4^1$
$x^2 - 3x - 4 = 0$
$(x-4)(x+1) = 0$
$x-4 = 0$ or $x+1 = 0$
$\quad x = 4$ or $\quad x = -1$
The solutions are –1 and 4.

7. $\dfrac{6}{x-2} \geq 3$
The denominator is 0 when
$x - 2 = 0$, or $x = 2$.

$\dfrac{6}{x-2} = 3$
$6 = 3(x-2)$
$6 = 3x - 6$
$12 = 3x$
$4 = x$

Region	Test Point	$\frac{6}{x-2} \geq 3$; Result
A: $(-\infty, 2)$	0	$\frac{6}{-2} \geq 3$; False
B: $(2, 4)$	3	$\frac{6}{1} \geq 3$; True
C: $(4, \infty)$	5	$\frac{6}{3} \geq 3$; False

The solution is $(2, 4]$.

9. $\log_3(2x+1) - \log_3 x = 1$

$$\log_3 \frac{2x+1}{x} = 1$$

$$\frac{2x+1}{x} = 3^1$$

$$2x+1 = 3x$$

$$1 = x$$

The solution is 1.

Appendix C Exercise Set

1. $V = lwh$
$= (6 \text{ in.})(4 \text{ in.})(3 \text{ in.}) = 72$ cubic inches
$SA = 2lh + 2wh + 2lw$
$= 2(6 \text{ in.})(3 \text{ in.}) + 2(4 \text{ in.})(3 \text{ in.})$
$\quad + 2(6 \text{ in.})(4 \text{ in.})$
$= 36$ sq. in. $+ 24$ sq. in. $+ 48$ sq. in.
$= 108$ sq. in.

3. $V = s^3$
$= (8 \text{ cm})^3 = 512$ cu. cm
$SA = 6s^2 = 6(8 \text{ cm})^2 = 384$ sq. cm

5. $V = \frac{1}{3}\pi r^2 h$
$= \frac{1}{3}\pi(2 \text{ yd})^2(3 \text{ yd}) = 4\pi$ cu. yd
$= 4\left(\frac{22}{7}\right)$ cu. yd
$= 12.56$ cu. yd
$SA = \pi r\sqrt{r^2 + h^2} + \pi r^2$
$= \pi(2 \text{ yd})\sqrt{(2 \text{ yd})^2 + (3 \text{ yd})^2} + \pi(2 \text{ yd})^2$
$= \pi(2 \text{ yd})\left(\sqrt{13} \text{ yd}\right) + \pi(4 \text{ sq. yd})$
$= \left(2\sqrt{13} + 4\right)\pi$ sq. yd
$\approx 3.14\left(2\sqrt{13} + 4\right)$ sq. yd
≈ 35.20 sq. yd

7. $V = \frac{4}{3}\pi r^3$
$= \frac{4}{3}\pi(5 \text{ in.})^3$
$= \frac{500}{3}\pi$ cu. in.
$= \frac{500}{3}\left(\frac{22}{7}\right)$ cu. in.
$= 523\frac{17}{21}$ cu. in.
$SA = 4\pi r^2$
$= 4\pi(5 \text{ in.})^2$
$= 100\pi$ sq. in.
$\approx 100\left(\frac{22}{7}\right)$ sq. in.
$= 314\frac{2}{7}$ sq. in.

9. $V = \frac{1}{3}s^2 h$
$= \frac{1}{3}(6 \text{ cm})^2(4 \text{ cm})$
$= 48$ cu. cm
$SA = B + \frac{1}{2}pl$
$= (6 \text{ cm})^2 + \frac{1}{2}(24 \text{ cm})(5 \text{ cm})$
$= 36$ sq. cm $+ 60$ sq. cm
$= 96$ sq. cm

11. $V = s^3$

$\quad = \left(1\frac{1}{3} \text{ in.}\right)^3$

$\quad = \frac{64}{27} \text{ cu. in.}$

$\quad = 2\frac{10}{27} \text{ cu. in.}$

13. $SA = 2lh + 2wh + 2lw$

$\quad = 2(2 \text{ ft})(1.4 \text{ ft}) + 2(2 \text{ ft})(3 \text{ ft})$
$\quad\quad + 2(1.4 \text{ ft})(3 \text{ ft})$

$\quad = 5.6 \text{ sq. ft} + 12 \text{ sq. ft} + 8.4 \text{ sq. ft}$

$\quad = 26 \text{ sq. ft}$

15. $V = \frac{1}{3}s^2 h$

$\quad = \frac{1}{3}(5 \text{ in.})^2 (1.3 \text{ in.})$

$\quad = 10\frac{5}{6} \text{ cu. in.}$

17. $V = \frac{1}{3}s^2 h$

$\quad = \frac{1}{3}(12 \text{ cm})^2 (20 \text{ cm})$

$\quad = 960 \text{ cu. cm}$

19. $SA = 4\pi r^2 = 4\pi(7 \text{ in.})^2 = 196\pi \text{ sq. in.}$

21. $V = (2 \text{ ft})\left(2\frac{1}{2} \text{ ft}\right)\left(1\frac{1}{2} \text{ ft}\right) = 7\frac{1}{2} \text{ cu. ft}$

23. $V = \frac{1}{3}\pi r^2 h$

$\quad = \frac{1}{3}\left(\frac{22}{7}\right)(2 \text{ cm})^2 (3 \text{ cm})$

$\quad = \frac{88}{7} \text{ cu. cm}$

$\quad = 12\frac{4}{7} \text{ cu. cm}$

Appendix D

Viewing Window and Interpreting Window Exercise Set

1. Yes, since every coordinate is between -10 and 10.

3. No, since -11 is less than -10.

5. Answers may vary. Any values such that Xmin < -90, Ymin < -80, Xmax > 55, and Ymax > 80.

7. Answers may vary. Any values such that Xmin < -11, Ymin < -5, Xmax > 7, and Ymax > 2.

9. Answers may vary. Any values such that Xmin < 50, Ymin < -50, Xmax > 200, and Ymax > 200.

11. Xmax $= -12$ Ymax $= -12$
     Xmin $= 12$ Ymin $= 12$
     Xscl $= 3$ Yscl $= 3$

13. Xmax $= -9$ Ymax $= -12$
     Xmin $= 9$ Ymin $= 12$
     Xscl $= 1$ Yscl $= 2$

15. Xmax $= -10$ Ymax $= -25$
     Xmin $= 10$ Ymin $= 25$
     Xscl $= 2$ Yscl $= 5$

17. Xmax $= -10$ Ymax $= -30$
     Xmin $= 10$ Ymin $= 30$
     Xscl $= 1$ Yscl $= 3$

19. Xmax $= -20$ Ymax $= -30$
     Xmin $= 30$ Ymin $= 50$
     Xscl $= 5$ Yscl $= 10$

Graphing Equations and Square Viewing Window Exercise Set

1. Setting A:

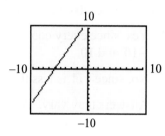

Setting B:

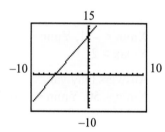

Setting B shows all intercepts.

3. Setting A:

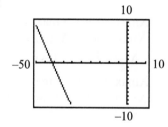

Setting B:

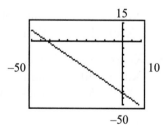

Setting B shows all intercepts.

5. Setting A:

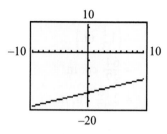

Setting B:

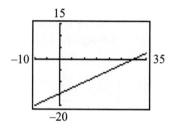

Setting B shows all intercepts.

7. $3x = 5y$

$y = \dfrac{3}{5}x$

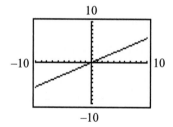

9. $9x - 5y = 30$

$\quad -5y = -9x + 30$

$\qquad y = \dfrac{9}{5}x - 6$

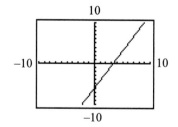

11. $y = -7$

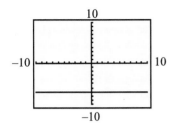

13. $x + 10y = -5$
$\quad\quad 10y = -x - 5$
$\quad\quad\quad y = -\dfrac{1}{10}x - \dfrac{1}{2}$

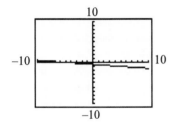

15. $y = \sqrt{x}$

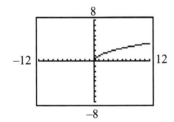

17. $y = x^2 + 2x + 1$

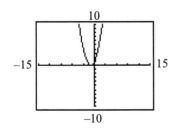

19. $y = |x|$

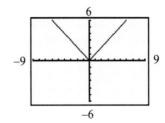

21. $x + 2y = 30$
$\quad\quad 2y = -x + 30$
$\quad\quad\quad y = -\dfrac{1}{2}x + 15$

Standard window:

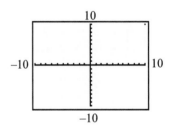

Adjusted window:

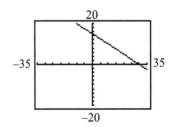